SCHAUM'S OUTLINE OF

THEORY AND PROBLEMS

OF

HUMAN ANATO

AND

PHYSIOLOGY

Second Edition

•

Kent M. Van De Graaff, Ph.D.
Professor of Zoology
Weber State University

R. Ward Rhees, Ph.D.
Professor of Zoology
Brigham Young University

•

SCHAUM'S OUTLINE SERIES
McGRAW-HILL

New York San Francisco Washington, D.C. Auckland Bogotá Caracas Lisbon London Madrid Mexico City Milan Montreal New Dehli San Juan Singapore Sydney Tokyo Toronto

KENT M. VAN DE GRAAFF is currently a Professor of Zoology at Weber State University in Ogden, Utah. He received his B.S. (1965) in Zoology at Weber State College, his M.S. (1969) at University of Utah, and his Ph.D. (1973) at Northern Arizona University. He completed a postdoctorate course in neuromyology (1974), and taught anatomy at University of Minnesota and Brigham Young University. Van De Graaff is the author or coauthor of several college textbooks, including *Human Anatomy, Concepts of Human Anatomy and Physiology,* and *Synopsis of Human Anatomy and Physiology.*

R. WARD RHEES is a Professor of Zoology at Brigham Young University. He received his B.S. (1967) in Pharmacy at the University of Utah and his Ph.D. (1971) in Physiology at Colorado State University. He taught at Weber State College and has been a Visiting Professor in the Department of Anatomy and the Brain Research Institute at the UCLA School of Medicine. His research on sexual differentiation of the brain has been published in numerous leading scholarly journals and presented at national and international conferences.

Schaum's Outline of Theory and Problems of
HUMAN ANATOMY AND PHYSIOLOGY

6 7 8 9 10 11 12 13 14 15 PRS PRS 9 0 2

ISBN 0-07-066887-6

Sponsoring Editor: Barbara Gilson
Production Supervisor: Pamela Pelton
Editing Supervisor: Maureen B. Walker

Library of Congress Cataloging-in-Publication Data

Van De Graaff, Kent M. (Kent Marshall), date
Schaum's outline of theory and problems of human anatomy and physiology / Kent M. Van De Graaff, R. Ward Rhees.
p. cm. -- (Schaum's outline series)
Includes index.
ISBN 0-07-066887-6
1. Human physiology--Outlines, syllabi, etc. 2. Human physiology--Problems, exercises, etc. 3. Human anatomy--Outlines, syllabi, etc. 4. Human anatomy--Problems, exercises, etc. I. Rhees, R. Ward. II. Title III. Series.
QP41.V36 1997
612--dc21

97-8163
CIP

McGraw-Hill

*A Division of The **McGraw-Hill** Companies*

To Karen and Karin

Preface

Mastery of the science of human anatomy and physiology is important for students who are planning careers in health-related fields such as medicine, nursing, dentistry, medical technology, physical therapy, and athletic training. The focus of the second edition of *Schaum's Outline of Human Anatomy and Physiology* is on presenting practical information that students will be able to apply to real-world situations they might encounter in their chosen discipline. In addition, numerous examples throughout this study outline reinforce the principle that learning anatomy and physiology helps students become better acquainted with themselves. The integration of anatomy and physiology in this study outline provides students with a focused perspective of body structure and function. The organization, level of rigor, and clinical focus of this study outline is especially appropriate for students preparing for health-related careers. In addition, this study outline provides students with an organized means of preparing for aspects of national MCAT, DAT, or allied health board certification examinations.

The topic sequence and content of this edition is designed to accompany any human anatomy and physiology textbook. If used as a supplement to a text and class notes, this study outline will improve a student's efficiency of study and performance on course examinations.

The organization of *Schaum's Outline of Human Anatomy and Physiology* is carefully designed to enhance learning. Each chapter is composed of objective–survey–problems modules. An objective represents a major topic and level of competency that a student should strive to achieve. A topic survey follows the objective and is identified with a magnifying glass icon. The survey is a carefully phrased body of information that gives the essence of the topic introduced in the objective. The problems and answers that follow the survey will test a student's understanding of the subject and provide additional information to meet the objective at the desired level.

Set off from the text narrative are short paragraphs highlighted by accompanying topic icons. This interesting information is relevant to the discussion that precedes it. The three icons used are as follows:

Clinical information is indicated by a physician's staff.

survey

Developmental information of practical importance is indicated by a human embryo.

Information relevant to the body processes that maintain homeostasis (a state of dynamic equilibrium) is indicated by a balance.

Schaum's Outline of Human Anatomy and Physiology is much more than just words. Because anatomy and physiology are visually oriented sciences, the preparation of an effective art program was a top priority in this edition. An abundance of carefully rendered figures supplements the text to maximize the learning effort. In addition to the anatomical renderings, flowchart figures are used throughout this study guide to clarify physiological processes. Each figure is placed as close as possible to its text reference.

New features to this edition include developmental, homeostatic, and additional clinical concepts related to the major topics. These new pedagogical features provide information concerning the development of body organs and facilitate the students comprehension of interactions of body systems. New figures, figure captions, and tables have been added to further complement the written material. Also, the labeling has been completely redone to enhance and improve the illustrations. At the close of each chapter is a set of review questions with complete answers by which students can measure their understanding of the concepts and information. Key clinical terms are defined at the end of chapters. The *Outline* is completed with a comprehensive index.

Several individuals assisted in the preparation of this outline, and sincere thanks is extended to each. Christopher H. Creek and Scott Schwendiman rendered the illustrations. Rendell Ashton and Joseph Ashton provide student input regarding the effectiveness of the questions. Special appreciation is expressed to Ann Mirels for her exceptional input as a copy editor. Michael W. Hancock and John L. Crawley were indispensible in the layout of the final product. Finally, we are appreciative of Maureen Walker of McGraw-Hill for her encouragement and editorial assistance in the completion of this project.

KENT M. VAN DE GRAAFF
R. WARD RHEES

Contents

Introduction to the Human Body 1

Objective A To describe *anatomy* and *physiology* as scientific disciplines and to explain how they are related.

Anatomy and physiology are subdivisions of the science of biology, which is the study of living organisms, both plant and animal. Human anatomy has to do with body structure and the relationships between body structures. Human physiology is concerned with the functions of the body parts. In general, function is determined by structure.

1.1 What are the subspecialties of human anatomy?

These include: *gross anatomy*, the study of structures observed with the unaided eye; *microscopic anatomy*, the study of structures observed with the aid of a microscope (*cytology* is the study of cells and their organelles, and *histology* is the study of tissues that make up organs); *developmental anatomy*, the study of structural changes from conception to birth; and *pathological anatomy (pathology)*, the study of structural changes caused by disease.

1.2 What are the subspecialties of human physiology?

These include *cellular physiology*, the study of the interactions of cell parts and the specific functions of the organelles and the cell in general; *developmental physiology*, the study of functional changes that occur as an organism develops; and *pathological physiology*, the study of the functional changes that occur as organs age or become diseased.

Objective B To describe the *human organism* with reference to a classification scheme and to list the *physical requirements for life*.

Homo sapiens, as we have named ourselves, is a biological organism that has features in common with all living animals. Because we have characteristics unique to us, we are a species within a classification scheme based on similarity of structural features.

1.3 Explain why humans are classed among the animals.

We breathe, eat and digest food, excrete body wastes, locomote, and reproduce our own kind, as do other animals. Being composed of organic materials, we decompose in death as other animals (chiefly microorganisms) consume our flesh. The processes by which our bodies produce, store, and utilize energy are similar to those used by all living organisms. The same genetic code that regulates our development is found throughout nature. The fundamental patterns of development observed in many animals are also seen in the formation of the human embryo.

1.4 What are the basic physical requirements for the survival of an organism?

Water, for a variety of metabolic processes; *food*, to supply energy, raw materials for building new living matter, and chemicals necessary for vital reactions; *oxygen*, to release energy from food materials; *heat*, to promote chemical reactions; and *pressure*, to allow breathing.

1.5 Classify human beings using taxonomic categories.

The descending series is shown in table 1.1. *Homo sapiens* is the only extant (existing) hominid.

Table 1.1 Classification of Human Beings

Taxon	*Grouping*	*Characteristics*
Kingdom	Animalia	Cells having a visible nucleus but lacking walls, plastids, and photosynthetic pigments
Phylum	Chordata	Notochord; dorsal hollow nerve cord; pharyngeal pouches
Subphylum	Vertebrata	Cartilaginous or bony endoskeleton; vertebral column
Class	Mammalia	Hair; mammary glands; three auditory ossicles; attached placenta; muscular diaphragm
Order	Primates	Prehensile hands with digits modified for grasping; large brains
Family	Hominidae	Large, well-developed cerebrum; flattened face; bipedal posture and locomotion; well-developed vocal structures; opposable thumb
Genus	*Homo*	
Species	*sapiens*	

Objective C To describe the *levels of organization* of the human body.

The chemical and cellular levels are respectively the basic structural and functional levels. Each level of body organization (fig. 1.1) represents an association of units from the preceding level. Although the cells in the adult body number in the trillions, there are only a few hundred specific kinds.

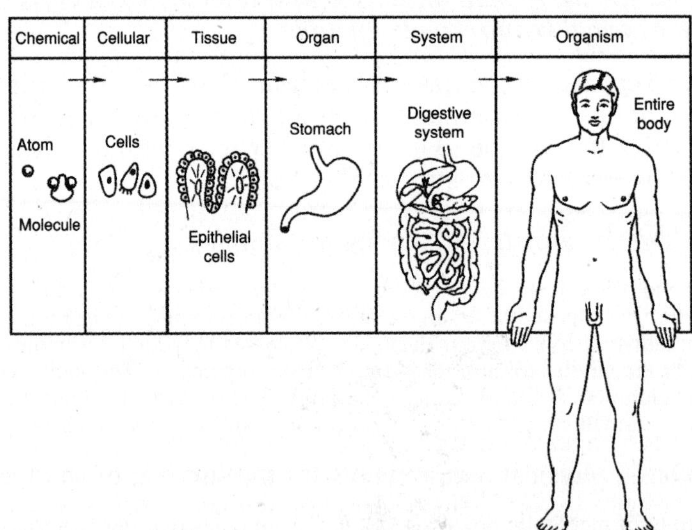

Figure 1.1 Levels of body organization. The chemical, cellular, and tissue levels are microscopic, whereas the organ, system, and organismic levels are macroscopic.

1.6 How are similar cells bound together?

Similar cells are uniformly spaced and bound together as tissue by nonliving *matrix,* which the cells secrete. Matrix varies in composition from one tissue to another and may take the form of a liquid, semisolid, or solid. Blood tissue, for example, has a liquid matrix, while bone cells are bound by a solid matrix. Not all similar cells, however, have a binding matrix; secretory cells, for instance, are solitary amidst a tissue of cells of another kind.

1.7 Define the term *tissue* and explain why the study of tissues is important.

A **tissue** is an aggregation of similar cells bound by supporting matrix that performs a specific function. *Histology* is the microscopic science concerned with the study of tissues. *Pathology* is the medical science concerned with the study of diseased tissues. Tissues are described in chapter 4.

1.8 List the four principal types of tissues and describe the functions of each.

Epithelial tissue (*epithelium*) covers body and organ surfaces, lines body cavities and *lumina* (hollow portions of body tubes), and forms various glands. Epithelial tissue is involved with protection, absorption, excretion, and secretion.

Connective tissue binds, supports, and protects body parts.

Muscle tissue contracts to produce movement of body parts and permit locomotion.

Nervous tissue initiates and transmits *nerve impulses* that coordinate body activities.

1.9 Use an example to define the term *organ* and describe the function of that organ.

A bone, such as the femur, is an **organ** because it is composed of several tissue types that are integrated to perform a particular function. The components of the femur include bone tissue, nervous tissue, vascular (blood) tissue, and cartilaginous tissue (at a joint). Not only does the femur, as part of the skeletal system, help to maintain body support, it also serves the muscular system by providing a place of attachment for muscles, and the circulatory system by producing blood cells in the bone marrow.

Vital body organs are those that are essential for critical body functions. Examples are the heart in pumping blood, the liver in processing foods and breaking down worn blood cells, the kidneys in filtering blood, the lungs in exchanging respiratory gasses, and the brain in controlling and correlating body functions. The reproductive organs are not vital body organs, nor are the organs within the appendages. Death of a person occurs when one or more of the vital body organs falters in its function.

1.10 Define the term *system* as it applies to body organization.

A **system** is an organization of two or more organs and associated structures working as a unit to perform a common function or set of functions, for example, the flow of blood through the body in the case of the circulatory system. Some organs serve more than one body system. The pancreas serves the digestive system in production and secretion of digestive chemicals (pancreatic juice) and the endocrine system in the production of hormones (chemical messengers, insulin, and glucagon). The basic structure and function of each of the body systems is presented in fig. 1.2 through fig. 1.11.

With the exception of the reproductive system, all of the organs that make up the body systems are formed within the 6-week embryonic period (from the beginning of the third week to the end of the eighth week) of prenatal development. Not only are the vital body organs and systems formed during this time, many of them become functional. For example, 25 days after conception the heart is pumping blood through the circulatory system. The organs of the reproductive system form between 10 and 12 weeks after conception, but they do not mature and become functional until a person goes through puberty at about age 12 or 13.

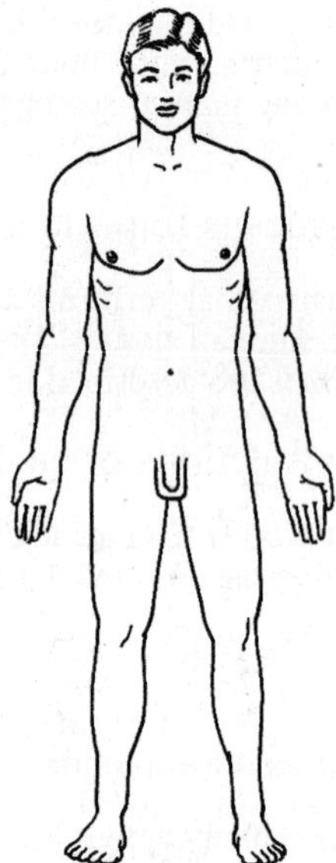

DEFINITION The integument (skin) and structures derived from it (hair, nails, and oil sweat glands).

FUNCTIONS Protects the body, regulates body temperature, eliminates wastes, and receives certain stimuli (tactile, temperature, and pain).

Figure 1.2 Integumentary system.

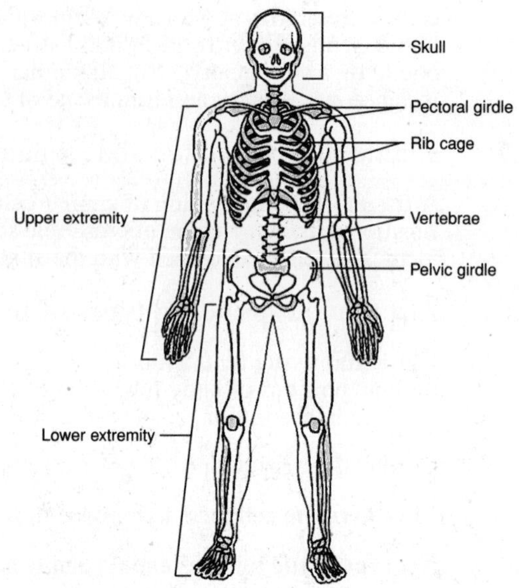

DEFINITION Bones, cartilage, and ligaments (which guy the bones at the joints).

FUNCTIONS Provides body support and protection, permits movement and leverage, produces blood cells (hemopoiesis), and stores minerals.

Figure 1.3 Skeletal system.

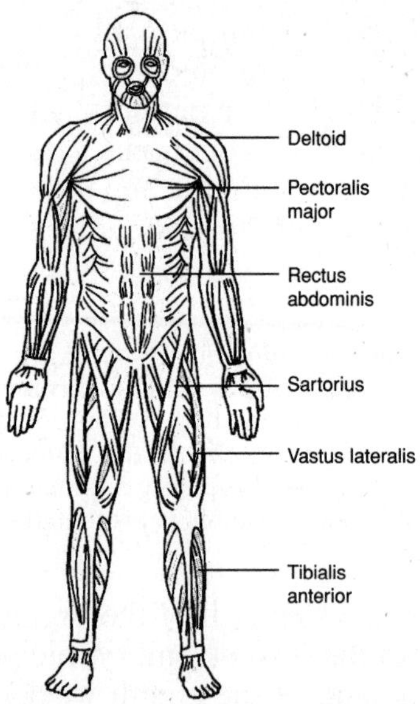

DEFINITION Skeletal muscles of the body and their tendinous attachments.

FUNCTIONS Effects body movements, maintains posture, and produces body heat.

Figure 1.4 Muscular system.

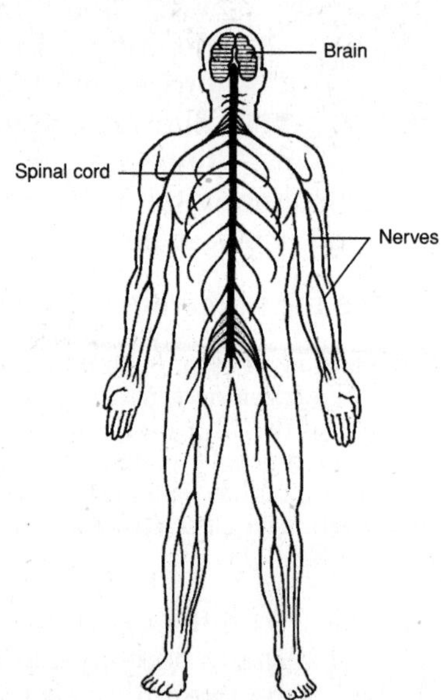

DEFINITION Brain, spinal cord, nerves, and sensory organs such as the eye and the ear.

FUNCTIONS Detects and responds to changes in internal and external environments, enables reasoning and memory, and regulates body activities.

Figure 1.5 Nervous system.

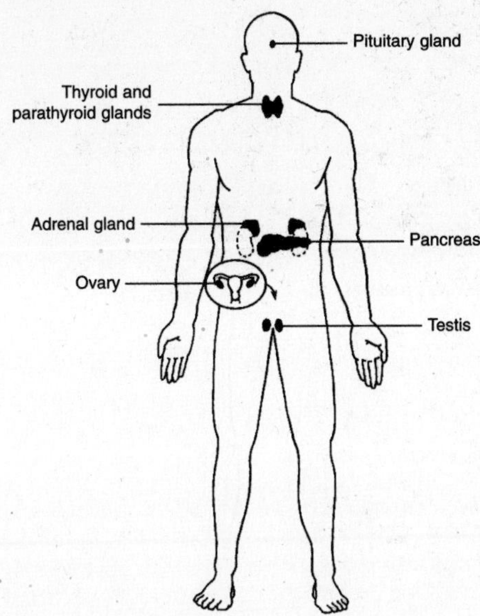

DEFINITION The hormone-producing glands.

FUNCTIONS Controls and integrates body functions via hormones secreted into the bloodstream.

Figure 1.6 Endocrine system.

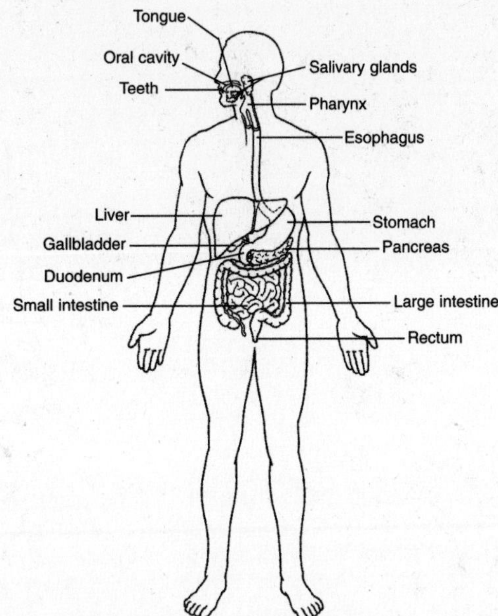

DEFINITION The body organs that render ingested foods absorbable.

FUNCTIONS Mechanically and chemically breaks down foods for cellular use and eliminates undigested wastes.

Figure 1.7 Digestive system.

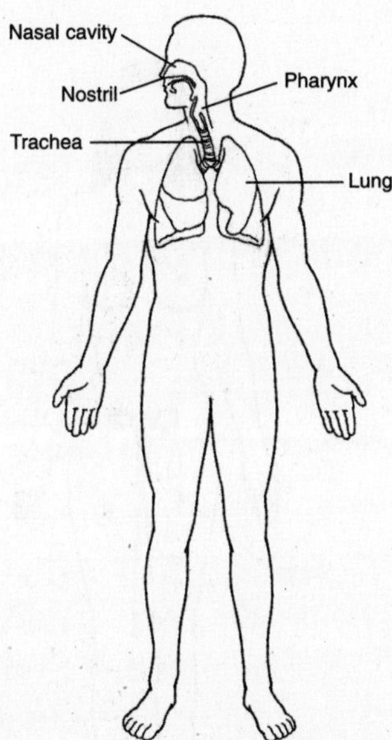

DEFINITION The body organs concerned with movement of respiratory gases (O_2 and CO_2) to and from the pulmonary blood (the blood within the lungs).

FUNCTIONS Supplies oxygen to the blood and eliminates carbon dioxide; also helps to regulate acid-base balance.

Figure 1.8 Respiratory system.

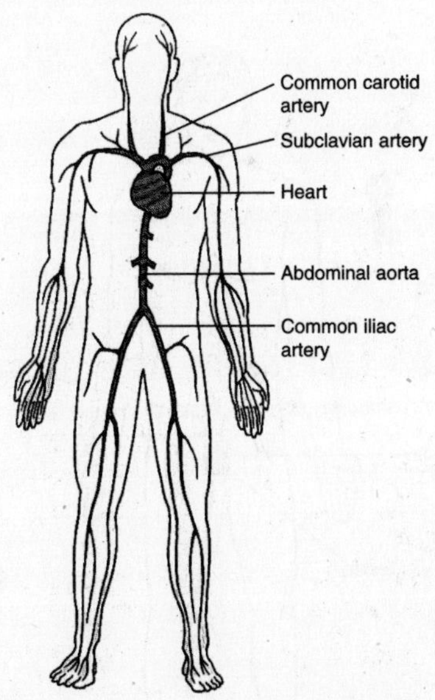

DEFINITION The heart and the vessels that carry blood or blood constituents (lymph) through the body.

FUNCTIONS Transports respiratory gases, nutrients, wastes, and hormones; protects against disease and fluid loss; helps regulate body temperature and acid-base balance.

Figure 1.9 Circulatory system.

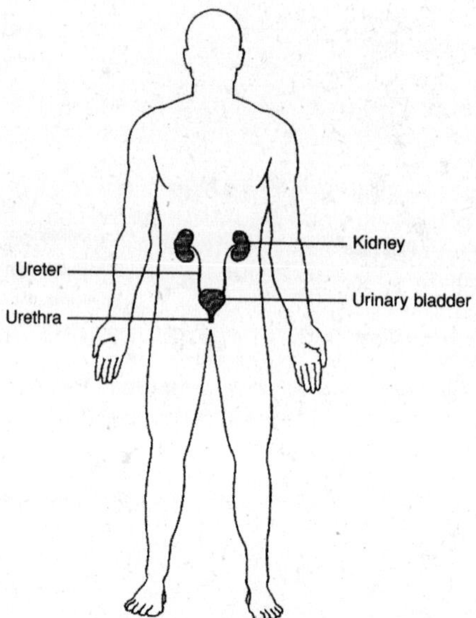

DEFINITION The organs that operate to remove wastes from the blood and to eliminate urine from the body.

FUNCTIONS Removes various wastes from the blood; regulates the chemical composition, volume, and elecrolyte balance of the blood; helps maintain the acid-base balance of the body.

Figure 1.10 Urinary system.

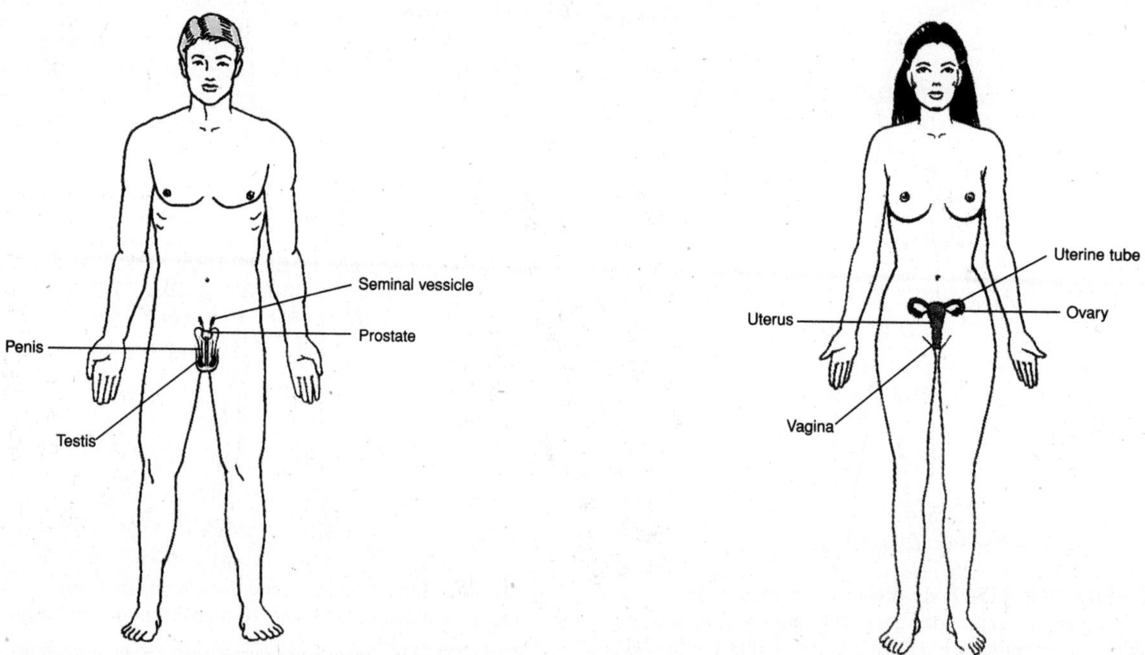

DEFINITION The body organs that produce, store, and transport reproductive cells (*gametes,* or sperm and ova).

FUNCTIONS Reproduce the organism, produce sex hormones.

Figure 1.11 Male and female reproductive systems.

Objective D To list the *body systems* and to describe the general functions of each.

1.11 Which body systems function in support and movement?

The muscular and skeletal systems are frequently referred to as the *musculoskeletal system* because of their combined functional role in body support and locomotion. Both systems, along with the movable (synovial) joints, are studied extensively in *kinesiology* (the mechanics of body motion). The integumentary system also provides some support, and its flexibility permits movement.

1.12 Which body systems function in integration and coordination?

The *endocrine system* and *nervous system* maintain consistency of body functioning, the former by secreting *hormones* (chemical substances) into the bloodstream and the latter by producing nerve impulses (electrochemical signals) carried via *neurons* (nerve cells).

1.13 Which body systems are involved with processing and transporting body substances?

Nutrients, oxygen, and various wastes are processed and transported by the *digestive, respiratory, circulatory, lymphatic,* and *urinary systems*. The lymphatic system, which is generally considered part of the circulatory system, is composed of lymphatic vessels, lymph fluid, lymph nodes, the spleen, and the thymus. It transports lymph from tissues to the bloodstream, defends the body against infections, and aids in the absorption of fats.

Diseases or functional problems of the circulatory system are of major clinical importance because of the potential for disruption of blood flow to a vital organ. *Arteriosclerosis*, or hardening of the arteries, is a generalized degenerative vascular disorder that results in the loss of elasticity and thickening of the arteries. *Atherosclerosis* is a type of arteriosclerosis in which plaque material called *atheroma* forms on the inside lining of vessels. A *thrombus* is a clot within a vessel. An *aneurysm* is an expansion or bulging of an artery, whereas a *coarctation* is a constriction of a segment of a vessel.

Objective E To explain what is meant by *homeostasis*.

Homeostasis is the process by which a nearly stable internal environment is maintained in the body so that cellular metabolic functions can proceed at maximum efficiency. Homeostasis is maintained by effectors (generally muscles or glands), which are regulated by sensory information from the internal environment.

1.14 What major regulatory process does the body use to maintain homeostasis?

Essentially all the control systems of the body are regulated by *negative feedback*. If a factor of the internal environment deviates from a set point, then the system that monitors that factor initiates a counterchange (hence "negative") that returns the factor to its normal state. A specific example is presented in fig. 1.12.

1.15 What is the relationship between homeostasis and pathophysiology?

They are opposed in meaning in the sense that health reflects homeostasis, whereas abnormal function—i.e., pathophysiology—marks a deviation from homeostasis. Pathophysiology is the basis for diagnosing disease and instituting treatment intended to restore normal function.

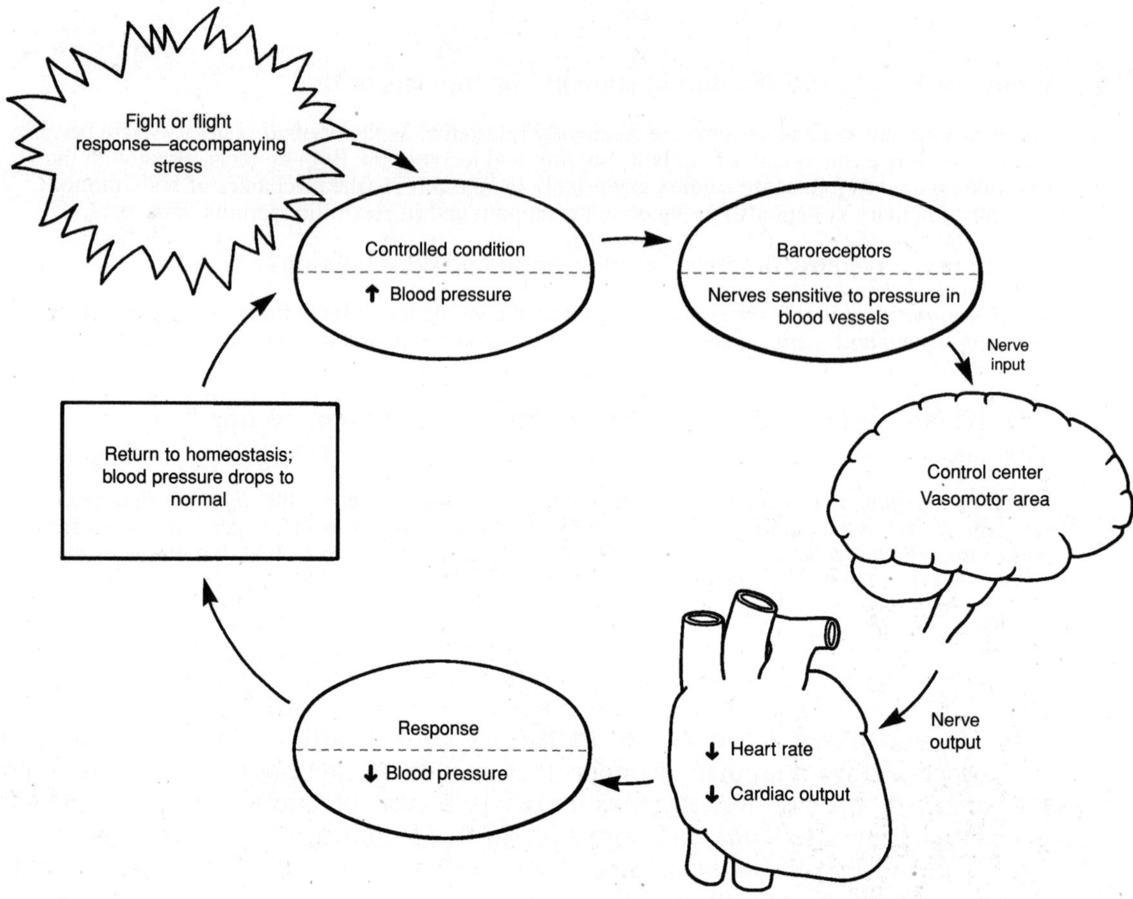

Figure 1.12 Homeostasis of blood pressure. Feedback mechanisms in the form of input (stimulus), a monitoring center, and output (response) maintain dynamic constancy.

Objective F To describe the *anatomical position.*

All terms of direction that describe the relationship of one body part to another are made in reference to a standard anatomical position (fig. 1.13). In the *anatomical position,* the body is erect, feet are parallel and flat on the floor, eyes are directed forward, and arms are at the sides of the body with the palms of the hands turned forward and the fingers pointing downward.

1.16 Why are the palms given an orientation that seems unnatural?

During early embryonic development, the palms are *supine* (facing forward or upward). Later, an axial rotation of each forearm puts the palms in a *prone* position (facing backward or downward). Thus, the anatomical position orients the upper extremities as in early development.

Objective G To identify the *planes of reference* used to locate and describe structures within the body.

A set of three planes (imaginary flat surfaces) passing through the body is frequently used to depict structural arrangement. The three planes are termed the *midsagittal, coronal,* and *transverse planes.*

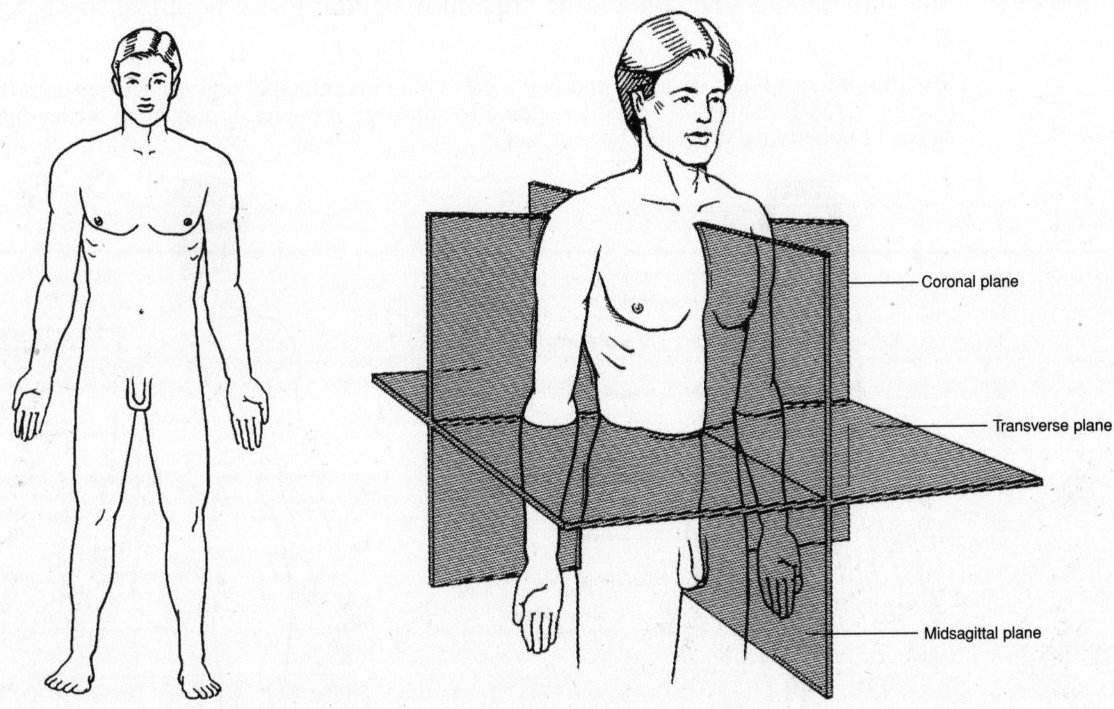

Figure 1.13 For descriptive purposes, the anatomical position provides a standard reference framework for the body.

Figure 1.14 Planes of reference through the body.

1.17 Distinguish between the principal body planes.

The **midsagittal plane** is the plane of symmetry of the body, dividing it into right and left halves. **Sagittal** (*parasagittal*) **planes** run parallel to the midsagittal plane; they divide the body into unequal right and left portions. **Coronal** (*frontal*) **planes** divide the body into front and back portions. **Transverse** (*horizontal* or *cross-sectional*) **planes** divide the body into superior (upper) and inferior (lower) portions. These planes are shown in fig. 1.14.

1.18 With reference to the planes of the body, discuss the advantage of CT (CAT) scans and MRIs over conventional X rays.

Conventional *radiographs (X rays)* are of limited clinical value because they are taken on a vertical plane, and thus images of various structures are often superimposed. One major advantage of *computerized tomographic images (CT scans)* and *magnetic resonance images (MRIs)* is that they can display images along transverse or sagittal planes. These images are similar to those that could otherwise be obtained only in actual sections through the body.

Objective H To identify and locate the principal *body regions.*

The principal body regions are the *head, neck, trunk, upper extremity* (two), and *lower extremity* (two). The *trunk* (torso) is frequently divided into the *thorax* and *abdomen.*

1.19 State the regions that contain the brachium, cubital fossa, popliteal fossa, and axilla.

Specific structures or clinically important areas within the principal regions have anatomical names (see fig. 1.15). Learning the specific regional terminology provides a foundation for learning the names of underlying structures later on.

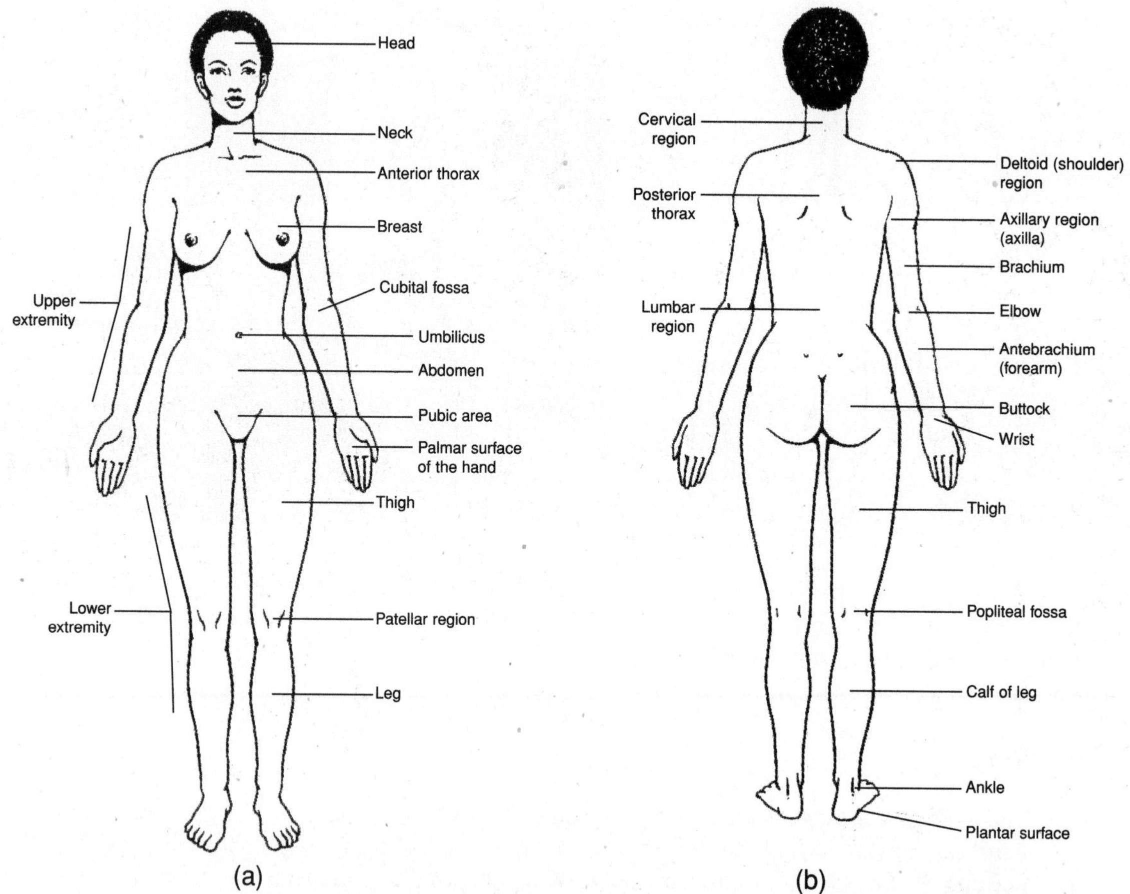

Figure 1.15 The principal body regions. (*a*) An anterior view and (*b*) a posterior view.

Objective I To identify and to locate the principal *body cavities* and the organs within them.

Body cavities are confined spaces in which organs are protected, separated, and supported by associated membranes. As shown in fig. 1.16, the **posterior (dorsal) cavity** includes the **cranial** and **vertebral cavities** (or **vertebral canal**) and contains the brain and spinal cord. The **anterior (ventral) cavity** includes the **thoracic, abdominal,** and **pelvic cavities** and contains visceral organs. The abdominal cavity and the pelvic cavity are frequently referred to collectively as the **abdominopelvic cavity**. Body cavities serve to segregate organs and systems by function. The major portion of the nervous system occupies the posterior cavity; the principal organs of the respiratory and circulatory systems are in the thoracic cavity; the primary organs of digestion are in the abdominal cavity; and the reproductive organs are in the pelvic cavity.

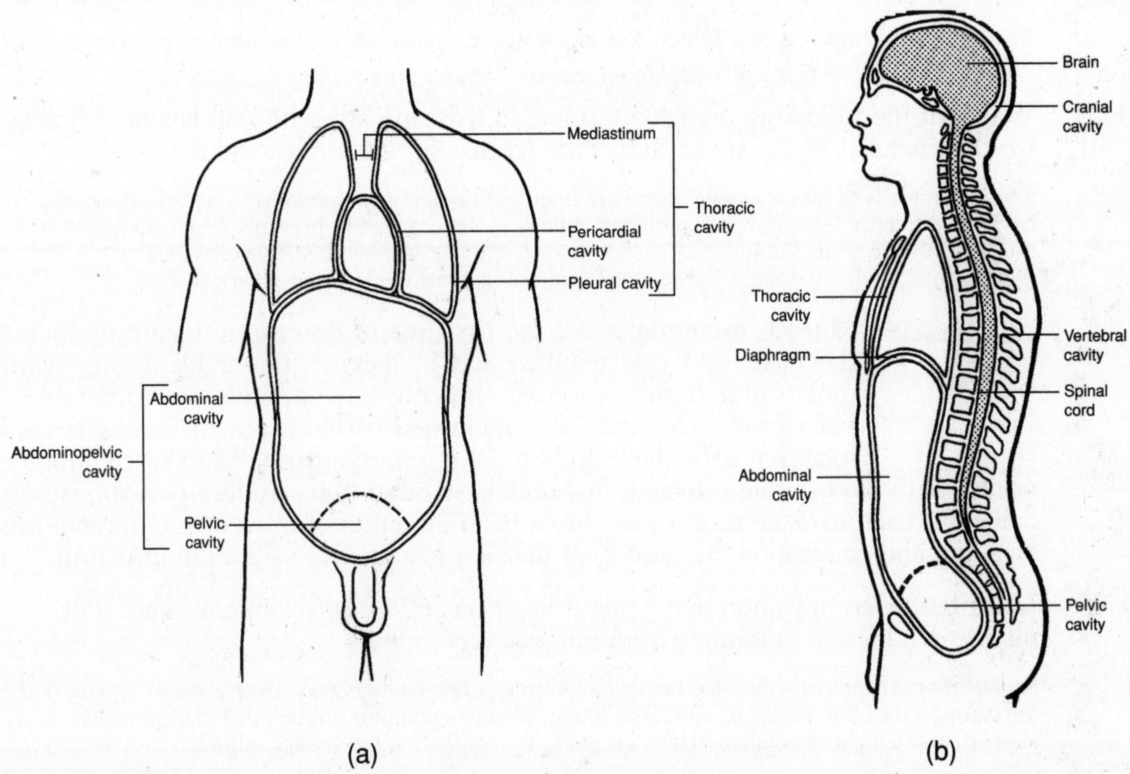

Figure 1.16 The principal body cavities. (*a*) An anterior view and (*b*) a midsagittal view.

1.20 What are visceral organs?

Visceral organs, or **viscera**, are those that are located within the anterior body cavity. Viscera of the thoracic cavity include the heart and lungs. Viscera of the abdominal cavity include the stomach, small intestine and large intestine, spleen, liver, and gallbladder.

1.21 Where are the pleural and pericardial cavities?

The thoracic cavity is partitioned into two **pleural cavities**, one for each lung, and the **pericardial cavity**, surrounding the heart. The area between the two lungs is known as the **mediastinum**.

1.22 What is the clinical significance of the thoracic organs being in separate compartments?

Because each thoracic organ is positioned in its own compartment, trauma is minimized and the risk of disease spreading from one organ to another is reduced. Although the lungs function together, they also work independently. Trauma may cause one lung to collapse, but the other will remain functional.

Objective J To discuss the types and functions of the various *body membranes.*

Body membranes are composed of thin layers of connective and epithelial tissue. They serve to cover, protect, lubricate, separate, or support visceral organs or to line body cavities. The two principal types are *mucous membranes* and *serous membranes.*

1.23 What are the functions of mucous membranes?

Mucous membranes secrete a thick, viscid substance, called *mucus,* that lubricates and protects the body organs where it is secreted.

1.24 Which of the following organs are lined, at least in part, with mucous membranes? (*a*) the trachea, (*b*) the stomach, (*c*) the uterus, (*d*) the mouth and nose

The inside walls of all the organs listed are lined with mucous membranes. Mucus in the nasal cavity and trachea traps airborne particles; mucus in the oral cavity prevents desiccation (drying); mucus coats the epithelial lining of the stomach to protect against digestive enzymes and hydrochloric acid; and mucus in the uterus protects against the entry of pathogens.

Mucous membranes are the first line of defense in locations such as the nasal and oral cavities and in the uterine cavity. Being warm, moist, and highly vascular, mucous membranes are vulnerable to pathogens. However, the acidic pH of the secreted mucus in these locations effectively kills most microorganisms. Mucous membranes occasionally do become infected, in which case other body immunity responses are called into action. A cold or a sore throat is an infection of mucous membranes, and swelling and congestion are among of the first responses to fight the infection.

1.25 Describe the composition and general locations of the serous membranes and distinguish these membranes from mucous membranes.

Serous membranes line the thoracic and abdominopelvic cavities and cover visceral organs. They are composed of thin sheets of epithelial tissue (simple squamous epithelium) that lubricate, support, and compartmentalize visceral organs. *Serous fluid* is the watery lubricant they secrete.

1.26 Give the specific locations of the individual serous membranes.

See table 1.2 and fig. 1.17.

Table 1.2 Serous Membranes and Their Locations

Cavity	*Serous Membrane*	*Location*
Thoracic	Visceral pleura	Adhering to outer surface of lungs
	Parietal pleura	Lining thoracic walls and thoracic surface of diaphragm
	Visceral pericardium (epicardium)	Covering outer surface of heart
	Parietal pericardium	Durable covering surrounding heart
Abdominopelvic	Visceral peritoneum	Covering abdominal viscera
	Parietal peritoneum	Lining abdominal wall
	Mesentery	Double fold of peritoneum connecting parietal to visceral peritoneum

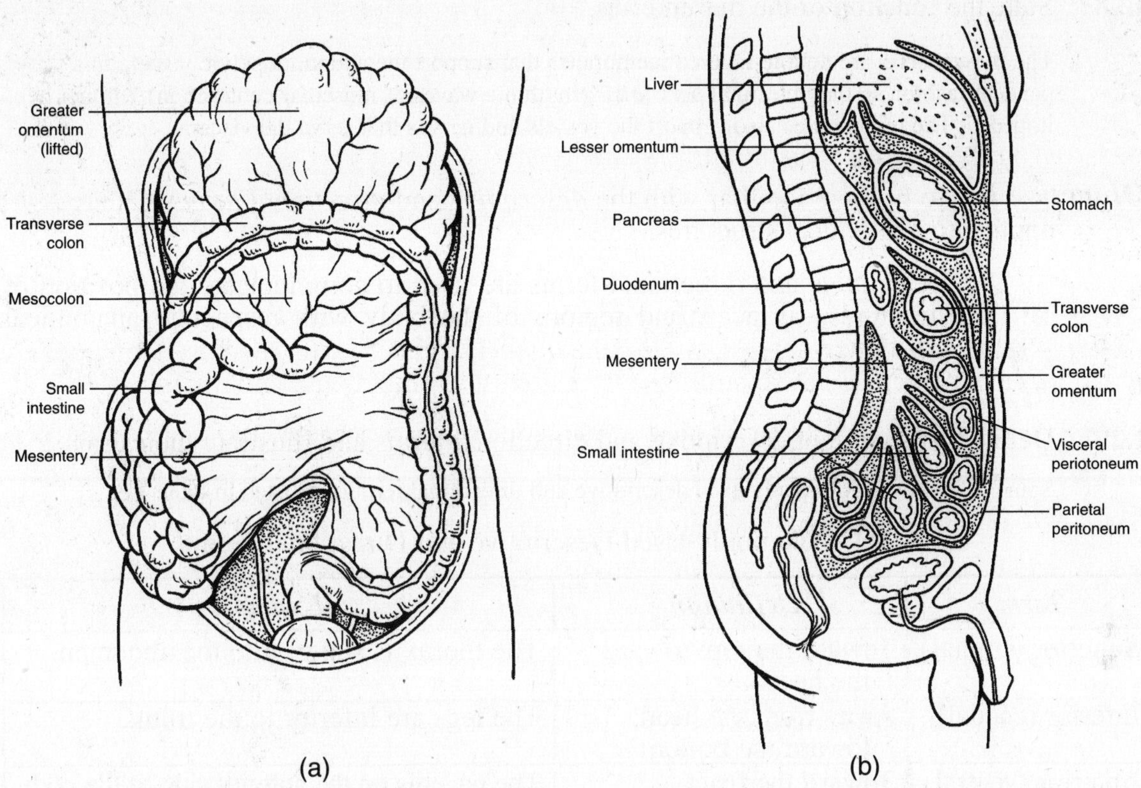

Figure 1.17 The serous membranes and their associated visceral organs. (*a*) An anterior view and (*b*) a midsagittal view.

Pleurisy is an inflammation of the pleural membranes associated with a lung. The infection is generally confined to just one of the pleural cavities. Trauma to a pleural cavity (such as from a crushed rib cage or a bullet or knife wound) may permit air to enter the pleural cavity—a condition known as a *pneumothorax*. Blood in a pleural cavity is known as a *hemothorax*. A pneumothorax causes the lung on the affected side to collapse. The compartmentalization of thoracic organs, however, ensures that one of the lungs will remain functional.

1.27 Define *peritoneal cavity* and explain what is meant by a retroperitoneal organ.

The parietal peritoneum is a thin membrane attached to the inside of the abdominal wall. It is continuous around the intestinal viscera as the visceral peritoneum. The **peritoneal cavity** is the space between the parietal and visceral portions of the peritoneum. *Retroperitoneal organs*, such as the kidneys, adrenal glands, and a portion of the pancreas, are positioned behind the parietal peritoneum, but are still within the abdominopelvic cavity.

Peritonitis is an inflammation of the peritoneal membrane. The infection is confined to the peritoneal cavity. Normally this cavity is aseptic, but it can become contaminated by trauma, rupture of a visceral organ (e.g., a ruptured appendix), an ectopic pregnancy (abnormal pregnancy site), or postoperative complications. Peritonitis is usually extremely painful and life threatening. Treatment usually involves the injection of massive doses of antibiotics and perhaps peritoneal intubation to permit drainage.

1.28 State the function of the mesenteries.

The **mesenteries** are double-layered membranes that support the abdominopelvic viscera in a pendent fashion so that intestinal *peristalsis* (rhythmic waves of muscular contraction) will not be impeded. The mesenteries also support the vessels and nerves that serve the viscera.

Objective K To become familiar with the *descriptive* and *directional terms* that are applied to anatomical structures.

Descriptive and directional terms are used to communicate the position of structures, surfaces, and regions of the body with respect to anatomical position.

1.29 Define the important descriptive and directional terms and illustrate their usage.

Some of the more commonly used descriptive and directional terms are listed in table 1.3.

Table 1.3 Commonly Used Descriptive and Directional Terms

Term	*Definition*	*Example*
Superior (cranial)	Toward the top; toward the head	The thorax is superior to the abdomen.
Inferior (caudal)	Away from the head; toward the bottom	The legs are inferior to the trunk.
Anterior (ventral)	Toward the front	The navel is on the anterior side of the body.
Posterior (dorsal)	Toward the back	The kidneys are posterior to the intestines.
Medial	Toward the midline of the body	The heart is medial to the lungs.
Lateral	Toward the side of the body	The ears are lateral to the head.
Internal (deep)	Away from the surface of the body	The brain is internal to the cranium.
External (superficial)	Toward the surface of the body	The skin is external to the muscles.
Proximal	Toward the main mass of the body	The knee is proximal to the foot.
Distal	Away from the main mass of the body	The hand is distal to the elbow.
Visceral	Related to internal organs	The lungs are covered by a thin membrane called the visceral pleura.
Parietal	Related to the body walls	The parietal pleura is the inside lining of the thoracic cavity.

Review Exercises

Multiple Choice

1. Production of secretory materials within cells would be studied as part of the science of (*a*) histology, (*b*) cytology, (*c*) developmental biology, (*d*) absorption, (*e*) anatomy.

2. A fingernail is a structure belonging to what body system? (*a*) skeletal, (*b*) circulatory, (*c*) integumentary, (*d*) lymphatic, (*e*) reticuloendothelial

3. Which two body systems are regulatory? (*a*) endocrine, (*b*) nervous, (*c*) muscular, (*d*) skeletal, (*e*) circulatory

4. The region of the body between the head and thorax is most appropriately referred to as (*a*) the lumbar region, (*b*) the throat region, (*c*) the trunk region, (*d*) the cervical region, (*e*) the gullet region.

5. A person in the anatomical position would be (*a*) lying face down; (*b*) lying face up; (*c*) standing erect facing forward; (*d*) in a fetal position.

6. In anatomical position, the thumb is (*a*) lateral, (*b*) medial, (*c*) proximal, (*d*) horizontal, (*e*) superficial.

7. Which is *not* one of the four principal tissue types? (*a*) nervous tissue, (*b*) bone tissue, (*c*) epithelial tissue, (*d*) muscle tissue, (*e*) connective tissue

8. Which is *not* a serous membrane? (*a*) parietal peritoneum, (*b*) mesentery, (*c*) visceral pleura, (*d*) lining of the mouth, (*e*) pericardium

9. The relationship between structure and function of an organ is best described as (*a*) a negative feedback system, (*b*) one in which function is determined by structure, (*c*) important only during homeostasis of the organ system, (*d*) nonexistent, except in certain parts of the body.

10. Which is *not* a chordate characteristic? (*a*) a vertebral column, (*b*) a notochord, (*c*) pharyngeal pouches, (*d*) a dorsal hollow nerve cord.

11. The abdominal cavity contains (*a*) the heart, (*b*) the lungs, (*c*) the spleen, (*e*) the trachea.

12. The ventral body cavity comprises all of the following cavities *except* (*a*) the spinal cavity, (*b*) the pleural cavity, (*c*) the thoracic cavity, (*d*) the pelvic cavity, (*e*) the abdominal cavity.

13. The *antebrachium* is (*a*) the chest area, (*b*) the hand, (*c*) the shoulder region, (*d*) the armpit, (*e*) the forearm.

14. Which is positioned retroperitoneally? (*a*) stomach, (*b*) kidney, (*c*) heart, (*d*) appendix, (*e*) liver

15. The foot is to the thigh as the hand is to (*a*) the brachium, (*b*) the shoulder, (*c*) the palm, (*d*) the digits.

16. Which term best defines the position of the knee relative to the hip? (*a*) lateral, (*b*) medial, (*c*) distal, (*d*) posterior, (*e*) proximal

17. The thoracic cavity is separated from the abdominopelvic cavity by (*a*) the mediastinum, (*b*) the abdominal wall, (*c*) the sternum, (*d*) the abdominal septum, (*e*) the diaphragm.

18. Long-distance regulation is accomplished via bloodborne chemicals known as (*a*) blood cells, (*b*) hormones, (*c*) ions, (*d*) motor impulses (*e*) neurotransmitters.

19. Which serous membrane would be cut first as a physician removes an infected appendix? (*a*) parietal peritoneum, (*b*) dorsal mesentery, (*c*) visceral pleura, (*d*) parietal pleura

20. If an anatomist wanted to show the structural relationship of the trachea, esophagus, neck muscles, and a vertebra within the neck, which body plane would be most appropriate? (*a*) sagittal plane, (*b*) coronal plane, (*c*) transverse plane, (*d*) vertical plane, (*e*) parasagittal plane

21. Which pairing of directional terms most closely approximates opposites? (*a*) medial and proximal, (*b*) superior and posterior, (*c*) proximal and lateral, (*d*) superficial and deep

22. A lung is located within (*a*) the mediastinal, pleural, and thoracic cavities; (*b*) the thoracic, pleural, and ventral cavities; (*c*) the peritoneal, pleural, and thoracic cavities; (*d*) the pleural, pericardial, and thoracic cavities; (*e*) none of the preceding.

23. Which of the following serous membrane combinations lines the diaphragm? (*a*) visceral pleura—visceral peritoneum, (*b*) visceral pleura—parietal peritoneum, (*c*) parietal pleura—parietal peritoneum, (*d*) parietal pleura—visceral peritoneum

24. In a negative feedback system, (*a*) input is always maintained constant (homeostatic), (*b*) input serves no useful purpose, (*c*) output is partially put back into the system, (*d*) output is always maintained constant.

25. What is the proper sequence of body cavities or areas traversed as blood flows from the heart to the uterus through the aorta and the uterine artery? (*a*) thoracic, pericardial, pelvic, abdominal; (*b*) pericardial, mediastinal, abdominal, pelvic; (*c*) pleural, mediastinal, abdominal, pelvic; (*d*) pericardial, pleural, abdominal, pelvic.

True or False

____ **1.** Histology is the microscopic examination of tissues.

____ **2.** The function of an organ is predictable from its structure.

____ **3.** A group of cells cooperating in a particular function is called a tissue.

____ **4.** In anatomical position the subject is standing erect, the feet are together, and the arms are relaxed to the side of the body with the thumbs forward.

____ **5.** A sagittal plane divides the body into right and left halves.

____ **6.** The thumb is lateral to the other digits of the hand and distal to the antebrachium.

____ **7.** The lungs are kept moist through the secretion of mucus from mucous membranes.

____ **8.** Increased body temperature during exercise is an example of a homeostatic feedback mechanism.

____ **9.** Mesenteries tightly bind visceral organs to the body wall so that they are protected from excessive movement.

____ **10.** A 6-inch knife wound lateral to the left nipple of a male would puncture the parietal pleura and cause a pneumothorax.

____ **11.** All of the visceral organs are contained within the abdominopelvic cavity.

____ **12.** A computerized tomographic (CT) scan permits an image to be displayed along a transverse plane.

____ **13.** The term *parietal* refers to the body wall, and the term *visceral* refers to internal body organs.

____ **14.** Humans are the only living members of the family Hominidae.

____ **15.** In the scientific name *Homo sapiens*, *Homo* is the genus designation and *sapiens* is the species designation.

Completion

1. Animals within the phylum ____________________ possess a notochord, dorsal hollow nerve cord, and pharyngeal pouches during some stage of their development.

2. ____________________ ____________________ is our scientic name.

3. A(n) ____________________ is an aggregation of similar cells bound by a supporting matrix.

4. The ____________________ system includes the skin, hair, nails, and oil and sweat glands.

5. The nervous system and the ____________________ system control and integrate other systems of the body.

6. ____________________ is the dynamic maintenance of a nearly stable internal environment in the body so that metabolism can occur.

7. ____________________ feedback mechanisms provide input to controlling organs in the process of maintaining homeostasis.

8. All terms of direction that describe the relationship of one body part to another are made in reference to a standard ____________________ position.

9. The ____________________ plane divides the body into equal right and left portions.

10. The armpit is technically known as the ____________________ .

11. The anterior portion of the elbow known as the ____________________ fossa is an important site for withdrawal of venous blood.

12. A lung is contained within a ____________________ cavity which, in turn, is contained within the thoracic cavity.

13. Mucus is secreted by ________________ membranes, and serous fluid is secreted by ________________ membranes.

14. ________________ support abdominopelvic viscera in a pendent fashion, thus enabling peristalsis.

15. ________________ is a directional term meaning "away from the head" or "toward the lower portion of the body."

Matching

Match the descriptions with the body planes or directional terms.

____	1. Toward a central reference point	(*a*) dorsal
____	2. Perpendicular to the craniocaudal axis	(*b*) cranial or superior
____	3. Divides the body into right and left halves	(*c*) transverse plane
____	4. Toward the back	(*d*) distal
____	5. Toward the head	(*e*) lateral
____	6. Away from the midsagittal plane	(*f*) anterior
____	7. Upper surface of the body	(*g*) posterior
____	8. Toward the front	(*h*) caudal or inferior
____	9. Divides the body into anterior and posterior portions	(*i*) medial
____	10. Toward the feet	(*j*) proximal
____	11. Away from a central reference point	(*k*) coronal plane
____	12. Toward the midsagittal plane	(*l*) midsagittal plane

Labeling

Label the body regions indicated on the figure to the right.

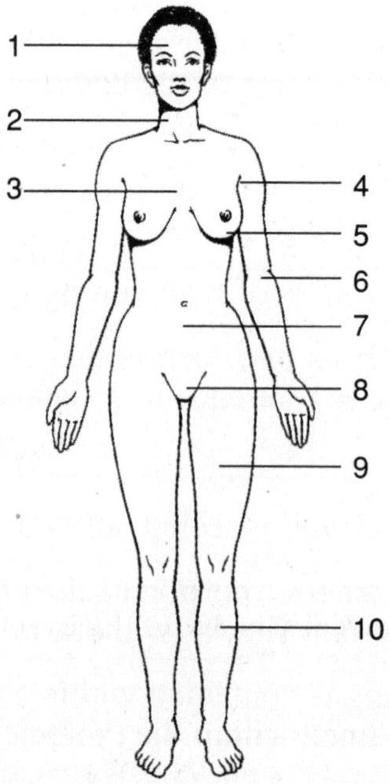

1. ________________
2. ________________
3. ________________
4. ________________
5. ________________
6. ________________
7. ________________
8. ________________
9. ________________
10. ________________

1
2
3
4
5
6
7
8
9
10

Table Completion

From the information provided, complete each row of the following table.

System	*Principal organs*	*Functions*
Circulatory system		
	Nose, pharynx, larynx, trachea, lungs	
		Processes ingested foods for cellular use; eliminates undigested wastes
	Kidney, urinary bladder, ureters, urethra	
		Supports, protects, and permits body movement; sites of hemopoiesis (manufacture of blood cells)
Muscular system		
	Brain, spinal cord, nerves, sense organs	
		Chemically controls and integrates many body activities
Reproductive system		

Answers and Explanations for Review Exercises

Multiple Choice

1. (*b*) Cytology is the study of cells and their functions. Since the production of secretory products involves cellular metabolic functions, it is considered an aspect of cytology.
2. (*c*) The integumentary system includes all of the outer surface structures of the body: the epidermis and the epidermal structures (hair, nails, and glands).
3. (*a*), (*b*) Both the endocrine system and the nervous system participate in controlling and coordinating the functions of the body. The effect of the nervous system is quicker, but the effect of the endocrine system is longer lasting.
4. (*d*) The term *cervical* refers to anything pertaining to the neck or a necklike region of an organ.
5. (*c*) In addition, the person's palms would be forward, with the arms and legs straight.
6. (*a*) Since the palm is forward in anatomical position, the thumb is on the lateral, or radial, side of the upper extremity.
7. (*b*) Bone is a type of connective tissue (see chapter 4).
8. (*d*) The lining of the oral cavity (mouth) derives from ectoderm and is stratified squamous epithelium. All serous membranes derive from mesoderm and are simple squamous epithelium (see chapter 4).
9. (*b*) All body structures are adapted to the specific function they perform, and when the structure is severely damaged or malformed, the function often cannot be performed.

10. (*a*) All vertebrates (animals with vertebral columns) are chordates, but not all chordates develop vertebrae.
11. (*c*) The heart, lungs, and trachea are contained in the thoracic cavity, superior to the abdominal cavity.
12. (*a*) The spinal cavity is contained within the posterior cavity.
13. (*e*) The term *ante* means "before or preceding"; the term *brachium* means "arm."
14. (*b*) Retroperitoneal organs are located behind the serous lining of the abdominal cavity. The kidneys are within the abdominal cavity but behind the parietal peritoneum.
15. (*a*) The brachium within the upper extremity corresponds in position to the thigh of the lower extremity.
16. (*c*) Distal means "farther from the center body mass," as the knee is to the hip.
17. (*e*) The diaphragm is a muscular partition that moves up and down with expiration and inspiration of air. All the abdominal organs lie beneath the diaphragm, and only the lungs and organs of the mediastinum lie above it.
18. (*b*) Hormones are chemicals released into the blood by endocrine glands. They influence the metabolism of target tissues or organs that are usually relatively distant from the gland releasing the hormone.
19. (*a*) The parietal peritoneum lines the inner side of the abdominal cavity wall and would always be cut first in any abdominal surgery.
20. (*c*) A transverse plane would give a cross-sectional view of the organs in the neck, showing clearly the spatial relationship between the various structures.
21. (*d*) *Superficial* means "near the outer surface of the body"; *deep* means "internal with respect to the surface of the body."
22. (*b*) The pleural cavity is formed by the serous membrane surrounding the lungs (the visceral pleura). The pleural cavity is inside the thoracic cavity, which is part of the anterior cavity.
23. (*c*) Since the diaphragm forms the dividing wall between the two cavities, and since the parietal membranes always line the inner cavity walls, the parietal pleura lines the superior surface of the diaphragm and the parietal peritoneum lines the inferior surface of the diaphragm.
24. (*c*) The system's output is entered into the system, where it inhibits further output.
25. (*b*) Only the lungs are contained in the pleural cavity, and the aorta carrying the blood must pass through the abdominal cavity before reaching the pelvic cavity.

True or False

1. True
2. True
3. True
4. False; the palms are facing forward and the thumbs are lateral
5. False; a sagittal plane divides the body into right and left portions; a midsagittal plane divides the body into right and left halves
6. True
7. False; serous membranes secrete a lubricating serous fluid around a lung
8. False; but sweating following exercise is a feedback phenomenon
9. False; mesenteries loosely attach the viscera in a pendant fashion to permit peristalsis
10. True
11. False; visceral organs are also contained within the thoracic cavity
12. True
13. True
14. True
15. True

Completion

1. Chordata
2. *Homo sapiens*
3. tissue
4. integumentary
5. endocrine
6. Homeostasis

7. Negative
8. anatomical
9. midsagittal
10. axilla
11. cubital
12. pleural
13. mucous, serous
14. Mesenteries
15. Inferior (caudal)

Matching

1. (*j*)
2. (*c*)
3. (*l*)
4. (*g*)
5. (*b*)
6. (*e*)
7. (*a*)
8. (*f*)
9. (*k*)
10. (*h*)
11. (*d*)
12. (*i*)

Labeling

1. Head
2. Neck
3. Thorax
4. Axilla
5. Breast
6. Cubital region
7. Abdomen
8. Pubic area
9. Thigh
10. Leg

Table Completion

System	*Principal organs*	*Functions*
Circulatory system	Heart, blood vessels, spleen, lymphatics	Transports materials via blood; regulates acid-base balance; protects against disease and fluid loss
Respiratory system	Nose, pharynx, larynx, trachea, lungs	Supplies O_2 to the blood and eliminates CO_2; helps regulate acid-base balance
Digestive system	Tongue, teeth, pharynx, esophagus, stomach, small intestine and large intestine; liver and pancreas	Processes ingested foods for cellular use; eliminates undigested wastes
Urinary system	Kidney, urinary bladder, ureters, urethra	Filters blood; regulates chemical composition, fluid volume, and electrolyte balance of blood
Skeletal system	Bones, cartilage, joints and ligaments	Supports, protects, and permits body movement; sites of hemopoiesis (manufacture of blood cells)
Muscular system	Muscles and tendons	Causes body movement; maintains posture, produces body heat
Nervous system	Brain, spinal cord, nerves, sense organs	Responds to environmental changes; enables reasoning and memory; regulates body activities
Endocrine system	Endocrine glands (pituitary gland, thymus, pancreas, adrenal glands, gonads, etc.)	Chemically controls and integrates many body activities
Reproductive system	Gonads and genital organs	Produces gametes and sex hormones; reproduces the organism

Cellular Chemistry 2

Objective A To identify by name and symbol the principal *chemical elements* of the body.

All matter, living and nonliving, consists of building units called *chemical elements*. Of the 110 chemical elements, 92 are naturally occurring and 22 of these are present in significant amounts in most animal tissues. The chemical composition of the human body is summarized in table 2.1.

Table 2.1 Chemical Composition of the Body

Chemical elements		***% body composition***
Carbon (C)	Nitrogen (N)	96%
Oxygen (O)	Hydrogen (H)	
Calcium (Ca)	Phosphorus (P)	3%
Potassium (K)	Sulfur (S)	
Iron (Fe)	Chlorine (Cl)	Trace quantities
Iodine (I)	Sodium (Na)	
Magnesium (Mg)	Copper (Cu)	
Manganese (Mn)	Cobalt (Co)	
Zinc (Zn)	Chromium (Cr)	
Fluorine (F)	Molybdenum (Mo)	
Silicon (Si)	Tin (Sn)	

2.1 Define the terms *atom* and *molecule* and distinguish between these terms.

An **atom** is the smallest unit of an element that retains its chemical properties. Every pure element is composed of only one kind of atom. For example, carbon, a key element in a living system, is composed of only carbon atoms.

A **molecule** is a combination of two or more atoms, joined by *chemical bonds.* Molecules may consist of atoms of the same element (as in the oxygen molecule, O_2) or of atoms of different elements (as in the hydrogen sulfide molecule, H_2S). Just as atoms are the smallest units of a chemical element, molecules are the smallest unit of a chemical compound. *Water* is a chemical compound that is essential for life. It consists of molecules, each containing one oxygen atom and two hydrogen atoms (H_2O).

Chemistry is sometimes called the *central science*, since its principles are central to understanding all aspects of science, including biology and physiology. Chemistry is vitally important in the training of health care workers. In order to understand the function and even the dysfunction of the body, a person must understand the component atoms and molecules and how they interact in the body. *Pharmacology* is the science of drugs, including their composition, uses, and effects on the body. *Drugs* are chemical compounds that have specific effects on the body's mechanisms.

Objective B To describe the structure of *atoms.*

An atom is composed of three kinds of elementary particles: **protons, neutrons,** and **electrons.** Particles are characterized by their weights (or masses) and their electric charges (table 2.2). The units for measuring weight and charge of the particles are such that a "normal" carbon atom has a weight of exactly 12 and an electron has a charge of –1. Protons and neutrons are bound in the nucleus of the atom.The number of protons in the nucleus is called the **atomic number (Z).** The atomic number is the same for all atoms of a given chemical element. Each chemical element has a consistent number of protons in the nucleus of each of its atoms. Surrounding the nucleus are precisely Z electrons, making the atom as a whole electrically neutral. Electrons orbit the nucleus, much as the planets of the solar system orbit the sun. However, because electrons have properties of waves as well as particles, it is more useful to speak of *energy levels* occupied by the electrons. If these energy levels are imagined as organized into successive shells, then the chemical properties of the element can be explained in terms of the distribution of the Z electrons among the shells.

Table 2.2 Subatomic Particles, Weights, and Charges

Particle (symbol)	*Weight (approximate)*	*Charge*
Proton (p^+)	1	+1
Neutron (n^0)	1	0
Electron (e^-)	$\frac{1}{1840}$	–1

2.2 Sketch structures for hydrogen (Z = 1), carbon (Z = 6), and potassium (Z = 19).

The shells of an element are often represented by concentric circles around the nucleus (fig. 2.1). The capacities of the first four shells are 2, 8, 8, and 18 electrons. The atom is built by one electron at a time, with a given shell entered only if all interior shells are full.

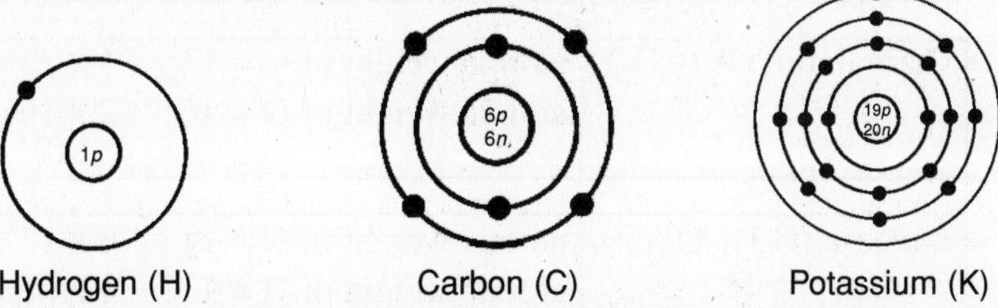

Figure 2.1 Atomic representation of energy levels, or shells.

2.3 What are *isotopes*?

Atoms of a given element (all containing the same number [Z] of protons) but with different numbers of neutrons are said to be *isotopes* of the element. For example, in addition to the standard 6-neutron variety of carbon, there exist 7-neutron and 8-neutron varieties. The atomic weight of an element, as given in the periodic table of chemical elements, is the average of the weights of all the isotopes of the element. For example, the weight of 6-neutron carbon is presented as 12.0000; however, the atomic weight of carbon is 12.01115. Since the number of neutrons in the nucleus tends to be close to the number of protons, it follows from the information given in table 2.2 that the atomic weight of an element is roughly 2Z. This rule does not hold up as well for larger atoms, but is a fairly good estimate in the smaller atoms. Because the various isotopes of an element have a common electron shell structure, they behave identically in ordinary chemical reactions. However, the difference in weight often creates a difference in stability and other properties.

Isotopes have important medical uses. Although all isotopes of a particular element behave identically in chemical reactions, some are radioisotopes, whose radioactivity can be detected by radiographic instruments. Radioisotopes are frequently used by radiologists and oncologists to diagnose and treat diseases. Through injection or ingestion, a physician may introduce a radioisotope into the body of a patient and then track the movement, cellular uptake, tissue distribution, or excretion of the isotope in the body.

Objective C To describe the structure and bonds of *molecules.*

survey

Molecules are structures composed of atoms held together by attractive forces called **bonds**. **Ionic bonds** form when atoms give up or gain electrons and become either positively or negatively charged. These charged atoms are called **ions**, and those with negative charges are attracted strongly to those with positive charges. **Covalent bonds** form when atoms share electrons. *Chemical reactions* occur when molecules form, are broken, or rearrange their component atoms. In chemical notation, subscripts denote how many atoms of each element are in one molecule of the compound.

2.4 Compute the molecular weight of water (H_2O), carbon dioxide (CO_2), and glucose ($C_6H_{12}O_6$).

The **molecular weight (MW)** is the sum of the weights of the atoms composing the molecule (table 2.3).

Table 2.3 The Molecular Weight of Water, Carbon Dioxide, and Glucose

Water (H_2O)	atomic weight of H = 1 atomic weight of O = 16	2 x 1 = 2 1 x 16 = 16 MW = 18
Carbon dioxide (CO_2)	atomic weight of C = 12 atomic weight of O = 16	1 x 12 = 12 2 x 16 = 32 MW = 44
Glucose ($C_6H_{12}O_6$)	atomic weight of C = 12 atomic weight of H = 1 atomic weight of O = 16	6 x 12 = 72 12 x 1 = 12 6 x 16 = 96 MW = 180

2.5 What types of bonds hold atoms together in molecules?

Ionic bonds. An ion is a charged atom that results from the loss or gain of one or more electrons from the atom's outer shell, causing it to lose its electrical neutrality. Atoms that gain electrons acquire an overall negative charge and are called **anions**. Atoms that lose electrons acquire an overall positive charge and are called **cations**. An **ionic bond** is the electrical attraction that exists between an anion and a cation. It is not as strong as a covalent bond in which electrons are shared rather than transferred. The NaCl molecule is held together by ionic bonding (fig. 2.2). Like most ionic compounds, NaCl has a very high melting point because the molecules have a strong attraction for each other. Ionic bonds dissociate easily in water.

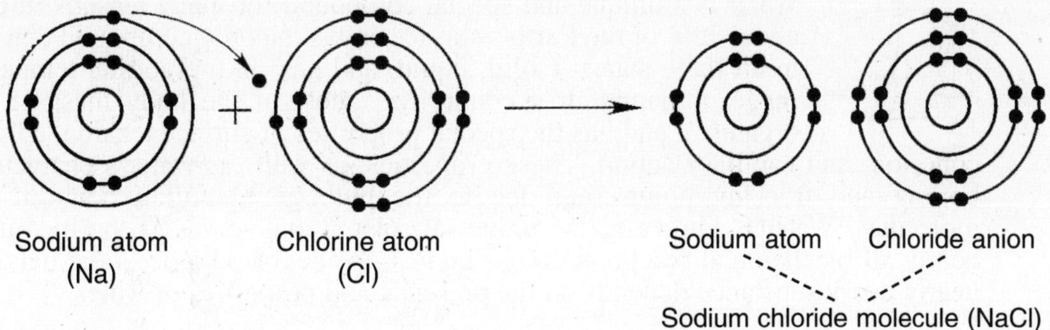

Figure 2.2 The formation of an ionic bond in the NaCl molecule.

Covalent bonds. Sometimes atoms share their electrons instead of transferring them completely. They may share one, two, or three pairs of electrons. Such a sharing of electrons between two atoms is called a **covalent bond**. Covalent bonds are extremely strong. A shared pair is indicated by a short line drawn between the chemical symbols. For instance, in the oxygen molecule, O_2, two pairs of electrons are shared (fig. 2.3), and so the molecule may be indicated as O=O.

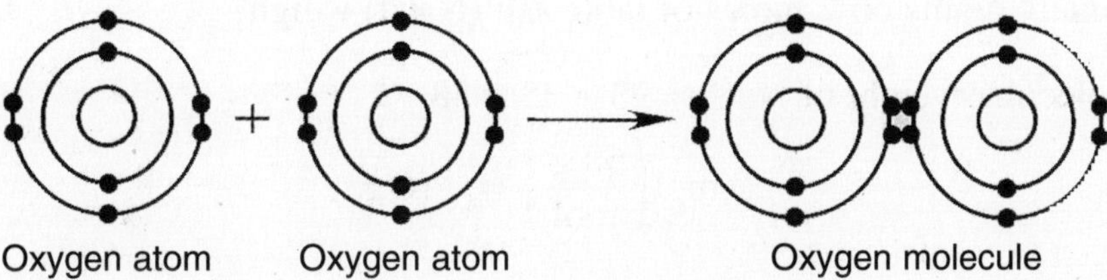

Figure 2.3 The formation of a covalent bond in the O_2 molecule.

Hydrogen bonds. When hydrogen forms a covalent bond with another atom, such as oxygen, the hydrogen atom often gains a slight positive charge as the larger oxygen atom exerts a stronger pull on the shared electron pair. The now slightly positive hydrogen atom has an affinity for the slightly negative oxygens of other molecules of the same compound, and this attraction is called a **hydrogen bond** (fig. 2.4). It is not a bond that forms new molecules, but rather a weak "bond" between molecules. Hydrogen bonding is not nearly as strong as covalent or ionic bonding, but it plays an important role in determining the properties of water and many other compounds that are vital to life.

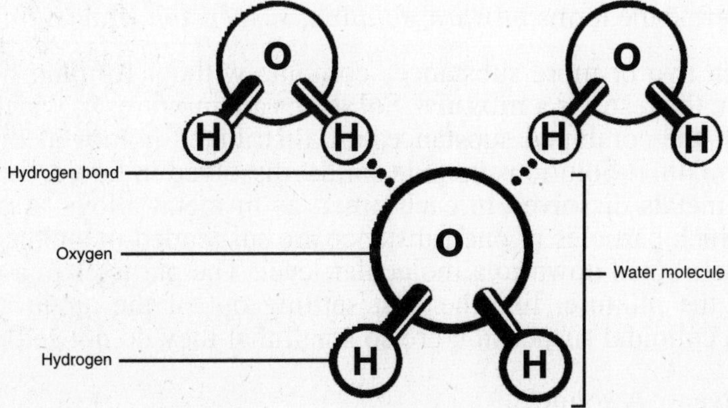

Figure 2.4 The configuration of hydrogen bonds between water molecules.

Water is a unique and special compound for many reasons. It covers about 70% of the Earth's surface and is the only compound that exists in all three states (solid, liquid, and gas) in the normal temperature range of nature. It accounts for most of the body mass of every organism, and has the special properties of surface tension, adhesion, cohesion, and capillary action. These properties, as well as water's characteristic boiling and freezing points, are due to the hydrogen bonding between water molecules. Water is known as the universal solvent and serves as the medium for nearly all biochemical reactions. In our bodies, the delicate homeostatic balance of nearly every substance depends on the presence and properties of water.

Objective D To understand the concept of *moles.*

A **mole** (mol) is a unit of measurement, just like a liter or a meter. It is a unit of weight, and it always contains 6.022 x 10^{23} molecules. A mole of water therefore contains 6.022 x 10^{23} molecules of water, and a mole of helium contains exactly 6.022 x 10^{23} helium atoms. A mole of any substance is equal to the same number of grams as the molecular weight of the substance.

2.6 How many grams do 2 moles of table salt (NaCl) weigh?

molecular weight of NaCl = 23 + 35 = 58

$$2 \text{ mol}\left(\frac{58 \text{ g}}{\text{mol}}\right) = 116 \text{ g}$$

2.7 How many water molecules are in 1 mL (milliliter) of water?

$$\begin{aligned} 1 \text{ mL } H_20 &= 1 \text{ g} \\ 1 \text{ mol } H_20 &= 18 \text{ g} \\ 1 \text{ mol } H_20 &= 6.022 \times 10^{23} \; H_20 \text{ molecules} \end{aligned}$$

$$(\text{ mL})\left(\frac{1 \text{ g}}{\text{mL}}\right)\left(\frac{1 \text{ mol}}{18 \text{ g}}\right)\left(\frac{6.022 \times 10^{23} \text{ molecules}}{\text{mol}}\right) = 3.34 \times 10^{23} \text{ molecules}$$

Objective E To define the terms *mixture, solution, suspension,* and *colloidal suspension.*

When two or more substances combine without forming bonds with each other, the result is a **mixture.** **Solutions** are mixtures in which the molecules of all the combined substances are distributed homogeneously throughout the mixture. Solutions include solids dissolved in liquid, as with salt water, and metals dissolved in each other, as in metal alloys. A **suspension** is a mixture in which particles of one substance are suspended in another substance, but not evenly distributed down to a molecular level. The particles in a suspension will settle out of the mixture, like the dust settling out of the air in a room, but the particles of a colloidal suspension are so small that they do not settle out.

2.8 What is a solvent? A solute?

Solutions are the most important kind of mixtures in organic chemistry, and most biological solutions consist of some solid substance dissolved in water. In this case, water serves as the

solvent of the solution, and the substance, be it a salt, sugar, or protein, is the solute. A practical definition of a solvent is that it is the substance of any solution present in greatest proportion, often water. All other substances are considered solutes. The distinction becomes less useful in solutions such as metal alloys, which may have equal amounts of two or more substances.

2.9 How are concentrations in solution measured?

Concentrations of solute in a solution may be measured in several ways, and the most appropriate way is determined by ease or need. For example, it is sometimes most useful to measure the *percentage of the solute in the solution.* **Molality** is a measure of the moles of solute per kilogram of solvent. **Molarity (M)** is a measure of the moles of solute per liter of solution. Molarity is by far the most frequently used measurement for biological solutions.

Objective F To describe *acids, bases,* and the *pH scale.*

In any sample of water, a certain minuscule proportion of water molecules exists in an ionized form, as H^+ (hydrogen ions) and OH^- (hydroxide ions). In pure water, the number of H^+ equals the number of OH^-, and the concentration of each is 10^{-7} M. Chemical substances that, when added to water solutions, increase the concentration of H^+ are called **acids**; those that increase the concentration of OH^- are called **bases**. The acidity of basicity of a solution is expressed as a value on the **pH scale**, which is a number derived from the logarithm of the concentration of hydrogen ions.

2.10 What is the pH of pure water?

Since pure water has a hydrogen ion concentration of 10^{-7} M, its pH is 7. This is found by taking the logarithm of the H^+ concentration, which is –7, and changing the sign to make it positive. Therefore, if the H^+ concentration of a solution is 10^{-2} M, the pH would be 2.

2.11 What is a strong acid? A weak acid?

Strong acids are acids that dissociate completely in water; or in other words, every one of the acid molecules loses its proton in the water solution. Examples of strong acids are hydrochloric acid (HCl) and sulfuric acid (H_2SO_4). *Weak acids* are acids that only partially dissociate; in other words, some but not all the molecules lose their protons in the water solution. Mole for mole, strong acids generally change the pH of a solution more significantly than do weak acids. However, weak acids and the salts they form are extremely important in organic chemistry, as they are the basis of *buffers*.

2.12 Define the term *salt*.

Salts are ionic compounds formed from the residue of an acid and the residue of a base. When an acid loses its proton and a base loses a hydroxyl group (OH^-), the remaining ions of the molecules, if both are present in the solution, will sometimes bind to each other, forming a salt. The reaction of HCl (an acid) with NaOH (a base) to form table salt (NaCl) is an example:

$$\underset{\text{acid}}{HCl} + \underset{\text{base}}{NaOH} \longrightarrow \underset{\text{water}}{H_2O} + \underset{\text{salt}}{NaCl}$$

Objective G To define *buffer*.

A **buffer** is a combination of a weak acid and its salt in a solution that has the effect of stabilizing the pH of the solution. If a solution contains a buffer, its pH will not change dramatically even when strong acids or bases are added. When acid is added to the solution, it is neutralized by the salt of the weak acid. When base is added to the solution, it is neutralized by the weak acid itself.

2.13 What is the pH of blood and how is it maintained at a constant level?

Blood has a pH of 7.4, which means it is slightly more basic than water. Blood maintains its pH in homeostasis (steady state) by means of the *bicarbonate buffer system*, which is regulated by the amount of carbon dioxide dissolved in the blood. The acid of the buffer system is carbonic acid, H_2CO_3, which forms from carbon dioxide and water. The salt is sodium bicarbonate, which exists in solution as bicarbonate ions, HCO_3^-.

2.14 List the most important buffer systems in the body and indicate their locations.

See table 2.4.

Table 2.4 Buffer Systems and Their Locations

Bicarbonate buffer	Blood, extracellular fluid (most easily adjusted body buffer)
Phosphate buffer	Kidneys, intracellular fluid
Protein buffer	All tissues (most plentiful body buffer)

Objective H To distinguish between *inorganic* and *organic compounds.*

Inorganic compounds do not contain carbon (exceptions include CO and CO_2), and are usually small molecules. **Organic compounds** always contain carbon and are held together by covalent bonds. Organic compounds are usually large, complex molecules. Both inorganic and organic compounds are important in *biochemistry*, the study of chemical processes that are essential to life.

2.15 List some inorganic compounds important in living organisms.

Water, oxygen, carbon dioxide, salts, acids, bases, and electrolytes (such as Na^+, K^+, and Cl^-).

Electrolytes have tremendous clinical significance. They function in every body system and are often an essential link in a body process. Electrolytes form when certain solutes held together by ionic bonds dissolve in water, yielding free ions in the water solution. The most important of these ions include potassium (K^+), sodium (Na^+), chloride (Cl^-), and calcium (Ca^{2+}). Electrolytes are important in the transmission of nerve impulses, maintenance of body fluids, and functioning of enzymes and hormones. Many disorders, such as kidney failure, muscle cramps, and some cardiovascular diseases, involve imbalances in electrolyte levels.

2.16 List the four major families of organic compounds and give examples of each.

See table 2.5.

Table 2.5 Organic Compounds and Examples

Carbohydrates	Glucose, cellulose, glycogen, starch
Lipids	Phospholipids, steroids, prostaglandins
Proteins	Enzymes, insulin, albumin, hemoglobin, collagen
Nucleic acids	DNA, RNA

2.17 Describe how biochemical compounds are formed and broken down.

All large biochemical molecules are formed by connecting small units together into large *macromolecules* in a process called **dehydration synthesis.** In this process, two units are joined, creating one large molecule and a single molecule of water. **Hydrolysis** is the reverse of this reaction. It is the use of water to break down macromolecules into their component building blocks. Dehydration synthesis and hydrolysis are the most important biological reactions. In living organisms, these reactions are usually catalyzed by enzymes, which are proteins that enhance and speed up reactions.

Objective I To describe the three types of *carbohydrates.*

All **carbohydrates** are composed of carbon, hydrogen, and oxygen. The ratio of hydrogen to oxygen in carbohydrates is 2 to 1. Carbohydrates are classified as **monosaccharides** (simple sugars, such as glucose); **disaccharides** (double sugars, such as sucrose); and **polysaccharides** (complex sugars, usually composed of thousands of glucose units, such as glycogen).

2.18 How are carbohydrates used in the body?

1. They serve as the principal source of body energy.
2. They contribute to cell structure and synthesis of cell products.
3. They form part of the structure of DNA and RNA (deoxyribose and ribose are both sugars).
4. They are converted into proteins and fats.
5. They function in food storage (glycogen storage in the liver and skeletal muscles).

2.19 Describe the various forms a monosaccharide may take.

Trioses are three-carbon sugars; *tetroses* are four-carbon sugars; *pentoses* are five-carbon sugars; *hexoses* are six-carbon sugars; and *heptoses* are seven-carbon sugars. Structures for the hexose glucose are shown in fig. 2.5 and structures of two important pentoses are shown in fig. 2.6.

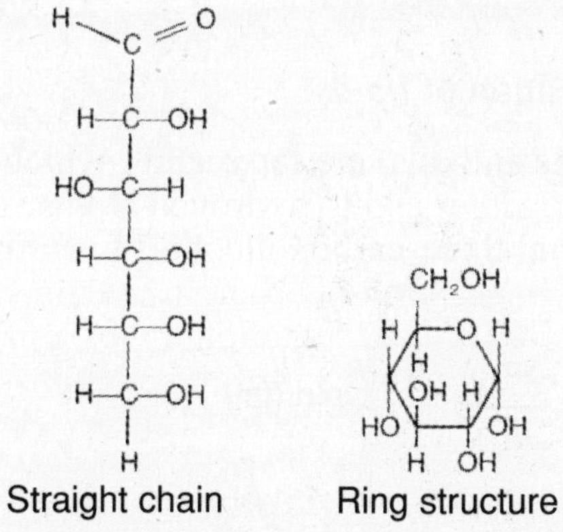

Straight chain | Ring structure | Ribose | Deoxyribose

Figure 2.5 Structures of glucose.

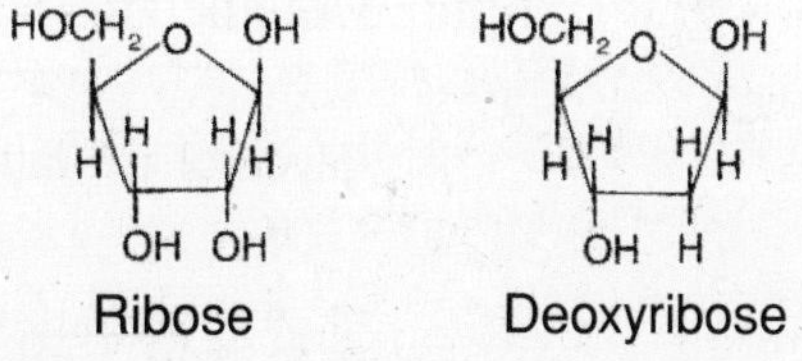

Figure 2.6 Sugars of RNA and DNA.

2.20 How are disaccharides built up from monosaccharides?

A *disaccharide* forms when two monosaccharides combine in a dehydration synthesis reaction, usually catalyzed by enzymes. The synthesis of maltose (a disaccharide composed of two bonded glucoses) is shown in fig. 2.7.

$-H_2O$ / $+H_2O$

Glucose + Glucose → Maltose

Figure 2.7 The formation of maltose (a disaccharide) from two glucoses (monosaccharides).

In a similar fashion: glucose + galactose = lactose
glucose + fructose = sucrose (table sugar)

The reverse of these dehydration synthesis reactions, the hydrolysis of the disaccharides, is the first step in the digestive process for these carbohydrates in the GI tract. Specific enzymes help to break down disaccharides into their component monosaccharides. Some common disorders of the body are due to the lack of these enzymes. The most notable is *lactose intolerance*, in which the enzyme lactase that breaks down lactose into glucose and galactose is lacking. Since lactose is the sugar in milk and other dairy products, a person unable to digest this sugar will experience gas pains and cramps, as well as diarrhea, after eating foods that contain milk. The lactose becomes food for bacteria in the GI tract. The person may be administered doses of the needed enzyme in order to digest the sugar.

2.21 In what ways do polysaccharides differ from monosaccharides and disaccharides?

Polysaccharides, or starches, are sometimes called *complex carbohydrates* because they contain many chemical bonds. The body is able to break them down in a more efficient and steady manner, supplying energy over a longer period of time, than is possible from the digestion of monosacchrides or disaccharides. Also, polysaccharides lack the characteristic sweet taste of monosaccharides and disaccharides.

Objective J To describe the chemical composition of *lipids*.

The building blocks of **lipids** (fats and oils) are fatty acids, which have long chains of carbon atoms bonded together and to hydrogen atoms. These fatty acids bond to a glycerol (a special three-carbon alcohol) to form the basic lipid molecule (fig. 2.8).

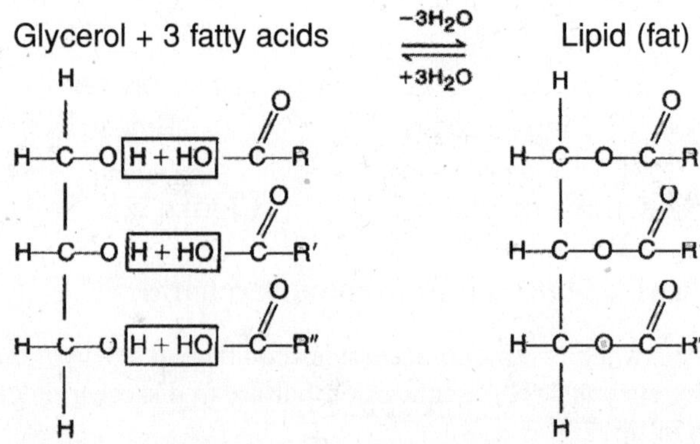

Figure 2.8 The formation of a basic lipid molecule (a triacylglycerol).

2.22 Distinguish between saturated and unsaturated fats, and give examples of each.

In *saturated fats*, each carbon in the molecule is bonded to as many hydrogens as possible; there are no double bonds between carbons. *Unsaturated fats* have at least one pair of carbons joined by a double bond.

Saturated:

butyric acid $CH_3(CH_2)_2COOH$
palmetic acid $CH_3(CH_2)_{14}COOH$

Unsaturated:

oleic acid $CH_3(CH_2)_7CH{=}CH(CH_2)_7COOH$
linolenic acid $CH_3(CH_2CH{=}CH)_3CH_2(CH_2)_6COOH$

Objective K To describe the chemical composition of *proteins*.

Proteins are large complex molecules formed by the dehydration synthesis of amino acids. The bonds between amino acids in a protein molecule are called peptide bonds and link the amino group (NH_2) of one amino acid to the acid carboxyl group (COOH) of another amino acid, which may be the same as or different from that of the first amino acid (fig. 2.9). If the molecular weight of the chain exceeds 10,000, the molecule is called a protein; smaller chains are called polypeptides. The function of the protein is determined by the character of the amino acids it contains. Proteins are the most diverse class of molecules and their functions vary widely.

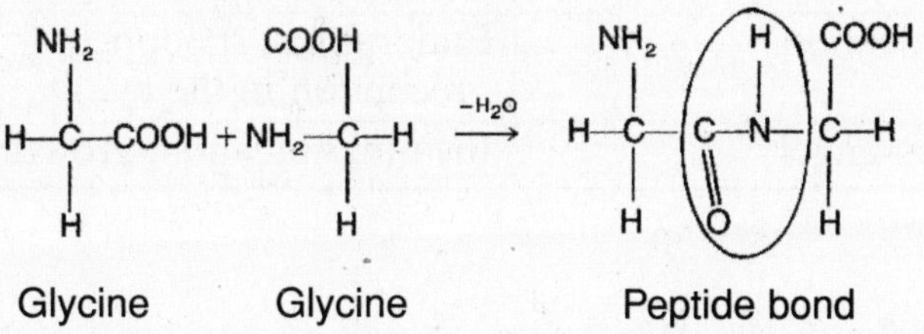

Glycine + Glycine $\xrightarrow{-H_2O}$ Peptide bond

Figure 2.9 The formation of a peptide bond between amino acids.

2.23 List the 20 amino acids and give their abbreviations.

See table 2.6.

2.24 What is the meaning of the term *essential amino acid*?

The body is able to convert certain amino acids to others; 12 of the 20 amino acids can be synthesized in this way. The remaining 8 are known as the essential amino acids because they must be supplied in the diet.

2.25 List some major functions of proteins and give some common examples.

See table 2.7.

Table 2.6 The 20 Amino Acids

Nonpolar	*Polar, uncharged*	*Polar, charged*
Glycine (Gly)	Serine (Ser)	Lysine (Lys)
Alanine (Ala)	Threonine (Thr)	Arginine (Arg)
Valine (Val)	Asparagine (Asn)	Histidine (His)
Leucine (Leu)	Glutamine (Gln)	Aspartic acid (Asp)
Isoleucine (Ile)	Tyrosine (Tyr)	Glutamic acid (Glu)
Methionine (Met)	Cysteine (Cys)	
Proline (Pro)		
Phenylalanine (Phe)		
Tryptophan (Trp)		

Table 2.7 Functions of Proteins and Examples

Function of proteins	*Examples*
Enzyme	Trypsin, chymotrypsin, sucrase, amylase
Transport and storage of molecules	Hemoglobin, myoglobin
Motion	Actin, myosin, tubulin (ciliary motion)
Structural support	Collagen, elastin
Immunity	Antibodies (immunoglobulins)
Neural communication	Endorphins, rhodopsin (pigment for light reception in the eye)
Intercellular messenger	Insulin, glucagon, growth hormones

Objective L To describe the chemical composition of *nucleotides*, the components of *nucleic acids.*

Survey

As indicated in fig. 2.10, **nucleotides** have three parts: a phosphate group (solid circle), a pentose sugar, and a nitrogenous base (oval). The pentose is always ribose in RNA and deoxyribose in DNA. The phosphate remains constant from one nucleotide to the next, but the base (in DNA) may be one of the following four: adenine (A) thymine (T), guanine (G), or cytosine (C). RNA substitutes uracil (U) for thymine. The nucleotides are joined together by dehydration synthesis into macromolecules. The structure and function of the DNA and RNA molecules are discussed in chapter 3.

2.26 Explain the difference between *purines* and *pyrimidines.*

Of the four nitrogenous bases of DNA, two are called **purine bases** and two are called **pyrimidine bases**. Fig. 2.11 shows two ring structures that contain nitrogen as well as carbon atoms. A comparison with fig. 2.12 shows that adenine and guanine are built on the purine ring, while cytosine and thymine are built on the pyrimidine ring.

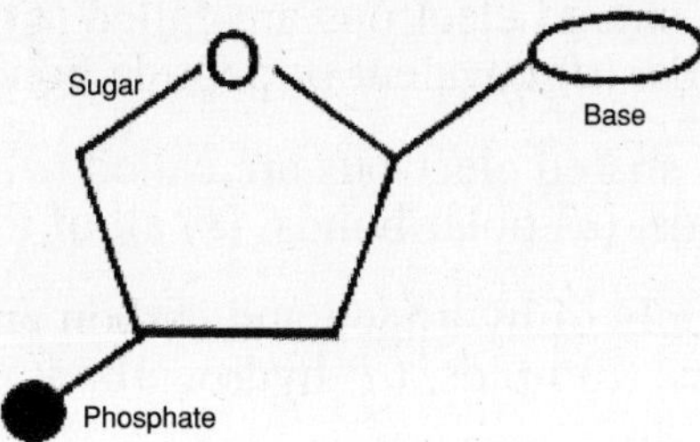

Figure 2.10 Components of a nucleotide.

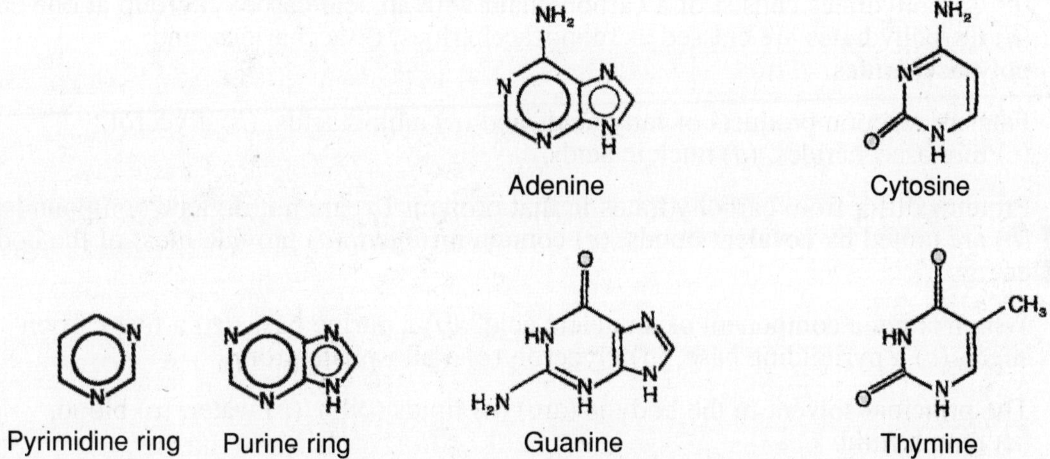

Figure 2.11 Basic ring structures.

Figure 2.12 Nitrogenous bases of DNA.

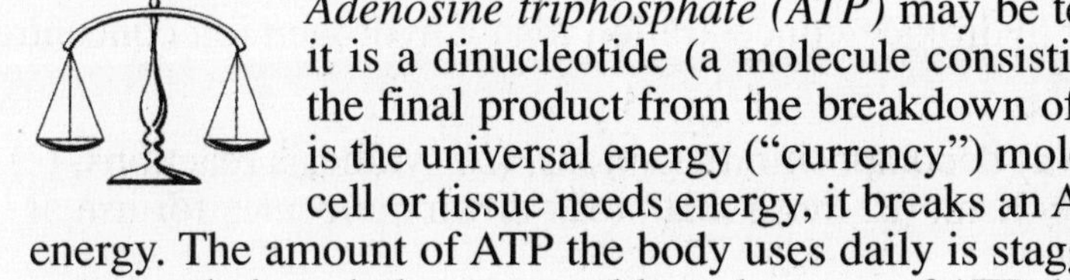

Adenosine triphosphate (ATP) may be termed a nucleic acid because it is a dinucleotide (a molecule consisting of two nucleotides). ATP, the final product from the breakdown of glucose and all other foods, is the universal energy ("currency") molecule of the body. Any time a cell or tissue needs energy, it breaks an ATP molecule apart to get that energy. The amount of ATP the body uses daily is staggering. If the molecules were not recycled, each day we would need a store of ATP that weighed approximately 50 pounds.

Review Exercises

Multiple Choice

1. A neutral atom contains (*a*) the same number of electrons as it does protons, (*b*) more protons than electrons, (*c*) the same number of electrons as it does neutrons, (*d*) more electrons than protons.

2. The number of protons in an atom is given by the (*a*) mass number, (*b*) atomic number, (*c*) difference between the atomic number and the mass number, (*d*) atomic weight.

3. A compound is a molecule (*a*) composed of two or more atoms, (*b*) composed of only one type of atom, (*c*) linked only by covalent bonds, (*d*) containing carbon.

4. Bonds that result from shared electrons are called (*a*) ionic bonds, (*b*) covalent bonds, (*c*) peptide bonds, (*d*) covalent or peptide bonds, (*e*) ionic or covalent bonds.

5. Bonds that result from shared electrons are called (*a*) ionic bonds, (*b*) covalent bonds, (*c*) peptide bonds, (*d*) polar bonds, (*e*) all of the preceding.

6. Molecules composed only of hydrogen and carbon are called (*a*) carbohydrates, (*b*) inorganic molecules, (*c*) lipids, (*d*) hydrocarbons.

7. Which of the following is a *false* statement?
(*a*) Carbohydrates are linked through dehydration reactions.
(*b*) Carbohydrates are composed of carbon, hydrogen, and oxygen.
(*c*) Carbohydrates consist of a carbon chain with an acid carboxyl group at one end.
(*d*) Carbohydrates are classed as monosaccharides, disaccharides, and polysaccharides.

8. Fats are reaction products of fatty acids and (*a*) amino acids, (*b*) glycerol, (*c*) monosaccharides, (*d*) nucleic acids.

9. Proteins differ from carbohydrates in that proteins (*a*) are not organic compounds, (*b*) are united by covalent bonds, (*c*) contain nitrogen, (*d*) provide most of the body's energy.

10. Which is *not* a component of a nucleic acid? (*a*) a purine base, (*b*) a five-carbon sugar, (*c*) a pyrimidine base, (*d*) glycerol, (*e*) a phosphate group

11. The principal solvent in the body is (are) (*a*) lipids (oils), (*b*) water, (*c*) blood, (*d*) lymph fluid.

12. Which of the following is a *false* statement?
(*a*) Acids increase hydrogen ion concentration in solution.
(*b*) Acids act as proton donors.
(*c*) Acids yield a higher hydroxide concentration than a hydrogen ion concentration.
(*d*) Acids have a low pH.

13. Anabolic reactions are (*a*) decomposition reactions, (*b*) synthesis reactions, (*c*) not part of the body's metabolism, (*d*) those that break down molecules for use as energy sources.

14. Deoxyribonucleotides are named according to (*a*) the base, (*b*) the sugar, (*c*) the phosphate group, (*d*) their position in the macromolecule.

15. Molecular weight is equal to (*a*) the sum of all the isotopic weights, (*b*) the sum of all the atomic weights, (*c*) the sum of the atomic numbers, (*d*) none of the preceding.

16. Phospholipids involve a phosphate group and (*a*) four or more fatty acids, (*b*) three fatty acids, (*c*) two fatty acids, (*d*) one fatty acid.

17. Of the following nitrogenous bases, which is found exclusively in RNA? (*a*) thymine, (*b*) guanine, (*c*) adenine, (*d*) uracil

18. Which represents the correct sequence in ascending order of size? (*a*) atom, amino acid, polypeptide, protein; (*b*) amino acid, atom, polypeptide, protein; (*c*) atom, amino acid, protein, polypeptide; (*d*) amino acid, atom, protein, polypeptide

19. Ions have (*a*) only positive charges, (*b*) only negative charges, (*c*) either positive or negative charges, (*d*) no charges.

20. Atoms of the same atomic number but of different mass numbers (different numbers of nuclear particles) are referred to as (*a*) ions, (*b*)isotopes, (*c*) cations, (*d*) tight atoms.

21. Which of the following is *not* an organic compound? (*a*) starch, (*b*) ribose, (*c*) carbon dioxide, (*d*) lipase

22. Which of the following is a disaccharide? (*a*) glucose, (*b*) ribose, (*c*) fructose, (*d*) lactose

23. The eight amino acids that cannot be formed in the body from other amino acids are referred to as (*a*) essential enzymes, (*b*) neutral amino acids, (*c*) normal amino acids, (*d*) essential amino acids.

24. Dehydration synthesis (*a*) requires water, (*b*) results in the splitting of molecules, (*c*) is the means for forming disaccharides, (*d*) occurs when glycogen stores are used by tissue cells.

25. Nucleotides lack (*a*) a phosphate group, (*b*) an amino group, (*c*) a nitrogenous base, (*d*) a five-carbon sugar.

True or False

____ **1.** Protons and electrons each have many times the mass of neutrons.

____ **2.** Of the 110 presently known elements, 75% are found in the body.

____ **3.** Sodium has atomic number 11 and mass number 23. Sodium, therefore, has 12 neutrons.

____ **4.** Positively charged ions are called cations.

____ **5.** Unsaturated fatty acids contain only single covalent bonds between carbon atoms.

____ **6.** Amino acids are linked by peptide bonds to form polypeptides.

____ **7.** The specific nature of a protein is determined mainly by its amino acid sequence and the properties of the respective amino acid R-groups.

____ **8.** Substances that increase the hydrogen ion concentration are called bases.

____ **9.** Covalent bonds are far more important in living organisms than ionic bonds.

____ **10.** Hydrogen, carbon, nitrogen, and oxygen account for about half of the body weight.

____ **11.** Nucleic acid molecules are small and unspecialized molecules.

____ **12.** Purine bases have a single ring of carbon and nitrogen atoms.

Completion

1. A ________________ is a combination of two or more atoms joined by chemical bonds.
2. ________________ ________________ form when atoms give up or gain electrons and become either positively or negatively charged.
3. A(n) ________________ bond is the strongest of the chemical bonds.
4. Composed of thousands of glucose molecules, the "food storage" polysaccharide in humans is called ________________.
5. ________________ are four-carbon sugars.
6. Three ________________ ________________ bound to one ________________ molecule form the basic lipid molecule.
7. ________________ fats contain no double bonds between carbon molecules but ________________ fats do.
8. The base ________________, unique to RNA, substitutes for the base thymine.
9. The nitrogenous bases ________________ and ________________ are built onto the purine ring.
10. ________________ ________________ are the building blocks of proteins.

Matching

Match the chemical component with its description.

____	**1.** Carbohydrates	(*a*)	proton acceptor
____	**2.** Protons and neutrons	(*b*)	adenine and guanine
____	**3.** Electrons	(*c*)	$C_n(H_2O)_n$
____	**4.** Covalent bonds	(*d*)	Cl^-
____	**5.** Nucleic acid	(*e*)	nucleus
____	**6.** Lipids	(*f*)	proton donor
____	**7.** Proteins	(*g*)	subshells
____	**8.** Hydrogen bonds	(*h*)	DNA and RNA
____	**9.** Peptide bonds	(*i*)	K^+
____	**10.** Purine base	(*j*)	cytosine and thymine
____	**11.** Pyrimidine bases	(*k*)	primary structure of proteins
____	**12.** Cation	(*l*)	secondary structure of proteins
____	**13.** Anion	(*m*)	water insoluble
____	**14.** Acid	(*n*)	shared electrons
____	**15.** Base	(*o*)	$H_2O - \underset{\displaystyle R}{\underset{\vert}{CH}} - COOH$

Answers and Explanations for Review Exercises

Multiple Choice

1. (*a*) Neutral refers to an absence of electrical charge or balance of opposite electrical charges. Since electrons carry a negative charge, the same number of protons are needed to balance the overall electrical charge.
2. (*b*) The atomic number represents the number of protons in an atom.
3. (*a*) A compound is a molecule composed of two or more atoms (e.g., H_2O, NaCl).
4. (*d*) Covalent and peptide bonds result from atoms sharing electrons.
5. (*a*) Ionic bonds result from the transfer of electrons.
6. (*d*) The prefix *hydro-* refers to hydrogen, and the suffix *-carbons* refers to carbon.
7. (*c*) A carbon chain with an acid carboxyl group at one end is a fatty acid.
8. (*b*) Three fatty acid molecules combine with one glycerol molecule to form one fat molecule (see fig. 2.7).
9. (*c*) Proteins contain nitrogen, whereas carbohydrates contain only carbon, hydrogen, and oxygen.
10. (*d*) A nucleic acid is composed of a five-carbon sugar bonded to a phosphate group and either a purine or a pyrimidine base (see fig. 2.10).
11. (*b*) Water is the principal solvent in the body.
12. (*c*) Acids yield a higher *hydrogen ion* concentration. Bases yield a higher *hydroxide* concentration.
13. (*b*) Anabolic reactions include the synthesis of large energy-storing molecules, such as glycogen, fat, and protein.
14. (*a*) The components of DNA nucleotides are identical except for the nitrogenous base,which determines the chemical nature of the entire molecule.
15. (*b*) The molecular weight is calculated by adding the atomic weights of all the atoms in the molecule.
16. (*c*) Phospholipids involve a phosphate group and two fatty acid molecules.
17. (*d*) The nitrogenous base uracil is found exclusively in RNA.
18. (*a*) Atoms are the building blocks for amino acids. Several amino acids strung together form a polypeptide. Several polypeptides strung together form a protein.
19. (*c*) Ions can be positively or negatively charged (e.g., Na^+, Cl^-, H^+, OH^-).
20. (*b*) Isotopes of an atom contain the same number of protons but a different number of neutrons.
21. (*c*) Carbon dioxide (CO_2) and carbon monoxide (CO) are the only notable exceptions to the rule that molecules containing carbon are organic molecules. These molecules form in natural processes that do not involve other organic molecules, as well as in organic systems.
22. (*d*) Lactose is composed of two monosaccharides: glucose and galactose bound together.
23. (*d*) The essential amino acids are those eight that cannot be formed within the body.
24. (*c*) A disaccharide forms when two monosaccharides combine in a dehydration synthesis reaction (see fig. 2.6).
25. (*b*) A nucleotide consists of a five-carbon sugar covalently bonded to a nitrogenous base and a phosphate group.

True or False

1. False; protons and neutrons are many times more massive than electrons
2. False; of the 106 known elements, only about 22 (21%) are found in the body
3. True
4. True
5. False; unsaturated fatty acids contain two covalent bonds between carbon atoms
6. True
7. True
8. False; bases increase the hydroxide ion (OH^-) concentration while acids increase the hydrogen ion (H^+) concentration
9. True
10. False; H, C, N, and O account for over 90% of human body weight
11. False; nucleic acids are large, highly specialized molecules
12. False; purine bases have a double ring of carbon and nitrogen atoms; pyrimidine bases have a single carbon and nitrogen ring

Completion

1. molecule
2. Ionic bonds
3. covalent bond
4. glycogen
5. Tetroses
6. fatty acids, glycerol
7. Saturated, unsaturated
8. uracil
9. adenine, guanine
10. Amino acids

Matching

1. (*c*)
2. (*e*)
3. (*g*)
4. (*n*)
5. (*h*)
6. (*m*)
7. (*o*)
8. (*l*)
9. (*k*)
10. (*b*)
11. (*j*)
12. (*i*)
13. (*d*)
14. (*f*)
15. (*a*)

Cell Structure and Function 3

Objective A To distinguish between *prokaryotic* and *eukaryotic cells.*

Prokaryotic cells (fig. 3.1*a*) lack a membrane-bound nucleus; instead they contain a single strand of *nucleic acid.* These cells contain few organelles. A rigid or semirigid *cell wall* surrounding *cell (plasma) membrane* gives the cell its shape.

Eukaryotic cells (fig. 3.1*b*) contain a true *nucleus* with multiple chromosomes. They also have several types of specialized membrane-bound *organelles.* Like prokaryotes, eukaryotic cells have a *cell (plasma) membrane.* Since all human cells are eukaryotic, most of this chapter will focus on eukaryotic cells and their functions.

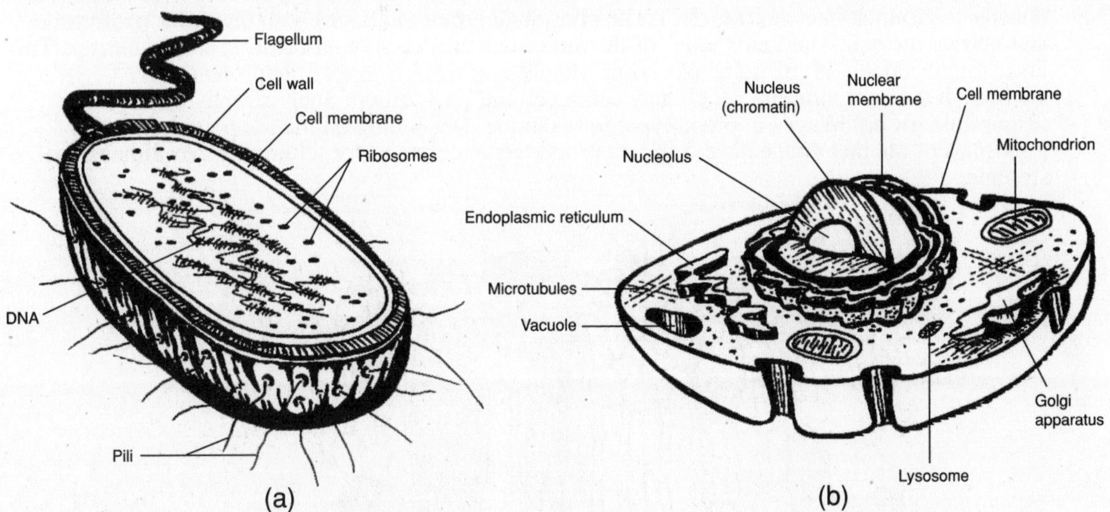

Figure 3.1 The structure of (*a*) a prokaryotic cell and (*b*) a eukaryotic cell.

3.1 Give examples of both prokaryotic and eukaryotic cells.

Bacteria are examples of prokaryotic single-celled organisms. Cyanobacteria (formerly called blue-green algae) are also prokaryotic organisms.

Organisms composed of eukaryotic cells include protozoa, fungi, algae, plants, and invertebrate and vertebrate animals.

Note that *viruses* are not classified as either prokaryotes or eukaryotes. Since they meet some, but not all, criteria for life, there is debate as to whether viruses are living organisms. They do carry DNA (or sometimes RNA), and pass it from generation to generation, but they do so only as parasites by making use of the reproductive process of a living host cell. They do not metabolize, respond to the environment, secrete or excrete. Still, viruses play an important role as pathogens, causing many common diseases, including the common cold, flu, polio, measles, chicken pox, and AIDS. Viruses are useful in genetic research, providing one means by which molecular biologists can introduce genetic material from one organism into another.

Objective B To describe the structures common to all cells.

Survey

Both prokaryotic and eukaryotic cells are bound by a *cell (plasma) membrane*. The genetic information of the cell is encoded in DNA, which is organized into bodies called *chromosomes*. The fluid matrix of the cell's interior is the *cytoplasm*. All cells express their genetic information by manufacturing proteins at small, granular organelles called *ribosomes*, which are made of RNA and proteins. Some prokaryotes, and some eukaryotes as well, have a *cell wall* in addition to the cell membrane. The cell wall gives extra protection and rigidity to the cell.

3.2 Describe the cell (plasma) membrane.

A major component of the cell membrane is **phospholipid**, a lipid (fat) molecule with a charged phosphate group at one end (fig. 3.2). The phosphate group interfaces with the water both inside and outside the cell. The lipid "tails" of the molecules face each other creating a *lipid bilayer*. This layer is embedded with **proteins** of various shapes and sizes. Some of the proteins have **carbohydrate components**, which may act in cellular recognition. Such carbohydrates are responsible for differences in blood type, for example. Depending on the temperature, the molecules of the membrane have different properties and may move within the two-dimensional structure.

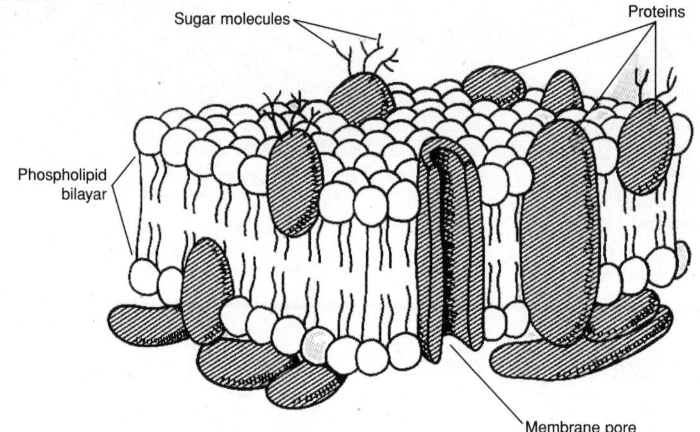

Figure 3.2 The structure of the cell membrane.

3.3 What is meant by *selectively permeable*, and why is this term applied to the cell membrane?

Permeable means that substances can pass through. *Selectively permeable* means that certain substances can pass through, but not others. One of the important functions of the cell membrane is to regulate which substances are admitted into and out of the cell. Water, alcohol, and gases readily pass through the cell membrane, but ions, large proteins, and carbohydrates do not.

3.4 What are the processes by which substances are transferred across cell membranes?

Diffusion – The movement of any substance from an area where it is more concentrated to an area where it is less concentrated. Oxygen enters the cell in this manner. It moves from the blood, where it is concentrated, to the inside of the cell, where it is not concentrated.

Osmosis – A type of diffusion, but involving only the movement of water across a membrane. The water moves to the side of the membrane that contains the most molecules of solute dissolved in it.

Facilitated transport – Accomplished by proteins that form gates, or channels, in the membrane, allowing the passage of large or charged molecules that would otherwise be restricted.

Active transport – The process of using energy to "pump" molecules across the membrane against the normal direction of diffusion. ATP is required.

Phagocytosis – The process by which a cell engulfs a foreign substance or body, like an amoeba engulfing its prey. The engulfed substance becomes contained in a membrane-bound vesicle before being digested or used otherwise.

Pinocytosis – The pumping of water across the membrane.

3.5 List the components of cytoplasm.

Cytoplasm is the fluid matrix within a cell. Consisting primarily of water, it suspends minute structures called *organelles*. Dissolved in the cytoplasm are

1. Gases, such as oxygen and carbon dioxide
2. Cellular wastes, such as urea
3. Building-block molecules, such as amino acids, fatty acids, and nucleotides
4. Food molecules, such as glucose
5. Ions, such as potassium (K^+), sodium (Na^+), chloride Cl^-), and calcium (Ca^{2+})
6. Proteins and RNA
7. Organelles of the cell, such as ribosomes and mitochondria (eukaryotes only)
8. ATP and other energy-carrying molecules
9. Hormones, drugs, or toxins transmitted by the blood

3.6 Explain the function of ribosomes.

Ribosomes are commonly called the "protein factories" of the cell. They are responsible for the process of translation, or taking the information from the DNA, encoded on RNA, and using it to create the proteins needed by the cell (see Objective D). Ribosomes are able to bond amino acids into long chains, in which the order of the different amino acids determines the properties and functions of the resulting protein. Ribsomes make proteins only when directed to do so by a piece of mRNA synthesized by the DNA in the nucleus. (Since prokaryotes have no nucleus, the ribosomes can work from the RNA before it has even left the DNA.)

3.7 Why is a cell wall needed in plants but not in animals?

The cells of all plants are surrounded by a rigid cell wall to the inside of the cell membrane. The pressure of the water inside the cell against the cell wall (turgor pressure) provides a rigidity to plants. This enables them to maintain their shape against gravity, wind, or other forces. Animals have adapted to the same forces by developing skeletal systems of various kinds, which generally allow more freedom of movement. The bony endoskeleton of humans and other vertebrates makes a cell wall unnecessary.

The cell walls of many kinds of bacteria are important in microbiology and in the use of antibiotics in clinical medicine. Whereas plant cell walls are made of cellulose, a glucose polymer, bacteria have cell walls made of peptidoglycan, a mixture of protein and carbohydrate. *Gram positive bacteria* are those that have a very simple cell wall and that readily accept chemical stains. They are also susceptible to antibiotics, which can readily penetrate the cell wall. *Gram negative bacteria* are more resistant to both stains and antibiotics. Gram staining provides the means for classifying the many kinds of bacteria. Some antibiotics, notably penicillin and its analogs, act by inhibiting the formation of the cell wall when bacterial cells divide.

3.8 What is the molecular composition of a chromosome?

DNA in the nucleus is wound tightly around proteins called **histones** (fig. 3.3). This wound strand is then coiled again many times around other proteins into large, rod-shaped molecules known as **chromosomes**. The DNA and proteins together are called **chromatin**. When a section of DNA is actively being expressed or replicated, that region of the chromosome is not tightly wound, but rather uncoiled to allow enzymes access to the DNA for replication or transcription.

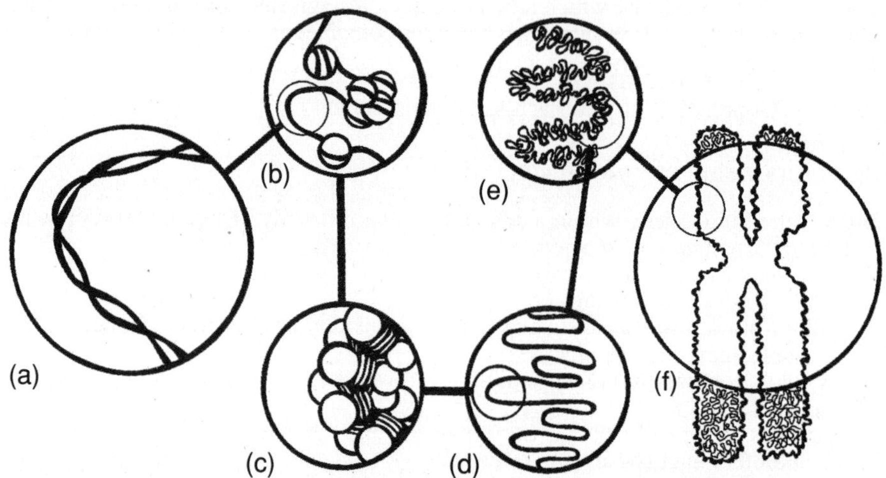

Figure 3.3 The structure of a chromosome. (*a*) A DNA double helix, (*b*) nucleosomes on a segment of DNA, (*c*) a chromatin fiber, (*d*) looped domains, (*e*) a portion of a chromatin (heterochromatin), and (*f*) a chromosome in metaphase.

3.9 How many chromosomes do humans have in each cell?

Every species of organism has a distinct number of chromosomes in each cell nucleus. Humans have 46 chromosomes, or 23 pairs (*diploid*) in each somatic (body) cell. The sex cells (*gametes*, sperm and eggs) have 23 chromosomes (*haploid*).

Some common diseases or developmental problems are related to chromosome count. *Down syndrome* is a condition caused by the presence of an extra chromosome (chromosome #21) in every nucleus. People with Down syndrome are usually mentally handicapped to some degree, exhibit characteristic developmental abnormalities, and appear to be extremely prone to developing other diseases, such as *Alzheimer's disease*. In addition, they generally have a shorter life span than other people.

Objective C To describe the *organelles* of eukaryotic cells.

An organelle is any subcellular structure having a specific function. In addition to the structures mentioned in Objective B, most eukaryotic cells have some or all of the organelles listed in table 3.1.

Table 3.1 Organelles of Eukaryotic Cells

Organelle	*Structure*	*Function*
Nucleus	Round or oval organelle; contains nucleolus and is surrounded by nuclear membrane; contains DNA organized into chromosomes	Storage of genetic material; control center for all cellular activities
Nucleolus	Round mass of RNA within nucleus	Center for organizing ribosomes and other products with RNA
Ribosomes	Granular particles composed of proteins	Synthesis of proteins
Endoplasmic reticulum (ER)	Membranous network through cytoplasm; continuous with cell and nuclear membranes	
Rough ER	Membranous network with attached ribosomes	Synthesis of proteins for use outside a cell
Smooth ER	Lacking ribosomes	Steroid synthesis; intercellular transport; detoxification
Golgi apparatus (complex)	Stacked membranes and vessels (cisternae)	Packaging of proteins produced at rough ER; formation of secretory vesicles and lysosomes
Mitochondria	Rodlike or oval organelles; membrane forms folds called cristae	ATP production (through Krebs cycle and oxidative phosphorylation)
Lysosomes	Dense vesicles filled with enzymes	Breakdown of worn cellular components or engulfed particles
Secretory vessicles	Membrane-bound sacs	Storage of proteins and other synthesized material destined for secretion
Microtubules	Long, hollow structures; made of polymerized tubulin (protein)	Structural support; involved in cell divisions, cell movement, and transport
Microfilaments	Long, solid fibers; made of polymerized actin (protein)	Structural support; involved in cell movement
Centrioles	Two short rods or granules, composed of nine sets of three fused microtubules; located near nucleus	Involved in cell division; movement of chromosomes during mitosis

3.10 What tissues would logically contain cells with large amounts of rough ER? With large amounts of smooth ER?

Cells that manufacture proteins for secretion contain large amounts of rough ER. For example, the acini cells in the pancreas contain many rough ER organelles because they secrete the digestive enzyme trypsin.

Cells that manufacture steroids, such as in the testes, ovaries, and adrenal glands, contain large amounts of smooth ER organelles. Smooth ER is also plentiful in the hepatocytes (liver cells), which are responsible for detoxifying blood and metabolizing toxins.

3.11 Describe the digestive action of a lysosome on a substance ingested by the cell through endocytosis.

See fig. 3.4.

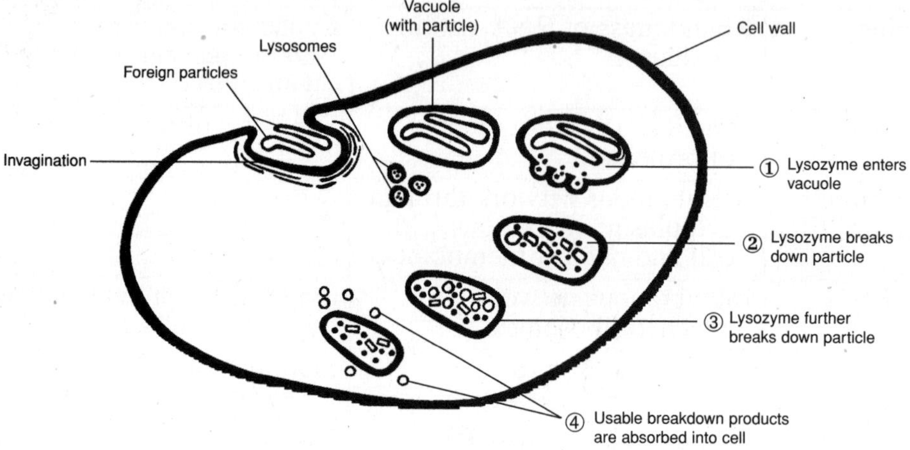

Figure 3.4 The digestive action of a lysosome on an ingested particle in the cell.

Many cell biologists believe that mitochondria, as well as chloroplasts in plants, evolved as independent prokaryotic cells which then developed an *endosymbiotic relationship* with larger cells. It is believed that these small prokaryotes gradually became dependent upon the larger host cells for food and protection as the host cells became dependent upon the smaller cells for energy production. Much evidence supports this *endosymbiont theory*, including similarities between genetic and protein material found in modern prokaryotes and organelles such as mitochondria and chloroplasts.

Objective D To describe the processes of *replication, transcription,* and *translation.*

Survey

Replication refers to the process in which DNA makes an identical copy of itself prior to cell division. **Transcription** refers to making mRNA from the DNA template. The mRNA then leaves the nucleus and joins with a ribosome in the cytoplasm to synthesize a protein in a process called **translation.** The three processes are sometimes referred to collectively as the *central dogma of biology*, since they constitute the method common to all life for the expression of genetically encoded information.

3.12 Diagram and describe the structure of the DNA molecule.

Each strand of DNA is composed of nucleotides linked together by phosphodiester bonds. The two strands are then wound around each other in a right-handed direction to form a double helix (fig. 3.5). The strands are complementary to each other, meaning that the bases of one strand are matched with their complementary bases on the other strand, A with T and C with G. (See chapter 2 for an explanation of the bases of DNA.)

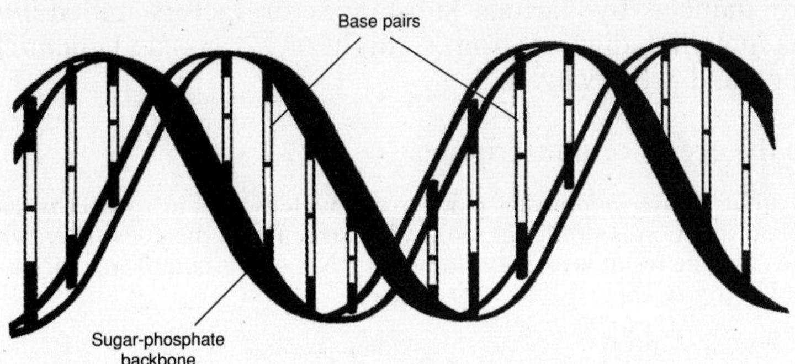

Figure 3.5 The double-helix structure of DNA.

3.13 Describe the events in replication.

Each step in the process of *replication* is accomplished by enzymes designed specifically for that step (fig. 3.6). An enzyme called *helicase* first unwinds the double helix into two parallel strands, then "unzips" the two strands. DNA polymerases I and III then move in between the separated strands and copy them both by adding complementary bases to each strand, one at a time, until finally there are two double strands, identical to each other. Other enzymes prime the DNA, check the copies for mistakes, and fix mistakes where they have occurred. The entire process is remarkably accurate. The various enzymes that function at different stages reduce the error rate to one in 10 billion bases copied.

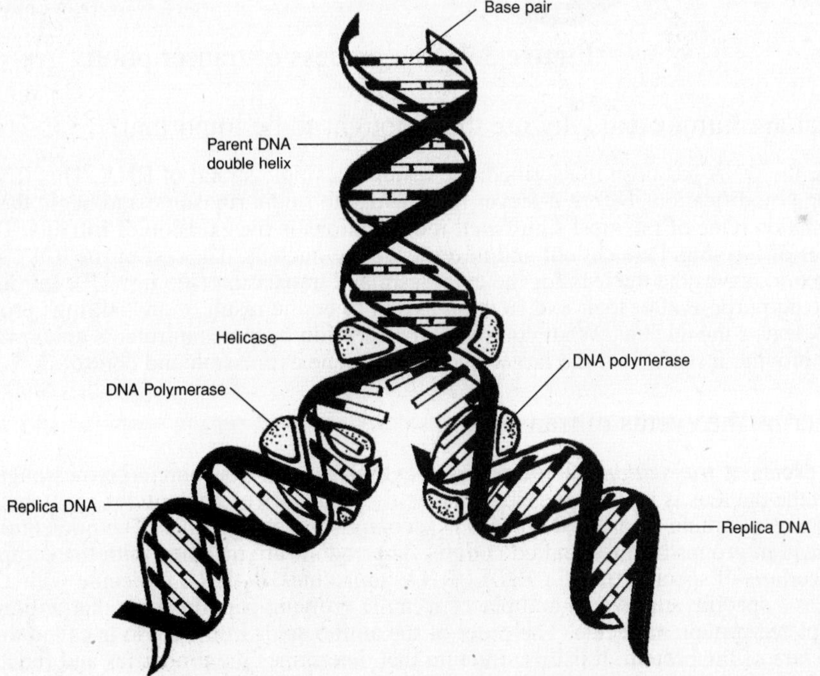

Figure 3.6 The steps of replication.

A *mutation* is an unrepaired mistake in replication. Mutations are surprisingly rare for the amount of replication that takes place. Still, they play an important role in creating diversity in the genetic makeup of a species. Most mutations are harmless or unnoticeable. Some may be harmful or even lethal to the organism, and others may be beneficial. Some mutations occur spontaneously, but many are induced by various substances or factors called *mutagens.* Common mutagens include radiation (from sunlight or X rays) and chemicals in certain dyes, sweeteners, and preservatives.

3.14 Describe the events of transcription.

Like replication, *transcription* takes place in the nucleus. It is instigated by specific enzymes. The process of transcription is similar to replication, except that the copying enzymes are RNA polymerases and the result is a single strand of RNA that is complementary to one of the DNA strands copied (fig. 3.7).

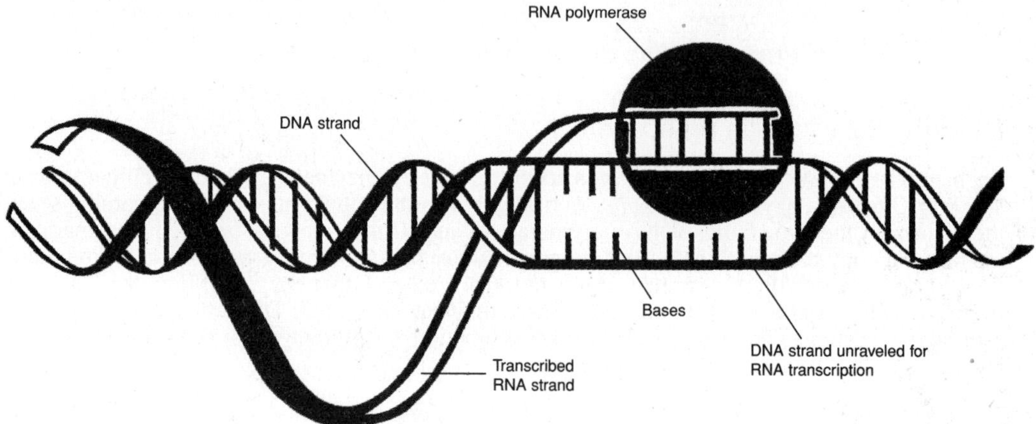

Figure 3.7 The process of transcription.

3.15 What are introns and why are they thought to be important?

After the DNA has been transcribed, the result is a single strand of RNA. This RNA undergoes several modifications before it leaves the nucleus to find a ribosome and begin the process of translation. One of the most significant modifications is the excision of **introns**. These noncoding pieces of the strand are cut out and never leave the nucleus. The rest of the RNA strand, known as the **exon**, leaves the nucleus for the cytoplasm and translation (see fig. 3.7). Introns apparently serve no purpose after their excision, and seem to be the result of an "editing" process before the RNA leaves the nucleus. What controls which section becomes introns is unknown, but the answer may provide insights into the larger question of gene expression and control.

3.16. Describe the events of translation.

The events of *translation* take place in the cytoplasm at a ribosome. The messenger RNA (mRNA) from the nucleus is held by the ribosome as its sequences of nucleotides are translated into a sequence of amino acids. The process is accomplished by enzymes. The nucleotides of the mRNA are read in groups of three, called **codons**. The codons are matched with the complementary **anticodons** of specific transfer RNA (tRNA) molecules. A tRNA molecule with a certain anticodon carries a specific amino acid and places it in the growing peptide chain that will eventually be a completed protein molecule. The order of the amino acids in the chain is called the *primary structure* of the protein. It is this structure that determines the properties and functions of the molecule.

Objective E To describe the processes of *mitosis* and *meiosis.*

Mitosis is the process of normal cell division (fig. 3.8). It occurs whenever body cells need to produce more cells for growth or for replacement and repair. The result of mitosis is two identical *daughter cells* with the same chromosomal content as the parent cell.

Meiosis is the process of *gamete* (sex cell) formation. It resembles mitosis in many ways, except that the end result is four daughter cells, each with half the chromosomal content of the parent cell (fig. 3.8).

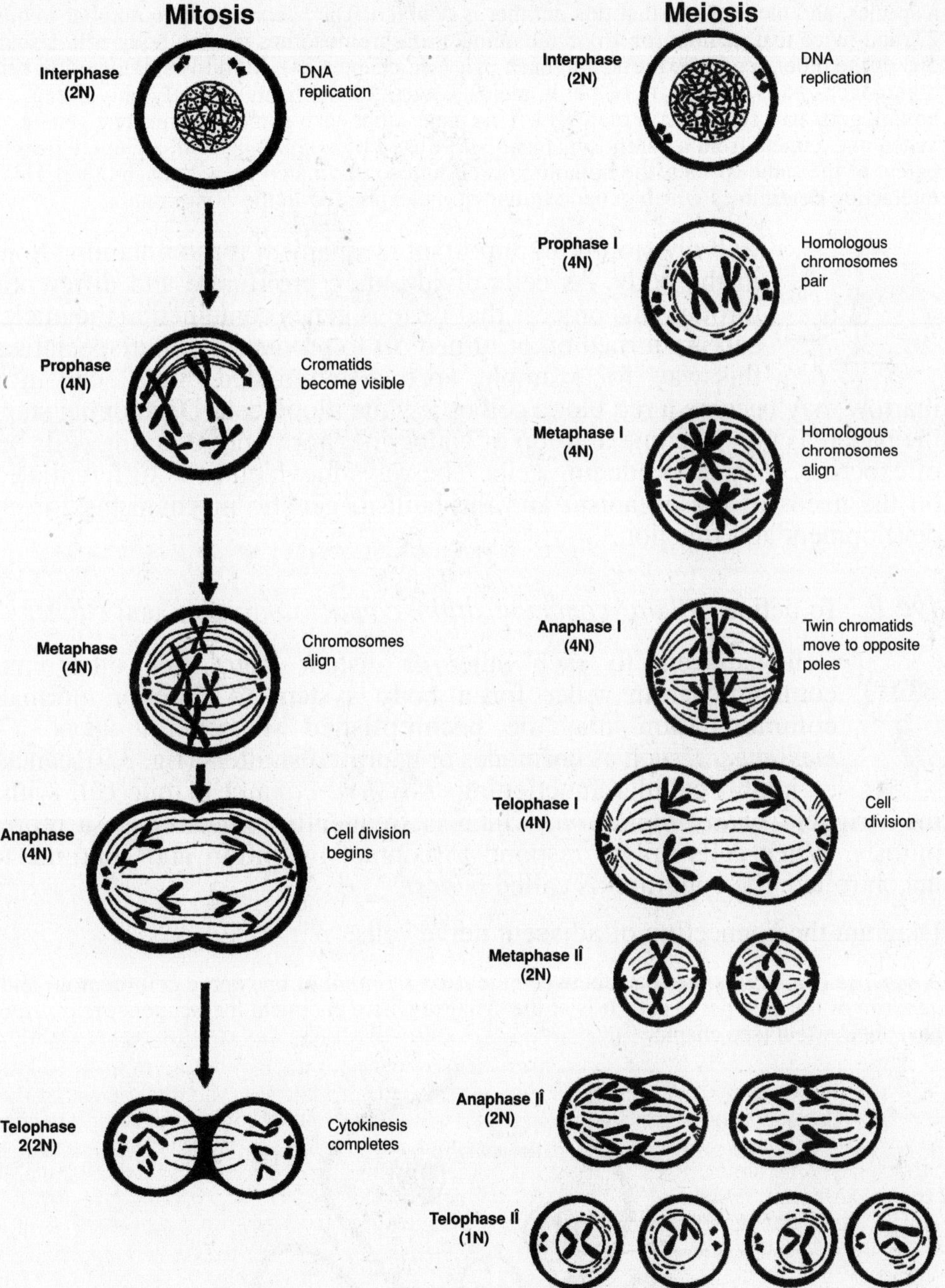

Figure 3.8 Stages of mitosis and meiosis.

3.17 Do the mutations that occur in mitosis and in meiosis have the same effect?

Mutations that occur during mitosis are not passed on to the next generation. Only the genetic information contained in gametes is passed to the offspring. Mutations that occur during meiosis, therefore, may establish traits or characteristics in all following generations.

3.18 Why are chromosomes organized into pairs in organisms that reproduce sexually?

In all cells, except for gametes, each chromosome has a homologous partner chromosome containing genes for the same traits. This pairing of chromosomes with similar traits is called a *diploid* (2*N*) condition. The letter *N* represents the specific number of chromosomes in each cell of a species, and the 2 means that this number is doubled. The *N* chromosome number in humans is 23, and twice that number (or 46) is the number of chromosomes in each body cell. Usually only one or the other gene is expressed in each pair (see chapter 24). In mitosis, interaction between the homologous pairs is not critical, but in meiosis, gametes can form correctly only if each homologous pair separates in meiosis I. This means that each gamete carries one gene per trait. When one gamete from a parent organism is fertilized by combining with a gamete from another parent of the same species, the homologous chromosomes match up with each other. This interaction determines which genetic traits will be expressed in the offspring.

Cell division is an important mechanism for maintaining homeostasis in the body. As cells divide, they proliferate and differentiate. It is through this process that various genes contained in the nucleus of the cell are turned on or turned off to produce cellular specializations. In this way, for example, an undifferentiated "stem" cell in red bone marrow may become a red blood cell or a white blood cell. Developing stem cells in the pancreas may become a group of endocrine, hormone-secreting cells or a group of exocrine, enzyme-producing cells. The ways in which cells differentiate depends on the needs of the organism and the built-in genetic mechanisms for control of development and function.

Objective F To define *cellular communication, contact inhibition,* and *cancer.*

Cells adjacent to each other or distant from each other must often communicate in order for a body system to function normally. This communication may be accomplished in several ways. *Chemical messengers,* such as hormones or neurotransmitters (fig. 3.9), can accelerate or inhibit cellular functioning. *Physical contact* of one cell with another may trigger *contact inhibition,* which is frequently expressed as a repression of mitosis. When cells fail to respond to contact inhibition and continue to divide uncontrolled, the condition is called *cancer.*

3.19 Diagram the connection of adjacent nerve cells.

A *synapse* (fig. 3.9) is the space between the *axon terminal* of one nerve cell (neuron) and the *dendrite* of the next nerve cell. It is at the synapse where chemical messengers (*neurotransmitters*) have their effect (see chapter 9).

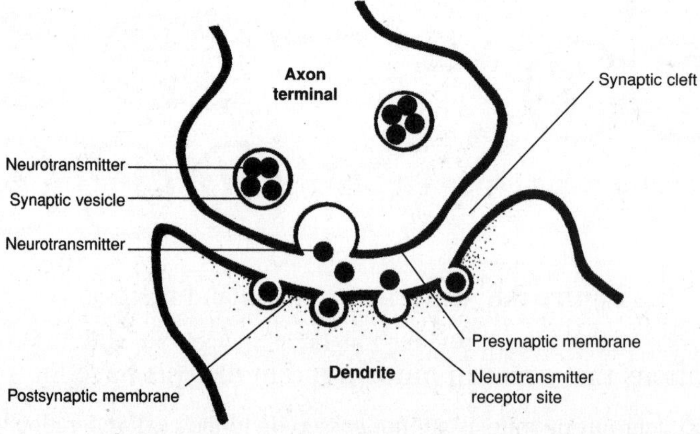

Figure 3.9 A synapse is the space between a presynaptic neuron and a postsynaptic neuron.

3.20 Describe the manufacture and secretion of a protein hormone.

Hormones are proteins, steroids, or other molecules (see chapter 13). Protein hormones are produced through the process of translation (as are other proteins) at a ribosome. Because the hormone is destined for secretion from the cell, the protein is contained in a membrane vesicle and packaged for secretion by the Golgi apparatus (see table 3.1). When a cell is stimulated to release hormones through a feedback mechanism, the membrane vesicle fuses with the cell membrane and the contents of the vesicle diffuse into the interstitial fluid surrounding the cell. Hormones are quickly carried by the blood to their target locations.

3.21 List some common kinds of cancer and their frequency of occurrence in the United States.

See table 3.2.

Table 3.2 Kinds of Cancer and Frequency of Occurrence

Lung cancer	20% of all cancers in men; 11% in women
Breast cancer	28% of all cancers in women
Prostate cancer	21% of all cancers in men
Skin cancer (melanoma)	3% of all cancers in both men and women
Colon/rectal cancer	15% of all cancers in both men and women
Ovarian cancer	4% of all cancers in women
Uterine cancer	9% of all cancers in women
Leukemia/lymphomas	17% of all cancers in men, 7% in women

3.22 What are some causes of cancer?

Finding out what causes cancer is the goal of a great deal of research. Some evidence indicates that certain viruses have tumor-inducing effects. Similar effects may result from genes carried normally in the genome that are activated by unknown stimuli. Many substances, collectively called *carcinogens*, are known to cause cancer. Generally, a critical level of exposure is needed for a substance to be carcinogenic. Known carcinogens include tobacco, alcohol, radiation from the sun and from X rays, industrial chemicals (such as asbestos and vinyl chloride), a high-fat diet, and certain drugs (such as steroids).

3.23 List the warning signs that may indicate the development of cancer.

The warning signs of cancer are listed below. People with any of these symptoms should consult a physician immediately.

- A lump or thickening in the breast or any other part of the body
- A sore that does not heal
- Rapid change in a birthmark, wart, or mole
- A hoarse voice or cough that does not go away
- Indigestion or difficulty swallowing
- Unusual bleeding
- A change in bowel or urinary habits

3.24 List and describe some methods of treating cancer.

Surgery – Especially effective in removing tumors that have not metastasized (spread to other locations). Surgery is usually coupled with other treatments.

Radiation therapy – Also known as *irradiation, cobalt therapy,* or *X-ray therapy*. Radiation therapy bombards the cancerous tissue with high-energy radiation, which kills the cells of the tumor.

Chemotherapy – The use of anticancer drugs. Chemotherapy is the most effective treatment of cancer that has metastasized.

All treatments for cancer have limitations, and some may have negative side effects. Treatment is most effective on cancers that have been detected early and that have not metastasized.

Review Exercises

Multiple Choice

1. The cell membrane (*a*) encloses components of the cell, (*b*) regulates absorption, (*c*) gives shape to the cell, (*d*) does all of the preceding.

2. The largest structure in the cell is (*a*) the Golgi apparatus, (*b*) the nucleus, (*c*) the ribosome, (*d*) the mitochondrion.

3. Which organelle contains hydrolytic enzymes? (*a*) lysosome, (*b*) ribosome, (*c*) mitochondrion, (*d*) Golgi apparatus

4. In question 3, which organelle is involved in protein synthesis?

5. Endoplasmic reticulum with attached ribosomes is called (*a*) smooth ER, (*b*) a Golgi apparatus, (*c*) nodular ER, (*d*) rough ER.

6. Engulfing of solid material by cells is called (*a*) pinocytosis, (*b*) phagocytosis, (*c*) active transport, (*d*) diffusion.

7. The cell membrane is a "sandwich" of (*a*) lipid-protein-lipid, (*b*) lipid-lipid-protein, (*c*) protein-protein-lipid, (*d*) protein-lipid-protein.

8. The function of the Golgi apparatus is (*a*) packaging of material in membranes for transport out of the cell, (*b*) production of mitotic and meiotic spindles, (*c*) excretion of excess water, (*d*) production of ATP by oxidative phosphorylation.

9. The function of mitochondria is (*a*) packaging of materials in membranes for transport out of the cell, (*b*) conversion of light energy to chemical energy in the form of ATP, (*c*) excretion of excess water from the cell, (*d*) synthesis of ATP by oxidative phosphorylation.

10. During protein synthesis, amino acids become linked together in a linear chain by (*a*) hydrogen bonds, (*b*) peptide bonds, (*c*) ionic bonds, (*d*) phosphate bonds, (*e*) amino bonds.

11. Which nucleotide base is absent from DNA? (*a*) adenine, (*b*) cytosine, (*c*) guanine, (*d*) thymine, (*e*) uracil, (*f*) none of the preceding

12. Messenger RNA is synthesized in (*a*) the nucleus, under the direction of DNA; (*b*) the cytoplasm, under the direction of the centrioles; (*c*) the centrioles, under the direction of DNA; (*d*) the Golgi apparatus, under the direction of DNA.

13. The sequence of nucleotides in a messenger RNA molecule is determined by (*a*) the sequence of the nucleotides in a gene, (*b*) the enzyme RNA polymerase, (*c*) the sequence of amino acids in a protein, (*d*) the enzyme ribonuclease (RNase), (*e*) the sequence of nucleotides in the anticodons of tRNA.

14. The flow of genetic information in most organisms may be indicated as (*a*) protein-DNA-mRNA, (*b*) protein-tRNA-DNA, (*c*) DNA-mRNA-protein, (*d*) four nucleotides.

15. The genetic code for a single amino acid consists of (*a*) one nucleotide, (*b*) two nucleotides, (*c*) three nucleotides, (*d*) four nucleotides.

16. In the DNA molecule, the nitrogenous base adenine always pairs with (*a*) uracil, (*b*) thymine, (*c*) cytosine, (*d*) guanine.

17. The "backbone" of DNA consists of repetitive sequences of phosphate and (*a*) sugar (glucose), (*b*) sugar (deoxyribose), (*c*) nucleic acids, (*d*) protein (ribose).

18. The molecule to which an amino acid is attached preparatory to protein synthesis is (*a*) ribosomal RNA, (*b*) messenger RNA, (*c*) transfer RNA, (*d*) viral RNA, (*e*) nucleolar RNA.

19. The sequence of amino acids in protein molecules is determined by the sequence of (*a*) amino acids in other protein molecules, (*b*) bases in transfer RNA, (*c*) bases in messenger RNA, (*d*) bases in ribosomal RNA, (*e*) sugars in DNA.

20. A certain gene has 1200 nucleotides (bases) in the coding portion of one strand. The protein coded for by this gene consists of (*a*) 400 amino acids, (*b*) 600 amino acids, (*c*) 1200 amino acids, (*d*) 2400 amino acids, (*e*) 3600 amino acids.

21. If one strand of a DNA molecule has the base sequence ACGGCAC, the other strand will have the sequence: (*a*) ACGGCAC, (*b*) CACGGCA, (*c*) CATTACA, (*d*) UGCCGUG, (*e*) TGCCGTG.

22. A transfer RNA has the anticodon sequence UAC. With which codon in the messenger RNA will it pair? (*a*) GGC, (*b*) UAC, (*c*) AUU, (*d*) CAU, (*e*) AUG

23. Chromosome duplication takes place during (*a*) telophase, (*b*) interphase, (*c*) metaphase, (*d*) anaphase.

24. In which phase of question 23 does cytokinesis (cleavage) occur?

25. The two daughter cells formed by mitosis have (*a*) identical genetic constitutions, (*b*) exactly half as many genes as the parent cell, (*c*) the same amount of cytoplasm as the parent cell, (*d*) none of the preceding.

True or False

____ 1. Eukaryotic cells lack a membrane-bound nucleus and have few organelles.

____ 2. Active transport does not require energy and is the mechanism by which O_2 enters a cell.

____ 3. A plant cell has a cell wall that maintains the cell's shape against gravity and other forces.

____ 4. DNA is surrounded by proteins called histones.

____ 5. In Down syndrome, there is an extra #21 chromosome.

____ 6. Cells that produce steroids, such as those within the testes, ovaries, and adrenal glands, contain large amounts of rough ER.

____ 7. A mutation is an unrepaired mistake in the replication of DNA.

____ 8. Transcription takes place in the cytoplasm of a cell.

____ **9.** Messenger RNA contains anticodons.

____ **10.** Meiosis is the process of gamete (sex cell) formation.

____ **11.** Agents that may cause cancer are called carcinogens.

____ **12.** Chemotherapy is most effective in the treatment of metastasized cancer.

Completion

1. ________________ are organisms that are not classified as prokaryotes or eukaryotes.

2. ________________ ________________ ________________ are resistant to both stains and antibiotics.

3. ________________ is diffusion that involves only the movement of water across a membrane.

4. DNA and proteins are together called ________________.

5. The ________________ is a rounded mass of RNA within the nucleus.

6. The sugar in RNA is ________________.

7. The *N* chromosome number in a human cell is ________________.

8. ________________ are long, hollow structures made of polymerized tubulin.

9. Protein synthesis takes place in the ________________ within the cytoplasm.

10. Each strand of DNA consists of ________________ linked together by phosphodiester bonds.

Matching

Match the organelle with its description or function.

____	**1.** Lysosome	(*a*)	control center of cell
____	**2.** Centriole	(*b*)	vesicle containing hydrolytic enzymes
____	**3.** Golgi apparatus	(*c*)	synthesis of steroids and detoxification
____	**4.** Ribosome	(*d*)	movement of chromosomes during mitosis
____	**5.** Nucleus	(*e*)	formation of secretory vesicles and lysomes
____	**6.** Smooth ER	(*f*)	synthesis of proteins

Answers and Explanations for Review Exercises

Multiple Choice

1. (*d*) The cell membrane is a dynamic component of a cell that has many functions.
2. (*b*) The nucleus is larger than any of the cytoplasmic organelles.

3. (*a*) Lysosomes contain hydrolytic enzymes that break down cellular components or engulfed particles.
4. (*b*) Ribosomes synthesize proteins. Free ribosomes produce proteins to be used by the cell. Ribosomes attached to the endoplasmic reticulum produce proteins that are used outside the cell.
5. (*d*) The term rough ER is used because of the rough appearance of this organelle.
6. (*b*) Phagocytosis is the mechanism by which large, solid materials are taken into a cell.
7. (*d*) The cell membrane is a lipid bilayer with protein embedded within and on the outer and inner surface.
8. (*a*) The Golgi apparatus packages secretory material and forms lysosomes.
9. (*d*) ATP is produced via the Krebs cycle and oxidative phosphorylation with mitochondria.
10. (*b*) Amino acids are linked by peptide bonds.
11. (*d*) Thymine is replaced by uracil in RNA.
12. (*a*) All three types of RNA (mRNA, tRNA, and rRNA) are formed in the nucleus under the direction of DNA.
13. (*a*) mRNA is produced under the direction of nucleotides in a gene.
14. (*c*) DNA > RNA (transcription) > protein (translation).
15. (*c*) The genetic code for a single amino acid consists of three nucleotides called a codon.
16. (*b*) A–T, C–G
17. (*b*) The sugar in DNA is deoxyribose. The sugar in RNA is ribose.
18. (*c*) tRNA transports an amino acid to the ribosome to be incorporated into protein.
19. (*c*) The bases in mRNA (three together = a codon) determine the amino acid sequence in protein.
20. (*a*) 1200/3 (three in a codon) = 400 amino acids.
21. (*e*) TGCCGTG
22. (*e*) AUG
23. (*b*) The chromosomes are duplicated during interphase.
24. (*a*) The cell divides (cytokinesis) during telophase.
25. (*a*) Each daughter cell has the same number and kind of chromosomes as the original parent cell.

True and False

1. False; prokaryotic cells lack a nucleus and have few organelles
2. False; active transport requires the energy provided by ATP
3. True
4. True
5. True
6. False; they contain large amounts of smooth ER
7. True
8. False; transcription (DNA > RNA) takes place in the cell nucleus
9. False; messenger RNA contains codons
10. True
11. True
12. True

Completion

1. Viruses
2. Gram negative bacteria
3. Osmosis
4. chromatin
5. nucleolus
6. ribose
7. 23
8. Microtubules
9. ribosomes
10. nucleotides

Matching

1. (*b*)
2. (*d*)
3. (*e*)
4. (*f*)
5. (*a*)
6. (*c*)

Tissues 4

Objective A To define *histology* and *tissue*, and to distinguish between the four major tissue types.

Histology is the microscopic study of the tissues that compose body organs. A tissue is an aggregation of similar cells that perform a specific set of functions. The body is composed of over 25 kinds of tissues, classified as *epithelial tissue*, *connective tissue*, *muscle tissue*, and *nervous tissue*.

4.1 What are the bases for the classification of tissues?

Classification of tissues is based on embryonic development, structural organization, and functional properties. **Epithelial tissue,** or **epithelium,** embryonically derives from ectoderm, mesoderm, and endoderm; it covers body and organ surfaces, lines body cavities and lumina (the hollow portions of body organs or vessels), and forms various glands. Epithelial tissue is involved in protection, absorption, excretion, and secretion. **Connective tissue** derives from mesoderm; it binds, supports, and protects body parts. **Muscle tissue** derives from mesoderm; it contracts to enable locomotion and movement within the body. **Nervous tissue** derives from ectoderm; it initiates and conducts nerve impulses that coordinate body activities.

4.2 What part do the tissues play in clinical diagnosis?

In many cases, a particular disease is indicated by the abnormal appearance of a tissue removed in biopsy or postmortem examination (autopsy) and microscopically examined.

Pathology is a branch of medicine that deals extensively with the study of tissues. A *pathologist* is a physician who examines the organs from a cadaver on both gross and microscopic levels in an attempt to determine the cause of death. Many diseases cause characteristic changes in the appearance and function of the cells making up tissues. When a pathologist examines a deceased person, the procedure is known as an *autopsy*. When tissues are taken from a living person for microscopic examination, the procedure is known as a *biopsy*.

Objective B To describe *epithelial tissue* on the cellular level and to differentiate between the various kinds.

An *epithelium* (plural, *epithelia*) consists of one or more cellular layers. The outer surface is exposed either to the outside of the body or to a lumen or cavity within the body. The deep inner surface of epithelium is usually bound by a *basement membrane* consisting of glycoprotein from the epithelial cells and a meshwork of collagenous and reticular fibers from the underlying connective tissue. Epithelial tissue is avascular (without blood vessels) and is composed of tightly packed cells. Epithelium composed of a single layer of cells is called *simple*; multilayered epithelium is *stratified*. According to the shape of the cells on the exposed surface, epithelial tissue is *squamous* (flattened surface cells–"scaly"), *cuboidal*, or *columnar*.

4.3 Catalog the five kinds of simple epithelia as to structure, function, and location within the body

See table 4.1 and fig. 4.1.

Table 4.1 Classification of Simple Epithelial Tissue

Type	*Structure and function*	*Location*
Simple squamous epithelium	Single layer of flattened, tightly bound cells; diffusion and filtration	Forming capillary walls; lining air sacs (alveoli) of lungs; covering visceral organs; lining body cavities
Simple cuboidal epithelium	Single layer of cube-shaped cells; excretion, secretion, or absorption	Covering surface of ovaries; lining kidney tubules, salivary ducts, and pancreatic ducts
Simple columnar epithelium	Single layer of nonciliated column-shaped cells; protection, secretion, and absorption	Lining digestive tract, gallbladder, and excretory ducts of some glands
Simple ciliated columnar epithelium	Single layer of ciliated column-shaped cells; transport role through ciliary motion	Lining uterine (fallopian) tubes and limited areas of respiratory tract
Pseudostratified ciliated columnar epithelium	Single layer of ciliated, irregularly shaped cells; protection, secretion, ciliary motion	Lining respiratory passageways and auditory (eustachian) tubes

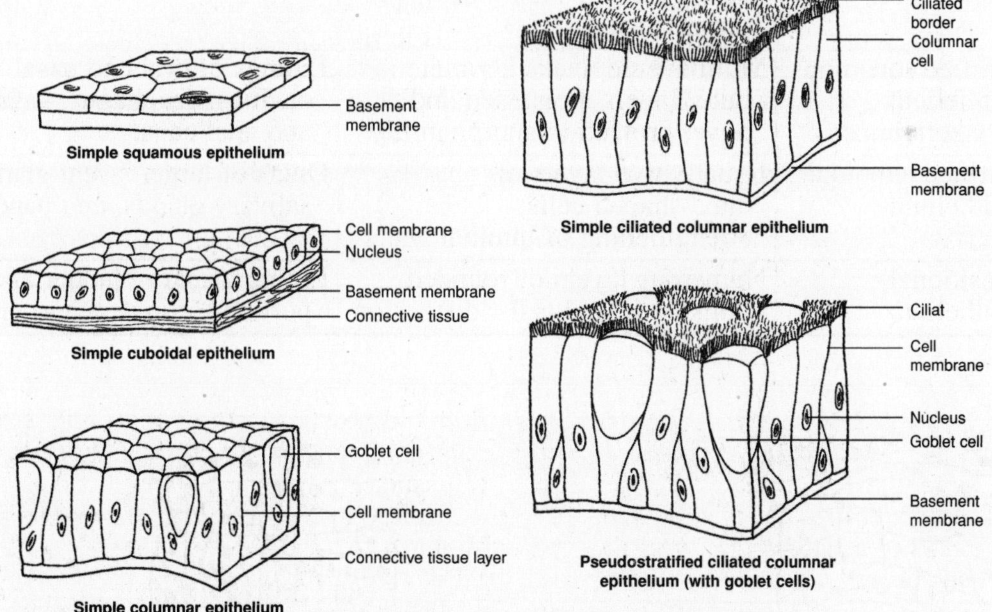

Figure 4.1 A comparison of simple epithelial tissues.

4.4 What is the basement membrane?

The **basement membrane** is a binding material of epithelial tissue in contact with the dividing layer of cells. Most epithelia have a basement membrane. It consists of glycoprotein from the epithelial cells and a meshwork of collagenous and reticular fibers from the underlying connective tissue.

4.5 *True or false*: Endothelium and mesothelium are types of simple epithelia.

True in the sense that the simple squamous epithelium lining blood and lymphatic vessels is frequently referred to as **endothelium**, while that covering visceral organs and lining body cavities is called **mesothelium**.

4.6 Which of the following epithelia contain goblet cells? (*a*) simple columnar epithelium, (*b*) simple ciliated columnar epithelium, (*c*) pseudostratified ciliated columnar epithelium

Specialized unicellular glands, called **goblet cells**, are dispersed throughout all types of columnar epithelial tissue; they are especially numerous in pseudostratified ciliated columnar epithelium. Goblet cells secrete a lubricative and protective *mucus* along the exposed surfaces of the tissues. The relative numbers of goblet cells in an epithelial lining depends on the need for mucus in the specific area of the lining. Since pseudostratified ciliated columnar epithelium is found in the respiratory tract, where abundant mucus is vital, this type of lining has large numbers of goblet cells.

4.7 Catalog the four kinds of stratified epithelia as to structure, function, and location within the body.

See table 4.2 and fig. 4.2.

Table 4.2 Classification of Stratified Epithelial Tissue

Type	*Structure and function*	*Location*
Stratified squamous epithelium (keratinized)	Multilayered, contains *keratin* (see problem 4.8), outer layers flattened and dead; protection	Epidermis of the skin
Stratified squamous epithelium (nonkeratinized)	Multilayered, lacks keratin, outer layers moistened and alive; protection and pliability	Linings of oral and nasal cavities, esophagus, vagina, and anal canal
Stratified cuboidal epithelium	Usually two layers of cube-shaped cells; strengthening of luminal walls	Ducts of larger sweat glands, salivary glands, and pancreas
Transitional epithelium	Numerous layers of rounded nonkeratinized cells; distension	Lining urinary bladder and portions of ureters and urethra

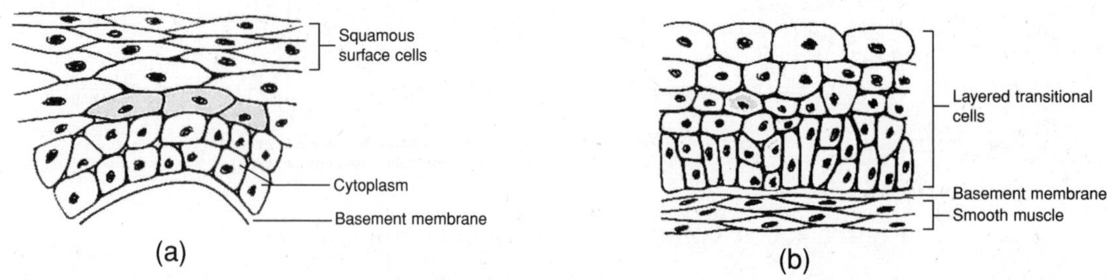

Figure 4.2 A comparison of (*a*) stratified squamous epithelium and (*b*) transitional epithelium.

4.8 Define *keratinization* and *cornification* and explain the value of these processes in stratified squamous epithelium.

The terms *keratinized* and *cornified* are frequently used interchangeably, although keratin and corneum are technically different. **Keratin** is the protein that forms during *keratinization* in conjunction with cellular death, as the layered cells are physically moved away from the life support of the vascular tissue underlying the stratified squamous epithelium (see problem 5.11). As the cells approach the exposed surface, they become flattened and dried during the process of *cornification*. The stratum corneum is the outer layer of the epidermis of the skin, where cornification occurs. Keratinization waterproofs the skin and cornification protects the skin from abrasion and entry of pathogens.

4.9 How does transitional epithelium differ from stratified squamous epithelium?

Transitional epithelium is similar to nonkeratinized stratified squamous epithelium, except that the surface cells of the former are large and round rather than flat, and they may have two nuclei. Transitional epithelium is specialized to permit distension of the ureters and urinary bladder and to withstand the toxicity of urine. Distension is possible because the transitional epithelial cells are able to change their shape, sometimes resembling cuboidal cells and sometimes squamous cells.

The appearance and relative numbers of cells in an epithelial lining can be very meaningful to a pathologist. Too many or two few cells of a certain type or abnormal levels of secreted products may signal that an organ is diseased or dysfunctional. In conducting an autopsy, a pathologist carefully examines the linings of cavities and organs in the body for signs of such irregularities. For example, cells with *pycnotic* (flattened) *nuclei* indicate certain diseases. The presence of excessive mucus or pus might indicate that a particular organ was combating an infection.

Objective C To define *glandular epithelial tissue* and to describe the formation, classification, and function of exocrine glands.

During prenatal development, certain epithelial cells invade the underlying connective tissue and form specialized secretory accumulations called *exocrine glands*. These glands retain a connection to the surface in the form of a duct. By contrast, *endocrine glands* lack ducts and secrete their products (*hormones*) directly into the bloodstream.

4.10 Give examples of exocrine glands and state the body systems with which they are associated.

Exocrine glands within the integumentary system include *sebaceous* (oil-secreting) *glands*, *sudoriferous* (sweat) *glands*, and *mammary glands*. Within the digestive system, exocrine glands include the salivary glands, gastric glands within the stomach, and the pancreatic gland.

Dysfunction of exocrine glands can result in a variety of symptoms and diseases. *Acne* is inflammation of sebaceous glands. *Ulcers* are stress related and are accompanied by excessive secretion of hydrochloric acid within the stomach by parietal cells. *Mumps* is an infectious disease of the parotid gland that secretes saliva.

4.11 Classify exocrine glands according to structure and give examples of the secretory product of each type.

See tables 4.3 and 4.4 and fig. 4.3.

Table 4.3 Structural Classification of Exocrine Glands

	Type	*Function*	*Examples*
Simple	Unicellular glands	Protect and lubricate	Goblet cells
	Tubular glands	Aid digestion	Intestinal glands
	Branched tubular glands	Protect; aid digestion	Uterine glands; gastric glands
	Coiled tubular glands	Regulate temperature	Eccrine sweat glands
	Acinar glands	Provide additive to spermatozoa	Seminal vesicles
	Branched acinar glands	Condition skin	Sebaceous glands
Compound	Tubular glands	Lubricate male urethra; aid digestion	Bulbourethral gland; liver
	Acinar glands	Provide infant nutrition; aid digestion	Mammary glands, salivary glands (submandibular and sublingual)
	Tubuloacinar glands	Aid digestion	Salivary gland (parotid); pancreas

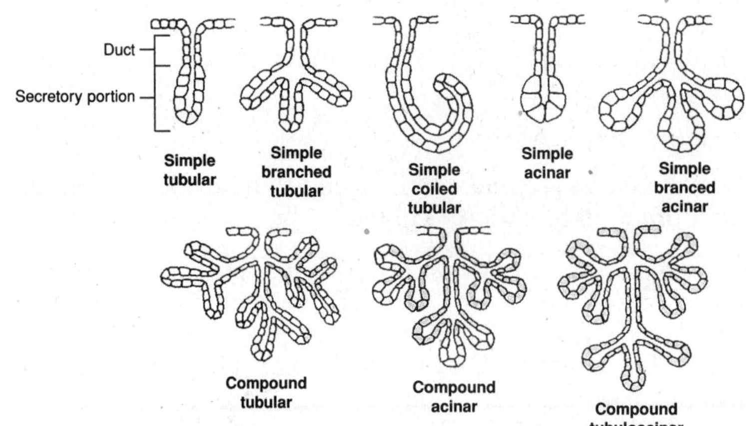

Figure 4.3 The structure of exocrine glands.

Table 4.4 Secretory Classification of Exocrine Glands

Type	*Function*	*Examples*
Merocrine	Anchored cell secretes water; regulates temperature, aids digestion	Salivary and pancreatic glands, certain sweat glands
Apocrine	Portion of secretory cell and secretion are discharged; provides nourishment to infant, assists in regulating temperature	Mammary glands, certain sweat glands
Holocrine	Entire secretory cell with enclosed secretion is discharged; conditions skin	Sebaceous glands of skin

Objective D To describe the characteristics, locations, and functions of *connective tissue.*

One of the most important components of connective tissue is the **matrix**, a bed of secreted organic material of varying composition that binds widely separated cells of the tissue. All connective tissue is derived from embryonic mesoderm. Connective tissue is found throughout the body. It supports and binds other tissues, stores nutrients, and/or manufactures protective and regulatory materials.

4.12 What are the various types of connective tissue? Describe their structures and functions, and state where they are located.

Throughout the embryo is found undifferentiated connective tissue called *mesenchyme* from which mature, differentiated connective tissues derive. These mature tissues fall into the four major categories indicated in fig. 4.4. Note that one of these, blood tissue, differs from the rest in having a fluid matrix. Further classification is given in table 4.5.

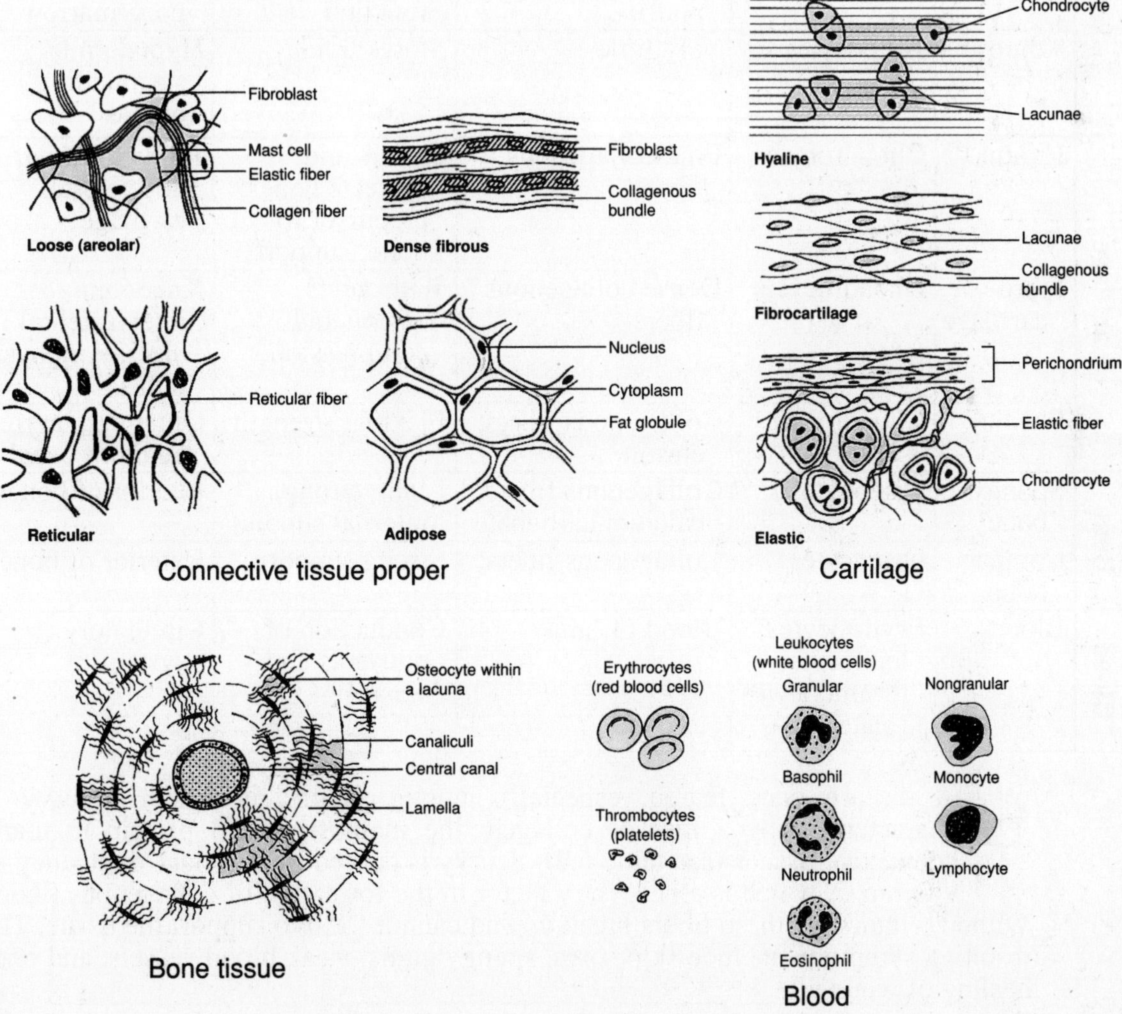

Figure 4.4 Types of connective tissue.

Table 4.5 Classification of Connective Tissue

	Tissue Type	*Cells*	*Matrix*	*Function*	*Location*
Connective Tissue Proper	Loose (areolar)	Fibroblasts; mast cells	Collagenous fibers; elastin	Binding and packing; protection and nourishment; holds fluids; secretes heparin	Deep to skin; surrounding muscles, vessels, and organs
	Dense fibrous	Fibroblasts	Densely packed collagenous fibers	Strong, flexible	Tendons, ligaments
	Elastic	Fibroblasts	Elastin fibers	Flexibility and distensibility	Arteries, larynx, trachea, bronchi
	Reticular	Phagocytes	Reticular fibers in jellylike matrix	Performs phagocytic function	Liver, spleen, lymph nodes, bone marrow
	Adipose	Adipocytes	Very little	Stores lipids	Hypodermis, surrounding organs
Cartilage	Hyaline	Chondrocytes	Fine collagenous fibers	Covers and protects bones; precursor to bone; support	Joints, trachea, nose, costal cartilage
	Fibro-cartilage	Chondrocytes	Dense collagenous fibers	Withstands tension and compression	Knee joint, intervertebral discs, symphysis pubis
	Elastic	Chondrocytes	Collagenous fibers; elastin	Flexible strength	Outer ear, larynx, auditory canal
Bone	Spongy bone	Osteocytes	Collagenous fibers; calcium carbonate	Light, strong, internal support	Interior of bones
	Compact bone	Osteocytes	Collagenous fibers;	Strong support	Exterior of bones
Blood	Blood	Erythrocytes, leukocytes, thrombocytes (platelets)	Blood plasma	Conduction of nutrients and wastes	Circulatory system

A disease once feared, especially among sailors, is *scurvy*. Scurvy is characterized by a loss of collagen, the main structural protein in many connective tissues (see table 6.4). Scurvy is caused by a dietary deficiency of vitamin C, which is a necessary factor in the formation of collagenous fibers. Without vitamin C, these fibers break up and cannot form to support the tissue. The resulting symptoms include skin sores, spongy gums, weak blood vessels, and poor healing of wounds.

4.13 Which of the following connective tissues are important in body immunity? (*a*) blood, (*b*) dense regular tissue, (*c*) fibrocartilage, (*d*) reticular tissue

Both the white blood cells (leukocytes) of the blood and the reticular tissue of lymphoid organs protect the body through *phagocytosis.*

4.14 Why is blood considered a connective tissue?

Because it contains cells (red blood cells, white blood cells, and platelets) and matrix (blood plasma), blood is considered a viscous connective tissue (see chapter 13).

4.15 Why are joint injuries involving cartilage slow to heal?

Cartilage is avascular and must therefore receive nutrients through diffusion from surrounding tissue. For this reason, cartilaginous tissue has a low rate of mitotic activity and, if damaged, heals slowly.

4.16 Distinguish between fat and adipose tissue.

The cells of adipose tissue contain large vacuoles adapted to store lipids, or fats. Overfeeding an infant during the first year, when **adipocytes** (adipose cells) are forming, causes excessive amounts of adipose tissue to develop. A person with a lot of adipose tissue is more susceptible to developing obesity later in life than a person with a lesser amount. Dieting eliminates the lipid stored within the tissue but not the tissue itself.

4.17 What do fibroblasts, reticular cells, mast cells, chondrocytes, and osteocytes all have in common? How do they differ?

All are specialized cells of different types of connective tissue; they are compared in table 4.6.

Table 4.6 Some Specialized Cells of Connective Tissue

Cell type	*Description*	*Location*	*Product*
Fibroblast	Large irregularly shaped cell	Throughout connective tissue proper	Collagenous, elastic, and reticular fibers
Reticular cell	Highly branched, interwoven cell	Reticular connective tissue; lymphoid organs	Phagocytes
Mast cell	Round, resembling a basophil	Loose connective tissue; surrounding blood vessels	Heparin (an anticoagulant)
Chondrocyte	Large ovoid cell	Cartilage tissue	Cartilaginous matrix
Osteocyte	Small ovoid cell	Bone tissue	Solid matrix

4.18 How is edema related to connective tissue?

Approximately 11% of the body fluid is found within loose connective tissue, where it is known as *tissue fluid* or *interstitial fluid.* Sometimes excessive tissue fluid accumulates, causing the swollen condition known as *edema.* The fluid surplus is generally symptomatic of other conditions.

4.19 What is the difference between compact bone tissue and spongy bone tissue?

Most bones of the skeleton are composed of compact (dense) bone tissue and spongy (cancellous) bone tissue (fig. 6.4). **Compact bone tissue** is the hard outer layer, whereas **spongy bone tissue** is the porous, highly vascular inner portion. Compact bone tissue is covered by the periosteum, which serves for attachment of tendons from muscles. Spongy bone tissue makes the bone lighter and provides a space for bone marrow where blood cells are produced. Bone tissue is further discussed in chapter 6.

4.20 What accounts for the hardness of bone tissue?

The hardness of bone is largely due to the calcium phosphate and calcium carbonate salts deposited within the intracellular (inorganic) matrix. Numerous collagenous fibers, also embedded within the matrix, give some flexibility to bone tissue.

Objective E To describe *muscle tissue* and to distinguish between the three types.

Through the property of *contractility*, muscle tissues cause movement of materials through the body, movement of one part of the body with respect to another, and locomotion. Muscle cells, also called *muscle fibers*, are elongated in the direction of contraction, and movement is accomplished through the shortening of the fibers in response to a stimulus. Derived from mesoderm, muscle cells are so specialized for contraction that, once the tissue formation has been completed prenatally, the cells can no longer replicate. There are three types of muscle tissue in the body: *smooth*, *cardiac*, and *skeletal*.

Skeletal muscle fibers begin to form about 4 weeks following conception. At this time, undifferentiated mesodermal cells, called *myoblasts*, begin migration to sites where the individual muscles will form. As the myoblasts arrive at these sites, they aggregate into *syncytial myotubes*. Myotubes grow in length by incorporating additional myoblasts, each with its own nucleus. As cell membranes break down within each myotube, multinucleated muscle fibers are formed. Muscle fibers are distinct at 9 weeks, and at 17 weeks muscles are sufficiently well developed for a pregnant woman to sense fetal movements known as *quickening*.

4.21 Describe the structure, function, and location of each type of muscle tissue.

See table 4.7 and fig. 4.5.

Table 4.7 A Comparison of the Three Types of Muscle Tissue

Type	*Location*	*Structure and function*
Smooth muscle	Walls of hollow internal organs	Elongated, spindle-shaped fiber with single nucleus; slow involuntary movements of internal organs
Cardiac muscle	Wall of heart	Branched, striated fiber with single nucleus and intercalated discs; rapid involuntary rhythmic contractions
Skeletal muscle	Spanning joints of skeleton via tendons	Multinucleated, striated, cylindrical fiber that occurs in fasciculi (slender bundles); rapid involuntary or voluntary movement of joints of skeleton

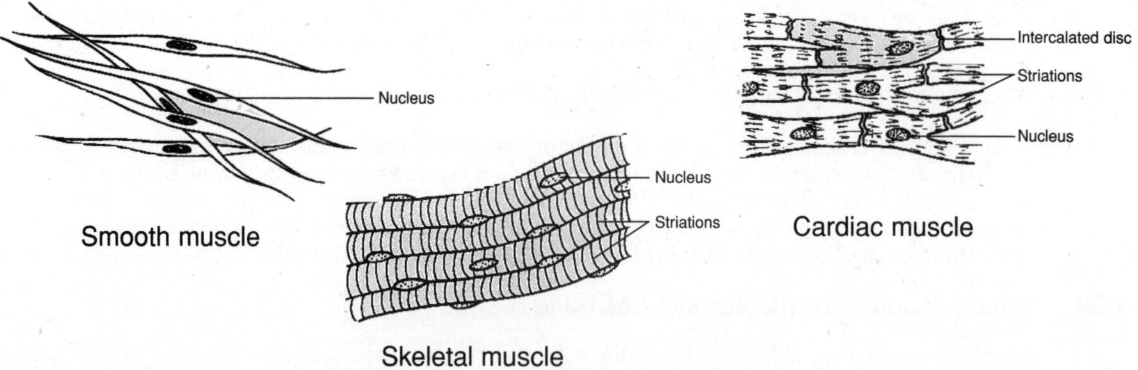

Figure 4.5 Types of muscle tissue.

4.22 Which of the following are characteristic properties of all muscle tissue? (*a*) irritability, (*b*) contractility, (*c*) extendibility, (*d*) elasticity

All are characteristic of muscle fibers. A muscle fiber exhibits irritability as it responds to a nerve impulse and contracts, or shortens. Once a stimulus has subsided and the muscle fiber is shortened but relaxed, it may passively stretch back or be extended by contracting fibers of opposing muscles. Each muscle fiber has innate tension, or elasticity, that causes it to assume a particular shape as it is relaxed.

Metabolism within cells releases heat as an end product. Muscles account for nearly one-half of the body weight, and even the fibers of resting muscles are in a continuous state of fiber activity (tonus). Thus, muscles are major heat sources. Maintaining a high body temperature is of homeostatic value in providing optimal conditions for metabolism. The rate of heat production increases immensely as a person exercises strenuously.

Objective F To describe the basic characteristics and functions of *nervous tissue.*

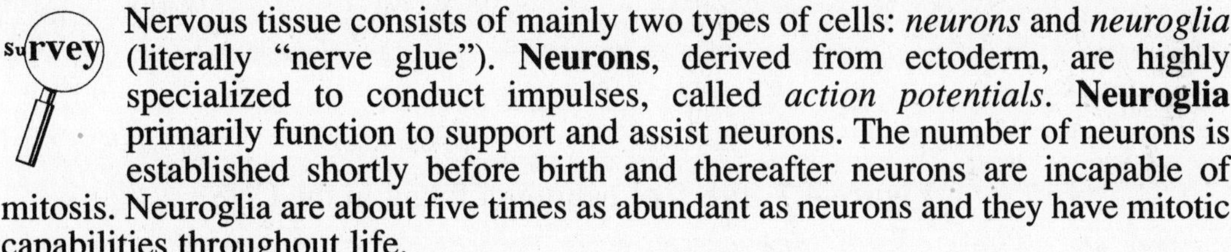

Nervous tissue consists of mainly two types of cells: *neurons* and *neuroglia* (literally "nerve glue"). **Neurons**, derived from ectoderm, are highly specialized to conduct impulses, called *action potentials.* **Neuroglia** primarily function to support and assist neurons. The number of neurons is established shortly before birth and thereafter neurons are incapable of mitosis. Neuroglia are about five times as abundant as neurons and they have mitotic capabilities throughout life.

4.23 How does the structure of a neuron reflect its function?

Branched **dendrites** (fig. 4.6) provide a large surface area for receiving stimuli and conducting impulses to the cell body. The elongated **axon** conducts the impulse away from the cell body to another neuron or to an organ that responds to the impulse.

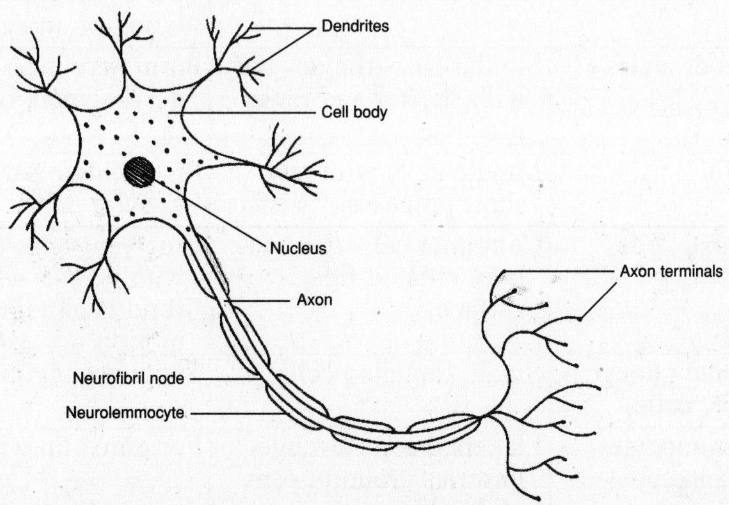

Figure 4.6 The structure of a neuron.

4.24 Describe the association of neurolemmocytes (Schwann cells) with certain neurons.

Neurolemmocytes (*Schwann cells*) are specialized neuroglial cells that support the axon (fig. 4.6) by ensheathing it with a lipid-protein substance called **myelin** (see chapter 8). This **myelin sheath** aids in the conduction of nerve impulses and promotes regeneration of a damaged neuron.

4.25 Describe the structure and function of neuroglia.

Besides neurolemmocytes, there are five kinds of *neuroglia*, four of which are illustrated in fig. 4.7. All six types of neuroglia are described in table 4.8 (CNS = central nervous system; PNS = peripheral nervous system).

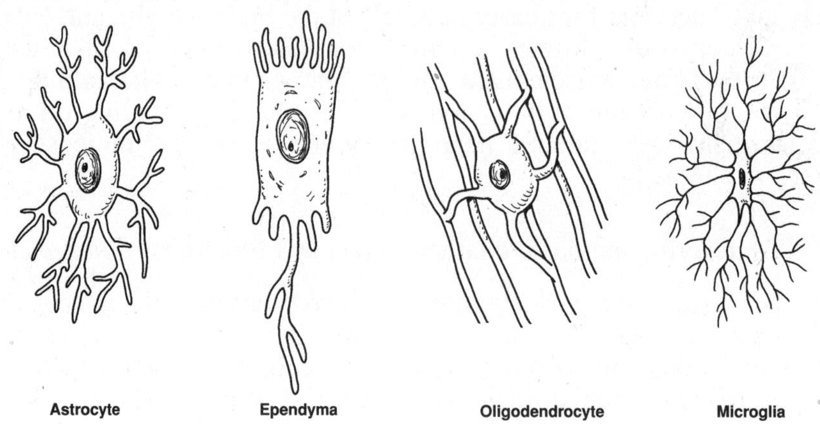

Figure 4.7 Types of neuroglia found in the CNS.

Table 4.8 Structure and Function of Neuroglia

Types	*Structure*	*Function*
Astrocytes	Stellate with numerous processes	Form structural support between capillaries and neurons within the CNS; contribute to blood-brain barrier
Oligodendrocytes	Similar to astrocytes but with shorter and fewer processes	Form myelin in CNS; guide development of neurons within the CNS
Microglia	Minute cells with few short processes	Phagocytize pathogens and cellular debris within CNS
Ependymal cells	Columnar cells that may have ciliated free surfaces	Line ventricles and central canal within CNS where cerebrospinal fluid is circulated by ciliary motion
Ganglionic gliocytes (satellite cells)	Small, flattened cells	Support ganglia within PNS
Neurolemmocytes (Schwann cells)	Flattened cells arranged in series around axons of dendrites	Form myelin within PNS

Because neurons no longer have the ability to undergo mitosis and proliferate once they have matured, destroying these cells can be permanently debilitating. Drugs and alcohol, as well as oxygen deprivation or trauma to the central or peripheral nervous system, can destroy neurons that can never be replaced. A number of diseases afflict either neurons or neuroglia. Three of these are *Alzheimer's disease*, *Parkinson's disease*, and *Huntington's disease*. Although neurons may seem fragile, if well nourished and kept free of drugs (including alcohol) they will endure and function a lifetime.

Review Exercises

Multiple Choice

1. Epithelia are involved in all the following *except* (*a*) protection, (*b*) secretion, (*c*) connection, (*d*) absorption, (*e*) excretion.

2. Which of the following is *not* a type of epithelium? (*a*) simple squamous, (*b*) transitional, (*c*) simple ciliated columnar, (*d*) complex stratified, (*e*) pseudostratified ciliated

3. Classification of epithelia is based on the number of layers of cells and on (*a*) shape, (*b*) staining properties, (*c*) size, (*d*) location, (*e*) ratio of living to nonliving cells.

4. The presence of a basement membrane is typical of most (*a*) epithelial tissues, (*b*) connective tissues, (*c*) nervous tissues, (*d*) muscle tissues, (*e*) cartilage tissues.

5. Simple squamous epithelium is *not* found in (*a*) blood vessels, (*b*) the lining of the mouth, (*c*) lymph vessels, (*d*) alveoli (air sacs) of the lungs, (*e*) linings of body cavities.

6. Goblet cells are a type of (*a*) multicellular gland, (*b*) intracellular gland, (*c*) unicellular gland, (*d*) intercellular gland, (*e*) salivary gland.

7. An example of a holocrine gland is (*a*) a sweat gland, (*b*) a salivary gland, (*c*) a pancreatic gland, (*d*) a sebaceous gland.

8. An exocrine gland in which a portion of the secretory cell is discharged with the secretion is termed (*a*) apocrine, (*b*) merocrine, (*c*) endocrine, (*d*) holocrine.

9. The inability to absorb digested nutrients may be due to damage of which type of epithelium? (*a*) ciliated columnar, (*b*) simple columnar, (*c*) simple squamous, (*d*) simple cuboidal, (*e*) stratified squamous

10. Which word combination applies to stratified squamous epithelium? (*a*) mesoderm–calcification, (*b*) ectoderm–keratinization, (*c*) mesoderm–ossification, (*d*) endoderm–cornification

11. Which statement best describes connective tissue?
(*a*) It is derived from endoderm and secretes metabolic substances.
(*b*) It is derived from mesoderm and conducts impulses.
(*c*) It is derived from mesoderm and contains abundant matrix.
(*e*) It is derived from ectoderm and is usually layered.

12. An infection would most likely increase phagocytic activity in (*a*) elastic tissue, (*b*) transitional tissue, (*c*) adipose tissue, (*d*) reticular tissue, (*e*) collagenous tissue.

13. Cartilage tissues are generally slow to heal following an injury because (*a*) cartilage is avascular, (*b*) cartilage does not undergo mitosis, (*c*) the matrix is semisolid, (*d*) chondrocytes are surrounded by fluids.

14. Which of the following is *not* a specialized type of cell found in connective tissues? (*a*) lymphocyte, (*b*) macrophage, (*c*) goblet cell, (*d*) mast cell, (*e*) fibroblast

15. The function of dense regular connective tissue is (*a*) elastic recoil, (*b*) binding and support, (*c*) encapsulation of blood vessels, (*d*) articulation.

16. Phagocytosis is a function of which type of connective tissue? (*a*) cartilage, (*b*) loose fibrous, (*c*) elastic, (*d*) reticular, (*e*) adipose

17. Adipose tissue forms (*a*) only during fetal development, (*b*) throughout life, (*c*) mainly during fetal development and the first postpartum year, (*d*) mainly at puberty.

18. Intervertebral discs are composed of (*a*) elastic connective tissue, (*b*) elastic cartilage, (*c*) hyaline cartilage, (*d*) fibrocartilage.

19. Intercalated discs are found in (*a*) cardiac muscle tissue, (*b*) movable joints, (*c*) the vertebral column, (*d*) bone tissue, (*e*) hyaline cartilage.

20. Tissue (interstitial) fluid would most likely be found in (*a*) loose connective tissue, (*b*) nervous tissue, (*c*) adipose tissue, (*d*) bone tissue, (*e*) muscle tissue.

True or False

____ 1. Connective tissues derive only from mesoderm and function to bind, support, and protect body parts.

____ 2. Simple ciliated columnar epithelium helps to move debris through the lower respiratory tract, away from the lungs.

____ 3. Cells of epithelia are tightly packed, mostly avascular, and without significant matrix.

____ 4. Nervous tissue is located only in the brain and spinal cord.

____ 5. Neurons are capable of mitosis to accommodate increased learning.

____ 6. Most bones in the body begin as fibrocartilage and then ossify to bone.

____ 7. Acinar glands have a flasklike secretory portion.

____ 8. Mast cells that produce the anticoagulant heparin are dispersed throughout loose connective tissue.

____ 9. Red blood cells are the only cellular component of blood tissue.

____ 10. Based on structure and method of secretion, mammary glands are classified as compound acinar and apocrine.

____ 11. Transitional epithelium occurs only in the urinary system.

____ 12. All stratified squamous epithelium is keratinized and cornified.

____ 13. Adipose tissue dies as a person diets, and new cells are formed as weight is gained.

____ 14. Skeletal and cardiac muscle fibers are striated.

____ 15. Neuroglia are specialized cells of nervous tissue that react to stimuli.

Completion

1. ______________________ is the scientific study of tissues.

2. Flattened, irregularly shaped cells that are tightly bound in a single-layered mosaic pattern compose ______________________ ______________________ epithelial tissue.

3. Epithelium consisting of two or more layers is classified as ______________________.

4. ______________________ is the name given to the simple squamous epithelium that lines the inside walls of blood vessels.

5. Rhythmic contractions of sheets of ______________________ muscle tissue in the intestinal wall results in involuntary movement of food materials.

6. ______________________ is a protein in the skin that strengthens the stratified squamous epithelium of the epidermis.

7. Pancreatic glands are classified as ______________________ glands because no portion of the gland is discharged with the secretion.

8. Bone tissue consisting of a latticework of thin plates of bone filled with bone marrow is termed ______________________ bone.

9. ______________________ is the matrix of blood tissue.

10. Alien matter is engulfed by leukocytes in the blood and in the ______________________ tissue of lymph nodes.

11. The abnormal pooling of fluid in tissues is called ______________________.

12. All connective tissue and muscle tissue is derived from the embryonic ______________________.

13. ______________________ muscle tissue is composed of multinucleated, striated, cylindrical fibers arranged into fasciculi.

14. The ______________________ of a neuron receive a stimulus and conduct the nerve impulse to the cell body.

15. The lipid-protein product of neurolemmocytes (Schwann cells) forms a cover of ______________________ around the axon of a neuron.

Matching

(*Set 1*) Match the epithelial tissues with their locations.

____	**1.** Simple squamous epithelium	(*a*)	lining the uterine tubes
____	**2.** Simple cuboidal epithelium	(*b*)	capillary walls
____	**3.** Simple columnar epithelium	(*c*)	lining the oral cavity
____	**4.** Pseudostratified ciliated columnar epithelium	(*d*)	lining pancreatic ducts
____	**5.** Stratified squamous epithelium	(*e*)	lining the digestive tract
____	**6.** Transitional epithelium	(*f*)	lining the respiratory tract
____	**7.** Simple ciliated columnar epithelium	(*g*)	lining the urinary bladder

(*Set 2*) Match the glands with their locations or descriptions.

____	**1.** Simple acinar gland	(*a*)	goblet cell
____	**2.** Compound tubular gland	(*b*)	parotid gland
____	**3.** Unicellular gland	(*c*)	seminal vesicle
____	**4.** Compound tubuloacinar gland	(*d*)	intestinal gland
____	**5.** Simple tubular gland	(*e*)	liver
____	**6.** Compound acinar gland	(*f*)	gastric gland
____	**7.** Simple branched tubular gland	(*g*)	mammary gland

(*Set 3*) Match the connective tissues or connective tissue structures with their locations or descriptions.

____	**1.** Hyaline cartilage	(*a*)	auricle of external ear
____	**2.** Spongy bone	(*b*)	minute canals
____	**3.** Canaliculi	(*c*)	intervertebral joint
____	**4.** Elastic cartilage	(*d*)	inner bone tissue
____	**5.** Compact bone	(*e*)	fetal skeleton
____	**6.** Fibrocartilage	(*f*)	covered by periosteum

Answers and Explanations for Review Exercises

Multiple Choice

1. (*c*) Epithelia are not involved with connection; this is the function of connective tissue. Epithelia are involved in protection (skin), secretion (exocrine glands), absorption (lining of GI tract), and excretion (glomerular capsule in kidney).
2. (*d*) "Complex" is not descriptive of epithelial tissue.
3. (*a*) Epithelial tissue is classified by (1) the number of layers (simple, stratified, pseudostratified), (2) the shape of cells (squamous, cuboidal, columnar), and (3) any surface modification (cornified, ciliated).
4. (*a*) Most epithelia have a basement membrane between the epithelial cells and the underlying connective tissue.
5. (*b*) The linings of ducts are composed of cuboidal cells, not squamous cells.
6. (*c*) Goblet cells are single cells that secrete a watery mucus into the lumen of the GI tract.
7. (*d*) Sebaceous glands are holocrine glands because they secrete by discharging entire cells full of product (sebum).
8. (*a*) Based on development and mode of secretion, mammary glands and certain sweat glands are classified as apocrine glands.
9. (*b*) The GI tract is lined with simple columnar epithelium, which allows a maximum number of cells to contact food particles.
10. (*b*) The epidermis of the skin derives from the ectoderm germ layer and is the primary site of keratinization in the body.
11. (*c*) All connective tissue is derived from mesoderm (mesenchyme cells). Connective tissues are classified according to the matrix the cells secrete and the arrangement of the components.
12. (*d*) Reticular tissue within lymphatic organs contains large numbers of phagocytic cells, which engulf invading pathogens.
13. (*a*) Lacking a capillary blood supply, cartilage tissue heals poorly.
14. (*c*) Goblet cells are found lining the respiratory tract and the GI tract, where they are needed to secrete their lubricating and protective mucus. Macrophages and lymphocytes are both found in connective tissue, where they aid the immune response.
15. (*b*) Dense regular connective tissue forms tendons and ligaments, as well as the capsules surrounding various organs.
16. (*d*) Reticular tissue contains large numbers of phagocytic cells. This tissue is present in lymphatic organs, such as the spleen, thymus, tonsils, and lymph nodes.
17. (*c*) The amount of lipid stored by adipocytes can vary throughout life, but the number of adipocytes remains about the same.
18. (*d*) Fibrocartilage is found in joints called symphyses, such as the symphysis pubis and between adjacent vertebrae. It is also found in the knee joint forming the menisci (see chapter 6).
19. (*a*) Intercalated discs are specialized junctions between adjacent cardiac muscle cells that allow the cells to conduct impulses, much like nerve cells.
20. (*a*) Tissue (interstitial) fluid fills the space between fibers and cells of connective tissue. Loose connective tissue has the most space for fluid to accumulate.

True or False

1. True
2. False; pseudostratified squamous epithelium is the characteristic epithelium in the respiratory tract
3. True
4. False; neurons and nerves of nervous tissue are found throughout the body
5. False; once formed prenatally, neurons cannot divide
6. False; most bones first form as hyaline cartilage
7. True
8. True
9. False; blood contains red blood cells, white blood cells, and platelets
10. True

11. True
12. False; epithelia lining the oral, anal, and vaginal cavities are nonkeratinized, as are epithelia on parts of the genitalia
13. False; only the lipid content is lost and gained as a person's weight fluctuates
14. True
15. False; neurons react to stimuli and neuroglia support and assist neurons

Completion

1. Histology
2. simple squamous
3. stratified
4. Endothelium
5. smooth
6. Keratin
7. merocrine
8. spongy
9. Plasma
10. reticular
11. edema
12. mesoderm
13. Skeletal
14. dendrites
15. myelin

Matching

(*Set 1*)

1. (*b*)
2. (*d*)
3. (*e*)
4. (*f*)
5. (*c*)
6. (*g*)
7. (*a*)

(*Set 2*)

1. (*c*)
2. (*e*)
3. (*a*)
4. (*b*)
5. (*d*)
6. (*g*)
7. (*f*)

(*Set 3*)

1. (*e*)
2. (*d*)
3. (*b*)
4. (*a*)
5. (*f*)
6. (*c*)

Integumentary System 5

Objective A To list the components of the *integumentary system* and to describe the characteristics and embryonic origin of the *skin*.

The *skin*, or *integument*, and associated structures (hair, glands, and nails) constitute the integumentary system. This system accounts for approximately 7% of the body weight and is a dynamic interface between the body and the external environment.

5.1 Why is the skin considered an organ?

The skin is an organ because it consists of several kinds of tissues that are structurally arranged to function together. It is the largest organ of the body, with a surface area of about 2 m^2 (22 ft^2) on the average adult. Its thickness ranges between 1.0 and 2.0 mm, but it is up to 6.0 mm thick on the palms and soles. The skin on the palms and soles is referred to as *thick skin*, as opposed to *thin skin* elsewhere on the body.

5.2 What is the embryonic origin of the skin?

The principal layers of the skin are established by the eleventh week of embryonic development. The epidermis and associated structures are derived from the *ectoderm germ layer*, and the dermis and the hypodermis are derived from the *mesoderm germ layer*. The principal layers of the skin are described in Objective C.

Dermatology is the specialty of medicine that deals with the skin. A dermatologist treats problems ranging from acne to severe burns and scarring. As we learn more about the dynamic nature of the skin and the many functional roles it plays, dermatology as a branch of medicine will continue to be of major importance.

Objective B To describe the basic functions of the integumentary system.

The functions of the integumentary system include physical protection, hydroregulation, thermoregulation, cutaneous absorption, synthesis, sensory reception, and communication. The skin is a physical barrier to most microorganisms, water, and most UV light. The acidic surface (pH 4.0–6.8) retards the growth of most pathogens. The skin protects the body from desiccation (dehydration) when on dry land and from water absorption when immersed in water. A normal body temperature of 37°C (98.6°F) is maintained by the antagonistic effects of shivering and sweating (see fig. 5.1). The skin permits the absorption of small amounts of UV light necessary for synthesis of vitamin D. It is important to note that certain toxins and pesticides also may enter the body through cutaneous absorption. The skin synthesizes **melanin** (a protective pigment) and **keratin** (a protective protein). Numerous sensory receptors are located in the skin, especially in parts of the face, palms and fingers of the hands, soles of the feet, and genitalia. Certain emotions, such as anger or embarrassment, may be reflected in changes of skin color.

5.3 Which body systems functionally interact with the integumentary system?

The *circulatory system* interacts extensively with the integumentary system in maintaining homeostasis. The sex hormones (androgens and estrogens) of the *endocrine system* influence the function and maintain the appearance of the integument. The white blood cells and the lymphatics of the circulatory system also provide body immunity within the skin. Furthermore, platelets that aid clotting provide a defense against excessive bleeding. Countless sensory receptors within the skin convey impulses to the *nervous system.* Various emotions are conveyed through facial expression, which involves the *muscular system.* Blushing is the result of vasodilation of cutaneous arterioles of the circulatory system.

5.4 List some defense mechanisms by which the skin helps to prevent infection.

(1) The thickness of the outer layer of the skin (epidermis) and its toughened exposed surface are physical barriers to microorganisms. (2) The acidic pH at the oily surface of the skin inhibits the growth of many microorganisms. (3) The skin is highly vascular; its huge network of blood vessels can quickly deliver the white blood cells and other protein factors necessary for inflammatory and immune responses.

5.5 Describe how skin helps to maintain a constant body temperature.

A relatively constant body temperature of 37°C (98.6°F) is maintained by the *hypothalamus* within the brain that functions like a thermostat. If the body temperature falls below 98°F, *cutaneous vasoconstriction* conserves heat, and additional heat is generated through shivering. If the body temperature rises above 99°F, heat loss is accelerated through *cutaneous vasodilation* and sweating. In each situation, a deviation from the normal state autonomically triggers a response in what is described as a *negative feedback mechanism.* The hypothalamus autonomically "switches on or off" the necessary physiological mechanisms to maintain homeostasis of body temperature.

Objective C To list the layers of the skin and to describe their structure.

The skin consists of two principal layers. The outer **epidermis** is stratified into five or six structural and functional layers. The thick and deeper **dermis** consists of two layers. Not considered a separate layer, the **hypodermis** (subcutaneous tissue) binds the skin to underlying structures. A diagram of the skin is shown in fig. 5.1.

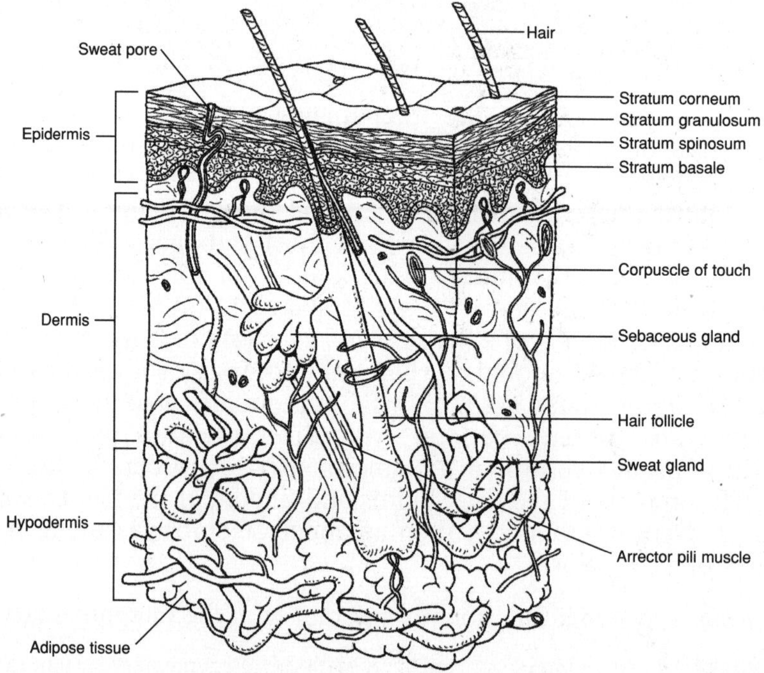

Figure 5.1 The skin.

5.6 In tissue composition, how does the epidermis differ from the dermis?

The protective epidermis is composed of *stratified squamous epithelium*, which averages 30 to 50 cells in thickness. The layered cells are avascular (without blood vessels) The outer cells of the epidermis are dead, keratinized, and cornified. By contrast, the considerably thicker dermis is highly vascular and consists of a variety of living cells. Numerous collagenous, elastic, and reticular fibers give support to the dermis. The dermis also has numerous sweat and oil glands, as well as nerve endings and hair follicles.

5.7 Describe the composition of the hypodermis.

The *hypodermis (subcutaneous tissue)* contains loose (areolar) connective tissue, adipose tissue, and blood and lymph vessels. Collagenous and elastic fibers reinforce the hypodermis, particularly on the palms and soles.

5.8 *True or false*: Females have a thicker hypodermis than do males.

True. The hypodermis of adults is approximately 8% to 10% thicker in females than in males. The greater thickness is due to a greater deposition of lipids within adipocytes (fat cells) and is apparently hormonally influenced. Studies confirm that extremely low fat reserves are typical of women who experience amenorrhea (absence of menstruation). Ovulation may also be disturbed in these women, impairing fertility.

5.9 What are the functions of the hypodermis?

The hypodermis binds the dermis to underlying organs; it also stores lipids, insulates and cushions the body, and regulates temperature. In mature females, this layer, through its softening of body contour, plays a part in sexual attraction.

Since it is rich in adipose tissue, many fat-soluble drugs and medications are designed to be injected into the hypodermis. A subcutaneous injection is often used when a patient is unable to take medication orally. Fat-soluble drugs often have a longer lasting effect than do water soluble drugs. A hypodermic needle is so named because it is used to inject the drug below the dermis into the tissue of the hypodermis.

Objective D To describe the strata, or structural layers, of the epidermis.

Survey

In table 5.1, the epidermal layers are listed in order from the stratum basale, in contact with the basement membrane, to the outer exposed stratum disjunction. These layers are illustrated in fig. 5.2.

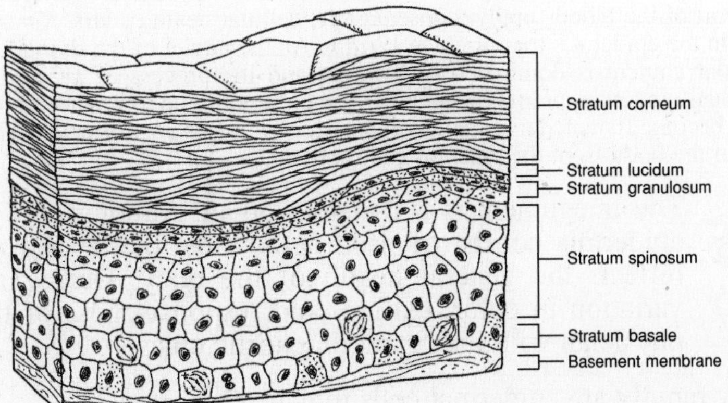

Figure 5.2 Layers of the epidermis.

Table 5.1 Layers of the Epidermis

Stratum (layer)	*Characteristics*
Stratum disjunction	Outermost layer of stratum corneum that continuously sloughs off in microscopic pieces
Stratum corneum	Several layers of keratinized corneum; a collagenous matrix composed of the products of dead cells
Stratum lucidum	Thin, clear layer present in the thick skin of the palms and soles; no remaining living cells
Stratum granulosum	One or more layers of granular cells with shriveled nuclei; contains keratin
Stratum spinosum (part of *stratum germinativum*)	Several layers of cells with large, oval, centrally located nuclei and spiny processes; limited mitosis; most cells dying and being moved toward surface
Stratum basale (part of *stratum germinativum*)	Single layer of well-nourished cells contacting the basement membrane and undergoing continuous mitosis; contains melanocytes

Both the ectodermal and mesodermal germ layers (see problem 5.2) participate in the formation of the skin. The epidermis and accessory integumentary structures (hair, glands, and nails) develop from ectoderm. The dermis develops from a thickened layer of undifferentiated mesoderm called *mesenchyme*. Likewise, the cutaneous blood vessels and smooth muscle fibers contained within the dermis are formed from mesoderm.

5.10 What is the basement membrane?

The **basement membrane** is a binding material of epithelial tissue in contact with the dividing layer of cells (see fig. 5.2). It consists of glycoprotein from the epithelial cells and a meshwork of collagenous and reticular fibers from the underlying connective tissue.

5.11 Why are the outer cells of the epidermis dead? What purpose does this serve?

Mitosis, or cell division, occurs primarily in the deep stratum basale and to a slight extent in the stratum spinosum. Mitosis occurs at these locations because of their proximity to blood vessels that provide nutrients and oxygen to the dividing cells. As the cells longitudinally divide, only half of them will remain in contact with the dermis. The other cells are physically pushed away from the life support of the blood supply; consequently, cellular death occurs. *Keratinocytes* are specialized cells within the epidermis that produce *keratin*. As the nuclei of the dying keratinocytes degenerate, their cellular content is dominated by keratin, and the process of *keratinization* is completed. Keratin toughens and waterproofs the skin. As the cells continue to be moved toward the surface of the skin, they become flattened and scalelike in a process called *cornification*. The dead cells of the epidermis buffer the body from the external environment.

The integument is a dynamic organ. Although the outermost layers of the epidermis consist of dead cells, most of the skin is very much alive and reflects the general health of the body. During a physical examination, variation in color, texture, and responsiveness of the skin can provide the physician with important diagnostic clues.

5.12 How rapidly are epidermal cells replaced?

The average time it takes for cells to be pushed from the stratum basale to the stratum disjunction is about 7 weeks. This time varies according to location on the body and the age of the person. As a person ages, the epidermis becomes thinner and the rate of mitosis decreases.

5.13 What is a callus and why does it form?

A **callus** is a localized hyperplasia (overdevelopment) of the stratum corneum of the palms or soles due to pressure on the skin or friction and the consequent increase in mitotic activity of the stratum basale in that area. A callus provides additional localized protection against mechanical abrasion.

A *blister* is a vesicle of interstitial fluid located between the stratum basale and the stratum spinosum. Developing in response to rapid and intense friction on the surface of the skin, it serves to cushion and protect the delicate basale layer. In a *blood blister*, a pinch or bruise results in confined and localized hemorrhage.

5.14 What accounts for the variation in normal skin color?

Normal skin coloration is genetically determined and reflects a combination of three pigments: *melanin*, *carotene*, and *hemoglobin*. Melanin is a brown-black pigment formed in cells called **melanocytes** that are found throughout the stratum basale and stratum spinosum. The number of melanocytes is virtually the same in all races, but the amount of melanin produced is variable. Carotene is a yellowish pigment found in epidermal cells and the fatty part of the dermis. Hemoglobin is an oxygen-binding pigment found in red blood cells. Oxygenated blood flowing through the vascular dermis and hypodermis gives the skin its pinkish tones.

A discoloration of the skin may be indicative of a particular body dysfunction. *Cyanosis* is a bluish discoloration that appears in people with certain cardiovascular or respiratory diseases. People also become cyanotic during an interruption of breathing. Jaundice is a yellowing of the skin, mucous membranes, and eyes due to an excess of bile pigment in the bloodstream. Jaundice may be symptomatic of liver dysfunction or gallstones. *Erythema* is a redness of the skin generally due to vascular trauma, such as from a sunburn.

5.15 What is the functional relationship between melanocytes, melanin, and tanning?

Melanin is a proteinaceous pigment that protects against the ultraviolet rays in sunlight. Gradual exposure to sunlight promotes increased production of melanin within melanocytes, and hence tanning of the skin. Excessive exposure, however, can result in a *melanoma*, a tumor composed of melanocytes.

5.16 Is albinism due to an absence of melanocytes, melanin, or both?

The skin of a genetically determined albino has the normal complement of melanocytes but lacks the enzyme tyrosinase that converts the amino acid tyrosine to melanin.

Since an open wound on the skin is a potential entry site for pathogens, the skin is able to maintain homeostasis by healing itself rapidly. An abrasion or a superficial cut promotes mitotic activity in the area and healing is quick and efficient. A more serious problem results if the cells of the stratum basale sustain damage. In an open wound, blood vessels are broken and bleeding occurs. Through the action of blood *platelets* and the plasma protein *fibrinogen*, a clot forms and blocks the flow of blood. The dried clot covering the damaged area is known as a *scab*. Beneath the scab, mechanisms are activated to destroy bacteria, dispose of dead or injured cells, and isolate the injured area. Collectively, these mechanisms are referred to as *inflammation* and include such responses as redness, heat, edema, and pain. Inflammation promotes healing. If the wound is severe, granulation tissue forms from fibroblasts at the site and eventually develops into *scar tissue*. The collagenous fibers of scar tissue are denser than those of normal skin and scar tissue has no epidermal layer. In addition, scar tissue has fewer blood vessels than normal skin and may lack hair, glands, and sensory receptors.

Objective E To describe the structure and function of the dermis.

survey

The upper *papillary layer* of the dermis is in contact with the epidermis. The deeper and thicker *reticular layer* is in contact with the hypodermis (see fig. 5.1). Both dermal layers are highly vascular and nourish the stratum basale of the epidermis. The dermis supports sudoriferous (sweat) glands, hair follicles, and sebaceous (oil-secreting) glands. In addition, numerous sensory receptors located within the dermis respond to heat, cold, touch, pressure, and pain.

5.17 Which of the following connective tissue fiber types is *not* usually found within the dermis? (*a*) reticular, (*b*) elastic, (*c*) fibrous, (*d*) collagenous

(*c*). Elastic fibers are abundant within the papillary layer and provide skin tone; reticular fibers are abundant within the reticular layer and lend a strong meshwork to the skin; collagenous fibers, along with elastic fibers, course in definite directions and are imaged as lines of tension on the surface of the skin (fig. 5.3). The lines of tension are of clinical concern because if a surgical cut is made in the direction of these lines, healing will be faster with less scarring.

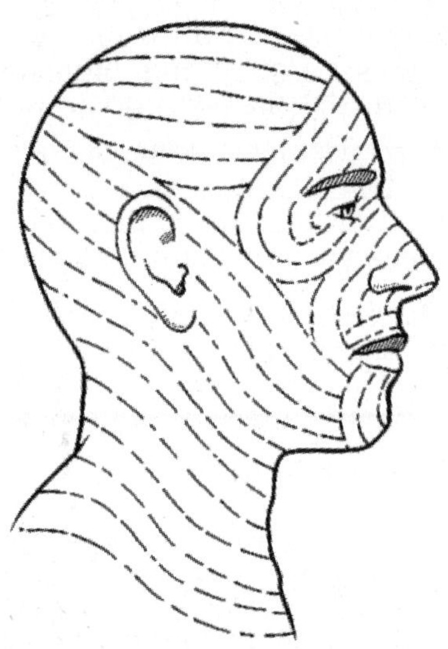

Figure 5.3 Tension lines of the skin covering the head and neck.

Figure 5.4 Print patterns are unique to the individual.

5.18 Define *friction ridges* and explain how these surface marks arise.

Friction ridges are *print patterns* that occur on the anterior surface of the hands and the plantar surface of the feet. They are especially prominent on the skin covering the digits, where they are known as *fingerprints* or *toeprints* (fig. 5.4). Friction ridges are individualistic and are established prenatally in response to the pull of the elastic fibers of the dermal papillary layer upon the epidermis. As the name implies, friction ridges prevent slippage when grasping objects or in locomotion.

5.19 Describe the innervation of the skin.

Specialized *effectors* consist of the muscles or glands within the dermis that respond to *motor (efferent) impulses* transmitted through the autonomic nervous system. Several types of **cutaneous sensory** (*afferent*) **receptors** respond to tactile (touch), pressure, temperature, tickle, or pain stimuli. Certain areas of the body, such as palms, soles, lips, and external genitalia, have a high concentration of sensory receptors and are therefore particularly sensitive to touch.

The cutaneous sensory receptors include those listed in table 5.2.

Table 5.2 Cutaneous Sensory Receptors

Receptor	*Function*
Corpuscles of touch (Meissner's corpuscles)	Detect light motion against the skin
Free nerve endings	Detect changes in temperature; respond to tissue trauma (pain receptors)
Root hair plexuses	Detect movements of hair
Lamellated (pacinian) corpuscles	Detect deep pressure, high-frequency vibration
Organs of Ruffini	Detect deep pressure, stretch
Bulbs of Krause	Detect light pressure, low-frequency vibration

5.20 Why is the vascular supply to the skin important in maintaining homeostasis?

Dermal blood vessels influence body temperature and blood pressure. An autonomic *vasoconstriction* or *vasodilation* will, respectively, shunt blood away from the superficial dermal arterioles or permit it to flow more freely. Blood flow in response to thermoregulatory stimuli can vary from 1 to 150 mL/min for each 100 g of skin. Skin color and temperature also depend on the blood supply. A cold, bluish, or grayish skin occurs when the arterioles are constricted and the capillaries dilated; when both are dilated, the skin is warm and ruddy. Vasoconstriction increases the blood pressure.

Shock is a sudden disturbance of mental equilibrium accompanied by acute peripheral circulatory failure due to marked hypotension (low blood pressure). Shock may be caused by loss of blood (from hemorrhage), diffuse systemic vasodilation, and/or inadequate cardiac function.

5.21 What are decubitus ulcers?

Decubitus ulcers (bedsores) are ulcerated wounds that may occur in debilitated patients who lie in one position for extended periods of time. They are caused by vasocompression in the skin overlying bony prominences—such as at the hip, heel, elbow, or shoulder—making it difficult for the tissue to heal. Periodically changing the position of the patient and daily massaging will minimize the occurrence of bedsores.

Specific locations of the body have characteristic densities of sensory receptors in the dermis. A neurologist uses this knowledge to test nervous system response. A patient should be able to perceive two points of touch as separate when the points are very close together on the face or hands. The ability to distinguish two close points by touch is greatly reduced on the back, however. A lack of sensitivity in certain areas of the body may indicate nerve damage due to disease or injury.

Objective F To describe *hair, nails, sebaceous glands, sudoriferous glands,* and *ceruminous glands.*

Hair, nails, and the three kinds of exocrine glands (glands that secrete a product through a ductule) form from the epidermal skin layer and are therefore of ectodermal derivation. These structures develop as downgrowths of germinal epidermal cells into the vascular dermis, where they receive sustenance and mechanical support.

5.22 Define *hair follicle, shaft, root,* and *bulb*; also, describe the layers of a hair and the arrector pili muscle.

The **hair follicle** is the germinal epithelial layer that has grown down into the dermis (fig. 5.5). Mitotic activity of the hair follicle accounts for growth of the hair. The **shaft of the hair** is the dead, visible, projecting portion; the **root of the hair** is the living portion within the hair follicle; and the **bulb of the hair** is the enlarged base of the root of the hair that receives nutrients and is surrounded by sensory receptors. Each hair consists of an inner **medulla**, a median **cortex**, and an outer **cuticle layer**. The keratinized cuticle layer appears scaly under a dissecting microscope. Variation in the amount of melanin accounts for different hair colors. A pigment with an iron base (trichosiderin) causes red hair; gray or white hair is due to a decrease in pigment production and air spaces between the three layers of the shaft of the hair. Each hair follicle has an associated **arrector pili muscle** (smooth muscle) that responds involuntarily to thermal or psychological stimuli, causing the hair to be pulled into a more vertical position.

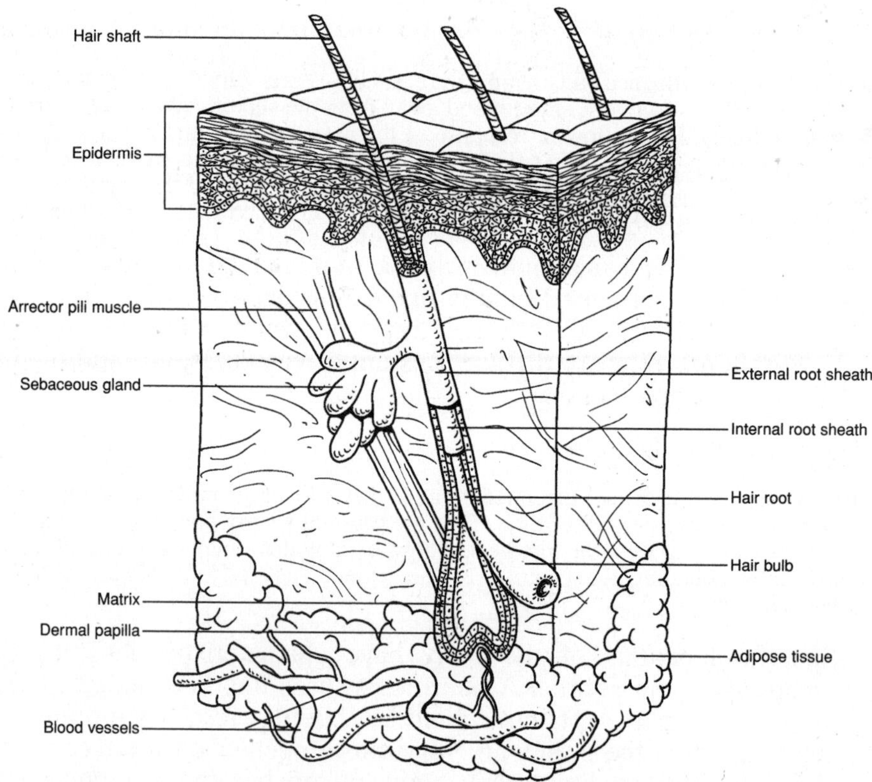

Figure 5.5 A hair within a hair follicle.

5.23 What are the functions of human hair?

The primary function of hair in humans is protection, although its effectiveness is limited. Hair on the scalp and eyebrows protects against sunlight, and hair in the nostrils and the eyelashes protects against airborne particles. An important secondary function of hair is as a means of individual recognition and sexual attraction.

5.24 What are the three distinct kinds of human hair?

Lanugo is fine, silky fetal hair that appears during the last trimester of development. Its function is unknown, but probably has something to do with the maturation of the hair follicles. **Angora hair** is continuously growing, as on the scalp, or on the face of sexually mature males. **Definitive hair** grows to a certain length and then stops growing. Examples are eyelashes, eyebrows, pubic hair, and axillary (armpit) hair.

5.25 Describe the structure and function of nails.

Nails are formed from the hardened transparent, stratum corneum of the epidermis. A parallel arrangement of keratin fibrils (see fig. 5.6) accounts for its hardness. Each nail consists of a **body;** a **free border**, which is attached to the undersurface by the *hyponychium*, and a **hidden border**, covered by the *eponychium* (cuticle). The body of the nail rests on a **nail bed**. The sides of the body of the nail extend into a furrow called a **nail groove**. The **matrix** is the growth area of the nail. A small part of the matrix, the **lunula**, is seen as a white, half-moon-shaped area at the base of the nail.

Fingernails grow about 1 mm per week, somewhat faster than toenails. They serve to protect the digits and aid in grasping small objects. All lizards, birds, and mammals have some sort of hardened sheath (claw, talon, hoof, or nail) protecting their terminal phalanges.

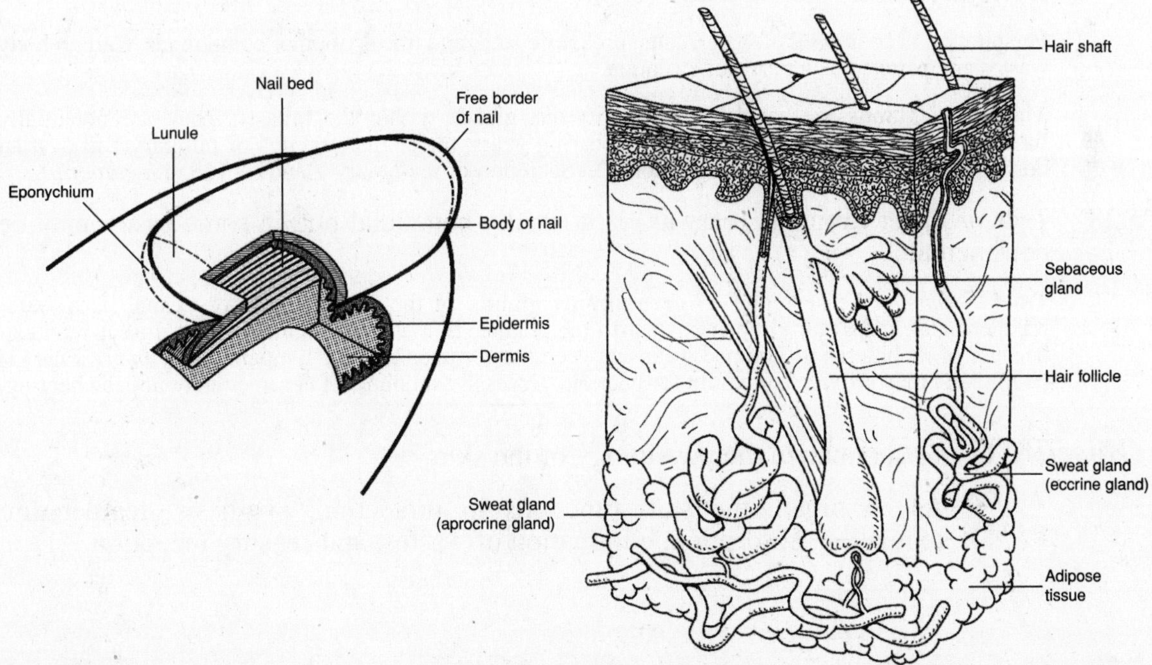

Figure 5.6 The structure of a nail. **Figure 5.7** Sudoriferous and sebaceous glands.

An *ingrown toenail* is a common condition in which the corner of the toenail grows into the skin on one or both lateral sides of the nail. Cutting the toenails straight across and not too close to the body of the nail is the best prevention. An infected ingrown toenail may require treatment by a *podiatrist*, a specialist who treats clinical conditions of the feet.

5.26 Describe the structure and function of sebaceous glands. Why are they of clinical significance?

Sebaceous glands are simple branched oil glands (fig. 5.7) that develop from the follicular epithelium of hair. They secrete acidic *sebum* (pH about 6.8) onto the shaft of the hair. The sebum is then dissipated to the surface of the skin where it protects, lubricates, and helps to waterproof the strata corneum and disjunction of the epidermis. Sebum consists mainly of lipids and some proteins. If the drainage ductule for sebaceous glands becomes blocked, the glands may become infected, resulting in *acne*. Sex hormones, particularly androgens, regulate the production and secretion of sebum.

5.27 Describe the structure and functions of sudoriferous glands and distinguish between eccrine and apocrine types.

Sudoriferous glands are *sweat glands*. As seen in figure 5.7, both eccrine and apocrine sweat glands are coiled, tubular structures that secrete perspiration (sweat) onto the surface of the skin. **Eccrine sweat glands** (also called *merocrine sweat glands*) are most abundant on the forehead, back, palms, and soles. These glands are formed prenatally and provide evaporative cooling in response to thermal or psychological stimuli. *Relaxed eccrine*, or *insensible, perspiration* accounts for 300 to 800 mL of water loss daily, depending upon the external temperature and humidity. *Active perspiration* in response to physical exercise may amount to as much as 5 L of water per day. Apocrine sweat glands are much larger than the eccrine type and are restricted to the axillary and pubic regions, in association with hair follicles. They are not functional until puberty, and their odoriferous secretion is thought to act as a sexual attractant.

Perspiration is composed of water, salts, urea, uric acid, and traces of other compounds. Certain body wastes are excreted as a result of sweating.

Mammary glands are specialized sudoriferous glands within the breasts. They are potentially functional in the female during her childbearing years, under the stimulus of pituitary and ovarian hormones following parturition (childbirth). Secretion of mammary glands is called *lactation*.

5.28 *True or false*: Cerumen (earwax) is normally beneficial but, in some cases, may be detrimental.

True: *Cerumen*, the secretion of **ceruminous glands** of the external auditory canal, is a water repellent and keeps the tympanum (eardrum) pliable. It is also thought to be an insect repellent because of its bitterness. Trapped water between the cerumen and the tympanum (*swimmer's ear*) is painful and may provide a medium for bacteria. Excessive amounts of cerumen may impede hearing.

Objective G To summarize the physiology of the skin.

As an organ, the skin functions in protection, synthesis, temperature regulation, absorption, elimination of wastes, and sensory reception.

5.29 Comment on each function of the skin and indicate where the function is performed.

The physiology of the skin is summarized in table 5.3.

Table 5.3 Summary of the Physiology of the Skin

	Function	*Site*	*Comments*
Protection against	Dehydration	Epidermis	Stratification forms dense barrier; sebum provides oily fibers; keratin toughens epidermis; basement membrane seals epidermis
Protection against	Mechanical injury	Epidermis	Stratification forms dense barrier; cornification of exposed layer; formation of calluses in response to friction; keratin toughens epidermis
Protection against	Pathogens	Epidermis	Stratification forms nearly impenetrable barrier; sebum is acidic (pH 4–6.8) and antiseptic, and lipid composition keeps epidermis from cracking; rapid rate of mitosis and shedding of cells from outer layer minimize entry of pathogens
Protection against	Ultraviolet (UV) light	Epidermis	Stratification forms dense barrier; scalp hair disperses light; melanin within melanocytes absorbs solar radiation
Protection against	Blood loss	Epidermis and dermis	Stratification forms dense barrier; process of *wound healing* (dermal vasoconstriction, blood coagulation, temporary scab, collagenous scar tissue)
	Synthesis	Epidermis and dermis	Keratin, melanin, and carotene synthesized in epidermis; dermis contains dehydrocholesterol, from which it synthesizes vitamin D in the presence of UV light
	Temperature regulation	Dermis and hypodermis	Cooling through vasodilation and sweating; warming through vasoconstriction and shivering; insulation provided by lipid content of hypodermis
	Absorption	Epidermis, dermis, and hypodermis	Limited by protective barriers, but some cutaneous absorption of O_2; CO_2; fat-soluble vitamins (A, D, E, and K); certain steroid hormones (cortisol); and certain toxic substances (insecticides)
	Elimination of wastes	Epidermis and dermis	Excessive water salt (NaCl) metabolic wastes (urea, uric acid)
	Sensory reception	Epidermis, dermis, and hypodermis	Lower layers of epidermis contain free nerve endings, responsive to temperature and pain; dermis contains *corpuscles of touch*, responsive to touch, and *lamellated corpuscles*, responsive to deep pressure; dermis and hypodermis contain *bulbs of Krause* and *organs of Ruffini*, responsive to pressure and stretch

Key Clinical Terms

Acne An inflammatory condition of sebaceous glands. Acne is affected by gonadal hormones and is therefore most common during puberty and adolescence. Pimples and blackheads on the face, chest, or back are expressions of this condition.

Alopecia Loss of hair, baldness. Baldness is usually due to genetic factors and may accompany old age. It is influenced by improper diet and poor circulation of blood.

Athlete's foot (*Tinea pedis*) A fungus disease of the skin of the foot.

Blister A collection of fluid between the epidermis and dermis, caused by excessive friction or a burn.

Boil A localized bacterial infection originating in a hair follicle or skin gland; also termed a *furuncle*.

Burn A lesion of the integument caused by heat, chemicals, electricity, or solar exposure. Classified as *first degree* (redness or hyperemia in superficial layers of skin), *second degree* (blisters involving deeper epidermal layers and dermis), or *third degree* (destruction of areas of the integument and damage to underlying tissue).

Callus A localized buildup of the stratum corneum on the palm or sole due to excessive friction.

Carbuncle Similar to a boil, except involving subcutaneous tissues.

Corn A localized buildup of the stratum corneum on the dorsal surface of the foot due to excessive friction.

Dandruff Common dandruff is the continual shedding of epidermal cells of the scalp; it can be controlled by normal washing and brushing of the hair. Abnormal dandruff may be due to certain skin diseases, such as seborrhea and psoriasis.

Decubitus ulcer A bedsore, or exposed ulcer from continual pressure on a localized portion of the skin, restricting the blood supply.

Dermatitis An inflammation of the skin.

Eczema A noncontagious inflammatory condition of the skin marked by red, itching, vascular lesions that may be crusty or scaly.

Gangrene Necrosis (death) of tissue due to obstruction of blood flow; may be localized or extensive, and perhaps secondarily infected with anaerobic microorganisms.

Melanoma A cancerous tumor of the melanocytes within the epidermis.

Nevus A mole or birthmark; congenital pigmentation of a certain area of the skin.

Psoriasis Inflammatory skin disease, usually expressed as circular scaly patches of skin.

Pustule A small, localized elevation of the skin containing pus.

Seborrhea A disease characterized by excessive activity of sebaceous glands and accompanied by oily skin and dandruff.

Shingles (*Herpes zoster*) A viral infection characterized by clusters of blisters along certain nerve tracts (dermatomes).

Urticaria (*hives*) A skin eruption consisting of reddish, itchy wheals; may arise from an allergic reaction or stress.

Wart A roughened projection of epidermal cells; caused by a virus.

Review Exercises

Multiple Choice

1. Which of the following word pairs is appropriately matched? (*a*) skin–gland, (*b*) skin–tissue, (*c*) skin–organ, (*d*) skin–body system

2. The skin is derived from (*a*) ectoderm and endoderm, (*b*) ectoderm and mesoderm, (*c*) mesoderm and endoderm, (*d*) ectoderm, mesoderm, and endoderm.

3. The skin accounts for what percentage of the body weight? (*a*) 2%, (*b*) 10%, (*c*) less than 2%, (*d*) 15%, (*e*) 7%

4. Which of the following is *not* a function of the skin? (*a*) prevention of body dehydration, (*b*) synthesis of vitamin A, (*c*) prevention of pathogen entry, (*d*) regulation of body temperature

5. Loss of body fluids through the integument is restricted by (*a*) keratin, (*b*) the stratum basale, (*c*) carotene, (*d*) melanocytes, (*e*) the thickness of the dermis.

6. Which epidermal layer is lacking within the skin of the head and trunk? (*a*) stratum spinosum, (*b*) stratum corneum, (*c*) stratum granulosum, (*d*) stratum lucidum, (*e*) stratum basale

7. Which of the following pairings is appropriate? (*a*) stratum basale–keratin, (*b*) stratum corneum–melanocytes, (*c*) stratum granulosum–keratin, (*d*) stratum lucidum–blood vessels, (*e*) stratum spinosum–cornified

8. Fingerprint patterns are established prenatally during development of (*a*) the stratum corneum, (*b*) the dermal papillary layer, (*c*) the stratum basale, (*d*) the dermal reticular layer, (*e*) the hypodermis.

9. It is *false* that the dermis (*a*) is highly vascular, (*b*) gives rise to sebaceous and sweat glands, (*c*) contains reticular, elastic, and smooth muscle fibers, (*d*) contains numerous nerve endings.

10. It is *false* that the epidermis (*a*) is highly vascular, (*b*) contains melanin and keratin, (*c*) is distinctly stratified, (*d*) gives rise to sebaceous and sweat glands.

11. Which grouping of terms is appropriate? (*a*) mesoderm, stratified squamous epithelium, epidermis; (*b*) epidermis, ectoderm, stratified squamous epithelium; (*c*) hypodermis, ectoderm, adipose tissue; (*d*) dermis, endoderm, vascular tissue

12. "Rapunzel, Rapunzel, let down your long ______________ scalp hair." (*a*) axillary, (*b*) lanugo, (*c*) definitive, (*d*) angora, (*e*) alopecia

13. What is the proper sequence of epidermal strata (layers) pierced as a sliver penetrates the epidermis on the palm of the hand?
    (*a*) spinosum, basale, granulosum, lucidum, corneum, disjunction
    (*b*) basale, spinosum, granulosum, disjunction, lucidum, corneum
    (*c*) disjunction, corneum, lucidum, granulosum, spinosum, basale
    (*d*) corneum, disjunction, lucidum, spinosum, granulosum, basale

14. Cells from the stratum basale reach the stratum disjunction in approximately (*a*) 15 to 20 days, (*b*) 6 to 8 weeks, (*c*) 8 to 10 days, (*d*) 12 to 15 weeks, (*e*) 4 to 6 months.

15. Which of the following is *not* a type of cutaneous sensory receptor? (*a*) lamellated corpuscle, (*b*) bulb of Krause, (*c*) free nerve ending, (*d*) organ of Ruffini, (*e*) Golgi apparatus

16. Produced in the epidermis of the skin, melanin (*a*) protects against ultraviolet light, (*b*) prevents infections, (*c*) helps regulate body temperature, (*d*) keeps the epidermis pliable, (*e*) reduces water loss.

17. Identify the mismatch (*a*) yellowish skin in people of Asian origin—carotene abundant, (*b*) tanning of skin in response to sunlight—increased synthesis of melanin, (*c*) bluish skin (cyanotic)—oxygenated blood, (*d*) lack of skin pigmentation (albinism)—heredity, (*e*) dark skin in people of African origin—greater synthesis of melanin.

18. The most probable cause of alopecia is (*a*) protein deficiencies, (*b*) dermal viral infection, (*c*) genetic inheritance, (*d*) stress.

19. Which of the following statements about sebaceous glands is *true*?
(*a*) They secrete sebum directly to the skin surface.
(*b*) They derive from specialized mesoderm.
(*c*) They are a type of oil-secreting gland.
(*d*) They are a compound saccular type.

20. Which of the following is *not* a function of the integument? (*a*) elimination of certain body salts, urea, and uric acid; (*b*) absorption of fat-soluble vitamins, steroid hormones, and certain toxic chemicals; (*c*) storage of lipids; (*d*) thermoregulation; (*e*) synthesis of proteins and carbohydrates; (*f*) prevention of desiccation and blood loss

True or False Questions

____ 1. Integument is synonymous with skin, and neither properly includes the hair or glands.

____ 2. Skin is the largest tissue of the body, accounting for approximately 7% of the body weight.

____ 3. Hair, nails, and integumentary glands are specializations of the epidermis and are derived from the embryonic ectodermal germ layer.

____ 4. A burn that damaged both the epidermis and dermis so that regeneration could occur only from the edges of the wound would be classified as a second-degree burn.

____ 5. The eponychium and lunula are both proximal to the hyponychium of a nail.

____ 6. The skin on the palm of the hand consists of six epidermal layers, and two dermal layers, the lowest of which is affixed to the hypodermis of the skin.

____ 7. Mitotic activity is characteristic of all tunics (layers) of the epidermis except the dead stratum disjunction, which is constantly being shed.

____ 8. People of African descent have more melanocytes in their skin than do lighter complexioned people.

____ **9.** Mammary glands are modified sebaceous glands that are hormonally prepared to lactate in association with the birth of a baby.

____ **10.** Stimulation of the free nerve endings within the skin would cause the perception of cold and may autonomically induce shivering.

____ **11.** Water-soluble substance would be more readily absorbed through the skin than fat-soluble substances.

____ **12.** All sudoriferous glands are formed and functional in a newborn.

____ **13.** The principal danger of a third-degree burn is excessive body fluid loss and disruption of homeostasis.

____ **14.** Alopecia is a disease that results in excessive loss of hair.

____ **15.** Warts, shingles, and acne are all viral infections of the integument.

Completion

1. The term ____________________ is synonymous with skin.

2. The epidermis of the skin consists of ____________________ ____________________ epithelial tissue.

3. The outermost layer of the epidermis of the skin is the stratum ____________________ and the deepest layer is the stratum ____________________.

4. Normal skin coloration reflects a combination of three pigments: hemoglobin, ____________________, and ____________________.

5. The dermis of the skin consists of an upper ____________________ layer and a deeper ____________________ layer.

6. ____________________ is a protein in the skin that strengthens the stratified squamous epithelium of the epidermis.

7. ____________________ glands secrete sebum into the hair follicles of skin.

8. ____________________ is silky fetal hair that appears during the last trimester of prenatal development.

9. Sudoriferous glands are of two types: ____________________ sweat glands are abundant on the forehead, back, palms, and soles; ____________________ sweat glands are abundant in the axillary and pubic regions of a sexually mature person.

10. ____________________ glands secrete cerumen (earwax) into the external auditory canal.

Labeling

Label the structures indicated on the figure to the right.

1. ______________________
2. ______________________
3. ______________________
4. ______________________
5. ______________________
6. ______________________
7. ______________________
8. ______________________
9. ______________________
10. ______________________

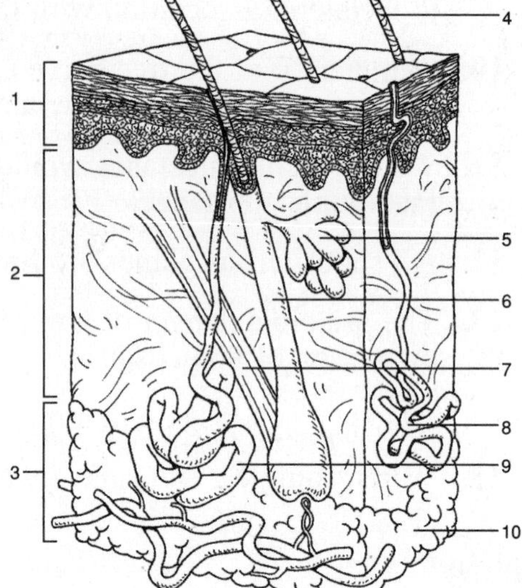

Answers and Explanations for Review Exercises

Multiple Choice

1. (*c*) "Skin–organ" is a match because the integument is an organ. An organ is a structure of the body composed of two or more types of tissues.
2. (*b*) The epidermis of the skin derives from embryonic ectoderm and the dermis of the skin derives from embryonic mesoderm.
3. (*e*) The skin and its accessory structures account for about 7% of a person's body weight, with some individual variation.
4. (*b*) The skin does synthesize vitamin D in the presence of ultraviolet light, but vitamin A can be obtained only from food.
5. (*a*) Keratin, a protein produced by dying epithelial cells within the epidermis, forms a waterproof barrier.
6. (*d*) The stratum lucidum is found only in areas of "thick skin," which are on the palms of the hands and the soles of the feet.
7. (*c*) The stratum granulosum is so named for the dark granules of keratohyalin within its cells. These granules contribute to the formation of keratin that permeates the upper layers of the epidermis.
8. (*b*) The contoured papillary layer of the dermis develops as a result of the genetically determined arrangement of the elastic and collagenous fibers that is established prenatally. Distinctly ridged, the fingerprints aid gripping. In criminology, they are also used as a means of identification.
9. (*b*) All integumentary glands have their origin as an invagination of the epidermis into the dermis, where they mature and become functional.
10. (*a*) The epidermis is avascular. Only the cells composing the stratum basale derive oxygen and nutrients necessary for mitosis. As the cells are moved away from the life-support of the dermis, they die and undergo the transformation of keratinization and cornification.
11. (*b*) Derived from the ectoderm germ layer, the epidermis is composed of stratified squamous epithelium.
12. (*e*) The lovely Rapunzel unloosed her scalp hair over the balcony to help her intrepid suitor scale the castle wall. Scalp hair is angora in that it grows continuously and has no predetermined length.
13. (*c*) The layers of the epidermis are in the same order throughout the body because they reflect the transitional changes that occur as they are moved away from the dividing stratum basale layer.

14. (*b*) The movement of cells in the epidermis varies in accordance with the rate of sloughing off from the outer surface and the rate of mitosis in the stratum basale.
15. (*e*) Lamellated corpuscles, bulbs of Krause, free nerve endings, and organs of Ruffini are all receptor types in the skin.
16. (*a*) Located in the stratum basale, melanin is a pigment produced in melanocytes that absorbs specific wavelengths of light. Ultraviolet light is a common radiation on Earth that is a potential health hazard.
17. (*c*) Oxygenated blood is bright red due to the formation of oxyhemolobin. Cyanosis, or blue blood, is the result of insufficient oxygen.
18. (*c*) Alopecia, or baldness, is usually genetically inherited, although viruses, stress, and protein deficiencies can influence the condition.
19. (*c*) Sebaceous glands secrete sebum into hair follicles. Derived from ectoderm, these exocrine glands are of the compound tubular type.
20. (*e*) The skin does not produce proteins but it does stores energy in the form of lipids. In order for lipids to be utilized as a source of food, however, they must be transported to the liver and converted to carbohydrates.

True or False

1. True
2. False; the skin is an organ
3. True
4. False; third-degree burn
5. True
6. True
7. False; mitosis occurs principally in the stratum basale, and to a slight degree in the stratum spinosum
8. False; all people have virtually the same number of melanocytes but vary in the ability to synthesize melanin
9. False; sudoriferous glands
10. True
11. False; the skin is virtually waterproof
12. False; apocrine sweat glands do not mature until puberty
13. True
14. False; alopecia is not a disease
15. False; acne is an inflammatory condition of sebaceous glands

Completion

1. integument
2. stratified squamous
3. disjunction, basale
4. melanin, carotene
5. papillary, reticular
6. Keratin
7. Sebaceous
8. Lanugo
9. eccrine, apocrine
10. Ceruminous

Labeling

1. Epidermis
2. Dermis
3. Hypodermis
4. Shaft of hair
5. Sebaceous gland
6. Hair follicle
7. Arrector pili muscle
8. Eccrine sweat gland
9. Apocrine sweat gland
10. Adipose tissue

Skeletal System 6

Objective A To describe the principal functions of the *skeletal system*.

survey

The skeletal system consists of **bones, cartilage,** and **joints**. The bones are the individual organs of the skeletal system and, in turn, are composed of bone tissue (see chapter 4).

The functions of the skeletal system fall into five categories. *Support*—the skeleton forms a rigid framework to which are attached the softer tissues and organs of the body. *Protection*—the skull, vertebral column, rib cage, and pelvic girdle enclose and protect vital organs; sites for blood cell production are protected within the hollow centers of certain bones. *Movement*—bones act as levers when attached muscles contract, causing movement about joints. *Hemopoiesis*—red bone marrow of an adult produces white and red blood cells and platelets (see problem 6.7). *Mineral storage*—the matrix of bone is composed primarily of calcium and phosphorus; these minerals can be withdrawn in small amounts if needed elsewhere in the body. Lesser amounts of magnesium and sodium are also stored in bone tissue.

6.1 How much of the body's calcium and phosphorus is contained in bones?

About 99% of the calcium within the body, and 90% of the phosphorus, is deposited in bones and teeth. These minerals give bone its rigidity, and they account for approximately two-thirds of the weight of bone. In addition, calcium is necessary for muscle contraction, blood clotting, and movement of molecules across cell membranes. Phosphorus is required for the activities of DNA and RNA, as well as for ATP utilization.

6.2 In addition to mineralization involving calcium and phosphorus, what other physiological mechanisms determine the stability of bone?

Body organs that perform regulatory functions have a direct effect on the stability of bone. The kidneys, for example, determine blood composition, which in turn affects bone. The digestive system—via proteins and vitamins A, D, and C—and the female reproductive system—via pregnancy—can cause alteration of bone. Enzymatic and metabolic controls (alkaline phosphatase, glycogen, etc.) of the liver affect bone structure. At least five hormones affect bone: pituitary growth hormone stimulates bone growth (osteogenesis); thyroid hormone promotes both osteogenesis and osteolysis (bone destruction); androgens and estrogens of the gonads stimulate bone growth and closure of the growth lines (epiphyseal plates); and unbalanced secretions of the adrenal cortisol and thyrocalcitonin may cause osteoporosis (bone atrophy).

Rickets and *osteomalacia* are metabolic diseases caused by a deficiency of vitamin D. Rickets occurs in children who have inadequate exposure to sunlight and a dietary deficiency of vitamin D. Without sufficient vitamin D, the body is unable to properly metabolize calcium and phosphorus. Children with rickets are irritable because of bone pain, and their bones are easily fractured. The bones within the legs are frequently bowed because of their inability to support the weight of the body. A deficiency of vitamin D in adults causes bone resorption, resulting in osteomalacia. Weakening of adult bones frequently leads to skeletal deformities, especially of the spine and legs. Both of these conditions are treated with supplements of vitamin D, calcium, and phosphorus.

Objective B To distinguish between the *axial* and *appendicular portions* of the skeletal system.

Survey

The **axial skeleton** consists of the bones that form the axis of the body and that support and protect the organs of the head, neck, and trunk. These bones include those of the *skull, vertebral column,* and *rib cage*. In addition, the *auditory ossicles* (ear bones) and the *hyoid bone* are included within the axial skeleton. The **appendicular skeleton** consists of the bones of the *pectoral* and *pelvic girdles* and the bones of the *upper* and *lower extremities*. The girdles anchor the appendages to the axial skeleton. The bones of the adult skeleton are illustrated in fig. 6.1 and listed in table 6.1.

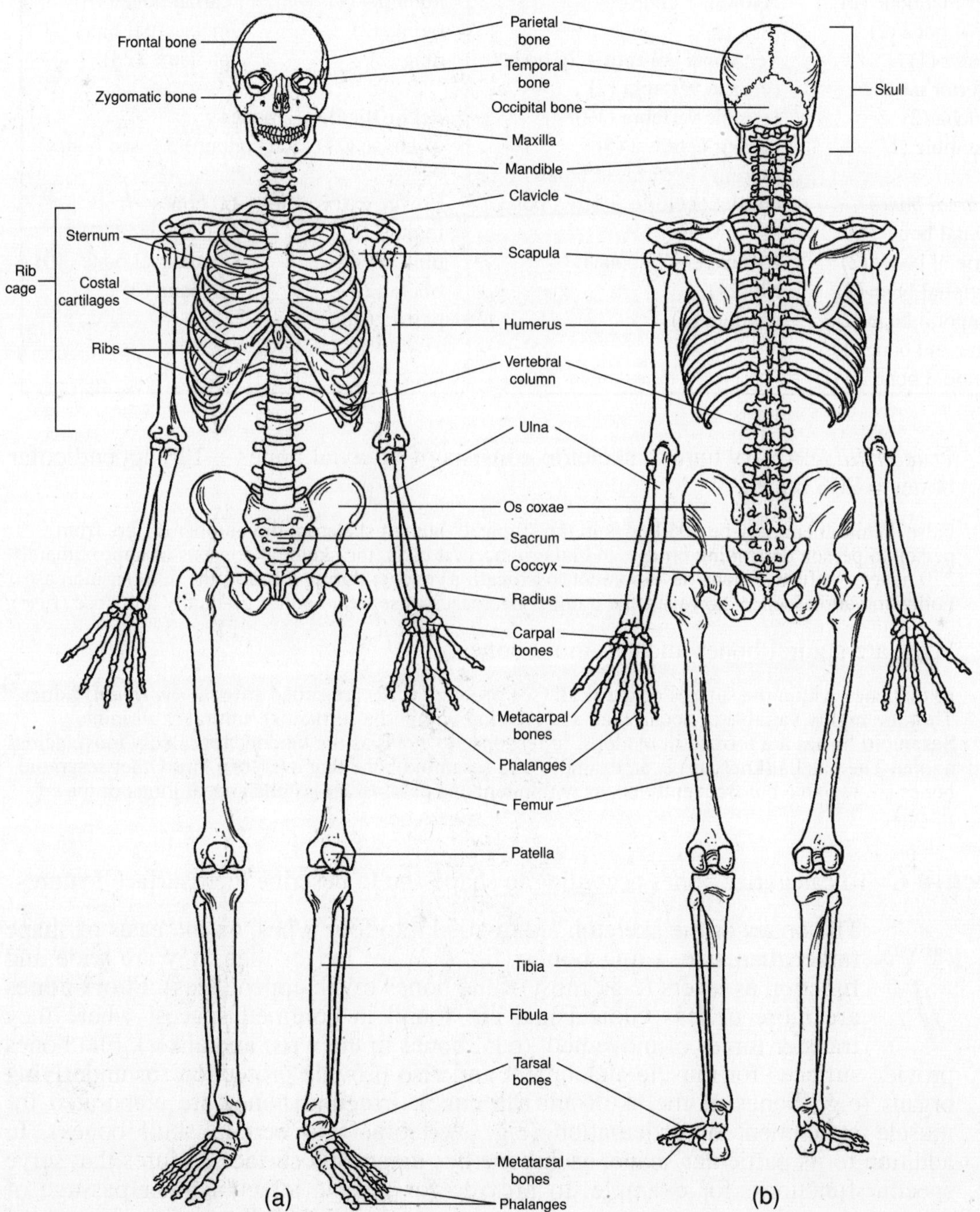

Figure 6.1 The skeleton. (*a*) An anterior view and (*b*) a posterior view.

Table 6.1 Classification of the Bones of the Adult Skeleton

Axial skeleton		**Appendicular skeleton**	
Skull—22 bones	**Auditory ossicles**—6 bones	**Pectoral girdle**—4 bones	
14 facial bones	malleus (2)	scapula (2)	
maxilla (2)	incus (2)	clavicle (2)	
palatine bone (2)	stapes (2)		
zygomatic bone (2)		**Upper extremities**—60 bones	
lacrimal bone (2)	**Hyoid**—1 bone	humerus (2)	carpal bone (16)
nasal bone (2)		radius (2)	metacarpal bone (10)
vomer (1)	**Vertebral column**—26 bones	ulna (2)	phalanx (28)
inferior nasal	cervical vertebra (7)		
concha (2)	thoracic vertebra (12)	**Pelvic girdle**—2 bones	
mandible (1)	lumbar vertebra (5)	os coxae (2) (each contains 3 fused bones)	
	sacrum (1) (5 fused bones)		
8 cranial bones	coccyx (1) (3 to 5 fused bones)	**Lower extremities**—60 bones	
frontal bone (1)		femur (2)	tarsal bone (14)
parietal bone (2)	**Rib cage**—25 bones	tibia (2)	metatarsal bone (10)
occipital bone (1)	rib (24)	fibula (2)	phalanx (28)
temporal bone (2)	sternum (1)	patella (2)	
sphenoid bone (1)			
ethmoid bone (1)			

6.3 *True or false*: Every human skeleton consists of 80 axial bones + 126 appendicular bones = 206 bones.

False. While there may be 206 bones in the "typical" human skeleton, the number differs from person to person depending on age and inheritance. At birth, the skeleton consists of approximately 270 bones. As further bone development (ossification) occurs during infancy, the number increases. Following adolescence, however, the number decreases as separate bones gradually ankylose (fuse).

6.4 What are sutural bones and sesamoid bones?

Extra bones within the sutures of the skull (see problem 6.12) are called **sutural** (*wormian*) **bones.** They are highly variable in occurrence and location within the serratelike sutural skull joints. **Sesamoid bones** are formed in tendons, in response to stress as the tendons repeatedly move across a joint. The patella (kneecap) is an example of a sesamoid bone that everyone has. Other sesamoid bones are variable but frequently occur within tendons passing across phalangeal joints of the fingers.

Objective C To categorize bones according to shape and to describe their surface features.

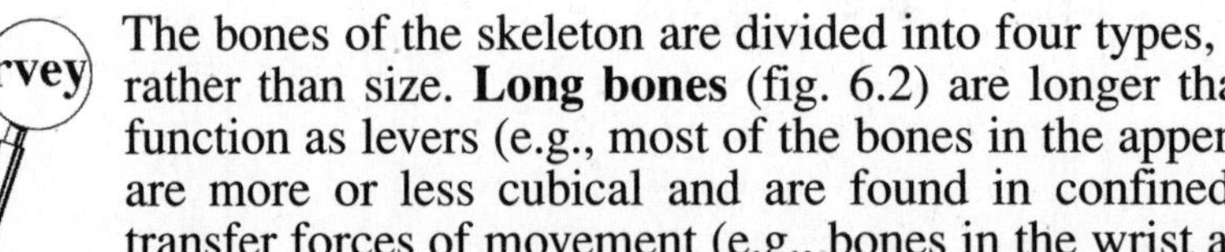

The bones of the skeleton are divided into four types, on the basis of shape rather than size. **Long bones** (fig. 6.2) are longer than they are wide and function as levers (e.g., most of the bones in the appendages). **Short bones** are more or less cubical and are found in confined spaces, where they transfer forces of movement (e.g., bones in the wrist and ankle). Flat bones provide surfaces for muscle attachment and also provide protection for underlying organs (e.g., bones of the skull and rib cage). Irregular bones are elaborated for muscle attachment or articulation (e.g., vertebrae and certain skull bones). In addition to its particular shape, each bone has diagnostic surface features that serve specific functions, for example, to provide for muscle attachment or passage of nerves or vessels, or to permit or restrict movement at joints. The surface features of bones are summarized in table 6.2.

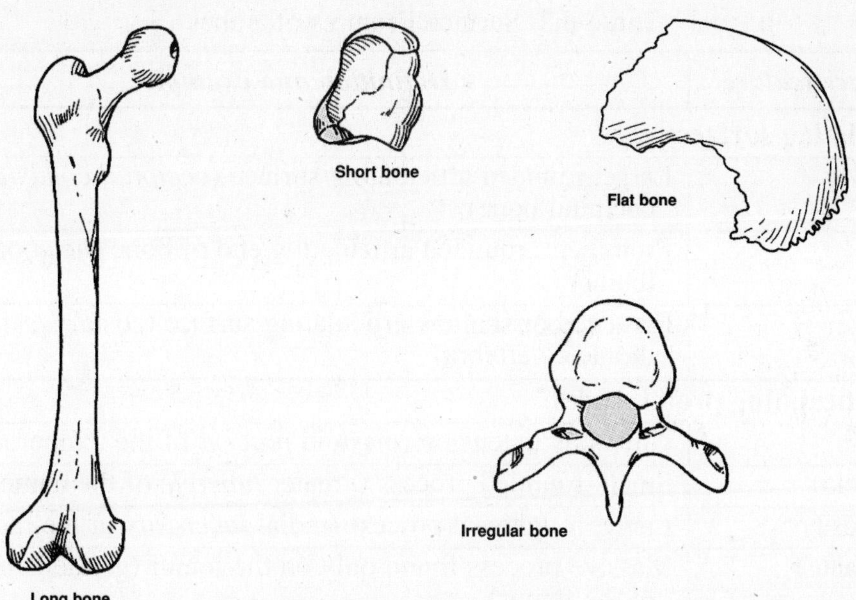

Figure 6.2 The shapes of bones.

Objective D To distinguish between *endochondral* and *intramembranous bone formation.*

Ossification (bone formation) begins during the fourth week of prenatal development. Bones develop either through *endochondral ossification*—going first through a cartilaginous stage—or through *intramembranous* (dermal) *ossification*—forming directly as bone.

6.5 Which bones are endochondral? Which are membranous?

The majority of bones are formed first as hyaline cartilage, which then undergoes endochondral ossification. The bones of the face (*facial bones*), however, and the bones surrounding the brain (*cranial bones*) are all membranous, except for the sphenoid and occipital bones, which are endochondral. Sesamoid bones are also membranous bone.

6.6 What are the fontanels, and why are they important?

During fetal development and infancy, the membranous bones of the top and sides of the cranium are separated by fibrous *sutures*. There are also six large membranous areas, called **fontanels** ("soft spots"), that permit the skull to undergo changes in shape (molding) during parturition (childbirth); four of these are illustrated in fig. 6.3. The fontanels also permit rapid growth of the brain during infancy. Ossification of the fontanels is normally complete by 20 to 24 months of age.

Table 6.2 Surface Features of Bones

Surface feature	*Definition and example*
Articulating surfaces	
Condyle	Large, rounded articulating surface (*occipital condyle* of the occipital bone)
Head	Prominent, rounded articulating end of bone (*head* of the femur)
Facet	Flattened or shallow articulating surface (costal *facet* of a thoracic vertebra)
Nonarticulating prominences	
Process	Any bony extension (mastoid *process* of the temporal bone)
Tubercle	Small, rounded process (greater *tubercle* of the humerus)
Tuberosity	Large, roughened process (radial *tuberosity* of the radius)
Trochanter	Massive process found only on the femur (greater *trochanter* of the femur)
Spine	Sharp, slender process (*spine* of the scapula)
Crest	Narrow, ridgelike projection (iliac *crest* of the os coxae)
Epicondyle	Projection above a condyle (medial *epicondyle* of the femur)
Depressions and openings	
Fossa	Shallow depression (mandibular *fossa* of the temporal bone)
Sulcus	Groove that accommodates a vessel, nerve, or tendon (intertubercular *sulcus* of the humerus)
Fissure	Narrow, slitlike opening (superior orbital *fissure* of the sphenoid bone)
Meatus, or canal	Tubelike passageway (external acoustic *meatus* of the temporal bone
Alveolus	Deep pit or socket (maxillary *alveoli* for teeth)
Foramen (pl., *foramina*)	Rounded opening through a bone (*foramen* magnum of the occipital bone
Sinus	Cavity or hollow space (frontal *sinus* of the frontal bone)
Fovea	Small pit or depression (*fovea* capitis femoris of the femur)

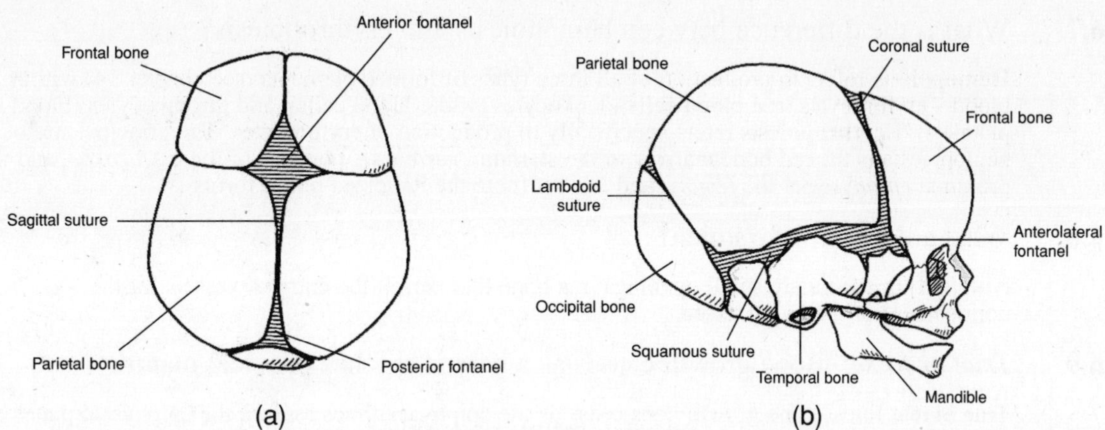

Figure 6.3 The fontanels of the fetal skull and the principal sutures. (*a*) a superior view and (*b*) a lateral view.

Objective E To describe the gross structure of a typical long bone.

Survey

Within the **diaphysis** (shaft) of a long bone is a **medullary cavity** that is lined with a thin layer of connective tissue called the **endosteum** (fig. 6.4). The medullary cavity contains fatty *yellow bone marrow*. On either end of the diaphysis is an **epiphysis**, consisting of *spongy bone* surrounded by *compact bone*. *Red bone marrow* is found within the pores of the spongy bone. Separating the diaphysis and epiphysis is an **epiphyseal plate**, a region of mitotic activity responsible for linear bone growth (elongation); an **epiphyseal line** replaces the plate when bone growth is completed. A **periosteum** of dense regular connective tissue covers the bone and is the site of tendon-muscle attachment and diametric bone growth (widening).

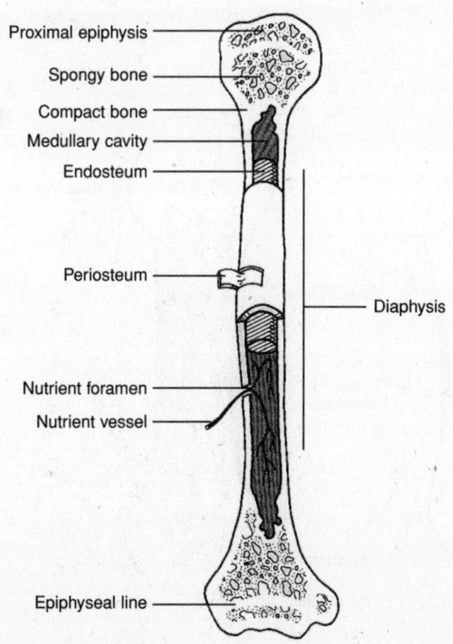

Figure 6.4 The structure of a long bone.

6.7 What is the difference between hemopoiesis and erythropoiesis?

Hemopoiesis refers to production of all three types of formed elements (see chapter 14) within blood—erythrocytes (red blood cells), leukocytes (white blood cells), and thrombocytes (blood platelets). **Erythropoiesis** refers specifically to production of erythrocytes. The principal site of hemopoiesis is the red bone marrow of the sternum, *vertebrae*, portions of the *ossa coxae*, and the proximal *epiphyses* of the *femora* and *humeri* (note the italicized plural forms).

6.8 What are nutrient foramina?

Nutrient foramina are small openings in a bone that permit the entry of vessels for the nourishment of the living tissue.

6.9 *True or false*: Bone growth ceases as a person reaches physical maturity.

True in that linear bone growth does cease as the epiphyseal lines replace the epiphyseal plates and ossification occurs between the epiphyses and diaphyses. However, diametric bone growth and enlargement of bony processes may occur at any time to accommodate an increase in body mass (as with a weight lifter).

6.10 Where is articular cartilage found?

Articular cartilage is thin hyaline cartilage that caps each epiphysis to facilitate joint movement. Technically, bones do not articulate; rather, the articular cartilage of one bone articulates with the articular cartilage of another.

Objective F To describe endochondral bone formation.

survey

Endochondral ossification begins in a primary center (fig. 6.5), in the shaft of a cartilage model, with hypertrophy of chondrocytes (cartilage cells) and calcification of the cartilage matrix. The cartilage model is then vascularized, osteogenic cells form a bony collar around the model, and osteoblasts lay down bony matrix around the calcareous spicules. Ossification from primary centers occurs before birth; from secondary centers (in the epiphyses), it occurs during the first 5 years.

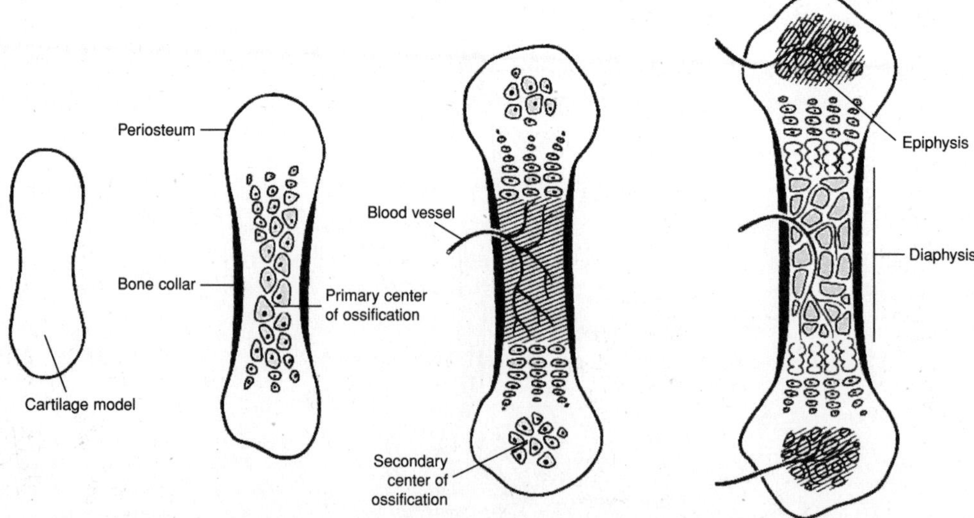

Figure 6.5 Ossification of a long bone.

Both the ectodermal and mesodermal germ layers (see chapter 4) participate in the formation of the skin. The epidermis and accessory integumentary structures (hair, glands, and nails) develop from ectoderm. The dermis develops from a thickened layer of undifferentiated mesoderm called *mesenchyme*. Likewise, the cutaneous blood vessels and smooth muscle fibers contained within the dermis are formed from mesoderm.

6.11 What are osteogenic cells?

There are several different types of bone cells, each with a particular function. **Osteogenic cells** are progenitor cells that give rise to all bone cells. **Osteoblasts** are the principal bone-building cells; they synthesize collagenous fibers and bone matrix, and promote mineralization during ossification. Once this has been accomplished, the osteoblasts, which are trapped in their own matrix, develop into **osteocytes** that maintain the bone tissue. **Osteoclasts** contain lysosomes and phagocytic vacuoles. These bone-destroying cells demineralize bone tissue.

Objective G To list the *cranial* and *facial bones* of the skull, to describe their locations and structural characteristics, and to identify the articulations that affix them together.

The skull is composed of 8 *cranial bones,* which articulate firmly with one another to enclose and protect the brain and associated sense organs, and 14 *facial bones,* which form the foundation for the face and anchor the teeth. These bones are listed in table 6.1 and illustrated in figs. 6.6 through 6.10.

6.12 Define the term *suture* and describe the locations of the principal sutures of the skull?

The bones of the skull are united by serrated immovable joints called **sutures** (see figs. 6.3, 6.7, and 6.9). The frontal bone is joined to the two parietal bones at the *coronal suture*; the parietal bones meet each other at the *sagittal suture*; the occipital bone meets the parietal bones at the *lambdoid suture*; and a parietal bone joins a temporal bone at the *squamous suture.*

6.13 List the cavities of the skull.

The **cranial cavity** is the largest cavity of the skull, with a capacity of 1300 to 1350 cubic centimeters. The **nasal cavity** is formed by both cranial and facial bones. Four sets of **paranasal sinuses** are located within the bones surrounding the nasal area. **Middle-** and **inner-ear chambers** are located within the temporal bones. The **oral**, or **buccal**, **cavity** (mouth) is only partially defined by bone. The two *orbits* for the eyeballs are formed by both facial and cranial bones.

6.14 What is a foramen? What are the major foramina of the skull, where are they located, and what structures pass through them?

A **foramen** (plural, *foramina*) is an opening through a bone for the passage of a vessel or a nerve. The principal foramina of the skull are summarized in table 6.3. Refer to figures 6.6 through 6.10 for illustrations of these foramina.

Table 6.3 Principal Foramina of the Skull

Foramen	*Location*	*Structures transmitted*
Carotid canal	Petrous part of temporal bone	Internal carotid artery and sympathetic nerves
Greater palatine foramen	Palatine bone of hard palate	Greater palatine nerve and descending palatine vessels
Hypoglossal foramen/canal	Anterolateral edge of occipital condyle	Hypoglossal nerve and branch of ascending pharyngeal artery
Incisive foramen	Hard palate, posterior to incisor teeth	Nasopalatine nerve and branches of descending palatine vessels
Inferior orbital fissure	Between maxilla and greater wing of sphenoid bone	Maxillary nerve of trigeminal cranial nerve, zygomatic nerve, and infraorbital vessels
Infraorbital foramen	Anterior surface of maxilla, inferior to orbit	Infraorbital nerve and artery
Jugular foramen	Between petrous part of temporal and occipital bones, posterior to carotid canal	Internal jugular vein; vagus, glossopharyngeal, and accessory nerves
Foramen lacerum	Between petrous part of temporal and sphenoid bones	Branches of ascending pharyngeal artery and internal carotid artery
Lesser palatine foramen	Hard palate, posterior to greater palatine foramen	Lesser palatine nerves
Foramen magnum	Occipital bone	Union of medulla oblongata and spinal cord; accessory nerves; vertebral and spinal arteries
Mandibular foramen	Medial ramus of mandible	Inferior alveolar nerve and vessels
Mental foramen	Below second premolar on lateral side of mandible	Mental nerve and vessels
Nasolacrimal canal	Lacrimal bone	Nasolacrimal (tear) duct
Olfactory foramen	Cribriform plate of ethmoid bone	Olfactory nerves
Optic foramen	Back of orbit in lesser wing of sphenoid bone	Optic nerve and ophthalmic artery
Foramen ovale	Greater wing of sphenoid bone	Mandibular nerve of trigeminal cranial nerve
Foramen rotundum	Body of sphenoid bone	Maxillary nerve of trigeminal cranial nerve
Foramen spinosum	Posterior angle of sphenoid bone	Middle meningeal vessels
Stylomastoid foramen	Between styloid and mastoid processes of temporal bone	Facial nerve and stylomastoid artery
Superior orbital fissure	Between greater and lesser wings of sphenoid bone	Oculomotor, trochlear, and abducens cranial nerves; ophthalmic nerve of trigeminal cranial nerve
Supraorbital foramen	Supraorbital ridge of orbit	Supraorbital nerve and artery
Zygomaticofacial foramen	Anterolateral surface of zygomatic bone	Zygomaticofacial nerve and vessels

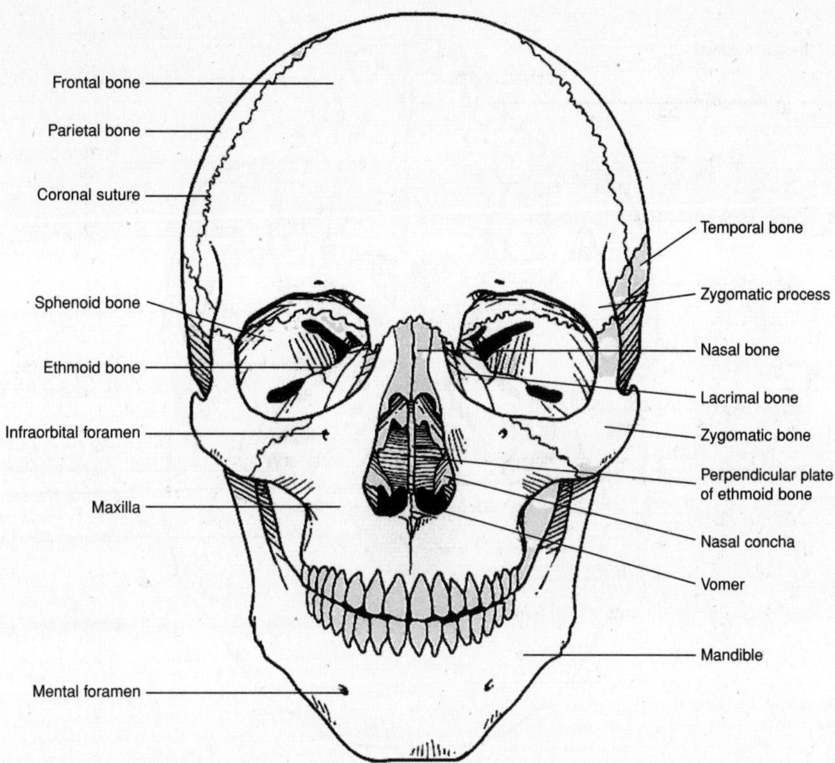

Figure 6.6 An anterior view of the skull.

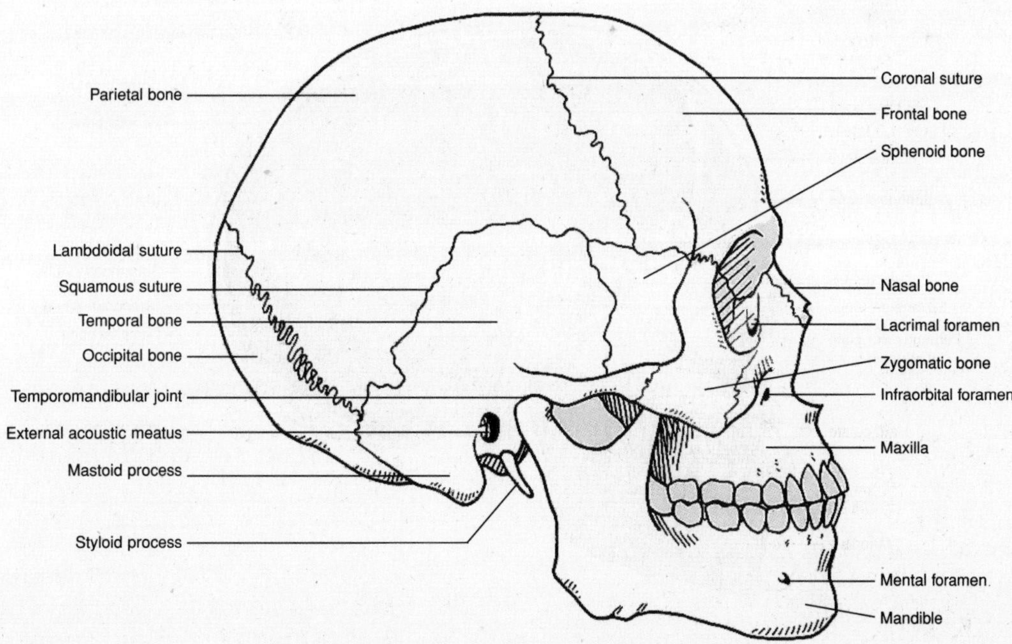

Figure 6.7 A lateral view of the skull.

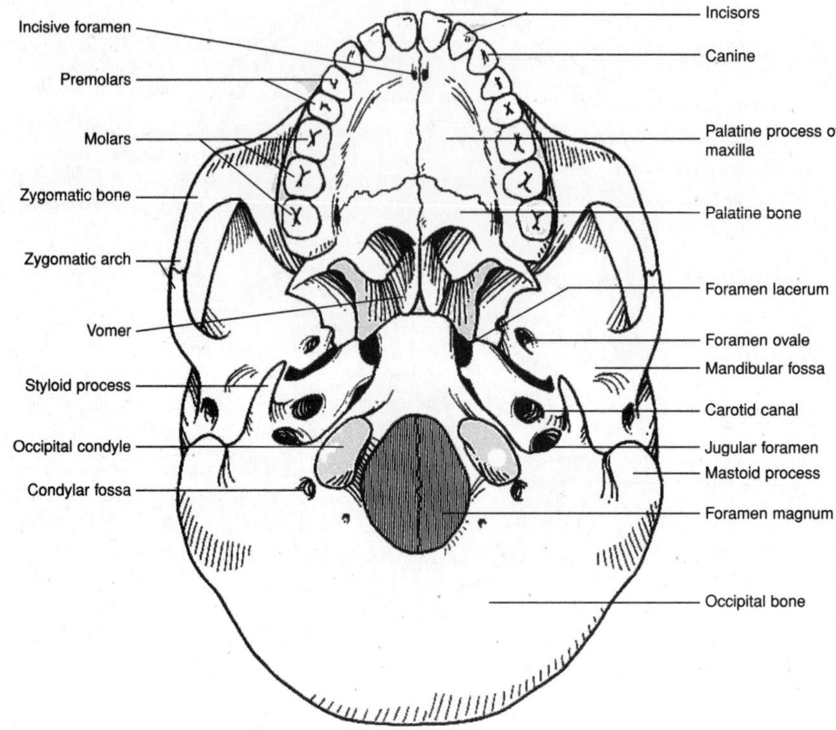

Figure 6.8 An inferior view of the skull.

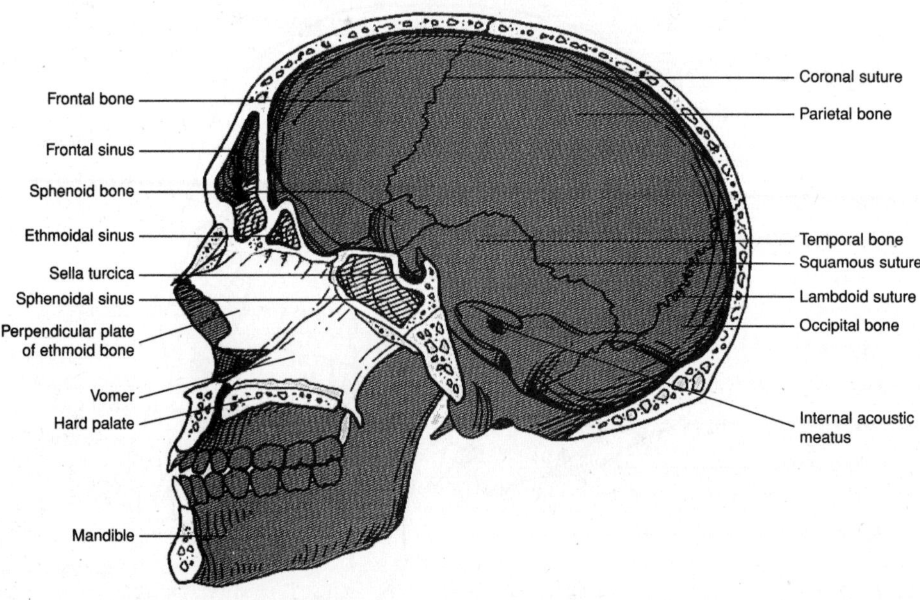

Figure 6.9 A sagittal view of the skull.

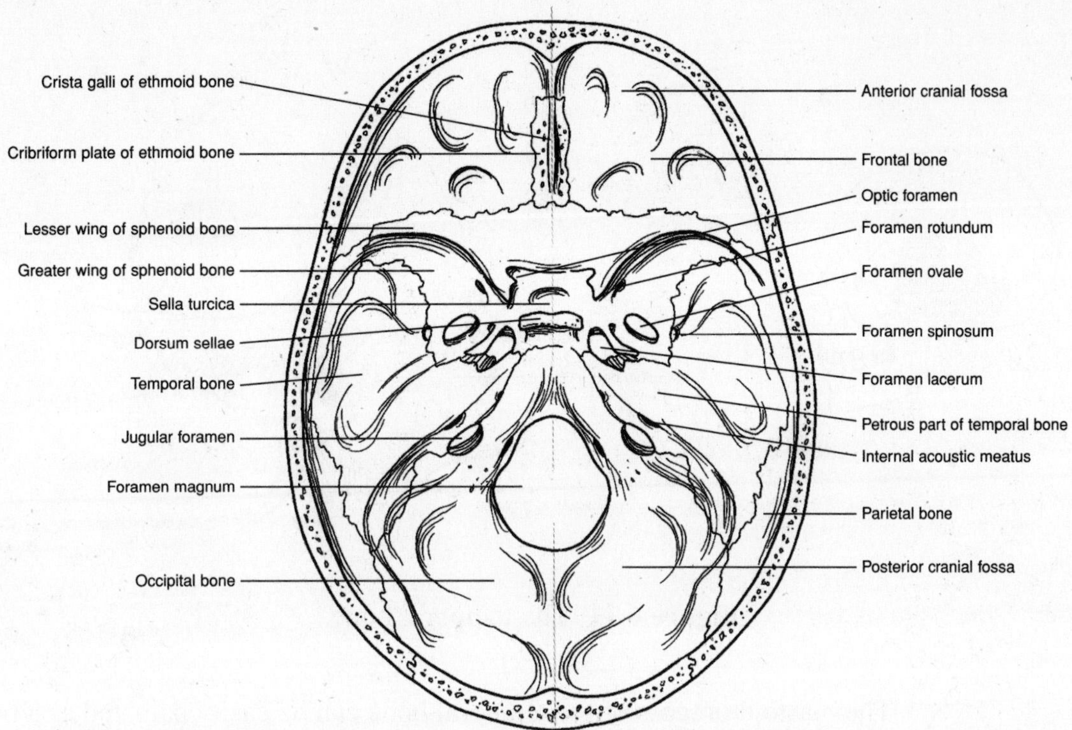

Figure 6.10 The floor of the cranial cavity.

6.15 Describe the anatomical features of the frontal bone.

The **frontal bone** forms the anterior roof of the cranium, the roof of the nasal cavity, and the *supraorbital margin* over the orbit of each eye (figs. 6.6, 6.7, and 6.9). The *supraorbital foramen* along the supraorbital margin is an opening for the small supraorbital nerve and artery. The frontal bone contains paired *frontal sinuses* (fig. 6.7) connected to the nasal cavity.

6.16 Identify the paranasal sinuses and state their function.

There are four **paranasal sinuses** that lessen the weight of the skull and act as sound chambers for voice resonance. These sinuses are named according to the bones in which they are found. Thus, there are the *frontal, maxillary, sphenoidal,* and *ethmoidal sinuses* (fig. 6.7).

Sinusitis is an inflammation of the mucous membrane that lines the paranasal sinuses. Since these sinuses connect to the nasal cavity, they are vulnerable to infections that originate in the nasal mucosa. Blowing the nose too hard may force microorganisms into the moist, warm environment of a paranasal sinus.

6.17 Describe the four parts of the temporal bone.

Each of the two **temporal bones** that form the lower sides of the cranium consists of four parts. The flattened **squamous part of the temporal bone** forms the posterior component of the *zygomatic arch* (see fig. 6.7) and has a **mandibular fossa** to receive the condyle of the mandible at the *temporomandibular joint* (fig. 6.11). The **tympanic part of the temporal bone** contains the *external acoustic meatus* (ear canal) and the *styloid process*. The **mastoid part of the temporal bone** consists of the *mastoid process*, which contains the mastoid and stylomastoid foramina. The dense and inferior **petrous part of the temporal bone** (see fig. 6.10) contains the middle and inner ears, as well as the three auditory ossicles (malleus, incus, and stapes) shown in fig. 6.15.

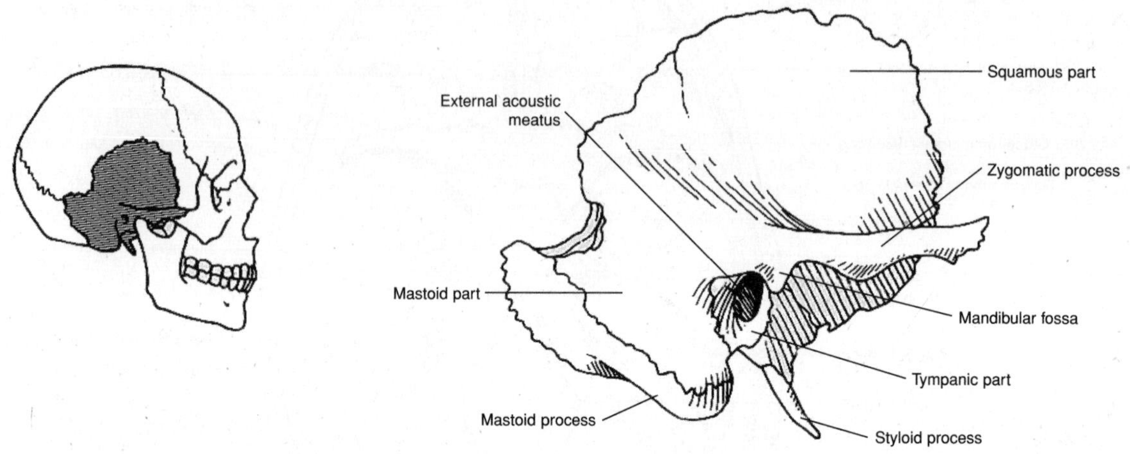

Figure 6.11 The temporal bone.

The mastoid process of the temporal bone can be easily palpated as a bony knob behind the earlobe. Although not present on a newborn, the mastoid process soon develops as the sternocleidomastoid muscle that attaches to it contracts, causing neck movement. As the process develops, a number of small air-filled spaces called *mastoid cells* form within the bone. These spaces are clinically important because they can become infected in mastoiditis. A tubular communication from the mastoid cells to the middle-ear cavity may allow ear infections to spread to this region.

6.18 What structures characterize the occipital bone?

The **occipital bone** forms the posterior and much of the inferior portion of the cranium. It contains the *foramen magnum,* through which the spinal cord attaches to the brain, and the *occipital condyles*, which articulate with the first cervical vertebra (see problem 6.27).

6.19 Which endocrine gland is supported by the sphenoid bone?

Located in the floor of the cranium (see fig. 6.10), the **sphenoid bone** resembles a butterfly with outstretched wings. The *sella turcica* is a bony depression in the sphenoid bone (fig. 6.12) that supports the pituitary gland. The sphenoid bone also contains the paired *optic foramina, foramina ovale, foramina spinosum, foramina lacerum, foramina rotundum,* and the *superior orbital fissures.*

The sphenoid bone is the most frequently fractured bone of the cranium (the bones supporting and surrounding the brain). Its broad, thin, platelike extensions are perforated by several foramina, weakening the sphenoid bone structurally. A blow to almost any portion of the skull causes the buoyed, fluid-filled brain to rebound against this vulnerable bone, often causing it to fracture. Because the bone is tightly confined, however, the fractured parts usually are not severely displaced and readily heal with no complications.

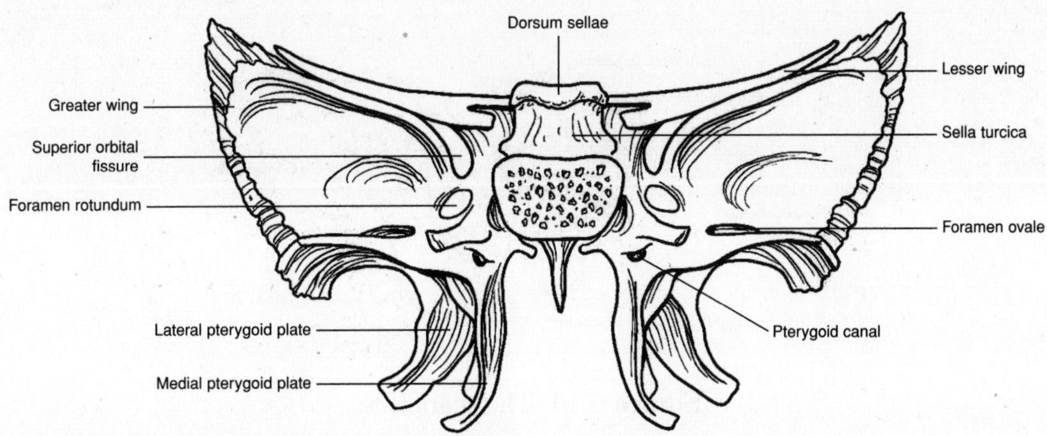

Figure 6.12 An anterior view of the sphenoid bone.

6.20 What do the perpendicular plate, crista galli, nasal conchae, and cribriform plate have in common?

All four structures are components of the **ethmoid bone** (fig. 6.13). The *perpendicular plate of the ethmoid bone* forms part of the *nasal septum*, which divides the *nasal cavity* into two *nasal fossae*. The *crista galli* attaches to the meninges covering the brain. The epithelium covering the scroll-shaped *superior* and *middle nasal conchae* warms and moistens inhaled air. The perforations in the *cribriform plate of the ethmoid bone* allow the passage of olfactory nerves.

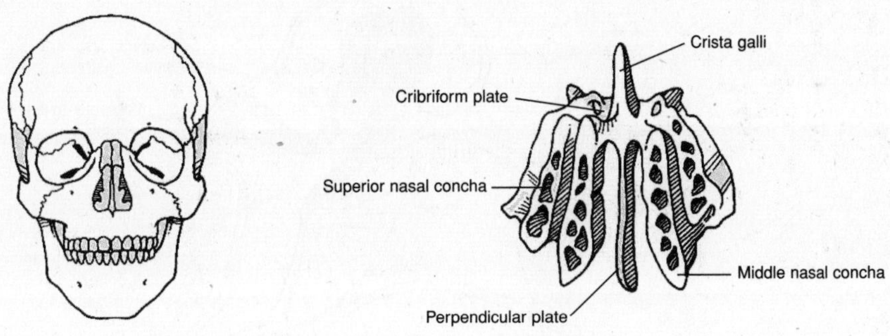

Figure 6.13 The ethmoid bone.

6.21 What is the hard palate?

The **hard palate** is the bony partition between the nasal and oral cavities formed by the union of the *palatine processes of the maxillae* and the *palatine bones*. The hard palate, along with the fleshy soft palate, forms the roof of the mouth.

6.22 What are the diagnostic features of the mandible?

The **mandible** (lower jawbone) (fig. 6.14) has *condylar processes* for attachment to the skull (at the temporomandibular joint). The *coronoid processes* are for attachment of the temporalis muscles. The *mandibular* and *mental foramina* are for passage of nerves (see table 6.2). Sixteen teeth are embedded in the adult mandible. The structure, function, and replacement sequence of teeth are discussed in chapter 19.

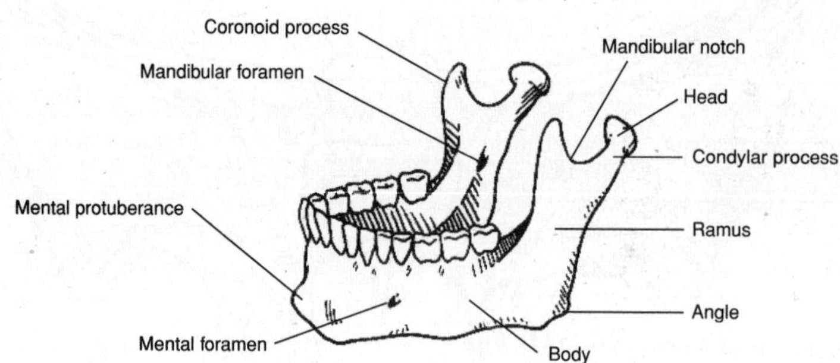

Figure 6.14 The mandible.

6.23 Why are the auditory (ear) ossicles, which are contained within the petrous parts of the temporal bones, not considered bones of the skull?

Each of the two middle-ear chambers contains three small **auditory ossicles**—the **malleus** (hammer), **incus** (anvil), and **stapes** (stirrup) (fig. 6.15). Because they originate in the pharyngeal region and then migrate into the position of the middle ear as the skull is forming, the auditory ossicles are not considered bones of the skull. In the functioning ear, the auditory ossicles amplify and transmit sound from the outer ear to the inner ear. A more detailed discussion of the structure and function of the auditory ossicles is included in chapter 12.

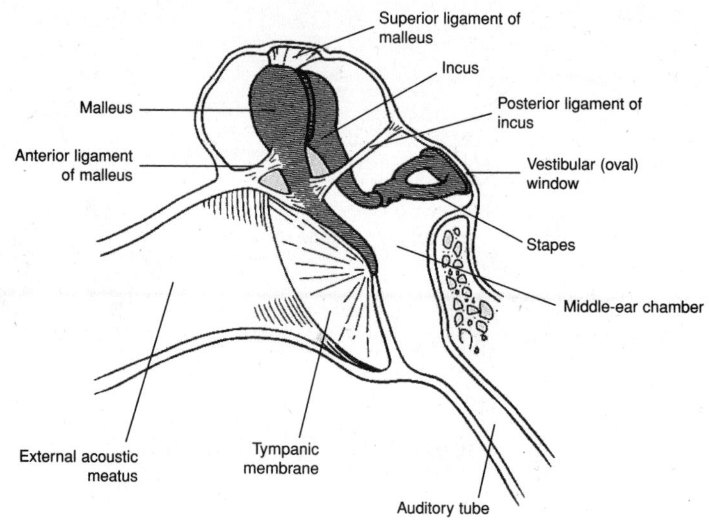

Figure 6.15 The auditory ossicles.

6.24 Where is the hyoid bone located and what are its functions?

The U-shaped **hyoid bone** (fig. 6.16) is located in the anterior neck, where it supports the tongue superiorly and the larynx (voice box) inferiorly. In addition, several anterior neck muscles attach to this bone. The hyoid bone plays a major role in swallowing. It is a unique bone in that it does not attach directly to any other bone. Instead, it is suspended from the styloid processes of the temporal bones by the stylohyoid ligaments.

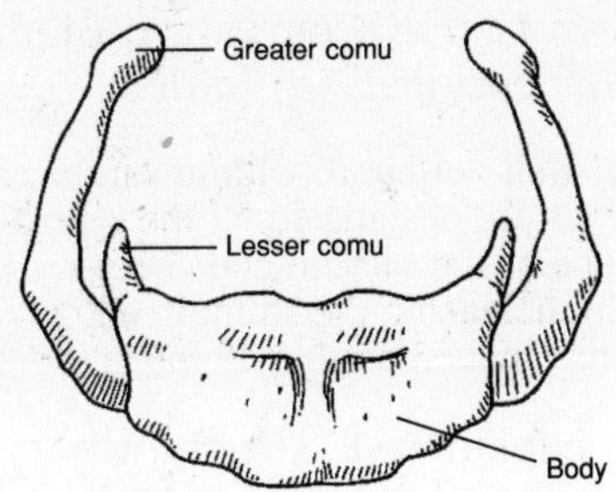

Figure 6.16 The hyoid bone.

Objective H To describe the structure and functions of the *vertebral column.*

As part of the axial skeleton, the **vertebral column** (backbone) supports and permits movement of the head and trunk and provides a site for muscle attachment. The vertebrae (bones of the vertebral column) also support and protect the spinal cord and permit passage of spinal nerves.

The vertebral column is composed of 33 individual vertebrae (singular, *vertebra*). There are 7 cervical, 12 thoracic, 5 lumbar, 4 or 5 fused sacral, and 4 or 5 fused coccygeal vertebrae; thus, the vertebral column is composed of a total of 26 movable parts (fig. 6.17). Vertebrae are separated by fibrocartilaginous **intervertebral discs** and are secured to one another by interlocking processes and binding ligaments. The structural arrangement of the vertebral column allows only limited movement between vertebrae but extensive movements of the vertebral column as a unit. Between the vertebrae are openings called **intervertebral foramina** that permit passage of spinal nerves.

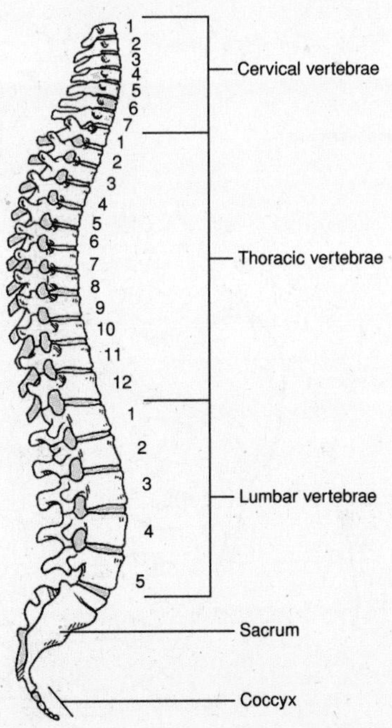

Figure 6.17 A lateral view of the vertebral column.

6.25 Which of the following is *not* a curvature of the vertebral column? (*a*) thoracic, (*b*) costal, (*c*) pelvic, (*d*) cervical, (*e*) lumbar

(*b*): Four curvatures of the adult vertebral column can be identified in a lateral view. The **cervical, thoracic,** and **lumbar curves** are designated by the type of vertebrae they include. The **pelvic curve** is formed by the shape of the sacrum and coccyx. The curves of the vertebral column play an important functional role in increasing the strength and maintaining the balance of the upper portion of the body; they also make possible a bipedal (two-footed) stance.

6.26 Could any vertebra be considered "structurally typical"?

While no single vertebra is typical, the various vertebrae shown in fig. 6.18 are representative of those from each of the five regions of the vertebral column. **Cervical vertebrae** have *transverse foramina* for the passage of vessels to the brain. **Thoracic vertebrae** are characterized by the presence of *facets* for articulation with the heads of ribs. The large **lumbar vertebrae** have prominent processes for muscle attachment. The *sacrum* consists of four or five fused **sacral vertebrae** and attaches to the pelvic girdle at the *sacroiliac joint*. The triangular *coccyx* ("tailbone") is composed of four or five fused **coccygeal vertebrae.**

Several generalities about vertebrae can be made. The drum-shaped **body** of a vertebra is in contact with the **intervertebral discs** on each end. The **neural arch** on the posterior surface of the body of the vertebra is composed of two supporting **pedicles** and two arched **laminae.** The hollow space formed by the *vertebral arch* and *body* is the **vertebral foramen**, or *vertebral canal*, that allows passage of the spinal cord. The **spinous process** extends posteriorly from the vertebral arch. Other processes of most vertebrae include paired **transverse processes**, paired **superior articular processes**, and paired **inferior articular processes.** The **intervertebral foramina** permit passage of spinal nerves.

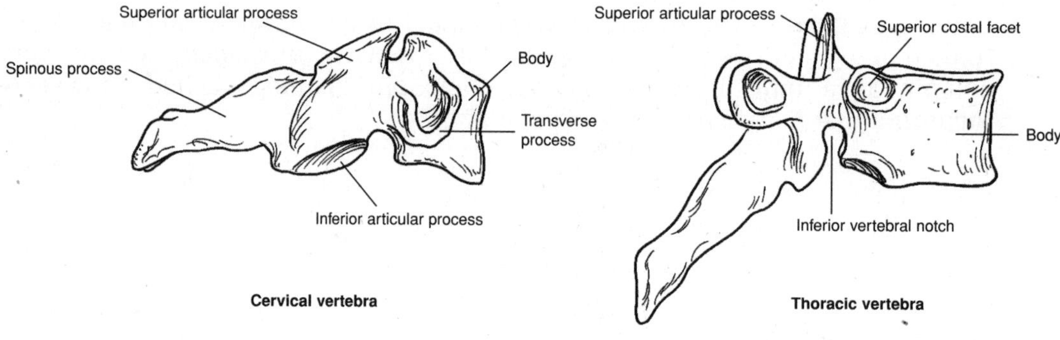

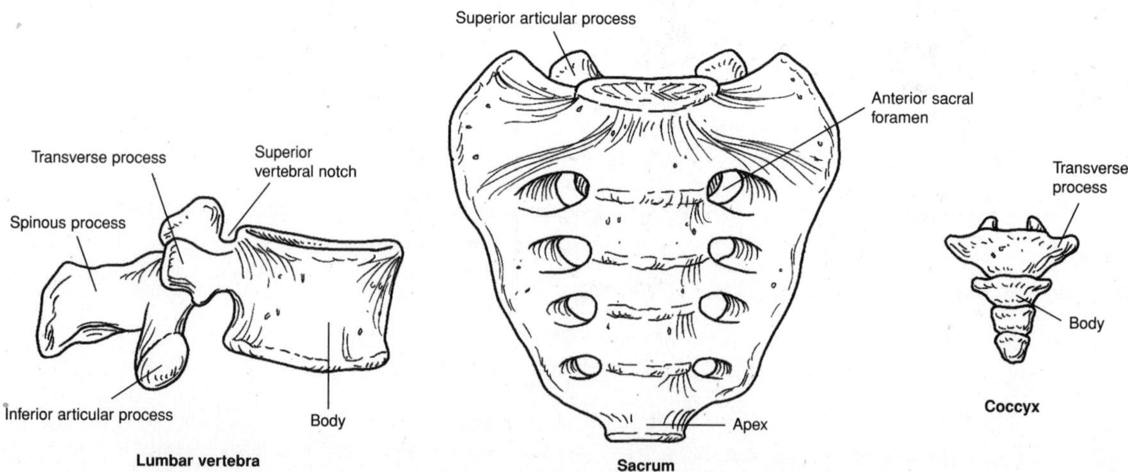

Figure 6.18 Examples of vertebrae from different vertebral regions.

6.27 Do any vertebrae have specific names besides a letter-number designation?

Only two. The **atlas**, which is the first cervical vertebra (C1) (fig. 6.19), is adapted to articulate with the *occipital condyles of the skull*, giving support and maneuverability to the head. (The *atlantooccipital joint* permits nodding of the head.) The **axis**, or second cervical vertebra (C2), has a peglike **dens**, or **odontoid process**, that provides a pivot for rotation with respect to the atlas, as in turning the head to the side.

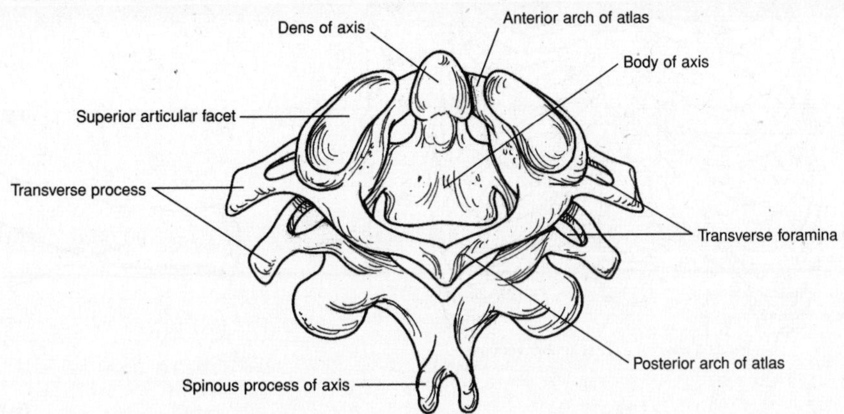

Figure 6.19 The atlas and axis cervical vertebrae.

Objective I To describe the structures of the *rib cage* and state their functions.

The **sternum**, **costal cartilages**, and **ribs** attached to the *thoracic vertebrae* form the **rib cage**, or *thoracic cage*, of the thorax. The anteroposteriorly compressed rib cage supports the pectoral girdle and upper extremities, protects and supports the thoracic and upper abdominal viscera, provides an extensive surface area for muscle attachment, and plays a major role in respiration.

6.28 Describe the structure of the sternum.

The elongated and flattened sternum is a compound bone, consisting of an upper **manubrium**, a central **body**, and a lower **xiphoid process** (fig. 6.20). On the lateral sides of the sternum are **costal notches,** where the costal cartilages attach.

6.29 *True or false*: Each of the 12 pairs of ribs attaches posteriorly to the thoracic vertebrae and anteriorly to the sternum via costal cartilages.

False. Only the first seven pairs, the **true ribs**, are anchored to the sternum by individual costal cartilages (fig. 6.18). The remaining five pairs are called **false ribs**. Ribs 8, 9, and 10 are attached to the costal cartilage of rib 7. The remaining two paired false ribs do not attach to the sternum and are frequently referred to as the **floating ribs**.

6.30 What features do ribs have in common?

Each of the first 10 paired ribs has a **head** and **tubercle** for articulation with a vertebra (fig. 6.20). The last two pairs have a head but no tubercle. All ribs have a **neck, angle,** and **shaft** (body).

Fractures of the ribs are relatively common injuries and most frequently occur between ribs 3 and 10. The first two pairs are protected by the clavicles, and the last two pairs move freely and will give with an impact. Little can be done to assist the healing of a broken rib other than binding it to restrict movement.

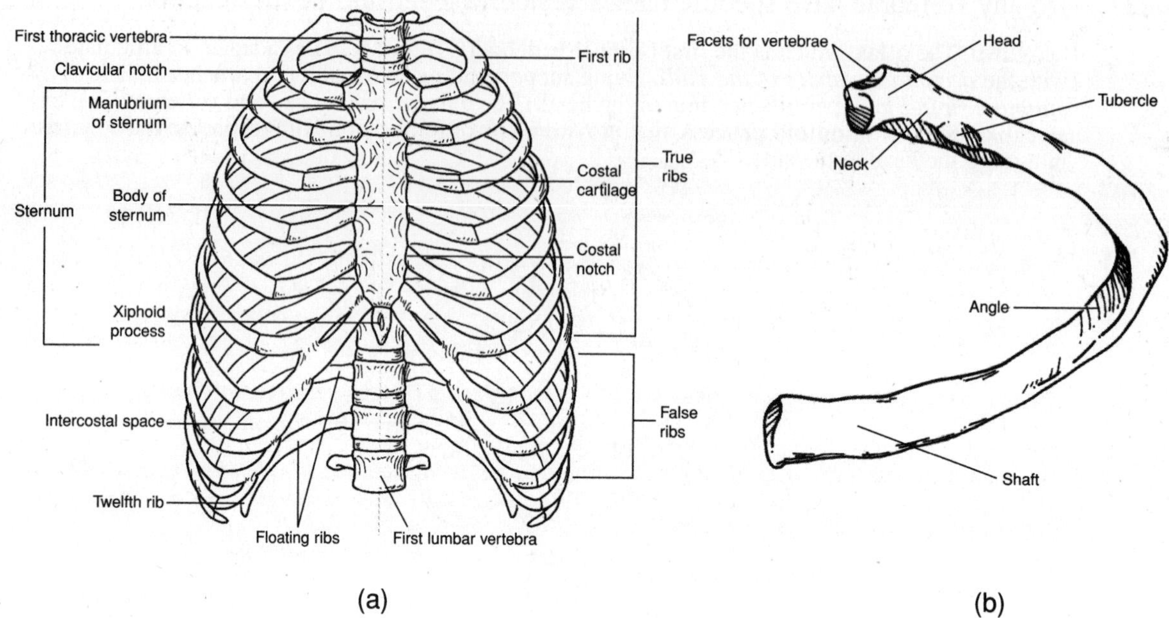

Figure 6.20 The rib cage consists of (*a*) the sternum, costal cartilages, and 12 paired ribs attached to the thoracic vertebrae. A typical rib (*b*) has facets for attachment to the sternum and a broadly flattened and rounded shaft (body) for protection of thoracic viscera and muscle attachment.

Objective J To describe the structure of the *pectoral girdle.*

The two scapulae and the two clavicles make up the **pectoral** (*shoulder*) **girdle** that attaches to the axial skeleton at the manubrium of the sternum. The pectoral girdle provides attachment for numerous muscles that move the brachium (arm) and antebrachium (forearm).

6.31 What are the functions of the clavicle?

The S-shaped **clavicle** (collarbone) (fig. 6.21) binds the upper extremity to the axial skeleton and positions the shoulder joint away from the trunk for freedom of movement. It also is the site of attachment for muscles of the trunk and neck.

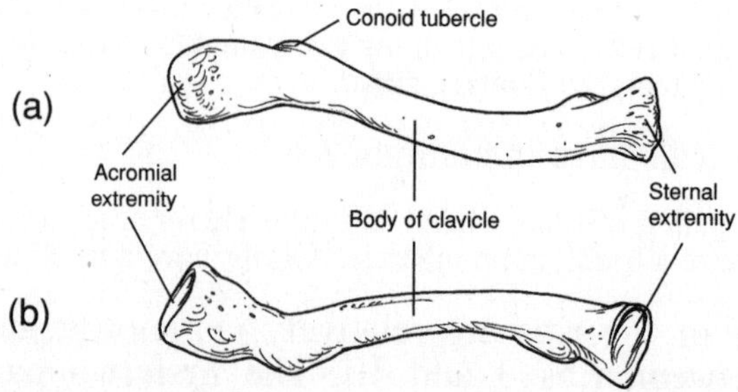

Figure 6.21 The right clavicle. (*a*) A superior view and (*b*) an inferior view.

The clavicle is the most frequently fractured bone in the body. Blows to the shoulder or an attempt to break a fall with an outstretched hand displaces the force to this long, delicate bone. Furthermore, the anterior border of the clavicle is directly subcutaneous and is not protected by fat or muscle. Because the clavicle is readily palpated, a fracture is usually easy to detect.

6.32 Identify the structural features of the scapula.

The flattened, triangular **scapula** (shoulder blade) has three borders, three angles, and three fossae (fig. 6.22). It also has diagnostic processes and other special features. The superior edge is called the **superior border**. The **medial border** (vertebral border) is nearest to the vertebral column, and the **lateral border** (axillary border) is directed toward the arm. The **superior angle** is located at the junction of the superior and medial borders, and the **inferior angle** is located at the junction of the medial and lateral borders. The **lateral angle** is located at the junction of the superior and lateral borders. Along the superior border, a distinct depression called the **scapular notch** serves as a passageway for a nerve. The **spine of the scapula** is a diagonal bony ridge on the posterior surface that separates the **supraspinous fossa** from the **infraspinous fossa**. The spine broadens toward the shoulder as the **acromion**. The **glenoid cavity** is a shallow depression into which the head of the humerus fits. The **coracoid process** lies superior and anterior to the glenoid cavity. On the anterior surface of the scapula is a slightly concave area known as the **subscapular fossa**.

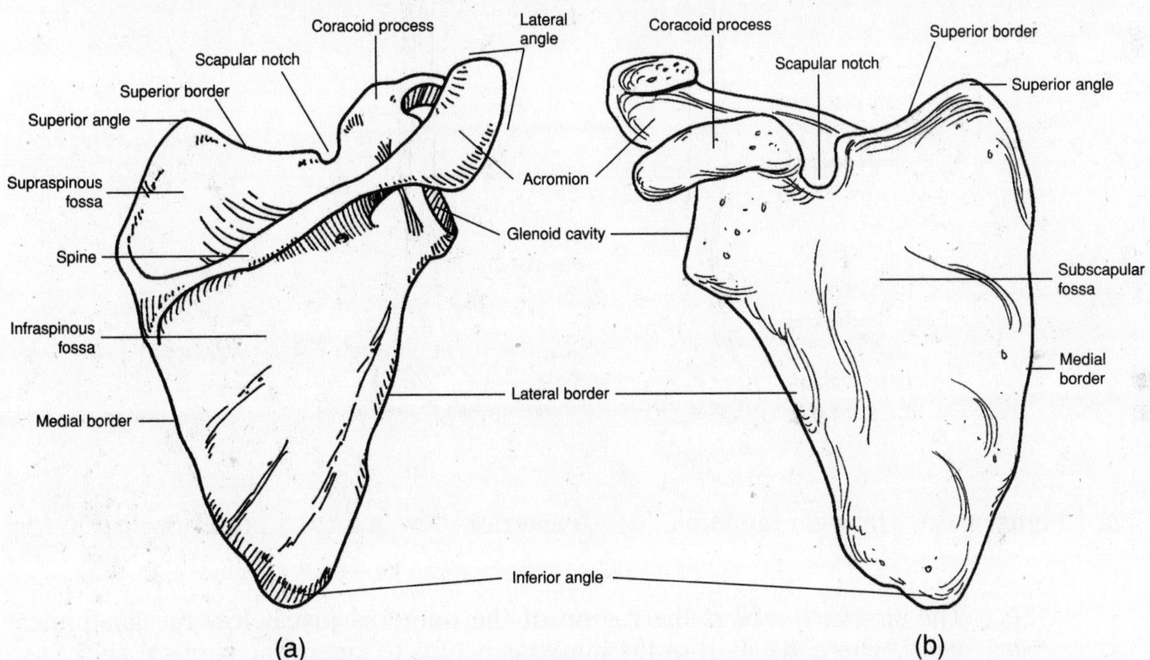

Figure 6.22 The right scapula. (*a*) A posterior view and (*b*) an anterior view.

Objective K To list the bones of the *upper extremity* and to describe the diagnostic features of the bones of the *brachium* (arm) and *antebrachium* (forearm).

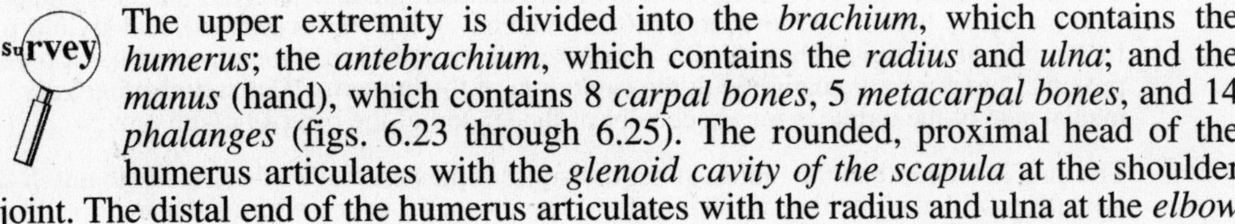

The upper extremity is divided into the *brachium*, which contains the *humerus*; the *antebrachium*, which contains the *radius* and *ulna*; and the *manus* (hand), which contains 8 *carpal bones*, 5 *metacarpal bones*, and 14 *phalanges* (figs. 6.23 through 6.25). The rounded, proximal head of the humerus articulates with the *glenoid cavity of the scapula* at the shoulder joint. The distal end of the humerus articulates with the radius and ulna at the *elbow*

joint. The distal ends of the radius and ulna articulate with the proximal row of carpal bones in the *wrist.* Numerous joints of various kinds occur within the hand.

6.33 Describe the structure of the humerus.

Located within the brachium, the **humerus** has a number of diagnostic features (fig. 6.23). The **anatomical neck** is an indented groove surrounding the margin of the **head of the humerus**. The **greater tubercle** is lateral to the head of the humerus. The **lesser tubercle** is slightly anterior to the greater tubercle and is separated from it by the **intertubercular groove**, through which passes the tendon of the biceps brachii muscle. The **shaft** (body) **of the humerus** is the long, cylindrical portion. Along its lateral midregion is a prominent ridge called the **deltoid tuberosity**. The **capitulum** at the distal end of the humerus is the lateral rounded condyle that receives the radius. The **trochlea** is the pulleylike medial surface that articulates with the ulna. On either side above the condyles are the **lateral** and **medial epicondyles**. The coronoid fossa is a depression above the trochlea on the anterior surface, and the **olecranon fossa** is a depression on the distal posterior surface.

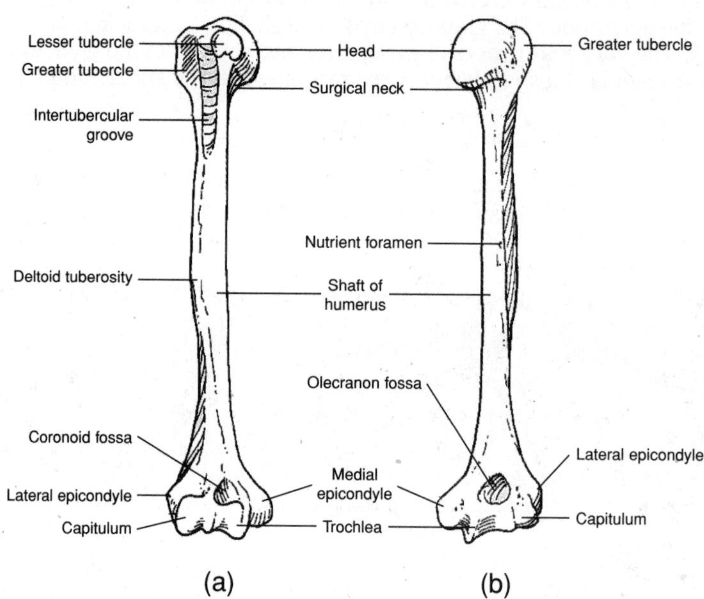

Figure 6.23 The right humerus. (*a*) An anterior view and (*b*) a posterior view.

The *surgical neck* is the region of the humerus just below the anatomical neck, where the shaft of the humerus begins to taper. The surgical neck is so named because of the frequency of trauma-induced fractures that occur at this location.

6.34 What do the radius and ulna have in common? How do they differ?

The lateral **radius** and the medial **ulna** both articulate proximally with the humerus and distally with the carpal bones. As shown in fig. 6.24, both have long **shafts** (*bodies*), and **styloid processes** for support of the wrist. The radius is shorter and more robust than the ulna and has a rounded proximal **head** for articulation with the capitulum of the humerus. The **radial tuberosity** on the medial side of the radius is for attachment of the tendon of the biceps brachii muscle.

The ulna is longer than the radius. It has a distinct depression called the **trochlear notch** that articulates with the trochlea of the humerus. The **coronoid process** forms the anterior lip of the **trochlear notch**, and the **olecranon** forms the posterior portion, or elbow. Lateral and inferior to the coronoid process is the **radial notch**, which accommodates the **head of the radius**.

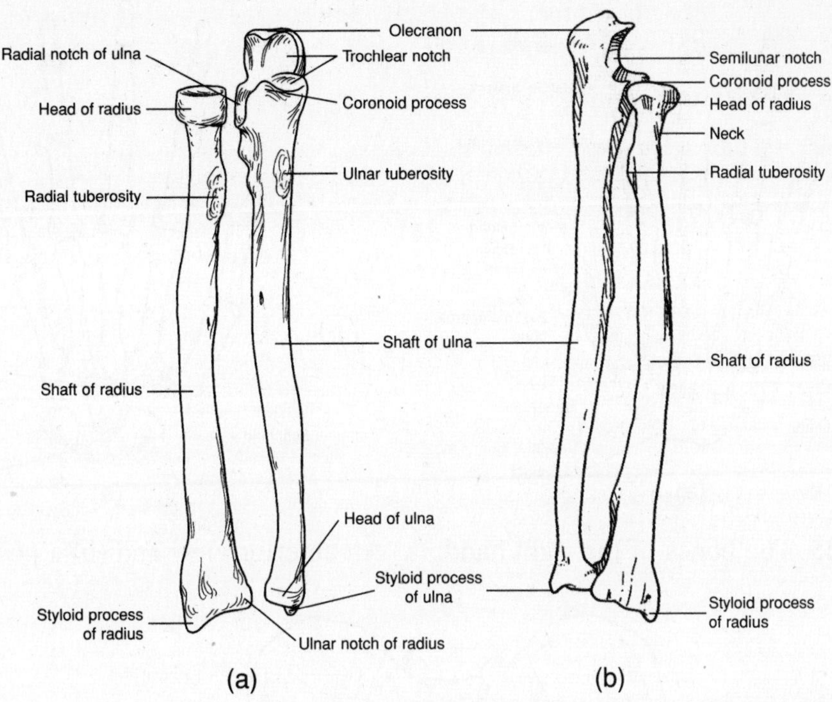

Figure 6.24 The right radius and ulna. (*a*) An anterior view and (*b*) a posterior view.

6.35 Describe the skeletal elements of the hand.

The 27 bones of the **manus**, or *hand*, are grouped into 8 *carpal bones*, 5 *metacarpal bones*, and 14 *phalanges* (fig. 6.25). The articulations (joints) between the cube-shaped carpal bones permit movement in a confined area, while the elongated metacarpal bones and phalanges act as levers about their freely movable joints.

The **carpal bones** are arranged in two transverse rows of four bones each. The proximal row, naming from lateral (thumb) to medial, consists of the *scaphoid bone, lunate bone, triquetral bone, and pisiform bone*. The distal row, from lateral to medial, consists of the *trapezium*, *trapezoid bone, capitate bone,* and *hamate bone*.

Each of the five **metacarpal bones** consists of a proximal *base*, a *shaft* (body), and a distal *head* that is rounded for articulation with the base of a proximal phalanx. The metacarpal bones are numbered I to V, the lateral, or thumb, side being I.

The 14 **phalanges** are the skeletal elements of the digits. A single finger bone is called a **phalanx**. The phalanges are arranged in a proximal row, a middle row, and a distal row. The thumb (pollex), however, has only a proximal and a distal phalanx.

Objective L To describe the structure and functions of the *pelvic girdle*.

The **pelvic girdle**, or *pelvis*, is formed by the two **ossa coxae** united anteriorly by the **symphysis pubis** (fig. 6.26). It is attached posteriorly to the *sacrum* of the vertebral column at the *sacroiliac joints*. The pelvic girdle and its associated ligaments support the weight of the body from the vertebral column. The pelvic girdle also supports and protects the lower viscera, including the urinary bladder, the reproductive organs, and in a pregnant woman, the developing fetus.

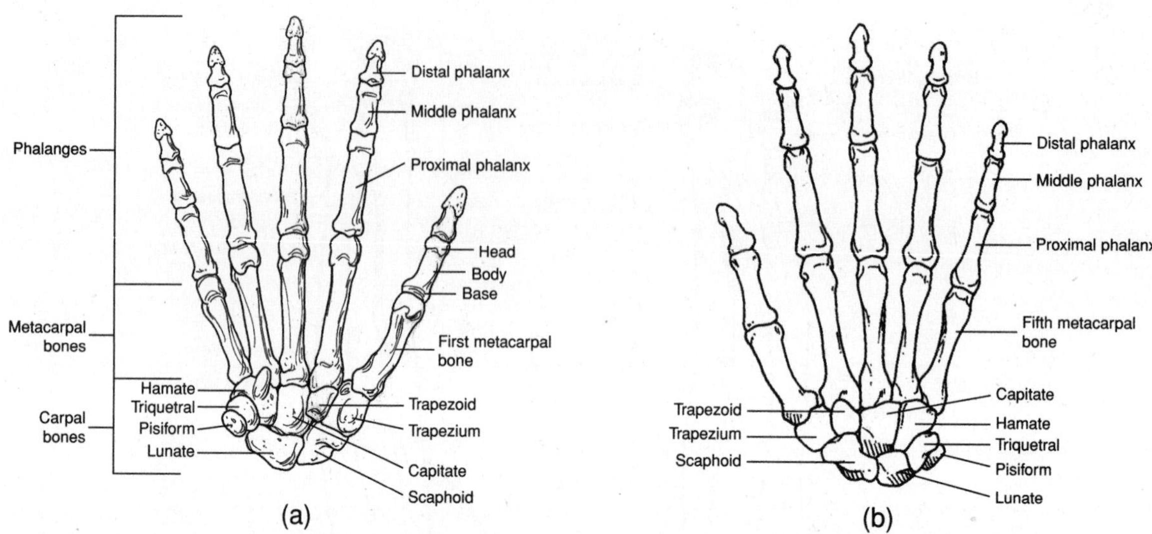

Figure 6.25 The bones of the right hand. (*a*) An anterior view and (*b*) a posterior view.

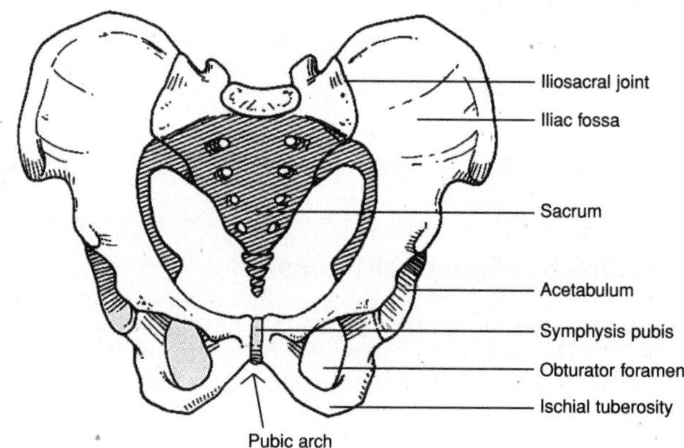

Figure 6.26 The pelvic girdle.

6.36 What three bones form the os coxae?

Each **os coxae** (hipbone) consists of an **ilium**, an **ischium**, and a **pubis**. In adults, these bones are firmly fused. On the lateral surface of the os coxae, where the three bones ossify, is a large circular depression, the **acetabulum** (fig. 6.27*a*), which receives the head of the femur. The **obturator foramen** is the large opening in the side of the os coxae. In a living person, the obturator foramen is covered by the obturator membrane, to which several muscles attach.

6.37 Describe the three bones of the os coxae.

The **ilium** is the largest and uppermost of the three bones of the os coxae. It is characterized by a prominent **iliac crest** that terminates anteriorly as the **anterior superior iliac spine** (fig. 6.27). Just below this spine is the **anterior inferior iliac spine**. The posterior termination of the iliac crest is the **posterior superior iliac spine**, and just below it is the **posterior inferior iliac spine**. Below the posterior inferior iliac spine is the **greater sciatic notch**. On the medial surface of the ilium is the roughened **auricular surface** that articulates with the sacrum. The **iliac fossa** is the smooth, concave surface on the anterior portion of the ilium.

The **ischium** is the posteroinferior component of the os coxae. The **spine of the ischium** is a prominent posterior projection from the bone. Inferior to the spine of the ischium is the **lesser sciatic notch**.

The **pubis** is the anterior component of the os coxae. It consists the **superior ramus, inferior ramus**, and the **body of the pubis**. The body of one pubis articulates with that of the other at the **symphysis pubis** of the pelvic girdle.

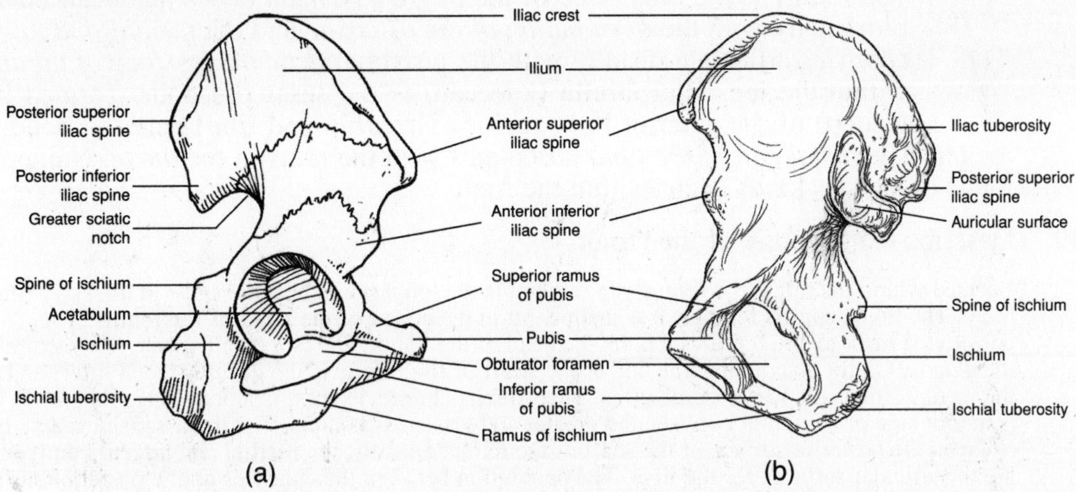

Figure 6.27 The right os coxae. (*a*) A lateral view and (*b*) a medial view.

The structure of the pelvic girdle and the way it is attached to the sacrum are adaptations for the *bipedal* (two-footed) *locomotion* characteristic of humans. An upright posture may cause problems, however. The sacroiliac joint may weaken with age, causing lower back pains. The weight of the viscera may weaken the lower abdominal walls and contribute to hernias. Some of the problems of childbirth are related to the structure of the mother's pelvis. Finally, the hip joints tend to deteriorate with age. Many elderly people suffer fractured hips and may need hip replacements.

6.38 *True or false*: There are sex-related differences in the adult pelvis.

True. Structural differences between the pelvis of an adult male and that of an adult female (table 6.4) reflect the female's role in pregnancy and parturition.

Table 6.4 A Comparison of the Male and Female Pelvic Girdles

Characteristic	*Male pelvis*	*Female pelvis*
General appearance	More massive; prominent processes	More delicate; processes not as prominent
Anterior superior iliac spines	Closer together	Wider apart
Pelvic inlet	Heart-shaped	Round or oval
Pelvic outlet	Narrower	Wider
Obturator foramen	Oval	Triangular
Symphysis pubis	Deeper, longer	Shallower, shorter
Pubic arch	Acute (less than 90°)	Obtuse (greater than 90°)

Objective M To list the bones of the *lower extremity* and to describe the diagnostic features of the bones of the *thigh* and *leg*.

survey

The **femur** is the only bone of the thigh. The *head of the femur* articulates proximally with the *acetabulum of the os coxae* and the *medial and lateral condyles* articulate distally with the proximal *articular surface of the tibia* within the leg. The **patella** (kneecap) is the sesamoid bone (formed in a tendon) of the anterior knee region. The **tibia** and **fibula** are the bones of the leg. The *distal end of the tibia* articulates with the *talus* in the *ankle*. Numerous joints of various kinds occur within the foot.

6.39 Describe the structure of the femur.

Located within the thigh, the **femur** (thighbone) is the longest and heaviest bone in the body (fig. 6.28). The **fovea capitis femoris** is a shallow pit in the center of the **head of the femur**. The constricted **neck of the femur** supports the head of the femur, and is a common site for fractures in the elderly. On the proximolateral side of the **shaft of the femur** is the **greater trochanter**, and on the medial side is the **lesser trochanter**. The **intertrochanteric crest** is a bony ridge on the posterior side of the femur between the greater and lesser trochanters. The **linea aspera** is a vertical ridge on the posterior surface of the shaft of the femur. Distally, the **medial** and **lateral condyles** are the articular surfaces for the tibia. The depression between the condyles on the posterior surface is called the **intercondylar fossa**, and the depression between the condyles on the anterior surface is called the **patellar surface**. On either side above the condyles are the **lateral** and **medial epicondyles**.

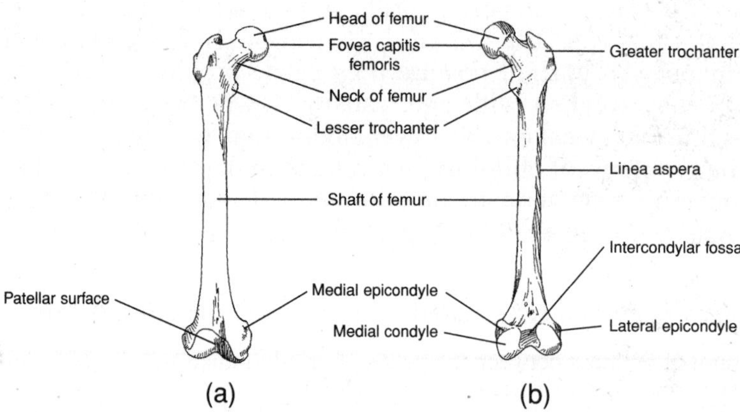

Figure 6.28 The right femur. (*a*) An anterior view and (*b*) a posterior view.

6.40 *True or false*. The only function of the patella is protection of the knee joint.

False. The functions of the **patella** (fig. 6.29) are to protect the knee joint and to strengthen the tendon of the quadriceps femoris muscle. It also increases the leverage of the quadriceps femoris muscle as it contracts to straighten (extend) the leg.

6.41 What do the tibia and fibula have in common? How do they differ?

The medial *tibia* (shinbone) and the lateral *fibula* are the two bones of the leg. As shown in figure 6.29, each has a long **shaft** (body) and a *malleolus* for support and protection of the ankle. The **tibia** is much more massive than the fibula. It has slightly concave surfaces, called the **medial** and **lateral condyles**, on the proximal end for articulation with the condyles of the femur. A sharp anterior border extends vertically along the anterior shaft of the tibia. The **tibial tuberosity**, for the attachment of the *patellar ligament* is located on the proximal portion of the anterior border. The **medial malleolus** is a prominent medial bony knob located on the distal end of the tibia.

The **fibula** is a thin, delicate bone that is more important for muscle attachment than for bearing weight. Proximally, the fibular articular facet articulates with the lateral epicondyle of the tibia (fig. 6.29). The **lateral malleolus** is a prominent lateral bony knob located on the distal end of the fibula.

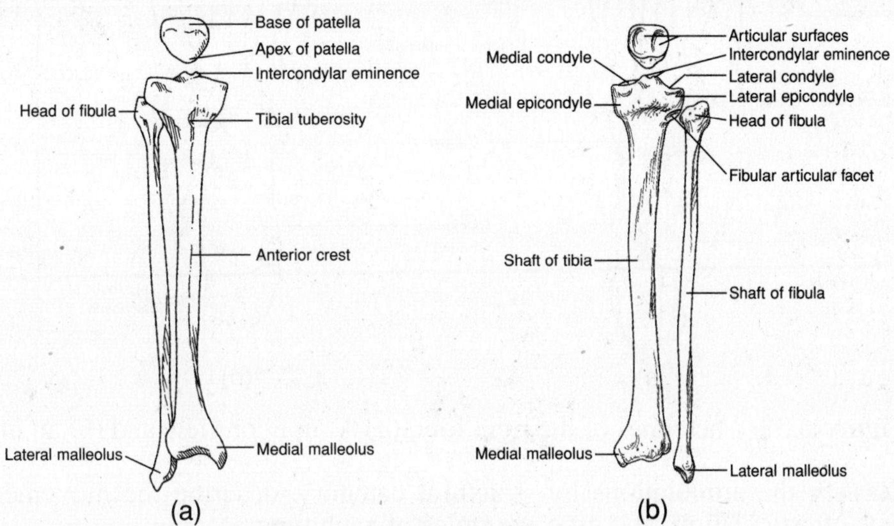

Figure 6.29 The right patella, tibia, and fibula. (*a*) An anterior view and (*b*) a posterior view.

6.42 Describe the skeletal elements of the foot.

The 26 bones of the **pes**, or *foot*, are grouped into 7 **tarsal bones**, 5 **metatarsal bones**, and 14 **phalanges** (fig. 6.30). The articulations (joints) between the cube-shaped tarsal bones permit movement in a confined area, while the elongated metacarpal bones and phalanges act as levers about their freely movable joints.

The **talus** is the tarsal bone that articulates with the tibia and fibula to form the *ankle joint*. The **calcaneous** is the largest of the tarsal bones and provides skeletal support for the heel of the foot. Anterior to the talus is the block-shaped **navicular bone**. The remaining four tarsal bones are from medial to lateral side, the **medial, intermediate,** and **lateral cuneiform bones** and the **cuboid bone**.

Each of the five **metatarsal bones** consists of a proximal **base**, a **shaft** (body), and a distal **head** that is rounded for articulation with the base of a *proximal phalanx*. The metatarsal bones are numbered I to V, the medial, or great toe, side being I.

The 14 **phalanges** are the skeletal elements of the digits. A single toe bone is called a **phalanx**. The phalanges are arranged in a proximal row, a middle row, and a distal row. The *great toe* (hallux), however, has only a proximal and a distal phalanx.

Objective N To describe the kinds of *articulations*, or joints, in the body and the range of movement permitted by each.

Joints may be classified according to structure or function. In the structural classification, a joint is *fibrous*, *cartilaginous*, or *synovial*. The functional classification distinguishes *synarthroses* (immovable joints), *amphiarthroses* (slightly movable joints), and *diarthroses* (freely movable joints). The following discussion applies only to the structural classification of joints.

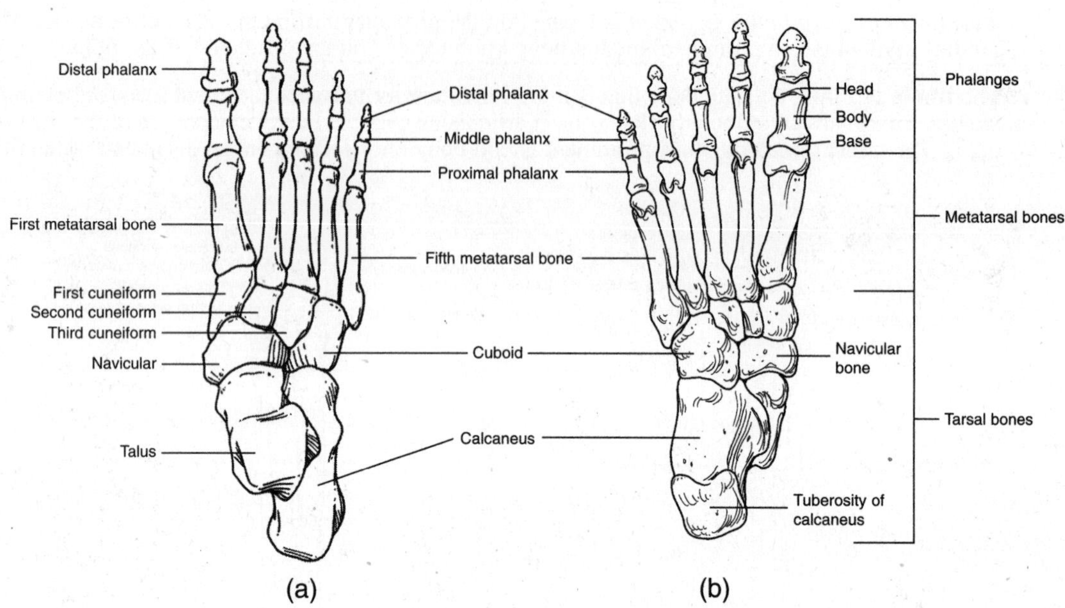

Figure 6.30 The bones of the right foot. (*a*) A superior view and (*b*) an inferior view.

6.43 Classify the articulations by structural category, describe the movements of each type of articulation, and give examples of each type.

See table 6.5.

6.44 Describe the structure of a synovial joint.

Synovial joints are enclosed by a fibroelastic **joint capsule**, which is lined by a thin **synovial membrane** (fig. 6.31). The synovial membrane secretes *synovial fluid*, which fills the joint capsule and lubricates the **articular cartilage** at the ends of the articulating bones. A few synovial joints, such as the knee joints, have cartilaginous pads, called **menisci**, that cushion and guide the articular cartilages.

Synovial fluid is also contained within small membranous sacs called **bursae** (singular, *bursa*) that cushion muscles and facilitate movements of tendons around synovial joints. Inflammation of the lining of a bursa is referred to as *bursitis*.

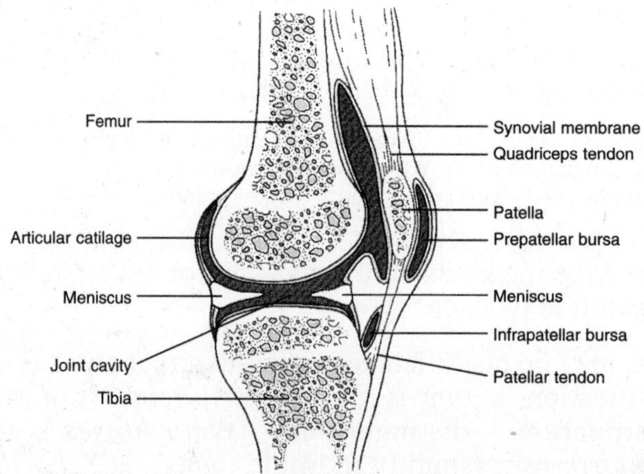

Figure 6.31 A synovial joint is represented by a sagittal view of the knee joint.

Table 6.5 Articulations of the Body

Classification	*Structure*	*Movements*	*Examples*
Fibrous joints	Articulating bones joined by fibrous connective tissue		
Sutures	Frequently serrated edges of articulating bones separated by thin layer of fibrous tissue	None	Sutures of skull
Syndesmoses	Articulating bones bound by interosseous ligaments	Slightly movable	Joints between tibia-fibula and radius-ulna
Gomphoses	Teeth bound into alveoli of bone	None	Teeth secured into alveoli (sockets)
Cartilaginous joints	Articulating bones joined by fibrocartilage or hyaline cartilage		
Symphyses	Articulating bones separated by pad of fibrocartilage	Slightly movable	Intervertebral joints; symphysis pubis and sacroiliac joint
Synchondroses	Mitotically active hyaline cartilage between bones	None	Epiphyseal plates within long bones
Synovial joints	Joint capsule containing synovial membrane and synovial fluid	Freely movable	
Gliding	Flattened or slightly curved articulating surfaces	Sliding	Intercarpal and intertarsal joints
Hinge	Concave surface of one bone articulates with convex surface of another	Bending motion in one plane	Knee joint; elbow joint; joints of phalanges
Pivot	Conical surface of one bone articulates with depression of another	Rotation about a central axis	Atlantoaxial joint; proximal radioulnar joint
Condyloid	Oval condyle of one bone articulates with elliptical cavity of another	Biaxial movement	Radiocarpal joint
Saddle	Concave and convex surface on each articulating bone	Wide range of movements	Carpometacarpal joint of thumb
Ball-and-socket	Rounded convex surface of one bone articulates with cuplike socket of another	Movement in all planes including rotation	Shoulder and hip joints

6.45 What are the technical terms for the types of movement permitted at synovial joints?

Flexion is a movement that decreases the angle between two bones; *extension* increases the angle (fig. 6.32). *Abduction* is movement away from the midline of the body or a body part; *adduction* is movement toward the midline or a body part. *Rotation* is the movement of a bone around its own axis, without lateral displacement. (*Pronation* is the forearm rotation that results in the palm of the hand being directed backward; the opposite rotation is called *supination*). *Circumduction* is a circular, conelike movement of a body segment.

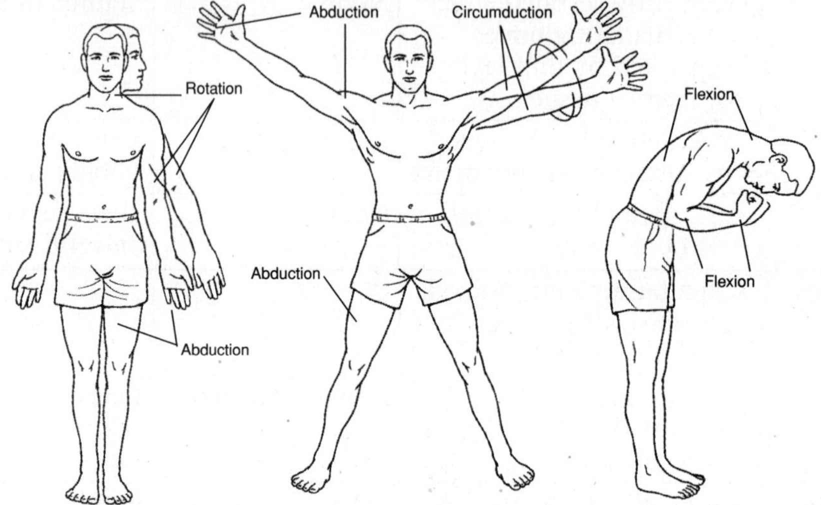

Figure 6.32 Movements at synovial joints.

Key Clinical Terms

Arthritis An inflammatory joint disease, usually associated with the synovial membrane and the articular cartilage. In certain types of arthritis, mineral deposits may form.

Bursitis Inflammation of a bursa.

Dislocation Displacement of one bone away from its natural articulation with another.

Fracture A cracking or breaking of a bone.

Kyphosis (*humpback*) An abnormal posterior convexity of the lower vertebral column.

Lordosis Excessive anteroposterior curvature of the vertebral column, generally in the lumbar region, resulting in a "hollow back" or "saddle back."

Osteoarthritis A localized degeneration of articular cartilage. (Not really an arthritis, since inflammation is not a primary symptom.)

Osteoporosis Atrophy of bone tissue, resulting in marked porosity in skeletal material. Causes include aging, prolonged inactivity, malnutrition, and an unbalanced secretion of hormones.

Scoliosis Excessive lateral deviation of the vertebral column.

Slipped disc Herniation of the *nucleus pulposus* of an intervertebral disc.

Spina bifida Developmental flaw in which the laminae of the vertebrae fail to fuse. The spinal cord may protrude through the opening.

Sprain Straining or tearing of the ligaments and/or tendons of a joint.

Review Exercises

Multiple Choice

1. Which of the following is *not* a function of the skeletal system? (*a*) production of blood cells, (*b*) storage of minerals, (*c*) storage of carbohydrates, (*d*) protection of vital organs

2. Mitosis resulting in elongation of bone occurs at (*a*) the articular cartilage, (*b*) the periosteum, (*c*) the epiphyseal plate, (*d*) the diploe.

3. Which hormone-bone cell combination may result in osteoporosis? (*a*) adrenal cortisol-osteoclast, (*b*) estrogen-osteoblast, (*c*) thyroid hormone-osteoclast, (*d*) thyrocalcitonin-osteoblast

4. Synovial fluid that lubricates a synovial joint is produced by (*a*) a meniscus, (*b*) the synovial membrane, (*c*) a bursa, (*d*) the articular cartilage, (*e*) the mucous membrane.

5. A flattened or shallow articulating surface of a bone is called (*a*) a tubercle, (*b*) a fossa, (*c*) a fovea, (*d*) a facet.

6. Which type of cartilage is the precursor to endochondral bone? (*a*) costal, (*b*) hyaline, (*c*) fibroelastic, (*d*) articular

7. Which suture extends from the anterior fontanel to the anterolateral fontanel? (*a*) coronal suture, (*b*) lambdoid suture, (*c*) squamous suture, (*d*) longitudinal suture.

8. A facial bone that is *not* paired is (*a*) the maxilla, (*b*) the lacrimal bone, (*c*) the vomer, (*d*) the nasal bone, (*e*) the palatine bone.

9. Hemopoiesis would most likely take place in (*a*) the hyoid bone, (*b*) a vertebra, (*c*) the maxilla, (*d*) the scapula.

10. Which of the following bones is *not* part of the axial skeleton? (*a*) hyoid bone, (*b*) sacrum, (*c*) sphenoid bone, (*d*) clavicle, (*e*) manubrium

11. The optic foramen is located within (*a*) the ethmoid bone, (*b*) the occipital bone, (*c*) the palatine bone, (*d*) the sphenoid bone.

12. An example of a gliding joint is (*a*) the intercarpal joint, (*b*) the radiocarpal joint, (*c*) the intervertebral joint, (*d*) the phalangeal joint.

13. The mandibular fossa is a feature of which part of the temporal bone? (*a*) squamous part, (*b*) petrous part, (*c*) tympanic part, (*d*) articular part

14. The superior and middle conchae are bony structures of which bone? (*a*) palatine bone, (*b*) nasal bone, (*c*) ethmoid bone, (*d*) maxilla

15. Which of the following bones does *not* contain a paranasal sinus? (*a*) frontal bone, (*b*) ethmoid bone, (*c*) vomer, (*d*) sphenoid bone, (*e*) maxilla

16. Teeth are supported by (*a*) the maxillae and mandible, (*b*) the mandible and palatine bones, (*c*) the maxillae and palatine bones, (*d*) the maxillae, mandible, and palatine bones.

17. The mastoid process is a structural prominence of (*a*) the sphenoid bone, (*b*) the parietal bone, (*c*) the occipital bone, (*d*) the temporal bone, (*e*) the ethmoid bone.

18. A joint characterized by an epiphyseal plate is called (*a*) a synovial joint, (*b*) a suture, (*c*) a symphysis, (*d*) a synchondrosis.

19. Which of the following bones is characterized by the presence of a diaphysis and epiphyses, articular cartilages, and a medullary cavity? (*a*) scapula, (*b*) sacrum, (*c*) tibia, (*d*) patella

20. Remodeling of bone is a function of (*a*) osteoclasts and osteoblasts, (*b*) osteoblasts and osteocytes, (*c*) chondrocytes and osteocytes, (*d*) chondroblasts and osteoblasts.

21. The cribriform plate is a specialized portion of which bone? (*a*) sphenoid bone, (*b*) maxilla, (*c*) temporal bone, (*d*) vomer, (*e*) ethmoid bone

22. Which of the following is *not* part of the os coxae? (*a*) acetabulum, (*b*) ischium, (*c*) pubis, (*d*) capitulum, (*e*) obturator foramen

23. A fractured coracoid process would involve (*a*) the clavicle, (*b*) the scapula, (*c*) the ulna, (*d*) the radius, (*e*) the tibia.

24. The false pelvis is (*a*) inferior to the true pelvis (*b*) found in the male only, (*c*) narrower in the male than in the female, (*d*) not really part of the skeletal system.

25. A fracture of the lateral malleolus would involve (*a*) the fibula, (*b*) the tibia, (*c*) the ulna, (*d*) a rib, (*e*) the femur.

26. Which of the following bones articulates distally with the talus in the foot? (*a*) navicular bone, (*b*) first metatarsal bone, (*c*) calcaneus, (*d*) first cuneiform bone, (*e*) cuboid bone

27. On a skeleton positioned in the anatomical position, which of the following structures faces anteriorly? (*a*) spinous process of the scapula, (*b*) subscapular fossa, (*c*) infraspinous fossa, (*d*) linea aspera of the femur, (*e*) spinous process of a thoracic vertebra

28. The sagittal suture is positioned between (*a*) the sphenoid and temporal bones, (*b*) the temporal and parietal bones, (*c*) the occipital and parietal bones, (*d*) the occipital and frontal bones, (*e*) the right and left parietal bones.

29. Which of the following bones lacks a styloid process? (*a*) sphenoid bone, (*b*) temporal bone, (*c*) ulna, (*d*) radius

30. Surgical entry through the roof of the mouth to remove a tumor of the pituitary gland would involve (*a*) the mastoid process, (*b*) the pterygoid process, (*c*) the styloid process, (*d*) the sella turcica.

True or False

____ 1. The tibia and fibula articulate with the femur at the knee joint.

____ 2. The proximal and distal ends of a long bone are referred to as diaphyses.

____ 3. Menisci occur only in certain synovial joints.

____ 4. Supination and pronation are specific kinds of circumductional movements.

____ **5.** Yellow bone marrow in certain long bones of an adult produces red blood cells, white blood cells, and platelets.

____ **6.** Bone matrix is composed primarily of calcium and magnesium, which may be withdrawn in small amounts as needed elsewhere in the body.

____ **7.** Thyroid hormone may promote either osteogenesis or osteolysis.

____ **8.** A furrow on a bone that accommodates a blood vessel, nerve, or tendon is known as a sulcus.

____ **9.** Cervical vertebrae are characterized by the presence of articular facets.

____ **10.** The two ossa coxae articulate anteriorly with each other at the symphysis pubis, and posteriorly with the sacrum.

____ **11.** The lateral malleolus of the tibia stabilizes the ankle joint.

____ **12.** Most of the bones of the skeleton form through intramembranous ossification.

____ **13.** There are a total of 56 phalanges in the appendicular skeleton.

____ **14.** Articular cartilage and synovial membranes are found only in synovial joints.

____ **15.** All joints or articulations in the body permit some degree of movement.

____ **16.** Flexion means "contraction of a skeletal muscle."

____ **17.** Osteoblasts actually destroy bone tissue in the process of demineralization.

____ **18.** A person has seven pairs of true ribs and five pairs of false ribs, the last two pairs of which are designated as floating ribs.

____ **19.** A stress fracture along the intertrochanteric line involves the femur.

____ **20.** Surgery of a meniscus could be performed only on either knee joint.

Completion

1. Red bone marrow produces blood cells in a process called ____________________.

2. The ________________________ skeleton consists of the skull, vertebral column, and rib cage; the ________________________ skeleton consists of the girdles and the appendages.

3. ________________________ bones, such as the patellae, are formed in tendons.

4. ________________________ bones are formed first as hyaline cartilage and ________________________ bones form directly as bone.

5. The ________________________ ____________________ is a diamond-shaped "soft spot" on the top of a newborn's skull that facilitates childbirth and permits brain growth.

6. Separating the diaphysis and epiphysis of a child's long bone is a(n) ______________ ____________________, which permits linear bone growth.

7. The ________________________ foramen is an opening in the mandible on the lateral side below the second premolar tooth.

8. The ________________ and the perpendicular plate of the ________________ bone compose the bony framework of the nasal septum.
9. In an adult, the ilium, ischium, and pubis are fused to form the ________________ ________________, or hipbone.
10. The foot contains ________ tarsal bones, ________ metatarsal bones, and ________ phalanges.

Labeling

Label the structures indicated on the figure to the right.

1. ________________________
2. ________________________
3. ________________________
4. ________________________
5. ________________________
6. ________________________
7. ________________________
8. ________________________
9. ________________________
10. ________________________

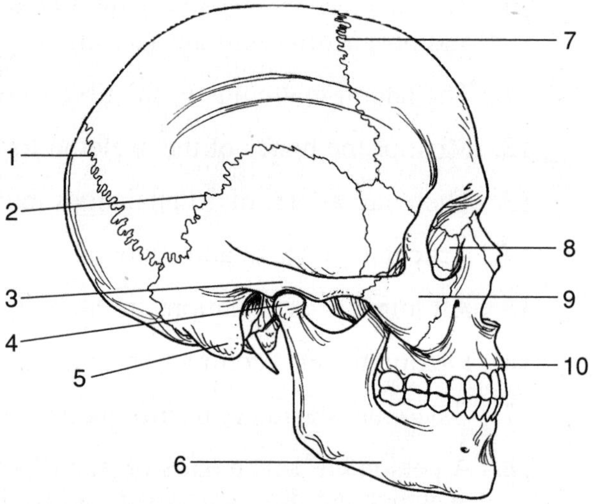

Answers and Explanations for Review Questions

Multiple Choice

1. (*c*) Carbohydrates are not stored within bone.
2. (*c*) Linear bone growth occurs at the epiphyseal plates through mitotic activity. Once adult height has been reached, cell division at these locations stops and the plates ossify.
3. (*a*) Both adrenal cortisol and osteoclasts break down bone tissue.
4. (*b*) The synovial membrane lining the inside of the joint capsule produces the lubricating synovial fluid.
5. (*d*) An example of a facet is the shallow depression on the side of a thoracic vertebra, where the head of a rib articulates.
6. (*b*) Most bones are endochondral, meaning that they began as a hyaline cartilage model before they ossified.
7. (*a*) Like a coronal plane (frontal plane) through the body, which divides the front from the back, the coronal suture appears to divide the skull from front to back.
8. (*c*) The only two unpaired facial bones are the vomer and the mandible.
9. (*b*) The principal sites for hemopoiesis are the sternum, vertebrae, ossa coxae, femora, and humeri.
10. (*d*) The clavicles are part of the pectoral girdle, a component of the appendicular skeleton.
11. (*d*) Contained within the sphenoid bone, the optic foramen is the passageway for the optic nerve from the eye.
12. (*a*) Each of the joints of the carpal bone (intercarpal joints) is of the gliding type.
13. (*a*) The squamous part of the temporal bone includes the zygomatic process and the mandibular fossa for articulation with the mandible at the temporomandibular joint.
14. (*c*) The nasal cavity contains three paired conchae. The superior and middle conchae are part of the ethmoid bone, and the inferior concha is a separate bone.

15. (*c*) The vomer is a flat bone that does not contain a sinus.
16. (*a*) In an adult who has all of his or her permanent teeth, 16 are supported in the maxillae and 16 are supported in the mandible.
17. (*d*) As a protrusion of the temporal bone, the mastoid process can be palpated as a bony knob directly behind the ear.
18. (*d*) Most synchondrotic joints ossify following the period of linear bone growth.
19. (*c*) Each of the long bones within the appendages of the body has a diaphysis, epiphyses, articular cartilage, and a medullary cavity.
20. (*a*) Osteoclasts break down bone tissue and osteoblasts build up bone tissue.
21. (*e*) With its numerous perforations, the cribriform plate of the ethmoid bone permits passage of the olfactory cranial nerves from the olfactory epithelium of the nasal cavity.
22. (*d*) The capitulum is a structure on the humerus.
23. (*b*) The coracoid process is an extension of the scapula from which several muscles attach.
24. (*c*) The false, or greater, pelvis is the distance between the two anterior superior iliac spines. What this means is that adult females have relatively wider hips than do adult males.
25. (*a*) The lateral malleolus is the knob of bone on the lateral side of the ankle. The lateral malleolus is on the distal end of the fibula, and the medial malleolus is on the distal end of the tibia.
26. (*a*) The navicular bone is sandwiched between the talus and the three cuneiform bones.
27. (*b*) The subscapular fossa is the slightly indented anterior surface of the scapula.
28. (*e*) The sagittal suture extends from the frontal bone to the occipital bone, between the two parietal bones.
29. (*a*) There are actually six styloid processes in the body—one on each of the paired ulna, radius, and temporal bones.
30. (*d*) The pituitary gland is supported inferiorly by the sella turcica of the sphenoid bone.

True or False

1. False; only the tibia
2. False; epiphyses
3. True
4. False; rotational
5. False; red bone marrow
6. False; calcium and phosphorus
7. True
8. True
9. False; transverse foramina
10. True
11. False; fibula
12. False; endochondral
13. True
14. True
15. False; some joints are immovable
16. False; "lessening the angle at a hinge joint"
17. False; osteoclasts
18. True
19. True
20. False; generally, but not always

Completion

1. hemopoiésis
2. axial, appendicular
3. Sesamoid
4. Endochondral, membranous
5. anterior fontanel
6. epiphyseal plate
7. mental
8. vomer, ethmoid
9. os coxae
10. 7, 5, 14

Labeling

1. Lambdoid suture
2. Squamous suture
3. Zygomatic process
4. Condyloid process
5. Mastoid process
6. Mandible (body)
7. Coronal suture
8. Lacrimal bone
9. Zygomatic bone
10. Maxilla

Muscle Tissue and Mode of Contraction 7

Objective A To review the classification of *muscle tissue.*

Recall from chapter 4, problem 4.21, that there are three types of muscle tissue: smooth, cardiac, and skeletal. Each type has a different structure and function, and each occurs in a different location in the body (see table 4.7 and fig. 4.5). Because they resemble tiny threads, muscle cells are called *muscle fibers.*

7.1 Which type of muscle constitutes the greatest portion of the body's total weight?

Skeletal muscle constitutes a body system by itself and accounts for about 40% of a person's body weight. Smooth and cardiac muscle tissues account for about 3% of the total body weight.

Muscle tissues are formed prenatally from undifferentiated mesoderm called *mesenchyme* that migrates throughout the body. Once in position and coalesced, the mesenchymal cells specialize into muscle fibers and lose their ability to mitotically divide. This means that a person at birth has all the muscle fibers he or she will ever have. With body growth and conditioning, the muscle fibers increase in size.

Objective B To describe the functions of muscles.

Motion. Contraction of skeletal muscles produces such body movements as walking, writing, breathing, and speaking. Movements associated with digestion and flow of fluids (lymphatic, urinary, and reproductive systems) require contraction of smooth muscles. Movements associated with the cardiovascular system require all three types of muscle tissue.

Heat production. All cells release heat as an end product of metabolism. Since a sizable portion of cells in the body are muscle cells, muscles are a major source of heat.

Posture and body support. The muscular system lends form and support to the body and helps to maintain posture in opposition to gravity.

7.2 How do the concepts of *synergism* and *antagonism* apply to skeletal muscles?

Synergistic muscles contract together and coordinate in effecting a particular movement. For example, the temporalis muscles and the masseter muscles both work together to elevate the jaw (close the mouth).

Antagonistic muscles perform tasks that oppose those of another group of muscles and are generally located on the opposite side of a limb or portion of the body. For example, the biceps brachii muscle flexes the elbow and the triceps brachii muscle extends it.

Objective C To identify the components of a *skeletal muscle fiber.*

Each skeletal muscle fiber is a multinucleated, striated cell containing a large number of rodlike **myofibrils** that extend, in parallel, the entire length of the cell. Each myofibril is composed of still smaller units, called **myofilaments**, that contain the contractile proteins **actin** and **myosin**.

7.3 Describe the protein structures involved in muscle contraction.

Each myofibril within a skeletal muscle fiber consists of several hundred protein strands called *myofilaments*. **Thin myofilaments** are about 6 nm in diameter and are composed primarily of the *actin proteins*. **Thick myofilaments** are about 16 nm in diameter and are composed primarily of *myosin proteins*.

Shaped like a golf club, each myosin protein has a long rod portion, the *light meromyosin (LMM) filaments*, and a globular head, the *heavy meromyosin (HMM) filaments*. Each myosin head contains an actin binding site and a myosin ATPase binding site. The myosin protein strands of the rod portion bind together with their globular heads projecting outward to form the thick filaments that lie between the thin filaments (fig. 7.1).

Three different proteins—*actin, tropomyosin,* and *troponin*—compose the thin myofilaments. Two long strands of spherical actin molecules, with binding sites for attachment with myosin cross bridges facing laterally, twist together like strings of pearls. This actin helix forms the backbone of the thin myofilaments. Long, thin, threadlike tropomyosin proteins spiral around and cover the binding sites on the actin helix. The *troponin molecule*, a small protein complex, fastens the ends of the tropomyosin molecule to the actin helix (fig. 7.2). The thick and thin myofilaments overlap within the myofibril like two halves of a deck of cards being shuffled, one layer of thin filament separating each layer of the deck. One thick myofilament, together with a thin filament above and one below, forms a myomere. The **sarcomere** (myomere) is the structural unit of the myofibril.

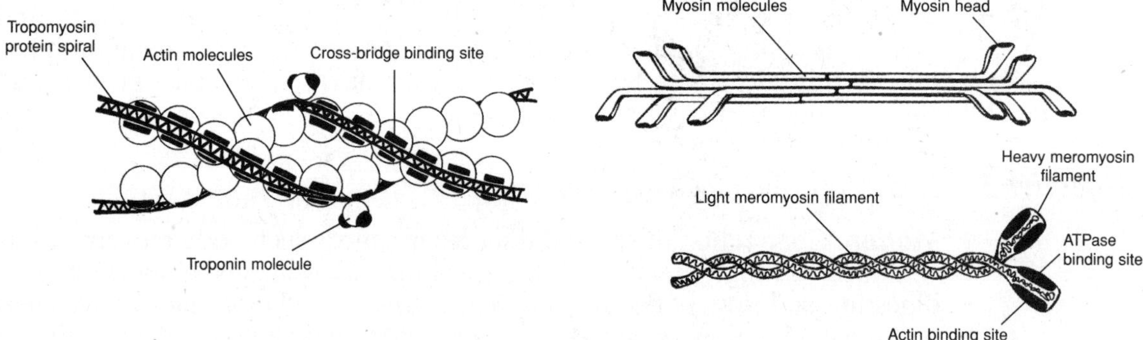

Figure 7.1 The structure of thin myofilaments.

Figure 7.2 The structure of thick myofilaments.

7.4 Why do skeletal and cardiac muscle fibers appear striated?

The regular spatial organization of the contractile proteins within the myofibrils is responsible for the cross-banding striations seen in skeletal and cardiac muscle fibers. The dark bands are called **A bands** (A = anisotropic bands) and the lighter bands are called **I bands** (I = isotropic bands). (Smooth muscle fibers contain the same contractile proteins, but in the absence of a regular spatial arrangement they lack the cross banding). The I bands are bisected by dark **Z lines**, where the actin filaments of adjacent sarcomeres join (fig. 7.3).

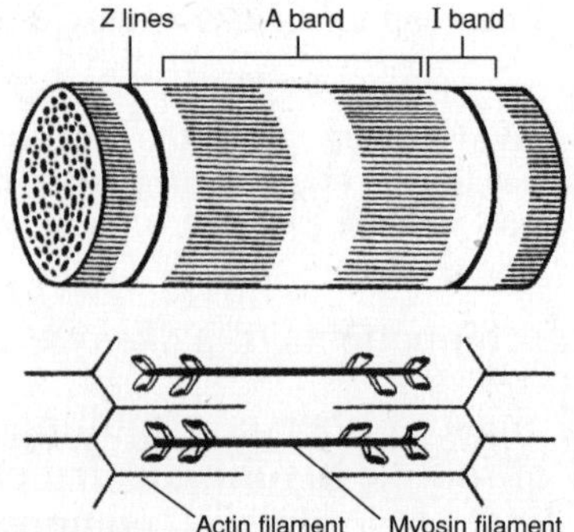

Figure 7.3 A sarcomere.

7.5 Describe the fine structure (electron micrograph) structure of a skeletal muscle fiber.

The **sarcolemma** (cell membrane) of a muscle fiber encloses the **cytoplasm** (sarcoplasm). The cytoplasm is permeated by a network of membranous channels, called the **sarcoplasmic** (endoplasmic) **reticulum**, which forms sleeves around the myofibrils. The longitudinal tubes of the sarcoplasmic reticulum empty into expanded chambers called **terminal cisternae**. Calcium ions (Ca^{2+}) are stored in the terminal cisternae and play an important role in regulating muscle contraction.

The **transverse tubules** (*T tubules*) are not part of the sarcoplasmic reticulum. Rather, they are internal extensions of the sarcolemma that extend perpendicular to the endoplasmic reticulum. The T tubules pass between adjacent segments of terminal cisternae and penetrate deep into the interior of the muscle fiber to allow the action potential from the cell surface to be delivered into the center of the fiber. A *muscle triad* consists of a T tubule and the cisternae on both sides (fig 7.4).

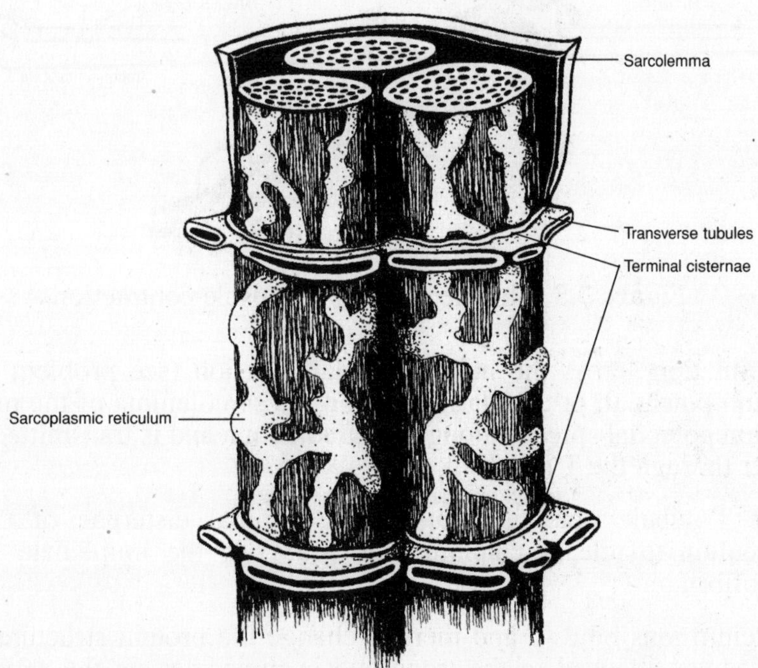

Figure 7.4 Configuration of a muscle triad.

Objective D To explain the sequence of events in *muscle contraction*

In the **sliding filament theory of contraction,** a skeletal muscle fiber, together with all of its myofibrils, shortens by movement of the insertion toward the origin of the muscle (see problem 7.17). Shortening of the myofibrils is caused by shortening of the sarcomeres, which is accomplished by sliding of the myofilaments. The A bands remain the same length during contraction, but are pulled toward the origin of the muscle. Adjacent A bands are pulled closer together as the I bands between them shorten. The mechanism that produces the sliding of the thin (actin) myofilaments over the thick (myosin) myofilaments during contraction is illustrated in fig. 7.5 and outlined in the steps that follow.

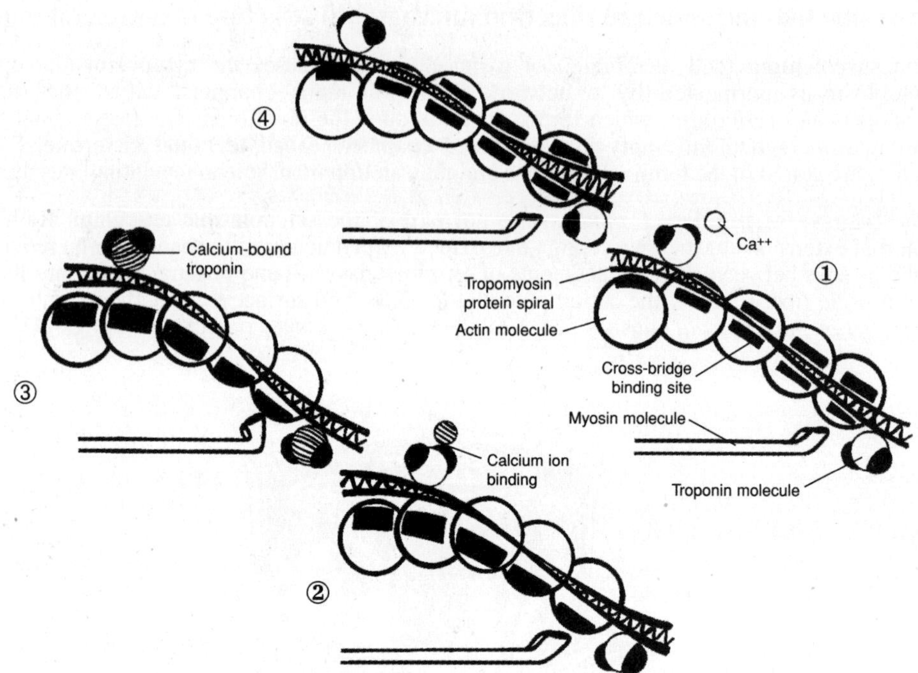

Figure 7.5 The mechanism of muscle contraction.

1. Stimulation across the neuromuscular junction (see problem 7.9) initiates an action potential, or depolarization, on the sarcolemma of the muscle fiber. This action potential spreads along the sarcolemma and is transmitted into the muscle fiber through the T tubules.
2. The T tubule potential causes the terminal cisternae of the sarcoplasmic reticulum to release calcium ions (Ca^{++}) in the immediate vicinity of each myofibril.
3. Calcium ions bind to and thereby change the protein structure of the troponin molecules attached to the tropomyosin molecules on the actin filaments. The resulting conformational change causes the tropomyosin to move aside, exposing the actin binding sites.
4. Myosin cross bridges bind to actin. Upon binding, the cocked (energized) HMM undergoes a conformational change, causing the head to tilt. This pulls the actin filament over the myosin filament in an action called a *power stroke.*
5. After the power stroke, ATP binds the HMM, causing detachment of the cross bridge from the actin binding sites. The enzyme ATPase within the HMM cleaves ATP to ADP + energy; the energy is used to recock the HMM. The HMM can then bind with another actin site (if these sites are sill exposed because of the presence of Ca^{2+}) and produce another power stroke.
6. Repeated power strokes successfully pull in the thin filaments, much like pulling in a rope hand over hand. This sliding-with-a-ratchet mechanism involves numerous actin binding sites and myosin cross bridges and constitutes a single muscle contraction.

7.6 How is muscle relaxation accomplished?

Just as an action potential sustains a muscle contraction, the cessation of an action potential causes the muscle to relax. Once the action potential ceases, the endoplasmic reticulum actively transports Ca^{2+} from the cytoplasm into the terminal cisternae. Without calcium ions, the troponin molecule resumes its original shape so that the tropomyosin is pulled back over the myosin binding sites of the actin molecule. With these sites covered, the myosin cross bridges can no longer bind to the actin molecule, and the actin filaments slide back to their noncontracted position.

Rigor mortis, "stiffness of death," demonstrates the importance of ATP in releasing the myosin head from the binding site on the actin molecule during muscle contraction. Following death, calcium ions leak through the cell membrane, initiating the contraction process that allows the myosin cross bridges to bind to the actin filaments. Without fresh stores of ATP, the myosin heads remain bound to the actin filaments, causing a stiffening of the muscle and thus immobility of the joints. This condition fades within days as the proteins involved degrade.

7.7 What causes muscle soreness following strenuous exercise?

For years it was believed that muscle soreness was simply caused by a buildup of lactic acid within the muscle fibers during exercise. Although lactic acid accumulation probably is a factor related to soreness, recent research has shown that there is also damage to the contractile proteins within the muscle. If a muscle is used to exert an excessive force to lift a heavy object or to run a distance farther than it is conditioned to, some of the actin and myosin filaments become torn apart. This microscopic damage causes an inflammatory response that results in swelling and pain. If enough proteins are torn, use of the entire muscle may be compromised.

Objective E To describe the *neuromuscular junction*.

A neuromuscular (myoneural) junction is the space between an axon terminal of a motor neuron and the cell membrane of a muscle fiber (fig. 7.6). The motor end plate is the combination of the axon terminal and the cell membrane as viewed histologically.

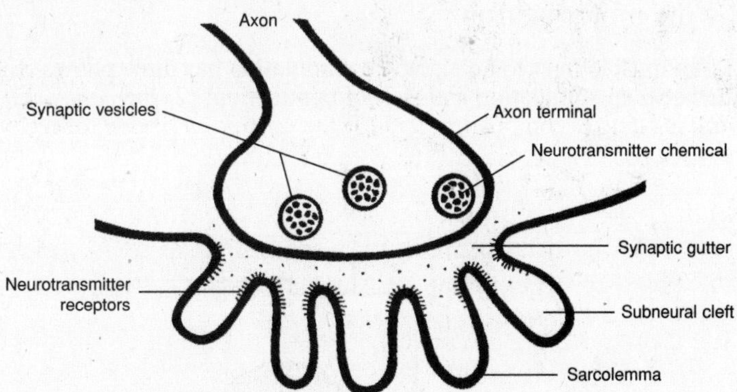

Figure 7.6 The neuromuscular junction.

7.8 The sarcolemma is invaginated forming a *synaptic gutter* at the site of the neuromuscular junction. At the bottom of the gutter are numerous folds called the *subneural clefts*. What is the function of these folds?

The **subneural clefts** of the sarcolemma (fig. 7.6) greatly increase the surface area over which the neurotransmitter (acetylcholine) can produce an action potential.

7.9 List the sequence of events occurring at the neuromuscular junction.

(1) The action potential travels along the motor neuron to the axon terminal, where it causes an influx of calcium ions. (2) The calcium ions cause synaptic vesicles (see fig. 7.6) to release acetylcholine, which diffuses across the synaptic gutter and combines with specific receptors on the sarcolemma. (3) An action potential radiates over the sarcolemma.

Myasthenia gravis is an autoimmune disease in which a person has developed antibodies that bind to and block the receptors for acetylcholine at the neuromuscular junction. The numbers of subneural clefts and acetylcholine receptors are also reduced. As a result, transmission of the signal across the neuromuscular junction is significantly reduced, causing muscle weakness.

Objective F To define *motor unit* and to describe how a motor unit works.

A **motor unit** consists of a single motor neuron together with the specific skeletal muscle fibers that it innervates. A large motor unit is one that serves many muscle fibers. A small motor unit is one that serves relatively few muscle fibers. Contraction of a skeletal muscle requires recruitment of motor units. Few motor units are recruited when fine, highly coordinated movements are being performed. Many motor units are recruited when a strength movement (e.g., lifting a heavy object) is being performed. Being "psyched-up" through sympathetic stimulation and secretion of adrenaline (epinephrine) facilitates motor unit recruitment.

The motor unit profile of a muscle is genetically determined, and each muscle in the body has its own motor unit profile. In some large muscles, such as in the back or thigh, a large motor unit may contain 200 to 500 muscle fibers. In some small muscles that are involved in precise movements, such as those in the face and hands, a small motor unit may contain 10 to 25 muscle fibers.

7.10 How do the individual muscle fibers of a motor unit respond to an electrical stimulus delivered by the motor neuron.

The response of a muscle fiber to an electrical stimulation has three phases (fig. 7.7): (1) the *latent period*, or time between stimulation and start of contraction; (2) the *contraction period*, or duration of time when work is being accomplished; and (3) the *relaxation period*, or recovery of the muscle fiber.

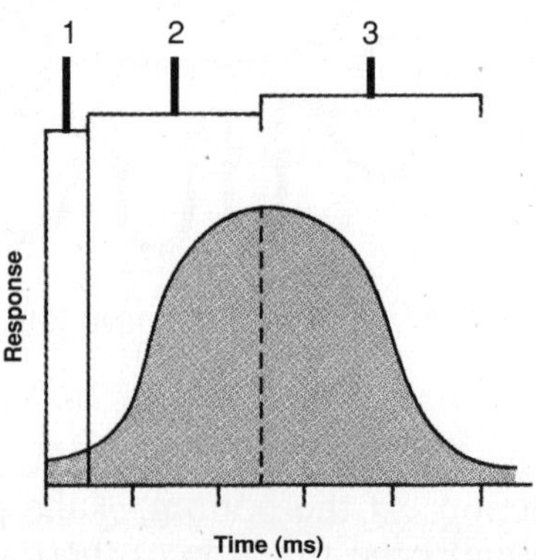

Figure 7.7 The activity of a muscle fiber in response to a stimulus.

7.11 Is the contraction time the same for all skeletal muscle fibers?

No. Skeletal muscle fibers are grouped according to biochemical performance characteristics into three different categories: *fast-twitch fibers*; *intermediate fibers*; and *slow-twitch fibers* (table 7.1). Each muscle contains a genetically determined percentage of fiber types. For example, one person may have more fast-twitch fibers in a particular muscle than another person. The effect that muscle conditioning has on changing the profile of muscle fiber types is uncertain. The percentages of fiber types, however, greatly influence muscular power and endurance. Anaerobic fast-twitch fibers (also called fast-glycolytic or Type IIb fibers) are able to contract very forcefully and rapidly. They are used primarily for power and speed. Aerobic slow-twitch fibers (also called slow-oxidative or Type I fibers) are highly resistant to fatigue. They are used primarily for endurance. The characteristics of intermediate fibers differ somewhat from fiber to fiber, but lie on the continuum between fast-twitch and slow-twitch fibers.

Table 7.1 A Comparison of Muscle Fiber Types

Fiber characteristic	*Fast-twitch fiber*	*Intermediate fiber*	*Slow-twitch fiber*
Fiber size	Large	Intermediate	Small
Glycogen content	High	Intermediate	Low
Myosin ATPase	High	High	Low
Myoglobin content	Low	High	High
Energy system	Anaerobic	Combination	Aerobic
Twitch	Fast	Fast	Slow
Primary use	Speed and power	Moderate activity	Endurance

7.12 How is the strength of a muscle contraction determined?

The strength of a muscle contraction is determined by the size and number of motor units that are recruited to perform the specific task. Motor units operate according to the *all-or-none* law. This means that when a motor unit is stimulated, all of the muscle fibers in that unit will contract. Therefore, the larger the motor unit being recruited, the greater the force that is generated.

The brain learns through experience about how many motor units it takes to perform a certain task. For instance, more motor units are recruited to smash a walnut than to crack an egg. Likewise, more motor units are recruited to pick up a book than a pencil. However, some objects appear heavier (or lighter) than they really are and thereby trick the mind into thinking that more (or fewer) motor units should be recruited than are actually needed. For example, if you reach out to pick up a carton of milk in your refrigerator that you assume is full when it is actually almost empty, you will recruit enough motor units to jerk the carton off the shelf and smack it against the surface above.

7.13 Explain muscle twitch, summation, and tetanus.

A single action potential to the muscle fibers of a motor unit produces a muscle **twitch**, or a very rapid (not sustained) contraction (fig. 7.8). If impulses are applied to a muscle in rapid succession through several motor units, one twitch will not have completely ended before the next begins. Therefore, since the muscle is already in a partially contracted state when the second twitch begins, the degree of muscle shortening in the second contraction will be slightly greater than the shortening that occurs with a single twitch. The additional shortening due to a rapid succession of two or more action potentials is termed **summation**. At sufficiently high stimulation frequencies, the overlapping twitches sum to one strong, steady contraction called **tetanus**. Relaxation of the muscle fiber is either partial (*incomplete tetanus*) or does not occur at all (*complete tetanus*). Most muscle contractions are short-term tetanic contractions, and are thus smooth and sustained.

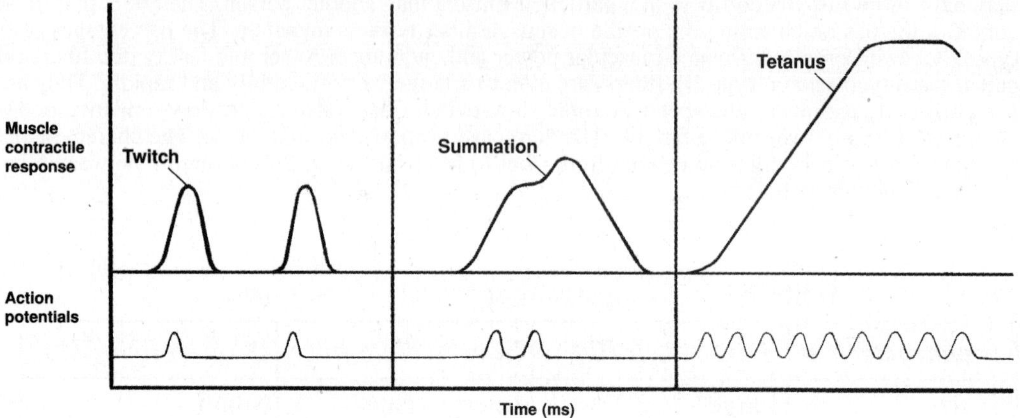

Figure 7.8 Patterns of various muscle contractions.

The bacterium *Clostridium tetani* is the causative agent of the disease *tetanus* (not to be confused with the normal manner of muscle contraction). The metabolic activity of this bacterium produces a toxin that interferes with the enzymes that break down neurotransmitters within the synaptic junctions. The presence of these neurotransmitters causes a constant action potential to be sent by the nerve to the muscle tissue, resulting in spasmodic contractions (tetany) of the muscle. When these painful, exhausting spasms occur in the masseter muscles (used for closing the jaw), the condition is commonly referred to as "lockjaw." Tetanus can be prevented with vaccines and treated with antibiotics.

7.14 Distinguish between isotonic and isometric contractions.

During **isotonic contraction**, the muscle shortens because the force of the contraction is greater than the resistance. During **isometric contraction** the length of the muscle stays the same because the antagonist force equals the force in the muscle being contracted. An isometric contraction becomes an isotonic contraction when increased force generated within the muscle overcomes the resistance, resulting in the muscle shortening.

Objective G To describe the *architecture of skeletal muscle.*

Skeletal muscle tissue, in association with connective tissue, is characterized by an organized pattern of muscle bundles (table 7.2). This muscle architecture determines the force and direction of the contracting muscle fibers.

7.15 Describe the principal fiber patterns in skeletal muscle.

See table 7.2.

Table 7.2 A Comparison of Muscle Fiber Arrangements

Appearance of fibers	*Types and Characteristics*	*Examples*
	Parallel fiber arrangement • straplike muscles with long excursions (contract over long distances) • few motor units • good endurance • not especially strong • relatively poor dexterity	*Sartorius muscle,* located along the anterior thigh region *Rectus abdominis muscle,* located along the anterior abdominal region
	Convergent fiber arrangement • fan-shaped muscles with moderate excursion • few motor units • moderate endurance • fairly strong • fairly dexterous	*Pectoralis major muscle*, located in anterior thoracic region *Temporalis muscle*, located over the temporal bone
	Pennate fiber arrangement • feather-shaped muscles with short excursions • many motor units • poor endurance • especially strong • excellent dexterity	*Antebrachial muscles*, located in anterior forearm and act on the hand *Crural muscles*, located in leg and act on the foot
	Sphincter muscles • fibers encircle a body orifice (opening) • many motor units • good endurance • moderately strong • good dexterity	*Orbicularis oris muscle* surrounding the mouth *Orbicularis oculi muscle* surrounding the eye

7.16 How are skeletal muscle fibers bound together and secured to a bone?

Loose fibrous connective tissues bind muscles at various levels to unify the force of contraction. A **fasciculus** is a bound group of individual muscle fibers (fig. 7.9). The *fasciculi* are the bundles of muscle fibers composing a muscle. In turn, each muscle is surrounded by a connective tissue called **fascia**. The fascia secures the muscle to a **tendon**. Composed of dense regular connective tissue (see chapter 4), tendons are strong, flexible structures that secure muscles to bones. More specifically, a tendon secures the fascia of a muscle to the periosteum of a bone. An *aponeurosis* is a sheetlike tendon.

The **endomysium** is the connective tissue in contact with the individual muscle fibers. The **perimysium** is the connective tissue binding the fasciculi together. The **epimysium** is the connective tissue surrounding the muscle and binding it to the fascia.

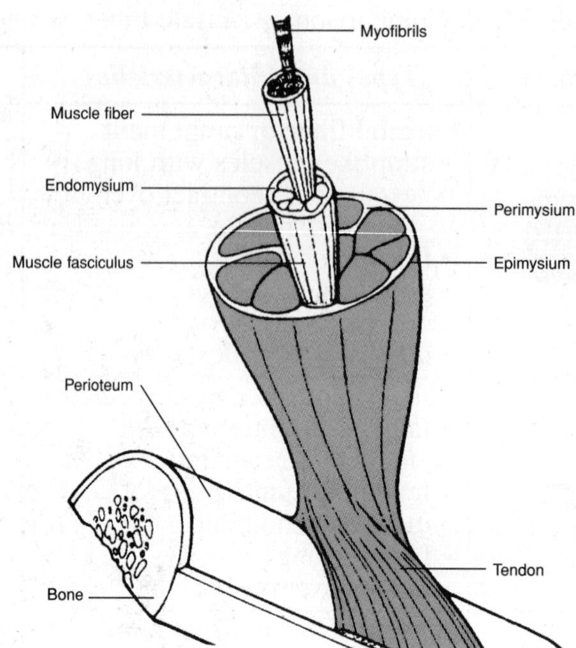

Figure 7.9 The associated connective tissues of a skeletal muscle.

7.17 How do the terms *origin* and *insertion* relate to muscles?

The **origin** of a muscle is the more stationary attachment of the muscle; the **insertion** is the more movable attachment. In the appendages, the origin is generally proximal in position, whereas the insertion is distal in position.

Review Exercises

Multiple Choice

1. Muscle fibers characterized by a lack of striations, a single centrally located nucleus in each cell, and involuntary contractions are referred to as (*a*) skeletal muscle fibers, (*b*) smooth muscle fibers, (*c*) cardiac muscle fibers, (*d*) autonomic muscle fibers.

2. The anisotropic dark bands of muscle fibers are called (*a*) Z bands, (*b*) I bands, (*c*) A bands, (*d*) D bands.

3. The structural unit of the myofibril is (*a*) the myofibril, (*b*) myosin, (*c*) the A band, (*d*) the myomere.

4. Muscle contraction is produced by shortening of all the following *except* (*a*) myofibrils, (*b*) sarcomeres, (*c*) A bands, (*d*) I bands.

5. Muscle contraction is initiated when (*a*) Ca^{2+} binds to the troponin, (*b*) actin is removed from troponin, (*c*) actin is made available to troponin, (*d*) Ca^{2+} is removed from the troponin.

6. The source of Ca^{2+} for the muscle is (*a*) the T tubule, (*b*) the central sac, (*c*) the terminal cisternae, (*d*) the sarcoplasmic reticulum.

7. In a relaxed muscle, (*a*) tropomyosin blocks attachment of myosin heads to actin, (*b*) the concentration of sarcoplasmic Ca^{2+} is low, (*c*) tropomyosin is moved out of the way so that the myosin heads can attach to actin, (*d*) myosin ATPase is activated.

8. Muscle relaxation occurs (*a*) as Ca^{2+} is released from the sarcoplasmic reticulum, (*b*) as long as Ca^{2+} is attached to troponin, (*c*) as action potentials are transmitted through the transverse tubules, (*d*) as the sarcoplasmic reticulum actively removes Ca^{2+} from the cytoplasm.

9. A muscle triad consists of (*a*) a T tubule and a sarcomere, (*b*) a T tubule and two terminal cisternae, (*c*) a T pump and two calcium pumps, (*d*) three myofibrils.

10. A single motor neuron and all the skeletal muscle fibers it innervates constitute (*a*) a motor unit, (*b*) a muscle triad, (*c*) a sarcounit, (*d*) a neuromuscular junction.

11. A muscle that develops tension against some load but that does not shorten is undergoing (*a*) isometric contraction, (*b*) isotonic contraction, (*c*) neither *a* nor *b*, (*d*) both *a* and *b*.

12. The channels that extend from the cell wall into the interior of a skeletal muscle cell form (*a*) the endoplasmic reticulum, (*b*) myofibers, (*c*) T tubules, (*d*) tropomyosin.

13. The globular heads on the myosin proteins of the myosin filament (*a*) are made up of troponin molecules; (*b*) are believed to be attached to ATP molecules, which are used to recock the myosin heads; (*c*) shorten during the contraction process; (*d*) have a high affinity for calcium ions released from the cisternae of the sarcoplasmic reticulum.

14. Troponin is a protein that (*a*) is bound to myosin to form a complex that is normally inhibited in the resting muscle fiber, (*b*) forms the binding site for the myosin heads when they attach to actin, (*c*) has a high affinity for calcium ions, (*d*) contains numerous molecules of ADP.

15. According to the all-or-none law, (*a*) all the contractile elements in a muscle fiber contract when the muscle fiber is stimulated, (*b*) all the muscle fibers in a muscle contract when the muscle is stimulated, (*c*) all the muscle fibers in a motor unit contract when the motor is stimulated, (*d*) none of the preceding are true.

True or False

____ 1. Muscle tissues account for approximately 40% of a person's weight.

____ 2. Actin is found only in the striated fibers of cardiac and skeletal muscle tissues.

____ 3. Slow-twitch muscle fibers are more resistant to fatigue than the other muscle fiber types.

____ 4. Fasciculi are enclosed in a covering of perimysium.

____ 5. A sarcomere is the region of a myofibril that lies between two consecutive Z lines.

____ 6. An action potential in a muscle fiber is initiated by stimulation across the neuromuscular junction.

____ **7.** Sustained contractions of skeletal muscle is known as tetanus.

____ **8.** Fast-twitch muscle fibers are primarily used in endurance activities.

____ **9.** A muscle triad consists of a sarcoplasmic reticulum, a T tubule, and a terminal cisternum.

____ **10.** A motor unit consists of a single motor neuron and the muscle fibers it innervates.

____ **11.** Lifting a dumbbell is an example of isometric contraction.

____ **12.** Thin myofilaments are primarily composed of myosin proteins.

____ **13.** An accumulation of lactic acid is the principal cause of sore muscles.

____ **14.** To initiate muscle contraction, calcium ions bind to and change the shape of the troponin protein molecules, which then pull the tropomyosin proteins off the myosin binding sites of the actin helix.

____ **15.** Synergistic muscles work together to perform a certain motion or action. Antagonistic muscles work in opposition to another group of muscles.

____ **16.** The strength of a muscle contraction is increased by recruiting more muscle fibers within a motor unit.

____ **17.** During muscle contraction, the I bands get smaller and the Z lines get closer together, but the A bands do not change in size.

____ **18.** The energy provided by ATP molecules allows the myosin head to bind to the exposed binding site on the actin molecule.

____ **19.** The transverse tubules (T tubules) store calcium ions needed for muscle contraction.

____ **20.** A tendon is a structure that binds the fascia of a muscle to the periosteum of a bone.

Labeling

Label the structures indicated on the figure to the right.

1. ______________________

2. ______________________

3. ______________________

4. ______________________

5. ______________________

2
3
1
4
5

Matching

Match the muscle fiber component with its description.

____	**1.** Z line	(*a*)	flat protein structure to which the thin filaments attach
____	**2.** Sarcomere	(*b*)	basic unit of a muscle fiber
____	**3.** A band	(*c*)	intramuscular saclike structures (tubules) derived from membranes
____	**4.** Sarcoplasmic reticulum	(*d*)	structure that binds calcium
____	**5.** Troponin	(*e*)	"trigger" or regulator of contraction
____	**6.** Calcium	(*f*)	composed mainly of myosin molecules
____	**7.** ATP-myosin complex	(*g*)	functions to release the energy in ATP

Answers and Explanations for Review Exercises

Multiple Choice

1. (*b*) Smooth muscle fibers lack visible striations because actin and myosin molecules are not regularly arranged. These fibers are under autonomic control, and each cell contains a single nucleus.
2. (*c*) They are called A bands because of their anisotropic property (they can polarize visible light).
3. (*d*) The sarcomere is the structural unit of the myofibril; it is the region of a myofibril between two successive Z lines.
4. (*c*) The myosin that forms the A bands does not shorten during contraction.
5. (*a*) Calcium ions bind to troponin and cause a conformational change in the tropomyosin, which exposes the actin binding site to myosin cross bridges.
6. (*c*) The terminal cisternae, or lateral sacs, store Ca^{2+}.
7. (*a*) Without the release of Ca^{2+}, the tropomyosin blocks the actin binding site.
8. (*d*) When there is no action potential, the calcium ions are actively returned and stored on the sarcoplasmic reticulum.
9. (*b*) A triad consists of a T tubule, which is an extension of the sarcolemma, and the terminal cisternae on both sides of the T tubule.
10. (*a*) A motor unit consists of a single motor neuron and the specific skeletal muscle fibers it innervates.
11. (*a*) During isometric contraction, the length of the muscle stays the same because the antagonistic force equals the force of the contracting muscle.
12. (*c*) T tubules are extensions of the sarcolemma.
13. (*b*) ATP binds to the myosin globular head. The enzyme ATPase within the head changes ATP to ADP and energy. The energy is used to recock the head.
14. (*c*) Calcium ions bind to troponin, which in turn causes tropomyosin to move aside so that the myosin cross bridge can attach to the actin binding site.
15. (*c*) Motor units operate on the all-or-none law; that is, when a motor unit is recruited, all of the muscle fibers in that motor unit contract.

True or False

1. True
2. False; actin is found in all muscle tissues, but in smooth muscle tissue it is not regularly arranged
3. True
4. True

5. True
6. True
7. True
8. False; fast-twitch muscle fibers are used primarily for resistance activities
9. False; a muscle triad consists of a T tubule and two cisternae
10. True
11. False; lifting a dumbbell is an example of isotonic contraction
12. False; thin myofilaments are composed chiefly of actin proteins; thick myofilaments are composed chiefly of myosin proteins
13. False; the primary cause of muscle soreness is damage to the thick and thin myofilaments
14. True
15. True
16. False; motor units follow the all-or-none law of physiological activity
17. True
18. False; energy released from the ATP molecule recocks the myosin head after the power stroke
19. False; terminal cisternae store calcium ions; T tubules conduct the action potential from the cell membrane into the center of the cell
20. True

Labeling

1. Muscle fasciculus
2. Myofibrils
3. Muscle fiber
4. Epimysium
5. Perimysium
6. Tendon

Matching

1. (*a*)
2. (*b*)
3. (*f*)
4. (*c*)
5. (*d*)
6. (*e*)
7. (*g*)

Muscular System 8

Objective A To become familiar with the nomenclature for muscles and their actions (table 8.1, 8.2).

Table 8.1 Examples of How Muscle Names Are Derived

Named according to:	*Examples*
Shape	*Rhomboideus* (like a rhomboid); *trapezius* (like a trapezoid); or, denoting the number or heads of origin, *biceps* (two heads)
Location	*Pectoralis* (chest region, or pectus); *intercostal* (between ribs); *brachii* (upper arm)
Attachment(s)	*Zygomaticus, temporalis, sternocleidomastoid*
Orientation	*Rectus* (straplike); *transverse* (across)
Relative position	*Lateralis, medialis, external*
Function	*Abductor, flexor, extensor, pronator*

Table 8.2 Examples of Muscle Actions (m. = muscle)

Action	*Definition*	*Example*
Flexion	Decreases a joint angle	Biceps brachii m.
Extension	Increases a joint angle	Triceps brachii m.
Abduction	Moves an appendage away from the midline	Deltoid m.
Adduction	Moves an appendage toward the midline	Adductor longus m.
Elevation	Raises a body structure	Levator scapulae m.
Depression	Lowers a body structure	Depressor labii inferioris m.
Rotation	Turns a bone around its longitudinal axis	Sternocleidomastoid m.
Supination	Rotates the hand so that the palm faces anteriorly	Supinator m.
Pronation	Rotates the hand so that the palm faces posteriorly	Pronator teres m.
Inversion	Turns the sole inward	Tibialis anterior m.
Eversion	Turns the sole outward	Peroneus tertius m.

Objective B To locate and learn the actions of the *muscles of the axial skeleton.*

The muscles of the axial skeleton include those used in facial expression, mastication, neck movement, and respiration; those that act on the abdominal wall; and those that move the vertebral column.

8.1 List the muscles of facial expression, along with their attachments and actions.

See fig. 8.1 and table 8.3.

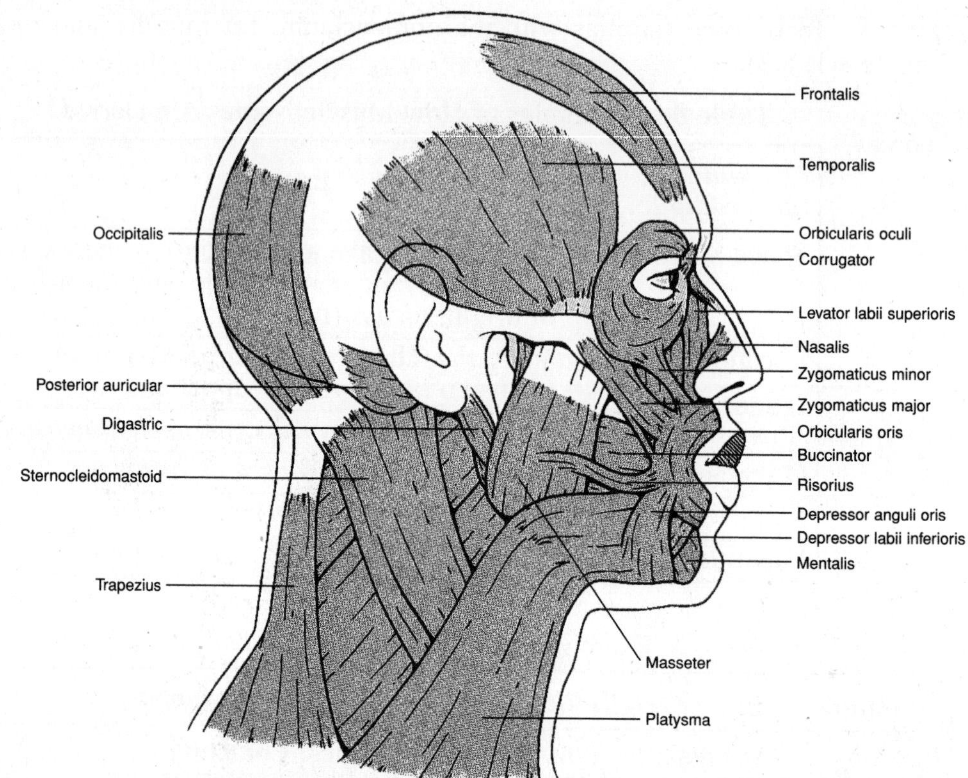

Figure 8.1 Muscles of facial expression.

Table 8.3 Muscles of Facial Expression

Facial muscle	*Origin(s)*	*Insertion(s)*	*Action(s)*
Frontalis	Galea aponeurotica	Skin of eyebrow	Wrinkles forehead; elevates eyebrow
Occipitalis	Occipital bone and mastoid process	Galea aponeurotica	Moves scalp backward
Corrugator	Fascia above eyebrow	Root of nose	Draws eyebrows toward midline, as in scowling
Orbicularis oculi	Bones of medial orbit	Tissue of eyelid	Closes eye, as in blinking
Nasalis	Maxilla and nasal bone	Aponeurosis of nose	Dilates nostrils
Orbicularis oris	Fascia surrounding lips	Mucosa of lips	Closes and purses lips, as in kissing
Levator labii superioris	Maxilla and zygomatic bone	Orbicularis oris	Elevates upper lip, as exposing upper teeth
Zygomaticus	Zygomatic bone	Orbicularis oris at lateral part of upper lip	Elevates corners of mouth, as in smiling

Table 8.3 (continued) Muscles of Facial Expression

Facial muscle	*Origin(s)*	*Insertion(s)*	*Action(s)*
Risorius	Fascia of cheek	Orbicularis oris at corner of lips	Draws corner of mouth laterally
Depressor anguli oris	Mandible	Inferolateral part of orbicularis oris	Depresses corner of mouth, as in frowning
Depressor labii inferioris	Mandible	Orbicularis oris and skin of lower lip	Depresses lower lip, as in exposing lower teeth
Mentalis	Mandible (chin)	Orbicularis oris	Elevates and protrudes lower lip, as in pouting
Platysma	Fascia of neck and clavicle	Inferior border of mandible	Depresses lower lip; tenses skin of neck
Buccinator	Maxilla and mandible	Orbicularis oris	Compresses cheek, as in sucking from a straw

8.2 List the muscles of mastication, along with their attachments and actions.

See fig. 8.2 and table 8.4.

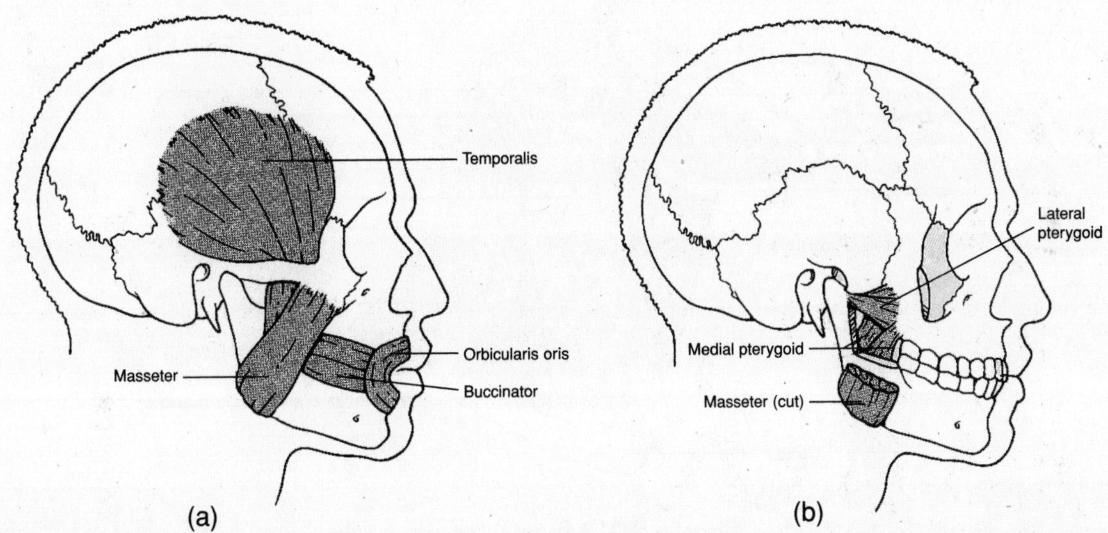

Figure 8.2 Muscles of mastication. (*a*) Superficial lateral view and (*b*) deep lateral view.

Table 8.4 Muscles of Mastication

Chewing muscle	***Origin(s)***	***Insertion(s)***	***Action(s)***
Temporalis	Temporal fossa	Coronoid process of mandible	Elevates jaw
Masseter	Zygomatic arch	Lateral ramus of mandible	Elevates jaw
Medial pterygoid	Sphenoid bone	Medial ramus of mandible	Depresses jaw; moves jaw laterally
Lateral pterygoid	Sphenoid bone and tuberosity of maxilla	Anterior side of mandibular condyle	Protracts jaw

8.3 List the muscles of neck movement, along with their attachments and actions.

See fig. 8.3 and table 8.5.

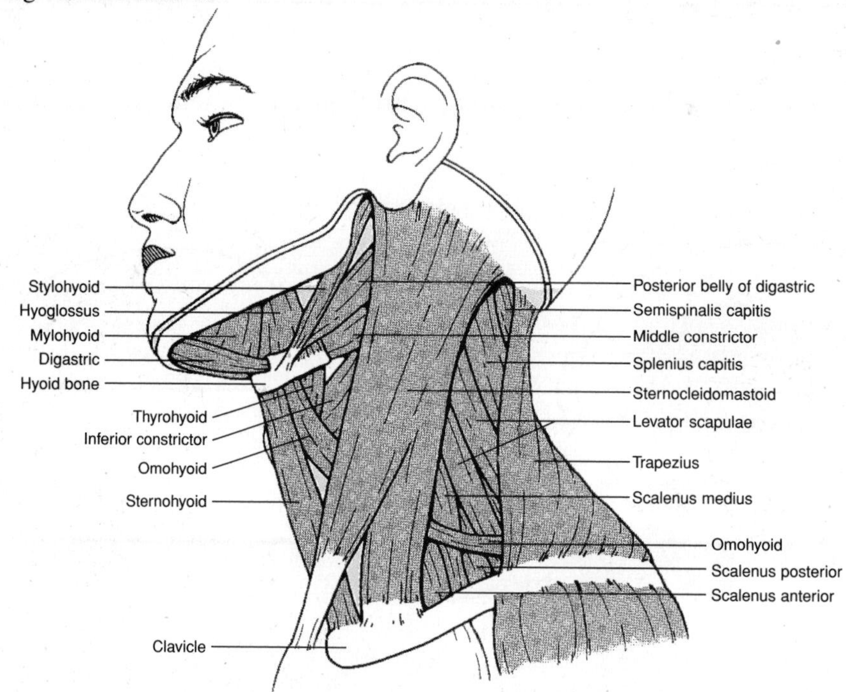

Figure 8.3 Muscles of the neck.

8.4 Describe the actions of the muscles involved in inhalation and the actions of those involved in exhalation.

During relaxed *inspiration* (inhalation), the important muscles are the **diaphragm**, the **external intercostal muscles**, and the **interchondral portion of the internal intercostal muscles** (fig. 8.4). A downward contraction of the dome-shaped diaphragm causes an increase in the vertical dimension of the thorax. A simultaneous contraction of the external intercostal muscles and the interchondral portion of the internal intercostal muscles produces an increase in the lateral dimension of the thorax. In addition, the **sternocleidomastoid** and **scalene muscles** may assist in inspiration through elevation of the first and second ribs, respectively.

Relaxed *expiration* (exhalation) is primarily a passive process, occurring as the muscles of

interosseous portion of the internal intercostal muscles contracts, causing the rib cage to be depressed. The **abdominal muscles** may also contract during forced expiration, which increases pressure within the abdominal cavity and forces the diaphragm superiorly, squeezing additional air out of the lungs.

Table 8.5 Muscles of the Neck

Neck muscle	*Origin(s)*	*Insertion(s)*	*Action(s)*
Sternocleidomastoid	Sternum and clavicle	Mastoid process of temporal bone	Flexes neck; turns head to side
Digastric	Inferior border of mandible and mastoid process of temporal bone	Hyoid bone	Depresses jaw to open the mouth; elevates hyoid bone
Mylohyoid	Inferior border of mandible	Hyoid bone and median raphe	Elevates hyoid bone and floor of mouth
Stylohyoid	Styloid process of temporal bone	Hyoid bone	Elevates and retracts tongue
Hyoglossus	Hyoid bone	Side of tongue	Depresses side of tongue
Sternohyoid	Manubrium	Hyoid bone	Depresses hyoid bone
Sternothyroid	Manubrium	Thyroid cartilage	Depresses thyroid cartilage
Thyrohyoid	Thyroid cartilage	Hyoid bone	Depresses hyoid bone; elevates thyroid cartilage
Omohyoid	Superior border of scapula	Clavicle and hyoid bone	Depresses hyoid bone

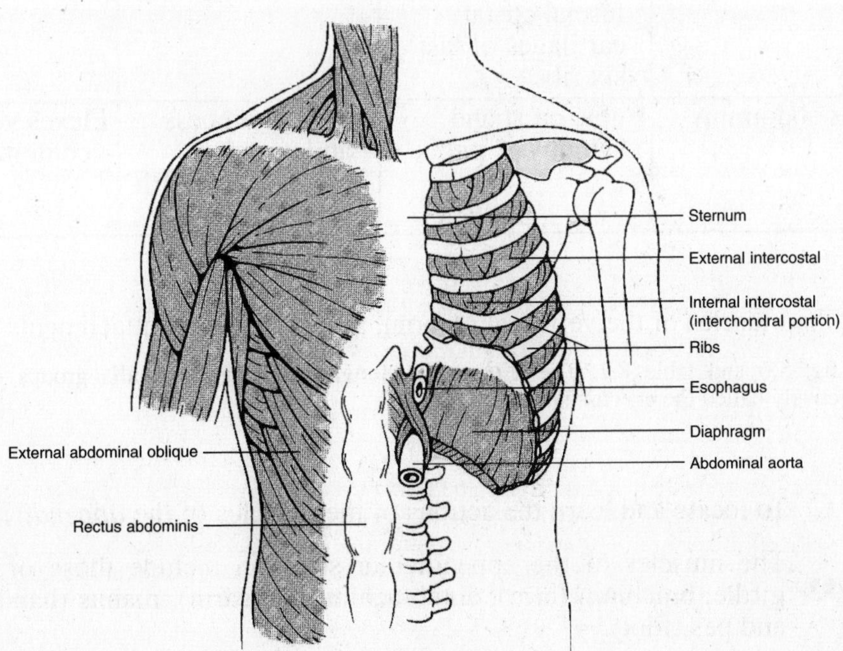

Figure 8.4 Muscles of respiration.

8.5 List the muscles of the abdominal wall, along with their attachments and actions.

See fig. 8.5 and table 8.6.

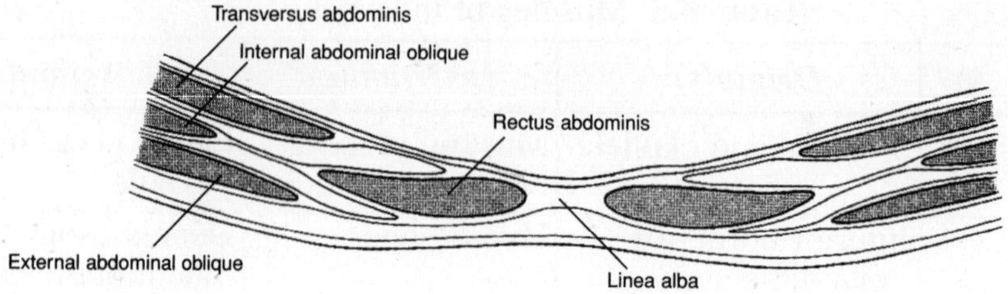

Figure 8.5 A cross section of the anterior abdominal wall.

Table 8.6 Muscles of the Abdominal Wall

Abdominal muscle	*Origin(s)*	*Insertion(s)*	*Action(s)*
External abdominal oblique	Lower eight ribs	Iliac crest and linea alba	Compresses abdomen; lateral rotation
Internal abdominal oblique	Iliac crest, inguinal ligament, lumbar fascia	Linea alba and costal cartilages of last three or four ribs	Compresses abdomen; lateral rotation
Transversus abdominis	Iliac crest, inguinal ligament, lumbar fascia, costal cartilages of last six ribs	Xiphoid process, linea alba, pubis	Compresses abdomen
Rectus abdominis	Pubic crest and symphysis pubis	Xiphoid process and costal cartilages of fifth to seventh ribs	Flexes vertebral column

8.6 List the muscles of the vertebral column, along with their attachments and actions.

See fig. 8.6 and table 8.7. The **iliocostalis**, **longissimus**, and **spinalis** groups of muscles are collectively called the **erector spinae** muscles.

Objective C To locate and learn the actions of the muscles of the *appendicular skeleton.*

The muscles of the appendicular skeleton include those of the pectoral girdle, brachium (arm), antebrachium (forearm), manus (hand), thigh, leg, and pes (foot).

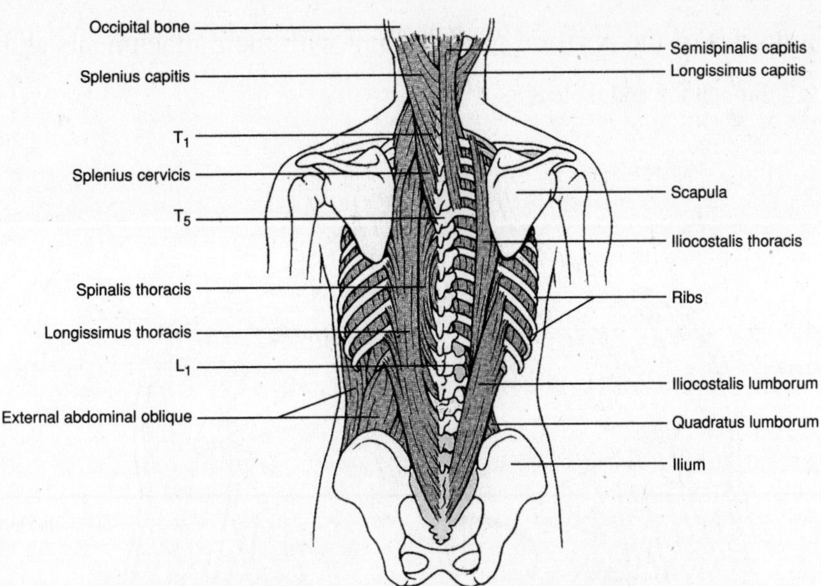

Figure 8.6 Posterior muscles of the vertebral column.

Table 8.7 Muscles of the Vertebral Column

Spinal muscle	*Origin(s)*	*Insertion(s)*	*Action(s)*
Quadratus lumborum	Iliac crest and lower three lumbar vertebrae	Twelfth rib and upper four lumbar	Extends lumbar region; laterally flexes vertebral column
Iliocostalis lumborum	Crest of ilium	Lower six ribs	Extends lumbar region
Iliocostalis thoracis	Lower six ribs	Upper six ribs	Extends thoracic region
Iliocostalis cervicis	Angles of three to six ribs	Transverse processes of fourth to sixth cervical vertebrae	Extends cervical region
Longissimus thoracis	Transverse processes of lumbar vertebrae	Lower nine ribs and transverse processes of all the thoracic vertebrae	Extends thoracic region
Longissimus cervicis	Transverse processes of upper five thoracic vertebrae	Transverse processes of second to sixth cervical vertebrae	Extends cervical region
Longissimus capitis	Transverse processes of upper four or five thoracic vertebrae	Transverse processes of second to sixth cervical vertebrae	Extends head; acting separately, turns face toward that side
Spinalis thoracis	Spinous processes of upper lumbar and lower thoracic vertebrae	Spinous processes of upper thoracic vertebrae	Extends vertebral column

8.7 List the muscles of the pectoral girdle, along with their attachments and actions.

See figs. 8.7 through 8.9 and table 8.8.

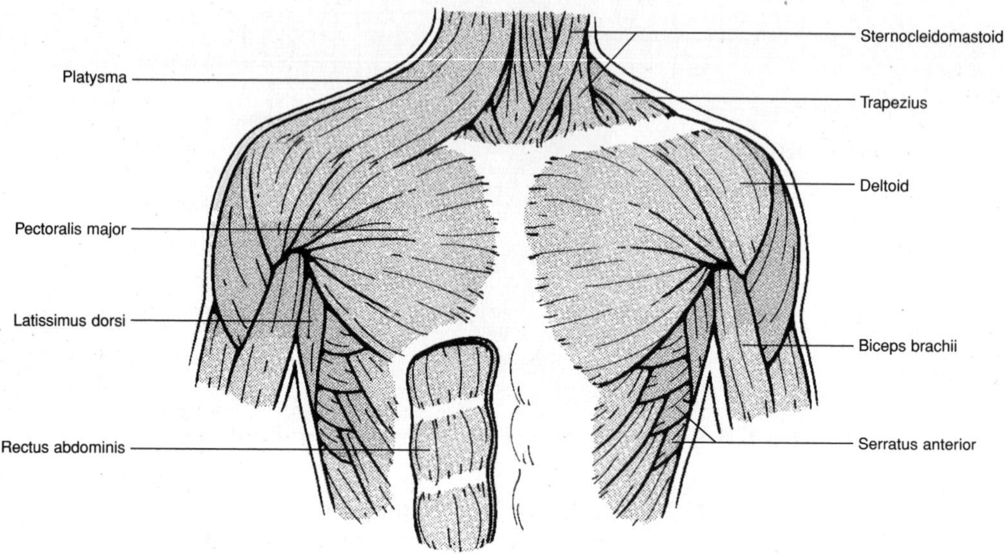

Figure 8.7 Anterior muscles of the thoracic and shoulder regions.

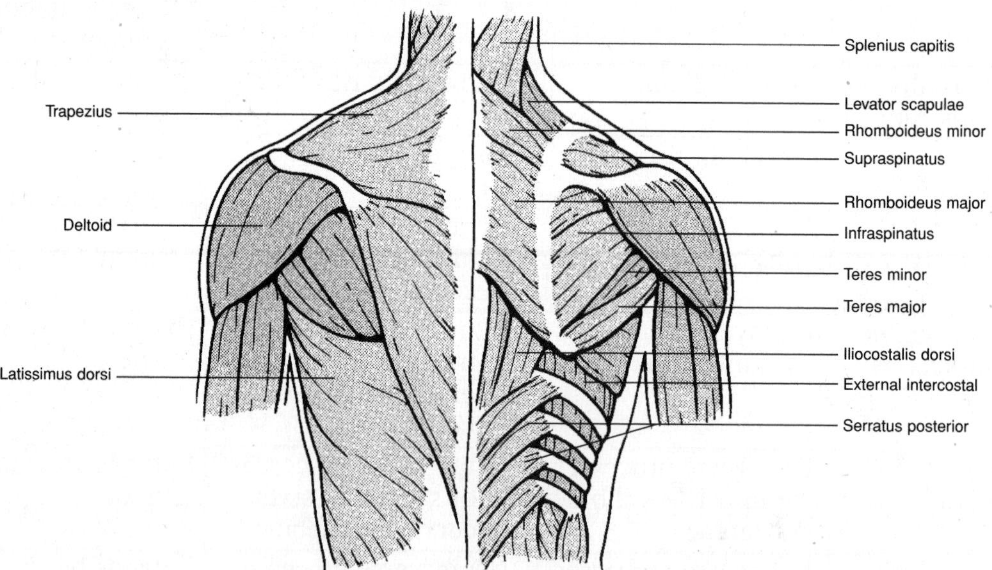

Figure 8.8 Posterior muscles of the thoracic and shoulder regions.

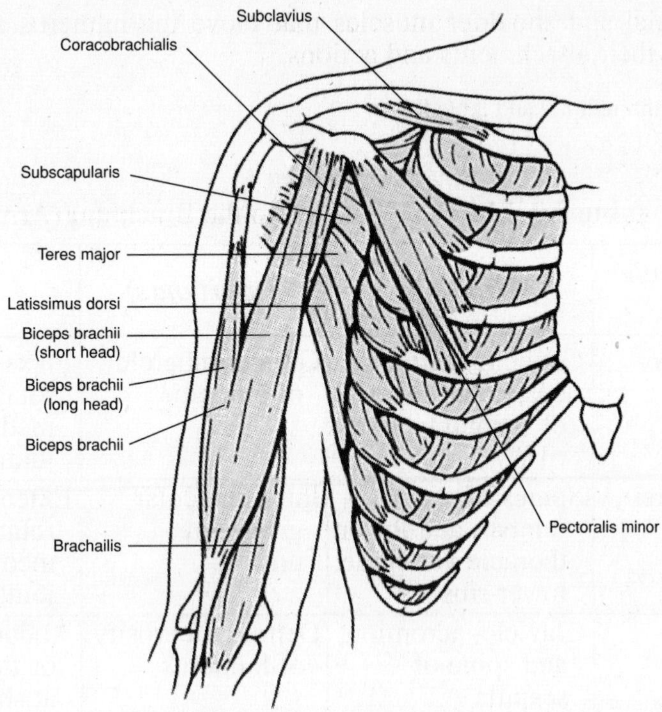

Figure 8.9 Deep anterior muscles of the thoracic and shoulder regions.

Table 8.8 Muscles That Act on the Pectoral Girdle

Pectoral muscle	*Origin(s)*	*Insertion(s)*	*Action(s)*
Serratus anterior	Upper eight or nine ribs	Anterior medial border of scapula	Pulls scapula forward and downward
Pectoralis minor	Sternal ends of third, fourth, and fifth ribs	Coracoid process of scapula	Pulls scapula forward and downward
Subclavius	First rib	Subclavian groove of clavicle	Draws clavicle downward
Trapezius	Occipital bone and spines of cervical and thoracic vertebrae	Clavicle, acromion, and spine of scapula	Elevates, depresses, and adducts scapula; hyperextends neck; braces shoulder
Levator scapulae	First to fourth cervical vertebrae	Superior border of scapula	Elevates scapula
Rhomboideus major	Spines of second to fifth thoracic vertebrae	Medial border of scapula	Elevates and adducts scapula
Rhomboideus minor	Seventh cervical and first thoracic vertebrae	Medial border of scapula	Elevates and adducts scapula

8.7 List the axial and shoulder muscles that move the humerus at the shoulder joint along with their attachments and actions.

See figs. 8.7 through 8.9 and table 8.9.

Table 8.9 Muscles That Act on the Brachium (Arm)

Axial or scapular muscle	*Origin(s)*	*Insertion(s)*	*Action(s)*
Pectoralis major	Clavicle, sternum, costal cartilages of second to sixth ribs	Greater tubercle of humerus	Flexes, adducts, and rotates humerus medially at shoulder joint
Latissimus dorsi	Spines of sacral, lumbar, and lower thoracic vertebrae; lower ribs	Intertubercular groove of umerus	Extends, adducts, and rotates humerus medially at shoulder joint; adducts arm
Deltoid	Clavicle, acromion, and spine of scapula	Deltoid tuberosity of humerus	Abducts arm; extends or flexes humerus at shoulder joint
Supraspinatus	Supraspinous fossa of scapula	Greater tubercle of humerus	Abducts and laterally rotates humerus at shoulder joint
Infraspinatus	Infraspinous fossa of scapula	Greater tubercle of humerus	Rotates arm laterally at shoulder joint
Teres major	Inferior angle and lateral border of scapula	Intertubercular groove of humerus	Extends, adducts, and rotates humerus medially at shoulder joint
Teres minor	Lateral border of scapula	Greater tubercle of humerus	Rotates humerus laterally at shoulder joint
Subscapularis	Subscapular fossa	Lesser tubercle of humerus	Rotates humerus medially at shoulder joint
Coracobrachialis	Coracoid process of scapula	Shaft of humerus	Flexes and adducts humerus at shoulder joint

8.8 List the muscles that act on the antebrachium (forearm) at the elbow joint, along with their attachments and actions.

See figs. 8.10 and 8.11 and table 8.10.

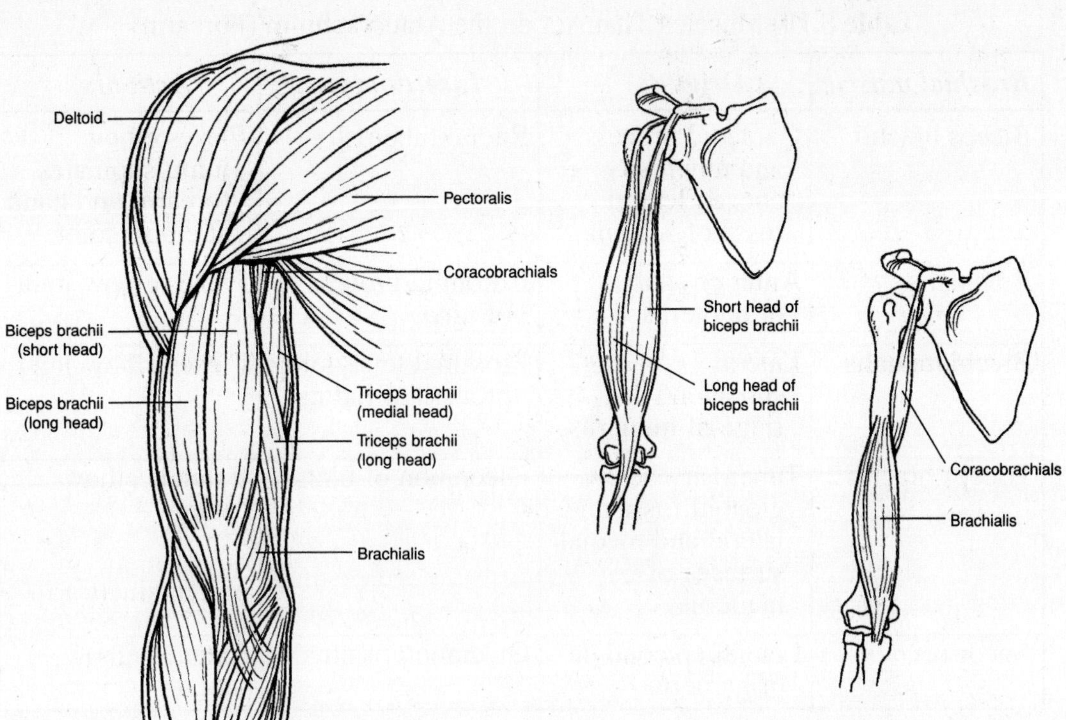

Figure 8.10 Anterior brachial muscles.

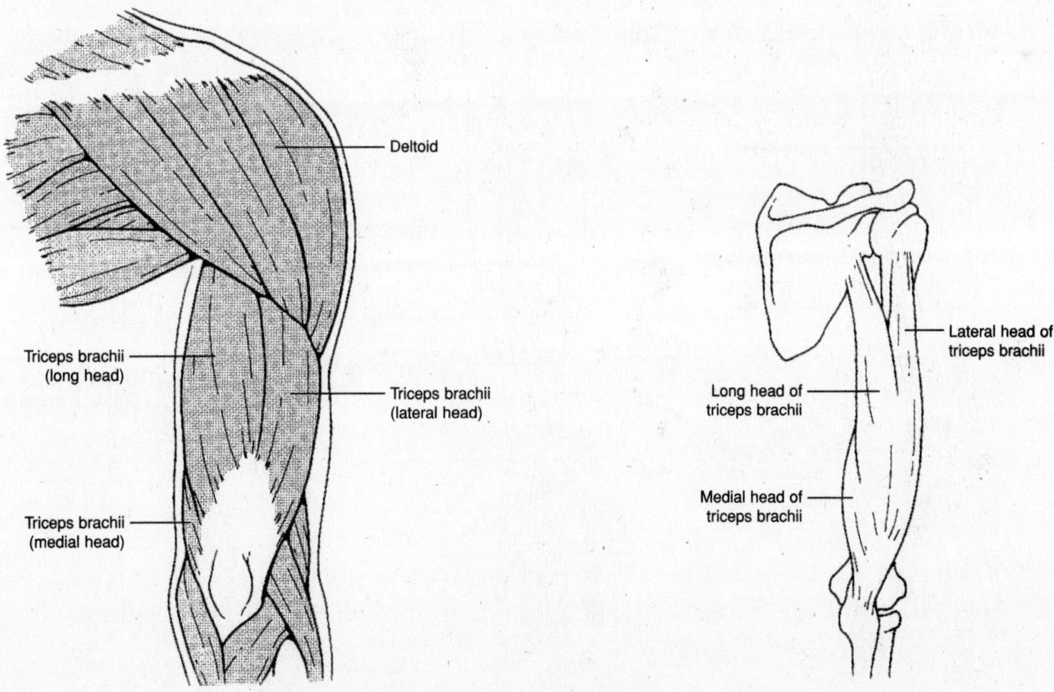

Figure 8.11 Posterior brachial muscles.

Table 8.10 Muscles That Act on the Antebrachium (Forearm)

Brachial muscle	*Origin(s)*	*Insertion(s)*	*Action(s)*
Biceps brachii	Coracoid process and tuberosity above glenoid fossa of scapula	Radial tuberosity	Flexes elbow joint; supinates forearm and hand at elbow joint
Brachialis	Anterior shaft of humerus	Coronoid process of ulna	Flexes elbow joint
Brachioradialis	Lateral supracondylar ridge of humerus	Proximal to styloid process of radius	Flexes elbow joint
Triceps brachii	Tuberosity below glenoid fossa and lateral and medial surfaces of humerus	Olecranon of ulna	Extends elbow joint
Anconeus	Lateral epicondyle of humerus	Olecranon of ulna	Extends elbow joint

8.9 List the muscles that act on the wrist, hand, and fingers, along with their attachments and actions.

See figs. 8.12 and 8.13 and table 8.11.

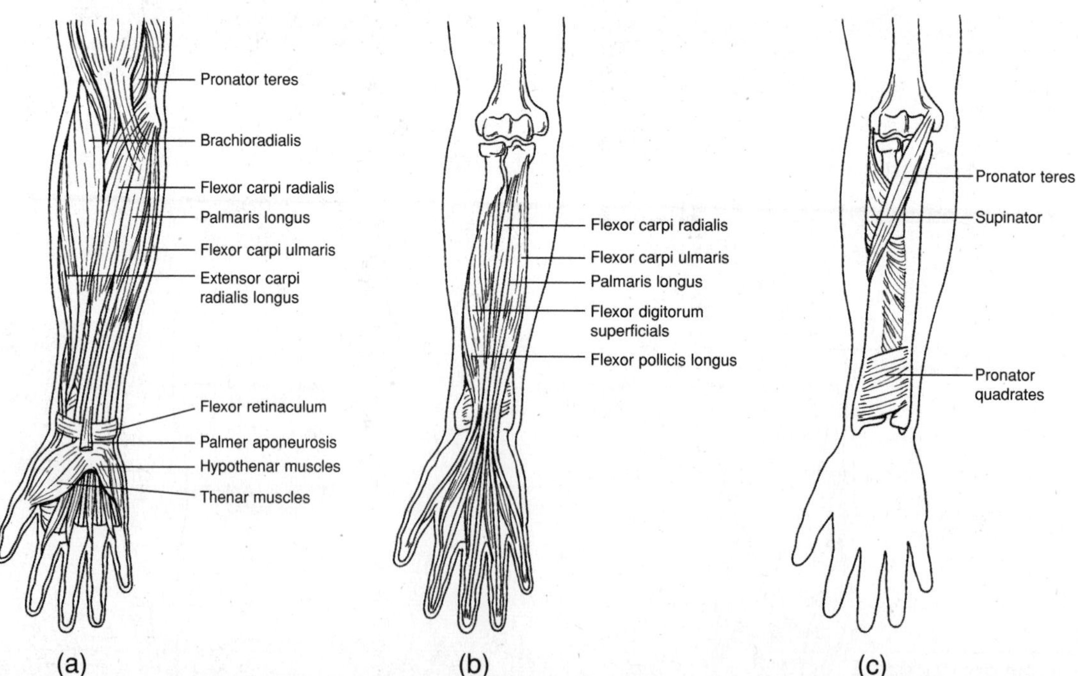

Figure 8.12 Anterior antebrachial (forearm) muscles that act on the wrist, hand, and fingers. (*a*) Superficial muscles, (*b*) deep muscles, and (*c*) deep rotator muscles.

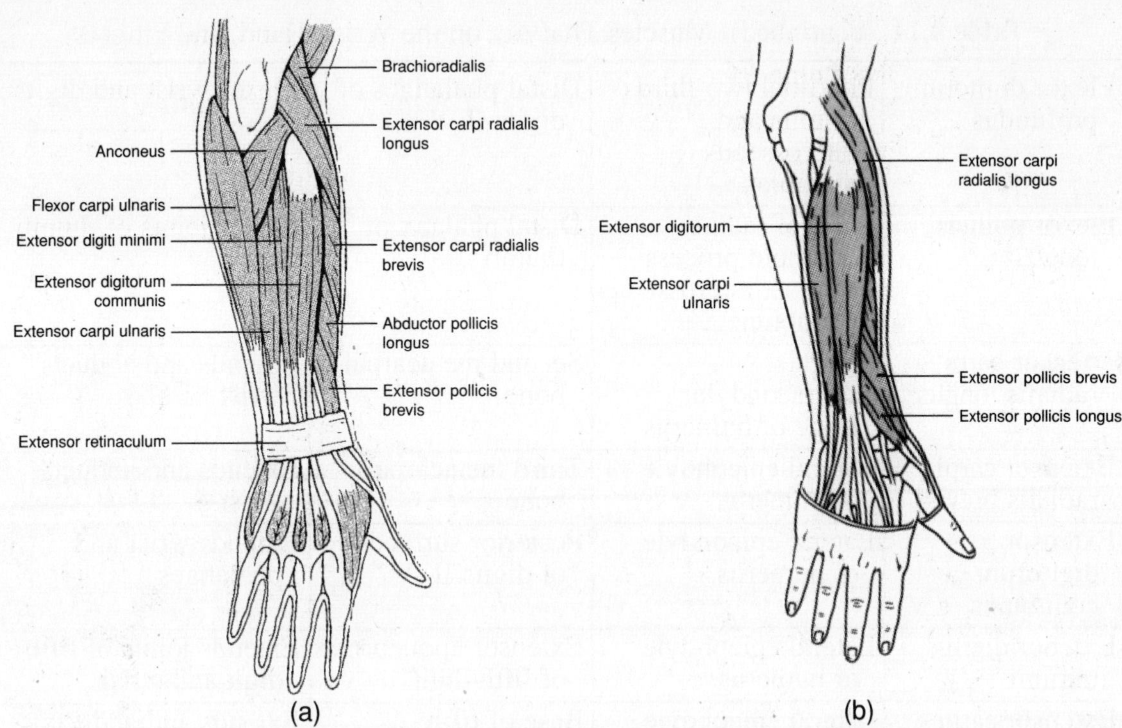

Figure 8.13 Posterior antebrachial (forearm) muscles that act on the wrist, hand, and fingers. (*a*) Superficial muscles and (*b*) deep muscles

Table 8.11 Muscles That Act on the Wrist, Hand, and Fingers

Antebrachial muscle	*Origin(s)*	*Insertion(s)*	*Action(s)*
Supinator	Lateral epicondyle of humerus and crest of ulna	Lateral surface of radius	Supinates hand
Pronator teres	Medial epicondyle of humerus	Lateral surface of radius	Pronates hand
Pronator quadratus	Distal fourth of ulna	Distal fourth of radius	Pronates hand
Flexor carpi radialis	Medial epicondyle of humerus	Base of second and third metacarpal bones	Flexes and abducts hand at wrist
Palmaris longus	Medial epicondyle of humerus	Palmar aponeurosis	Flexes wrist
Flexor carpi ulnaris	Medial epicondyle of humerus and olecranon of ulna	Carpal and metacarpal bones	Flexes and adducts wrist
Flexor digitorum superficialis	Medial epicondyle of humerus and coronoid process	Middle phalanges of digits II–V	Flexes wrist and digits

Table 8.11 (continued) Muscles That Act on the Wrist, Hand, and Fingers

Flexor digitorum profundus	Proximal two-thirds of ulna and interosseous membrane	Distal phalanges of digits II–V	Flexes wrist and digits
Flexor pollicis longus	Shaft of radius coronoid process of ulna, interosseous membrane	Distal phalanx of thumb	Flexes joints of thumb
Extensor carpi radialis longus	Lateral supracondylar ridge of humerus	Second metacarpal bone	Extends and abducts wrist
Extensor carpi radialis brevis	Lateral epicondyle of humerus	Third metacarpal bone	Extends and abducts wrist
Extensor digitorum communis	Lateral epicondyle of humerus	Posterior surfaces of digits II–V	Extends wrist and phalanges
Extenor digiti minimi	Lateral epicondyle of humerus	Extensor aponeurosis of fifth digit	Extends joints of fifth digit and wrist
Extensor carpi ulnaris	Lateral epicondyle of humerus and olecranon of ulna	Base of fifth metacarpal bone	Extends and adducts wrist
Extensor pollicis longus	Lateromedial shaft of ulna	Base of distal phalanx of thumb	Extends joints of thumb; abducts joints of hand
Extensor pollicis brevis	Distal shaft of radius and interosseous membrane	Base of first phalanx of thumb	Extends joints of thumb; abducts joints of hand
Abductor pollicis longus	Distal radius and ulna and interosseous membrane	Base of first metacarpal bone	Abducts joints of thumb and joints of hand

8.10 List the anterior and posterior muscles that move the thigh at the hip joint, along with their attachments and actions.

See fig. 8.14 and table 8.12.

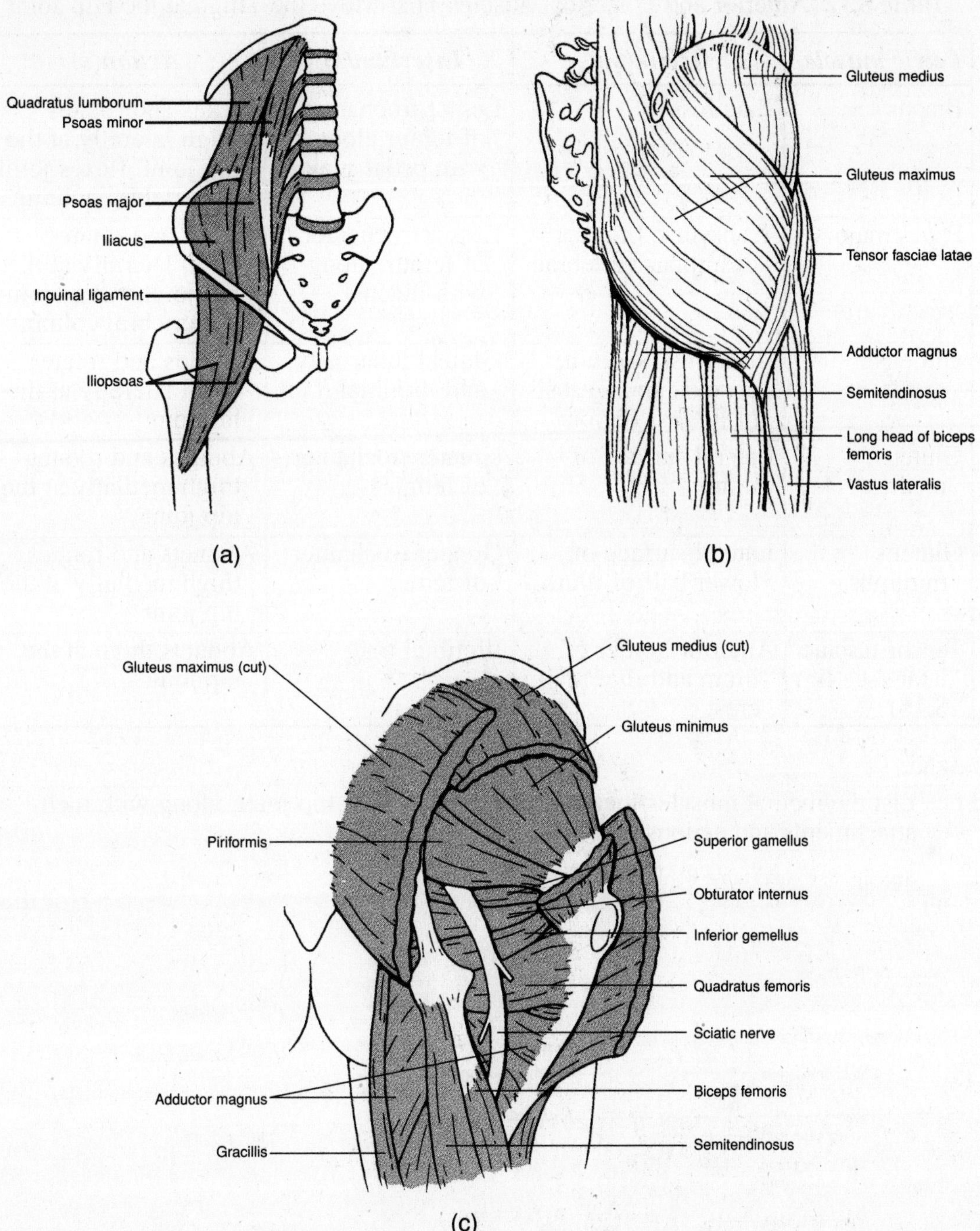

Figure 8.14 Muscles that move the thigh at the hip joint. (*a*) Anterior pelvic muscles, (*b*) superficial gluteal muscles, and (*c*) deep gluteal muscles.

Table 8.12 Anterior and Posterior Muscles That Move the Thigh at the Hip Joint

Pelvic muscle	*Origin(s)*	*Insertion(s)*	*Action(s)*
Iliacus	Iliac fossa	Lesser trochanter of femur along with psoas major	Flexes and rotates thigh laterally at the hip joint; flexes joints of vertebral column
Psoas major	Transverse process of lumbar vertebrae	Lesser trochanter of femur, along with iliacus	Flexes and rotates thigh laterally at the hip joint; flexes joints of vertebral column
Gluteus maximus	Iliac crest, sacrum, coccyx, aponeurosis of lumbar region	Gluteal tuberosity and iliotibial tract	Extends and rotates thigh laterally at the hip joint
Gluteus medius	Lateral surface of ilium	Greater trochanter of femur	Abducts and rotates thigh medially at the hip joint
Gluteus minimus	Lateral surface of lower half of ilium	Greater trochanter of femur	Abducts and rotates thigh medially at the hip joint
Tensor fasciae latae (see fig. 8.16)	Anterior border of ilium and iliac crest	Iliotibial tract	Abducts thigh at the hip joint

8.11 List the medial muscles that move the thigh at the hip joint, along with their attachments and actions.

See fig. 8.15 and table 8.13.

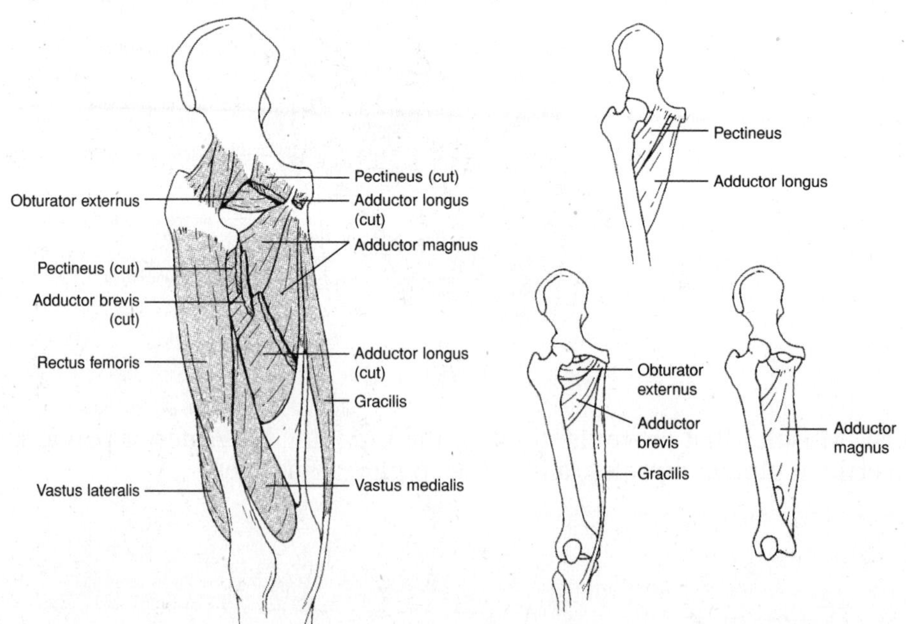

Figure 8.15 Medial (adductor) muscles that move the thigh at the hip joint.

Table 8.13 Medial Muscles that Move the Thigh at the Hip Joint

Adductor muscle	*Origin(s)*	*Insertion(s)*	*Action(s)*
Gracilis	Inferior edge of symphysis pubis	Proximomedial surface of tibia	Adducts thigh at hip joint; flexes and rotates leg at knee joint
Pectineus	Pectineal line of pubis	Distal to lesser trochanter of femur	Adducts and flexes thigh at hip joint
Adductor longus	Pubis, below pubic crest	Linea aspera of femur	Adducts, flexes, and laterally rotates thigh at hip joint
Adductor brevis	Inferior ramus of pubis	Linea aspera of femur	Adducts, flexes, and laterally rotates thigh at hip joint
Adductor magnus	Inferior ramus of ischium and inferior ramus of pubis	Linea aspera and medial epicondyle of femur	Adducts, flexes, and laterally rotates thigh at hip joint

8.12 List the muscles of the thigh that move the leg, along with their attachments and actions.

See fig. 8.16 and table 8.14.

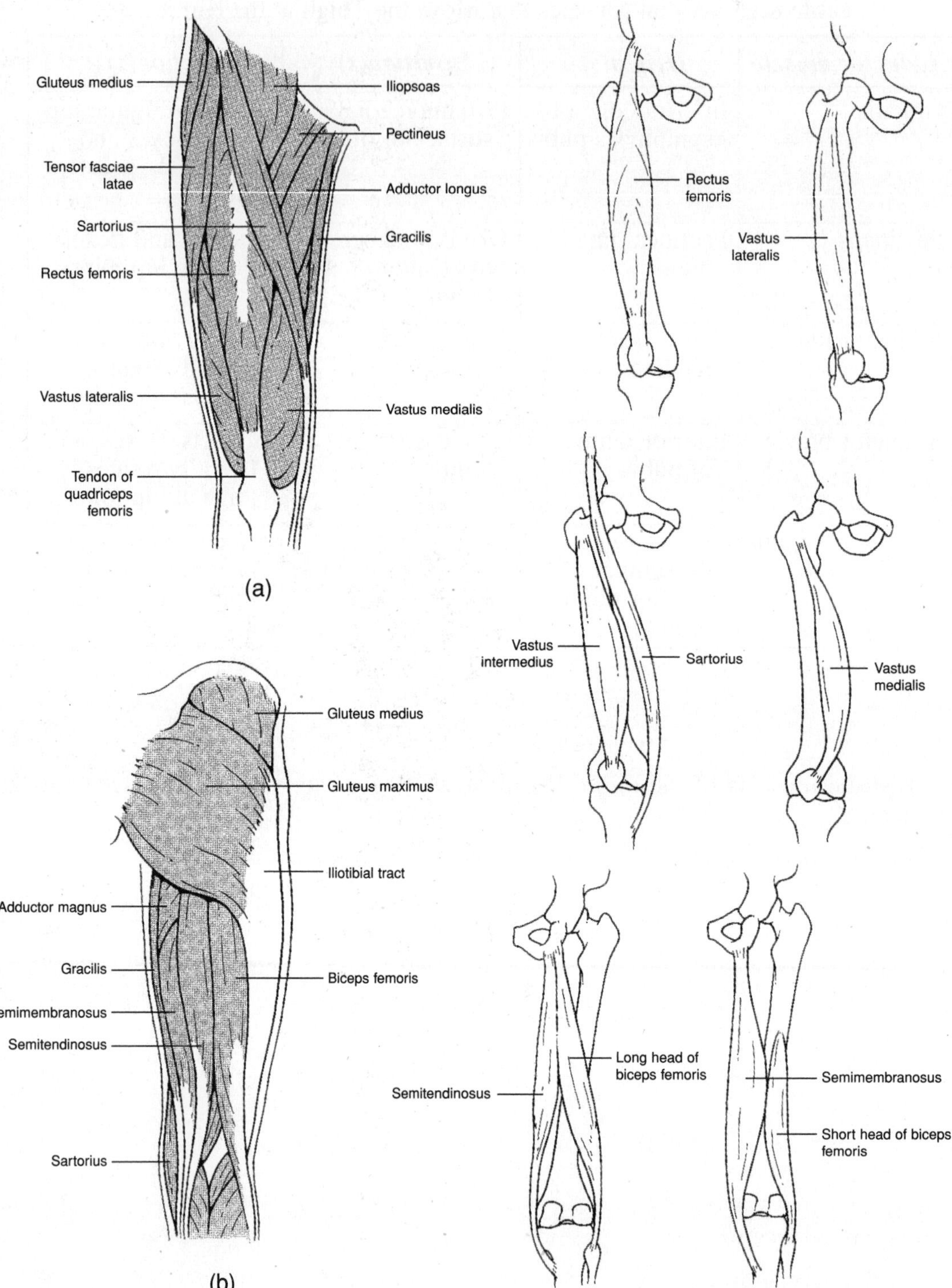

Figure 8.16 Muscles of the thigh that act on the leg. (*a*) Anterior thigh region and (*b*) posterior thigh region.

Table 8.14 Muscles of the Thigh That Act on the Leg

Thigh muscle	*Origin(s)*	*Insertion(s)*	*Action(s)*
Sartorius	Anterior superior iliac spine	Medial surface of tibia	Flexes leg and thigh; abducts and rotates thigh laterally; rotates leg medially at hip joint
Quadriceps femoris		Patella by common tendon, which continues as patellar ligament to tibial tuberosity	Extends leg at knee joint
Rectus femoris	Anterior inferior iliac spine		
Vastus lateralis	Greater trochanter and linea aspera of femur		
Vastus medialis	Medial surface and linea aspera of femur		
Vastus intermedius	Anterior and lateral surfaces of femur	Flexes leg at knee joint; extends and laterally rotates thigh at hip joint	
Biceps femoris	Long head—ischial tuberosity; short head—linea aspera of femur	Head of fibula and lateral epicondyle of tibia	
Semitendinosus	Ischial tuberosity	Proximal portion of medial surface of shaft of tibia	Flexes leg at knee joint; extends and medially rotates thigh at hip joint
Semimembranosus	Ischial tuberosity	Medial epicondyle of tibia	Flexes leg at knee joint; extends and medially rotates thigh at hip joint

8.13 List the muscles of the leg that move the ankle, foot, and toes, along with their attachments and actions.

See figs. 8.17 and 8.18 and table 8.15.

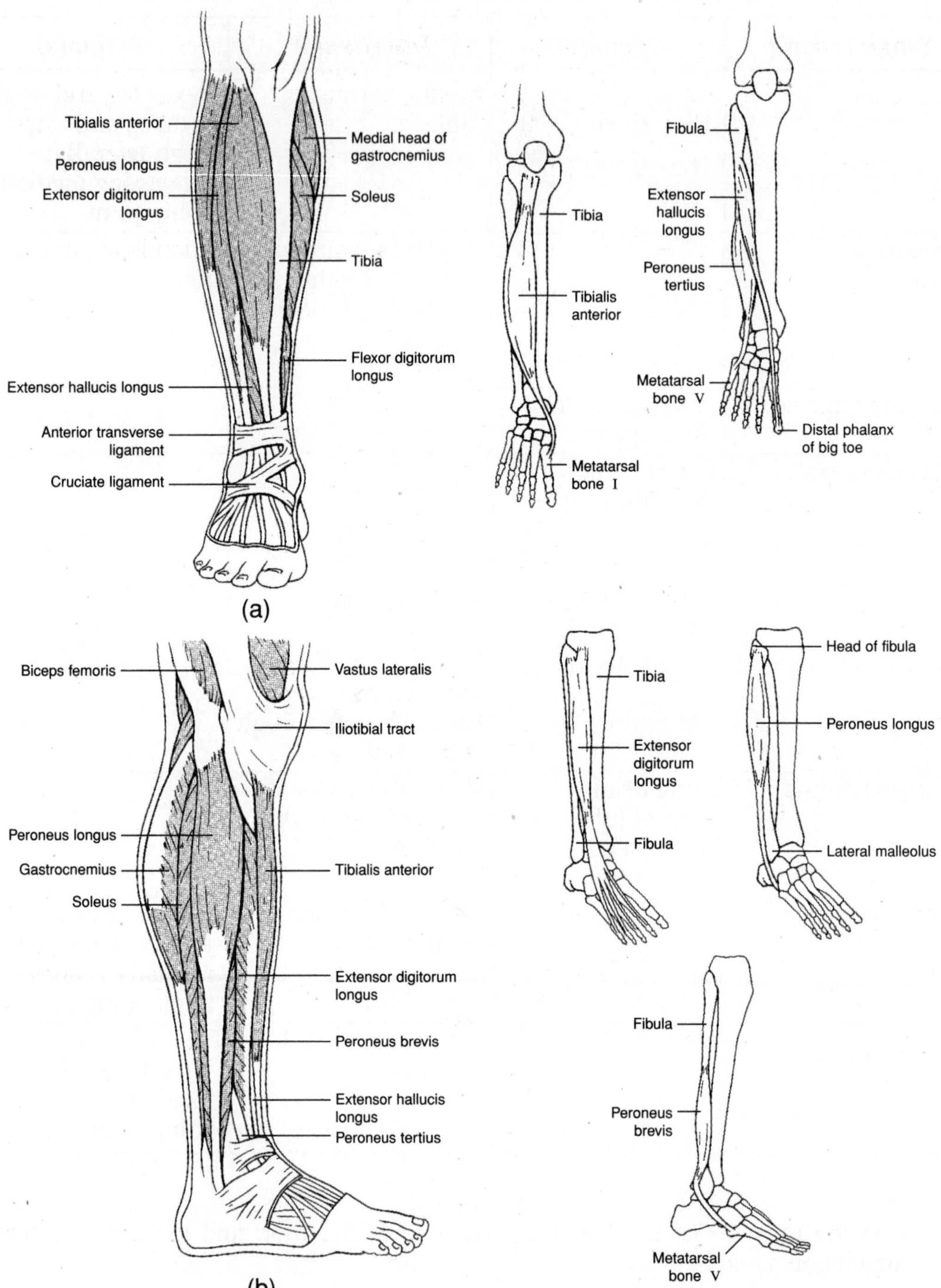

Figure 8.17 Muscles of the leg that move the ankle, foot, and toes. (*a*) Anterior leg region and (*b*) lateral leg region.

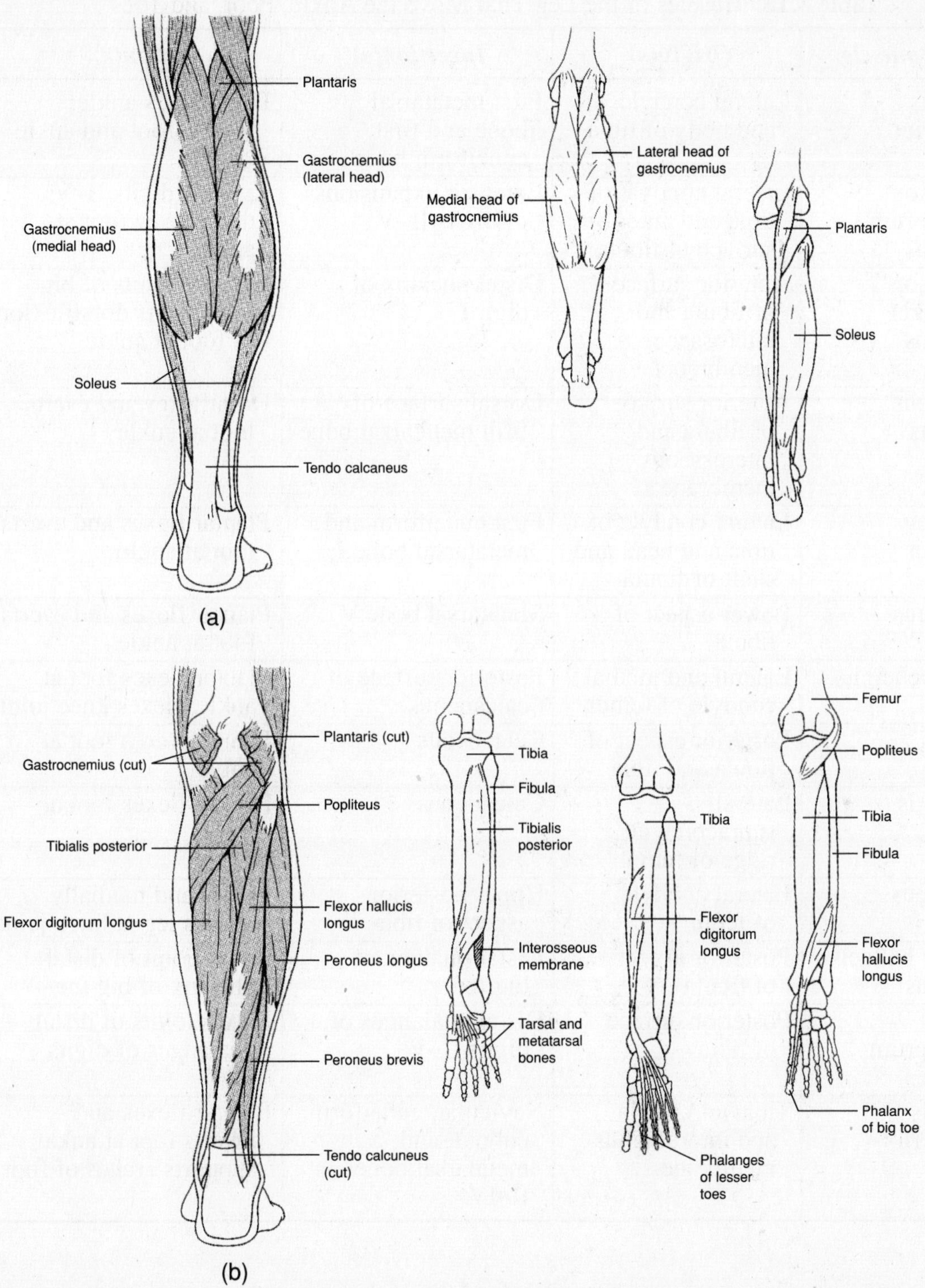

Figure 8.18 Posterior muscles of the leg that move the ankle, foot, and toes. (*a*) Superficial muscles and (*b*) deep muscles.

Table 8.15 Muscles of the Leg That Move the Ankle, Foot, and Toes

Leg muscle	*Origin(s)*	*Insertion(s)*	*Action(s)*
Tibialis anterior	Lateral condyle and body of tibia	First metatarsal bone and first cuneiform bone	Dorsiflexes ankle; inverts foot and ankle
Extensor digitorum longus	Lateral condyle of tibia and anterior surface of fibula	Extensor expansions of digits II–V	Extends digits II–V; dorsiflexes foot at ankle
Extensor hallucis longus	Anterior surface of fibula and interosseous membrane	Distal phalanx of digit I	Extends joints of big toe; assists dorsiflexion of foot at ankle
Peroneus tertius	Anterior surface of fibula and interosseous membrane	Dorsal surface of fifth metatarsal bone	Dorsiflexes and everts foot at ankle
Peroneus longus	Lateral condyle of tibia and head and shaft of fibula	First cuneiform and metatarsal bone I	Plantar flexes and everts foot at ankle
Peroneus brevis	Lower aspect of fibula	Metatarsal bone V	Plantar flexes and everts foot at ankle
Gastrocnemius	Lateral and medial condyle of femur	Posterior surface of calcaneous	Plantar flexes foot at ankle; flexes knee joint
Soleus	Posterior aspect of fibula and tibia	Calcaneous	Plantar flexes foot at ankle
Plantaris	Lateral supracondylar ridge of femur	Calcaneous	Plantar flexes foot at ankle
Popliteus	Lateral condyle of femur	Upper posterior aspect of tibia	Flexes and medially rotates leg at knee joint
Flexor hallucis longus	Posterior aspect of fibula	Distal phalanx of big toe	Flexes joint of distal phalanx of big toe
Flexor digitorum longus	Posterior surface of tibia	Distal phalanges of digits II–V	Flexes joints of distal phalanges of digits II–V
Tibialis posterior	Tibia and fibula and interosseous membrane	Navicular, cuneiform, cuboid, and metatarsal bones II–IV	Plantar flexes and inverts foot at ankle; supports arches of foot

Key Clinical Terms

Charley horse A cramp or stiffness in a muscle, especially in the back of the thigh, as a result of a sprain, tear, or bruise of the muscle.

Cramp A sustained spasmodic contraction of a muscle, usually accompanied by severe localized pain.

Fibromyositis An inflammation of both skeletal muscle tissue and the associated connective tissue. Lumbago, or rheumatism, is fibromyositis in the lumbar area of the back.

Graphospasm Writer's cramp.

Hernia Rupture, or protrusion through muscle tissue, of a portion of the underlying viscera. The most common hernias are the *femoral* (viscera passing through the femoral ring), the *inguinal* (viscera protruding through the inguinal canal), the *umbilical* (viscera protruding through the navel), and the *hiatal* (superior portion of the stomach protruding through the diaphragm).

Intramuscular injection Hypodermic injection at a heavily muscled area (most commonly the buttock), so that nerves will not be damaged.

Muscular atrophy A decrease in the size of a muscle that was previously fully developed, possibly as a result of disease, disuse, infections, nutritional problems, or aging.

Muscular dystrophy A genetic abnormality of muscle tissue, characterized by dysfunction and, ultimately, deterioration.

Myasthenia gravis Thought to be an autoimmune disease, myasthenia gravis is characterized by extreme muscle weakness and low endurance. There is a defective transmission of impulses at the neuromuscular junction.

Myopathy Any disease of the muscles.

Poliomyelitis A viral disease that often attacks and destroys the cell bodies of the somatic motor neurons of the skeletal muscles, causing paralysis.

Shin splints Tenderness and pain on the anterior surface of the leg, caused by a strain of the anterior tibial muscle or the extensor digitorum longus muscle.

Tetanus *(lockjaw)* A disease caused by the bacterium *Clostridium tetani*, which produces a toxin that causes painful muscle spasms. The jaw muscles are affected first.

Torticollis *(wryneck)* Persistent contraction of a sternocleidomastoid muscle, drawing the head to one side and distorting the face. Torticollis may be acquired or congenital.

Review Exercises

Multiple Choice

1. A flexor muscle of the shoulder joint is (*a*) the supraspinatus, (*b*) the trapezius, (*c*) the pectoralis major, (*d*) the teres major.

2. Which of the following muscles does *not* attach to the humerus? (*a*) teres major, (*b*) supraspinatus, (*c*) biceps brachii, (*d*) brachialis, (*e*) pectoralis major

3. Which of the following muscles does *not* insert upon the orbicular oris? (*a*) depressor labii inferioris, (*b*) zygomaticus, (*c*) risorius, (*d*) platysma, (*e*) levator labii superioris

4. The erector spinae muscle group does *not* include (*a*) the iliocostalis, (*b*) the longissimus, (*c*) the spinalis, (*d*) the semispinalis.

5. All of the following muscles are synergists in flexing the elbow joint *except* (*a*) the biceps brachii, (*b*) the brachialis, (*c*) the coracobrachialis, (*d*) the brachioradialis.

6. Which of the following muscles does *not* attach to the scapula? (*a*) deltoid, (*b*) latissimus dorsi, (*c*) coracobrachialis, (*d*) teres major, (*e*) rhomboideus major

7. Which of the following muscles attaches to the acromion of the scapula? (*a*) teres major, (*b*) deltoid, (*c*) supraspinatus, (*d*) rhomboideus major, (*e*) infraspinatus

8. Of the four quadriceps femoris muscles, which contracts over the hip and knee joints? (*a*) rectus femoris, (*b*) vastus medialis, (*c*) vastus intermedius, (*d*) vastus lateralis

9. Which of the following muscles plantar flexes and inverts the foot as it supports the arches? (*a*) flexor digitorum longus, (*b*) tibialis posterior, (*c*) flexor hallucis longus, (*d*) gastrocnemius

10. An eyebrow is drawn toward the midline of the face through contraction of which of the following muscles? (*a*) corrugator, (*b*) risorius, (*c*) nasalis, (*d*) frontalis

11. Which of the following is *not* used as a means of naming muscles? (a) location, (*b*) action, (*c*) shape, (*d*) attachment, (*e*) strength of contraction

12. Rotation of the hand so that the palm faces posteriorly is the action of which muscles? (*a*) supinators, (*b*) abductors, (*c*) adductors, (*d*) flexors, (*e*) extensors

13. The muscles that are synergistic to the diaphragm during inspiration are
(*a*) the external intercostal muscles
(*b*) the internal intercostal muscles (excluding the interchondral part)
(*c*) the abdominal muscles
(*d*) all of the above

14. A muscle of mastication is (*a*) the buccinator, (*b*) the temporalis, (*c*) the mentalis, (*d*) the zygomaticus, (*e*) the orbicularis oris.

15. Which of the following muscles does *not* originate on the lateral epicondyle of the humerus? (*a*) extensor carpi radialis brevis, (*b*) extensor digitorum, (*c*) extensor digiti minimi, (*d*) all of the preceding originate on the lateral epicondyle

16. The muscle that extends and laterally rotates the thigh is (*a*) the iliacus, (*b*) the gluteus medius, (*c*) the psoas major, (*d*) the gluteus maximus, (*e*) the gluteus minimus.

17. Which of the following muscles does *not* attach to the rib cage? (*a*) serratus anterior, (*b*) rectus abdominis, (*c*) pectoralis major, (*d*) serratus posterior, (*e*) latissimus dorsi

18. Which of the following muscles does *not* have its origin on the pubis? (*a*) gracilis, (*b*) adductor brevis, (*c*) pectineus, (*d*) sartorius

19. The gluteus minimus muscle originates on which bone? (*a*) coccyx, (*b*) ischium, (*c*) femur, (*d*) ilium, (*e*) pubis

20. Which of the following muscles is deep (beneath another muscle) in position? (*a*) platysma, (*b*) pectoralis major, (*c*) tensor fascia latae, (*d*) external abdominal oblique, (*e*) rhomboideus major

True or False

____ **1.** When contracted, the zygomaticus draws the angle of the mouth upward, as in a smile.

____ **2.** Contraction of the orbicularis oris compresses the lips together.

____ **3.** *Extension* and *abduction* are interchangeable terms in that both actions result in an appendage's being moved away from the body.

____ **4.** The digastric muscles are important in chewing because, when contracted, they lower the mandible and open the mouth.

____ **5.** Flexion of the vertebral column results when the iliocostalis muscles are contracted.

____ **6.** The triceps brachii originates from processes on the humerus and on the scapula.

____ **7.** When contracted, the semimembranosus flexes the leg at the knee joint and may also extend the thigh at the hip joint.

____ **8.** A pulled groin muscle could involve the gracilis.

____ **9.** The sartorius acts only on the hip joint.

____ **10.** The muscles of the quadriceps femoris are antagonistic to the hamstring muscles.

____ **11.** All three gluteal muscles insert on the greater trochanter of the femur.

____ **12.** When contracted, the pectoralis minor rotates and adducts the humerus.

____ **13.** Three muscles—the gastrocnemius, soleus, and plantaris—function synergistically in plantar flexion of the foot.

____ **14.** The palmaris longus is anterior in position and functions to flex the hand.

____ **15.** From superficial to deep, the anterior abdominal wall consists of the external abdominal oblique, internal abdominal oblique, and transverse abdominis muscles.

Completion

1. The ________________________ is synergistic to the temporalis muscle in closing the mouth.

2. Improper lifting may strain the ____________________ ____________________ complex of muscles of the vertebral column.

3. The ________________________ is a prominent muscle along the lateral surface of the forearm, where it flexes the elbow joint when contracted.

4. The ____________________ ____________________ muscle group is on the anterior thigh, and the ____________________ muscle group is on the posterior thigh.

5. The ______________________ is the largest of the posterior crural muscles.

6. The three most important muscles of relaxed inspiration are the dome-shaped ______________________, the ______________________ ______________________ muscles, and the interchondral portion of the ______________________ ______________________ muscles.

7. The ______________________ is a muscle that extends from the anterior superior iliac spine to the medial surface of the proximal portion of the tibia.

8. Of the three gluteal muscles, the ______________________ ______________________ is the one that extends and rotates the thigh laterally at the hip joint.

9. The iliacus and psoas major merge toward their point of insertion to form the ______________________.

10. The ______________________ muscle can open the mouth or elevate the hyoid bone.

Labeling

Label the muscles indicated on the figures that follow.

1.	______________________	**16.**	______________________
2.	______________________	**17.**	______________________
3.	______________________	**18.**	______________________
4.	______________________	**19.**	______________________
5.	______________________	**20.**	______________________
6.	______________________	**21.**	______________________
7.	______________________	**22.**	______________________
8.	______________________	**23.**	______________________
9.	______________________	**24.**	______________________
10.	______________________	**25.**	______________________
11.	______________________	**26.**	______________________
12.	______________________	**27.**	______________________
13.	______________________	**28.**	______________________
14.	______________________	**29.**	______________________
15.	______________________	**30.**	______________________

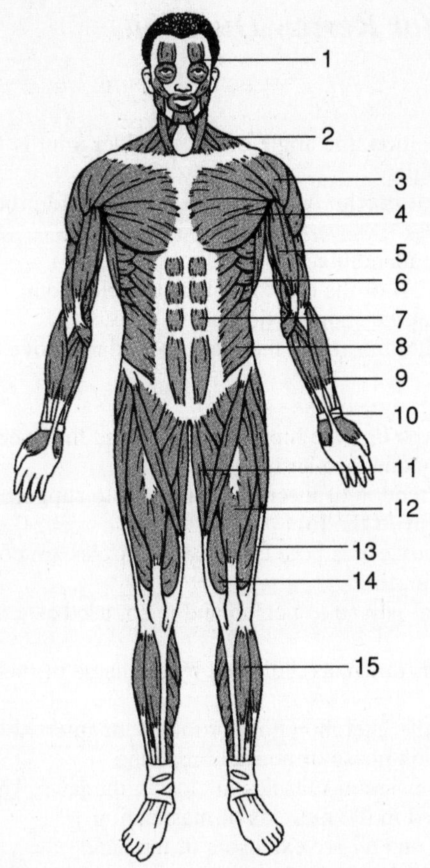

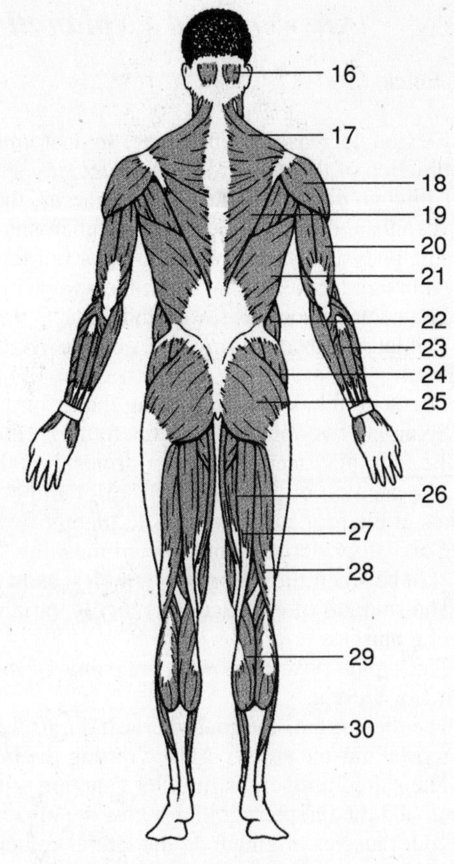

Matching

Match each of the following muscles with its action.

____	**1.** Deltoid	(*a*)	flexes and adducts the shoulder joint
____	**2.** Psoas major	(*b*)	flexes the joints of the vertebral column
____	**3.** Gracilis	(*c*)	extends the knee joint
____	**4.** Trapezius	(*d*)	adducts the scapula
____	**5.** Vastus lateralis	(*e*)	flexes the hip joint and joints of the vertebral column
____	**6.** Rectus abdominis	(*f*)	adducts and extends the shoulder joint
____	**7.** Semimembranosus	(*g*)	adducts the hip joint
____	**8.** Quadratus lumborum	(*h*)	abducts the shoulder joint
____	**9.** Latissimus dorsi	(*i*)	flexes the elbow joint
____	**10.** Coracobrachialis	(*j*)	extends the joints of the lumbar region of the vertebral column
____	**11.** Gluteus medius	(*k*)	flexes the knee joint
____	**12.** Brachialis	(*l*)	abducts and medially rotates the hip joint

Answers and Explanations for Review Questions

Multiple Choice

1. (*c*) Flexion decreases a joint angle. In anatomical position, the angle of the shoulder joint is 180°. Contraction of the pectoralis major decreases this angle.
2. (*c*) Although positioned along the humerus, the biceps brachii originates on the coracoid process of the scapula and inserts on the radial tuberosity.
3. (*d*) The platysma inserts on the inferior border of the mandible.
4. (*d*) Although located in the back, the spinalis is not part of the erector spinae muscle group.
5. (*c*) The coracobrachialis flexes and adducts the arm at the shoulder joint.
6. (*b*) The latissimus dorsi originates on the vertebrae and inserts on the intertubercular groove of the humerus.
7. (*b*) The deltoid has its origin along the acromion and spine of the scapula.
8. (*a*) Spanning two joints, the rectus femoris functions to flex the hip joint and extend the knee joint. Of the four quadriceps femoris muscles, it is the only one to span two joints.
9. (*b*) The position of the tibialis posterior and its long tendon of insertion permits it to support the arches of the foot as it functions to plantar flex and invert the foot.
10. (*a*) *Corrugator* derives from a word meaning "to wrinkle"; as both corrugator muscles are contracted, the skin between the eyebrows wrinkles, as in scowling.
11. (*e*) The strength of contraction is highly variable from person to person and is not used as a means of naming muscles.
12. (*a*) The supine position is with the palms of the hands down, as indicated by the name of the supinator muscle.
13. (*a*) The diaphragm, external intercostal muscle, and the interchondral portion of the internal intercostal muscle are synergistic during the inspiration phase of normal breathing.
14. (*b*) The paired temporalis muscles function with the masseter muscles in closing the jaws. The paired lateral and medial pterygoid muscles are also included in the muscles of mastication.
15. (*d*) Four muscles originate on the lateral epicondyle, and all are extensors of the hand.
16. (*d*) Of the three gluteal muscles, only the gluteus maximus extends and laterally rotates the hip joint.
17. (*e*) The latissimus dorsi originates on the vertebrae and inserts on the intertubercular groove of the humerus.
18. (*d*) The sartorius originates on the anterior superior iliac spine and inserts on the medial surface of the tibia
19. (*d*) Each of the three gluteal muscles has its origin on some part of the ilium.
20. (*e*) The rhomboideus major lies deep to the trapezius muscle.

True or False Questions

1. True
2. True
3. False; extension increases the angle of a joint; abduction moves an appendage away from the midplane of the body
4. True
5. False; the iliocostalis muscles extend the vertebral column and the rectus abdominis flexes the vertebral column
6. True
7. True
8. True
9. False; the sartorius acts on both the hip and knee joints
10. True
11. False; the gluteus maximus inserts on the gluteal tuberosity of the femur and iliotibial tract
12. False; the pectoralis minor does not attach to the humerus; it inserts on the coracoid process of the scapula and, when contracted, it pulls the scapula forward and downward
13. True
14. True
15. True

Completion

1. masseter
2. erector spinae
3. brachioradialis
4. quadriceps femoris/hamstring
5. gastrocnemius
6. diaphragm, external intercostal, internal intercostal
7. sartorius
8. gluteus maximus
9. iliopsoas
10. digastric

Labeling

1. Frontalis
2. Trapezius
3. Deltoid
4. Pectoralis major
5. Bicep brachii
6. Serratus anterior
7. Rectus abdominis
8. External abdominal oblique
9. Brachioradialis
10. Palmaris longus
11. Gracilis
12. Rectus femoris
13. Vastus lateralis
14. Vastus medialis
15. Tibialis anterior
16. Occipitalis
17. Trapezius
18. Deltoid
19. Infraspinatus
20. Triceps brachii
21. Latissimus dorsi
22. External abdominal oblique
23. Flexor carpi ulnaris
24. Gluteus medius
25. Gluteus maximus
26. Semimembranosus
27. Semitendinosus
28. Biceps femoris
29. Gastrocnemius
30. Soleus

Matching

1. (*h*)
2. (*e*)
3. (*g*)
4. (*d*)
5. (*c*)
6. (*b*)
7. (*k*)
8. (*j*)
9. (*f*)
10. (*a*)
11. (*l*)
12. (*i*)

Nervous Tissue 9

Objective A To distinguish between the *central nervous system, peripheral nervous system,* and *autonomic nervous system.*

survey On the basis of structure, the nervous system is divided into the **central nervous system (CNS)** and the **peripheral nervous system (PNS)**. The CNS is composed of the *brain* and the *spinal cord* (fig. 9.1). The PNS is composed of *cranial nerves* from the brain and *spinal nerves* from the spinal cord. In addition, *ganglia* and *plexuses* (table 9.1) are found within the PNS.

The **autonomic nervous system (ANS)** is a functional division of the nervous system. Structures within the brain are ANS control centers, and specific nerves are the pathways for conduction of autonomic nerve impulses. The ANS functions automatically to speed up or slow down body activities.

The nervous system develops very early in prenatal development. By day 20, *neuroectoderm* gives rise to the *neural groove*, which in turn becomes the *neural tube*. Once formed, the neural tube eventually develops into the brain and spinal cord. In addition, *neural crest cells* (from the crest of the enveloped neural tube) migrate throughout the body to give rise to various structures, including melanocytes, the adrenal medulla, some cranial nerve ganglia, and the neurolemmocytes (Schwann cells).

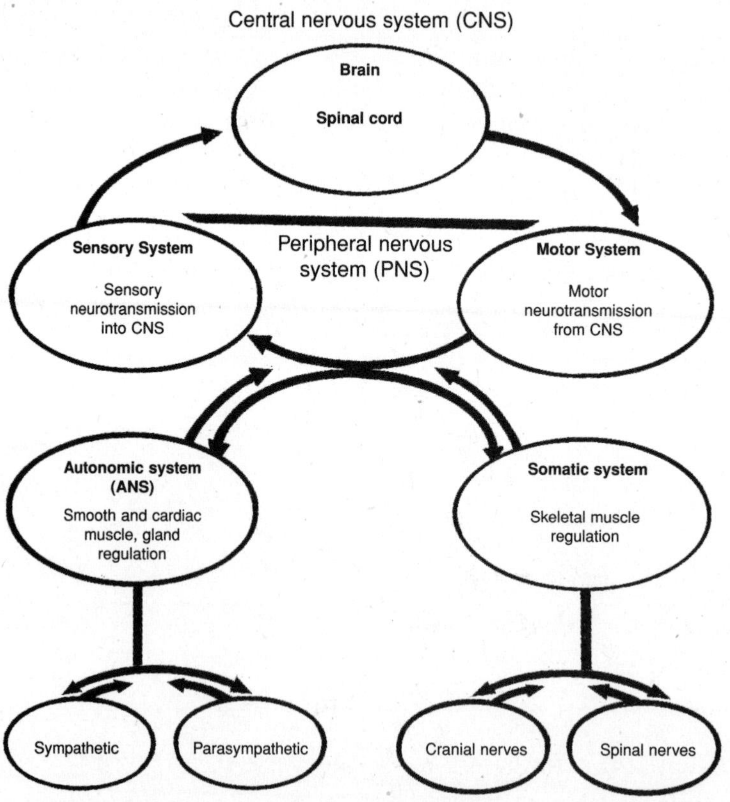

Figure 9.1 Organization of the nervous system.

Table 9.1 Divisions and Structures of the Nervous System

Division/Structure	*Description and location*	*Function*
Central nervous system (CNS)	Brain within the cranium and the spinal cord within the vertebral canal	Responds to nerve impulses (sensations) from sensory nerves; body control center
Peripheral nervous system (PNS)	Composed of sensory, motor, or mixed nerves	Conveys impulses to and from CNS
Autonomic nervous system (ANS)	Composed of specific structures of CNS and nerves of PNS; divided into sympathetic and parasympathetic divisions	Exerts involuntary (autonomic) control of vital body functions including heart rate, respiratory rate, blood pressure, digestion, body temperature, and so forth
Brain	Composed of gray and white matter within the cranium	Serves as control center for nervous system
Spinal cord	Composed of gray and white matter within the vertebral canal of the spinal column	Conveys messages (impulses) to and from brain; reflex center
Neuron	Cell within nervous tissue	Responds to stimuli and conveys nerve impulses
Sensory (afferent) neuron	Component of a sensory or a mixed nerve within PNS	Transmits impulses from sensory receptor to CNS
Motor (efferent) neuron	Component of a motor or a mixed nerve within PNS	Transmits impulses from CNS to effector organs (muscles or glands)
Neuroglium	Cell within nervous tissue	Supports neurons
Nerve	Bundle of nerve fibers within PNS	Conveys impulses
Tract	Bundle of nerve fibers within CNS	Interconnects structures of CNS; conveys impulses
Ganglion	Cluster of cell bodies of neurons within PNS	Serves as control center for a bundle of neurons
Nucleus	Cluster of cell bodies of neurons within white matter of CNS	Serves as control center for a bundle of neurons
Nerve plexus	Network of nerves within PNS	Provides overlapping innervation (nerve supply) to certain body regions

9.1 List the principal functions of the nervous system.

1. Responds to stimuli within the body and in the external environment.
2. Transmits nerve impulses to and away from the CNS.
3. Interprets nerve impulses arriving in the cerebral cortex of the brain.

4. Assimilates experiences as required in memory, learning, and intelligence.
5. Initiates glandular secretion and muscle contraction.
6. Programs instinctual behavior (more important in vertebrates other than humans).

9.2 Distinguish between the terms *stimulus, sensation,* and *perception.*

A **stimulus** is an energy source (chemical, pressure, light wave, etc.) that activates a *receptor cell* (specialized nerve cell) to transmit a **nerve impulse**, or **sensation**. If the sensation arrives in the conscious part of the brain, the cerebral cortex, a **perception** occurs. Perception is awareness of the stimulus.

Pricking one's finger, for example, is a stimulus that activates many receptor cells to send nerve impulses to the brain. Once these sensations reach the cerebral cortex, a person perceives (feels) pain. (See problem 11.13 for the role the reflex arc plays in a similar example.)

Objective B To describe the general structure of a *neuron* and to classify neurons.

Although neurons vary considerably in size and shape, they are generally composed of a **cell body**, **dendrites**, and an **axon**. While some neurons may be as long as a meter (3 ft), most are considerably shorter.

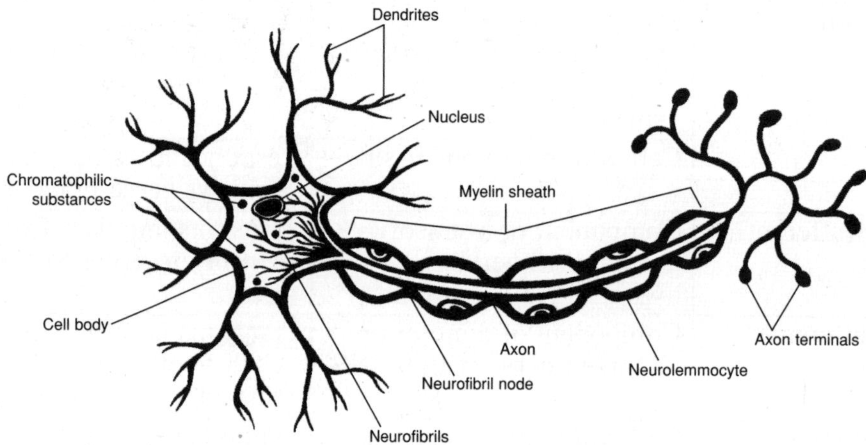

Figure 9.2 The structure of a neuron.

9.3 What are chromatophilic substances, neurofibrils, microtubules, collateral branches, and axon terminals?

See fig. 9.2.

Chromatophilic substances (*Nissl bodies*) are layers of rough ER (see chapter 3), whose function is protein synthesis. **Neurofibrils** are filamentous strands of protein that support the cell body. **Microtubules** are minute channels that transport material within the cell. **Collateral branches** are extensions from the axon that may also transmit impulses. **Axon terminals** are the slight enlargements at the ends of the branched axon. Axon terminals contain *synaptic vesicles* that produce and secrete *neurotransmitter chemicals* into the *synapses* (see Objective F).

9.4 Describe four ways of classifying neurons.

By direction of impulse conduction – *Sensory* (*afferent*) *neurons* transmit nerve impulses to the spinal cord or brain. *Motor* (*efferent*) *neurons* conduct impulses away from the spinal cord or brain. *Association neurons* (*interneurons* or *internuncial neurons*) conduct impulses from sensory to motor neurons. The term *innervation* means "nerve supply" and can be either motor or sensory. Motor neurons can be further classified as *alpha* and *gamma motor neurons*. Alpha motor neurons innervate and stimulate skeletal muscle. Gamma motor neurons innervate specialized muscle tissue called the *muscle spindle*. The muscle spindle is a small, highly differentiated part of muscle tissue, located deep within a muscle.

By area of innervation – *Somatic sensory neurons* are receptors within the skin, bones, muscles, and joints. They also include sensory receptors within the eyes and ears. *Somatic motor neurons* are effector neurons that innervate skeletal muscles and cause the muscle fibers to contract when stimulated. *Visceral sensory fibers* convey impulses from visceral organs and blood vessels. Most of these receptors convey autonomic sensations, but some respond to visceral stimuli, such as hunger pangs or intestinal aches. Sensory receptors within the tongue for taste and nasal epithelium for smell are also visceral sensory fibers. *Visceral motor fibers*, also called *autonomic motor fibers*, are part of the ANS. These fibers originate in the CNS and innervate cardiac muscle fibers, glands, and smooth muscles within visceral organs.

By number of processes – *Multipolar neurons* have one axon and two or more dendrites. *Bipolar neurons* have one axon and one dendrite. *Unipolar neurons* have a single process extending from the cell body, which divides into two branches. One branch extends to the spinal cord and serves as the axon; the other extends to the peripheral part of the body and serves as the dendrite.

By fiber diameter

Group	*Diameter*	*Function*
AA	12–20 μm	Proprioception
AB	5–12 μm	Pressure, touch
Aτ	3–6 μm	Motor-nerve-muscle-spindle junctions
Aσ	2–5 μm	Temperature, touch, pain
B	<3 μm	Preganglionic autonomic
C	0.3–1.3 μm	Postganglionic sympathetic

9.5 Describe the formation of the myelin sheath.

Myelin (Gk. *myelos*, marrow) is an insulating cellular membrane consisting of a fatlike lipid substance known as *sphingomyelin.* During *myelination* (fig. 9.3), the myelin wraps around the neuron creating a multilayered sheath. In the CNS, oligodendrocytes produce the sheath; in the PNS, *neurolemmocytes* (Schwann cells) assume this role. In the PNS, there are small gaps called *neurofibril nodes* (*nodes of Ranvier*) between segments of the sheath (see fig. 9.2). The sheath insulates nerve fibers and thereby inhibits the flow of ions between intracellular and extracellular fluid compartments.

Two fairly common diseases afflict the myelin sheaths. *Multiple sclerosis* (*MS*) is a chronic degenerative disease, marked by remission and relapse, that progressively destroys the myelin sheaths of neurons in multiple areas of the CNS. *Tay-Sachs disease* is an inherited disease in which the myelin sheaths are destroyed by excessive accumulation of lipids within the membrane.

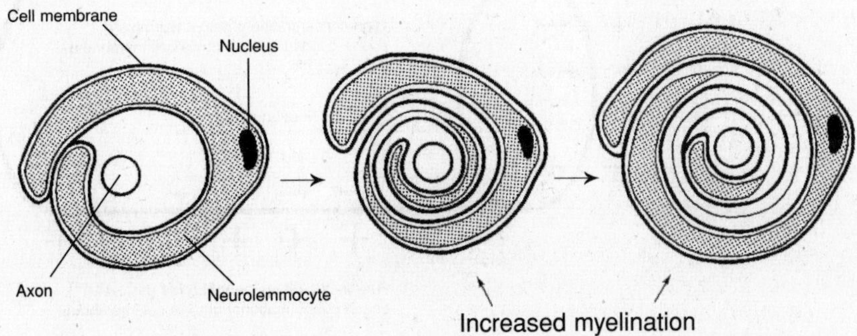

Figure 9.3 The process of myelination.

Objective C To classify *neuroglial.*

There are six categories of **neuroglia** (table 9.2). Also called *glia*, or *glial cells*, these specialized cells of the nervous system physically and physiologically support neurons by assisting in the transfer of nutrients and wastes to and from the neurons. Neuroglia mitotically divide and are estimated to be about five times more abundant than neurons.

9.6 List the different types of nueroglia, including their location and function.

Refer to table 4.8.

9.7 Why are microglia frequently considered part of the body's immune system?

Following trauma to the CNS, or during an infection of the brain or spinal cord, microglia respond by increasing in number, migrating to the site, and phagocytizing the bacterial cells or cellular debris.

Objective D To describe the *resting membrane potential.*

In a nonconducting ("resting") neuron, a voltage, or **resting potential,** exists across the cell membrane. This resting potential is due to an imbalance of charged particles (ions) between the extracellular and the intracellular fluids. The mechanisms responsible for the membrane having a net positive charge on its outer surface and a net negative charge on its inner surface (fig. 9.4) are as follows:

1. A *sodium-potassium pump* transports sodium ions (Na^+) to the outside and potassium ions (K^+) to the inside, with three Na^+ moved out for every two K^+ moved in.

2. The cell membrane is more permeable to K^+ than to Na^+, so that the K^+, which is relatively concentrated inside the cell, moves outward faster than the Na^+, relatively concentrated outside the cell, moves inward.

3. The cell membrane is essentially impermeable to the large (negatively charged) anions that are present inside the neuron, and therefore fewer negatively charged particles move out than positively charged particles.

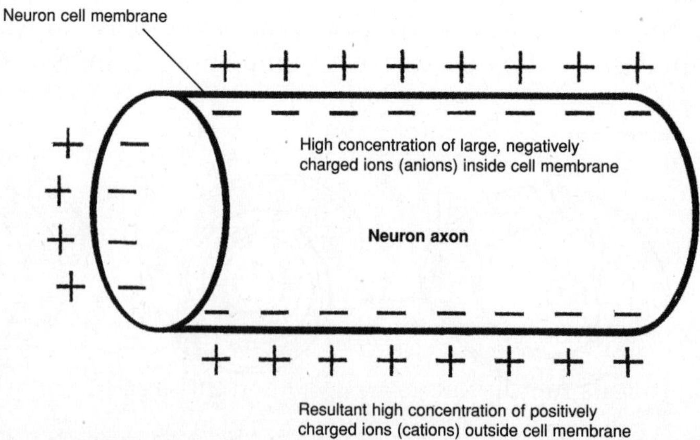

Figure 9.4 A segment of a neuron showing the locations of charges.

As electrical charges move across the membrane, a *physiological current* is induced. Physiological currents can be measured in some of the body's systems. An **EEG** (*electroencephalogram*) records brain activity by monitoring electrical currents in the brain. An **ECG** (*electrocardiogram*) records cardiac electrical activity. An **EMG** (*electromyogram*) records skeletal muscle activity. The electrical activity recordings can be used to diagnose various disorders of the body.

9.8 Since the membrane is 50 to 100 times more permeable to K^+ than Na^+, do these ions diffuse through different channels?

Yes. For example, tetrodotoxin—a poison obtained from puffer fishes—blocks the diffusion through Na^+ channels, but not through K^+ channels.

9.9 Is energy required to develop and maintain a resting membrane potential?

Yes. The sodium-potassium pump, like all other cellular active transport systems, requires the expenditure of metabolic energy derived from the hydrolysis of adenosine triphosphate (ATP).

Objective E To describe the chain of events associated with an *action potential*.

Nerve impulses, which carry information from one point of the body to another, may be described as the progression along the neuron membrane of an abrupt change in the resting potential. This "traveling disturbance," called an **action potential**, is illustrated in fig. 9.5.

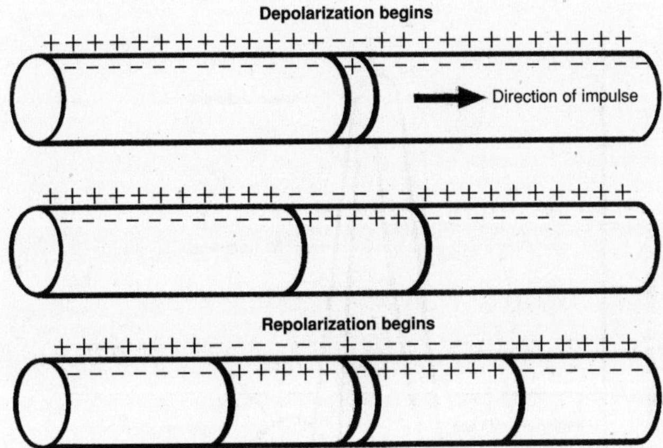

Figure 9.5 A schematic of an action potential, demonstrating the movement of charges.

The sequence of events is as follows:

1. Stimulus (chemical-electrical-mechanical) is sufficient to alter the resting membrane potential of a particular region of the membrane.
2. The membrane's permeability to sodium ions increases at the point of stimulation.
3. Sodium ions rapidly move into the cell through the membrane.
4. As sodium ions move into the cell, the transmembrane potential reaches zero (the membrane becomes locally depolarized).

5. Sodium ions continue to move inward, and the inside of the membrane becomes positively charged relative to the outside (reverse polarization).
6. Reverse polarization at the original site of stimulation results in a local current that acts as a stimulus to the adjacent region of the membrane.
7. At the point originally stimulated, the membrane's permeability to sodium decreases, and its permeability to potassium increases.
8. Potassium ions rapidly move outward, again making the outside of the membrane positive in relation to the inside (repolarization).
9. Sodium and potassium pumps transport sodium ions back out of, and potassium ions back into, the cell. (Now the cycle repeats at [1], relative to the advanced site.)

9.10 What determines whether a stimulus will be strong enough to produce an action potential in a nerve cell?

The resting membrane potential (fig. 9.6) is about –70 mV. This means that the potential of the inner surface is normally 70 mV below the potential of the outer surface. A *threshold stimulus* (just adequate) will sufficiently increase the permeability of the membrane to sodium ions to raise the membrane potential to about –55 mV. Once this *threshold potential* has been reached, complete depolarization and repolarization occur and an action potential is generated.

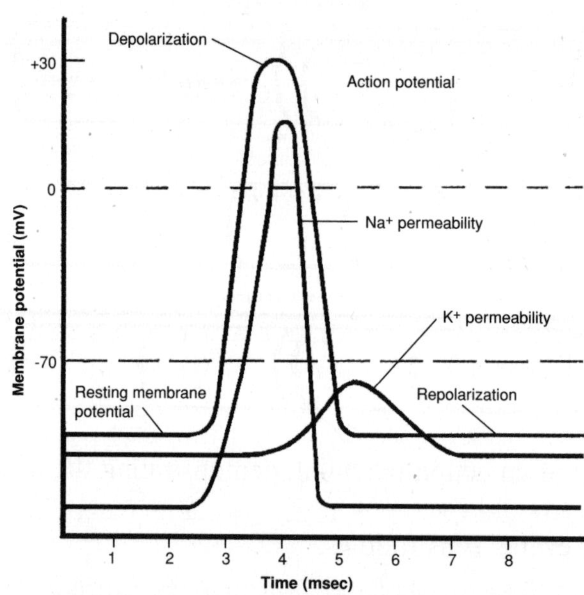

Figure 9.6 An action potential: Resting membrane potential, depolarization, and repolorization.

9.11 Is the size of the action potential related to the strength of the stimulus?

No. Nerve and muscle cells obey the *all-or-none law*, which states that a threshold stimulus evokes a maximal response and that a subthreshold stimulus evokes no response.

9.12 If a neuron has received a threshold stimulus and is undergoing depolarization and repolarization, how much time must pass before a second stimulus can produce an action potential?

In the interval from the onset of an action potential until repolarization is about one-third complete, no stimulus can elicit another response; the "dead phase" is referred to as the *absolute refractory period*. Following the absolute refractory period is an interval during which the neuron will not respond to a normal threshold stimulus but will respond to a suprathreshold stimulus; this is the *relative refractory period.*

9.13 What factors influence the speed at which impulses are conducted along excitable cell membranes?

Diameter of the conducting fiber. Conduction velocity is directly proportional to fiber diameter.

Temperature of the cell. Warmer nerve fibers conduct impulses at higher speeds.

Presence or absence of the myelin sheath. Myelinated fibers conduct impulses more rapidly than unmyelinated fibers. This is because action potentials "leap" from one neurofibril node to the next instead of progressing from point to point along the axon. This leaping or jumping of the impulse is called *saltatory conduction.* Saltatory conduction is not only faster but also consumes less energy, since the pumping of sodium and potassium ions need occur only at the nodes.

Objective F To define *synapse* and *synaptic transmission.*

A **synapse** is the specialized junction through which impulses pass from one neuron to another (*synaptic transmission*). With reference to fig. 9.7, the steps in the process are as follows:

1. An action potential spreads over the axon terminal.
2. An influx of calcium ions causes synaptic vesicles to fuse with the presynaptic membrane.
3. Neurotransmitter is released by exocytosis from the synaptic vesicles into the synaptic cleft.
4. The neurotransmitter diffuses across the synaptic cleft to the postsynaptic membrane.
5. The neurotransmitter combines with specific receptors on the postsynaptic membrane.
6. The permeability of postsynaptic membrane is altered, whereupon an impulse is initiated on the second neuron.
7. The neurotransmitter is removed from the synapse as a result of being enzymatically degraded, taken up in the presynaptic terminal, or diffused out of the synaptic region.

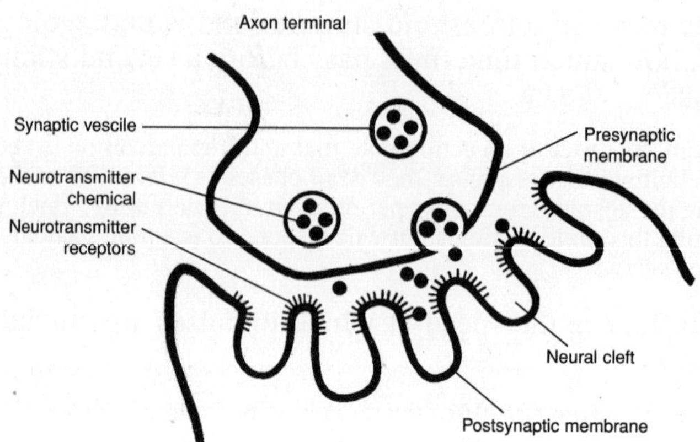

Figure 9.7 Synaptic transmission.

9.14 Briefly define *synaptic delay, synaptic fatigue,* and *one-way conduction.*

Synaptic delay. There is a delay of about 0.5 ms in the transmission of an impulse from the axon terminal of the presynaptic neuron to the postsynaptic neuron. The time is consumed in (1) the release of the neurotransmitter, (2) the diffusion of the neurotransmitter across the cleft, (3) the interaction of the neurotransmitter with receptors on the postsynaptic membrane, and (4) the initiation of the impulse in the postsynaptic neuron.

Synaptic fatigue. With repetitive stimulation there is a progressive decline in synaptic transmission due to depletion of the store of neurotransmitter in the axon terminal.

One-way conduction. Most synapses conduct impulses in one direction only because the neurotransmitter is usually present only on one side of the synapse.

9.15 Neurotransmitters may be *excitatory*, causing the postsynaptic neuron to become active, or *inhibitory*, preventing the postsynaptic neuron from becoming active. Briefly differentiate between excitatory and inhibitory mechanisms.

Excitatory neurotransmitters are those that increase the postsynaptic membrane's permeability to sodium ions. The increased but still subthreshold membrane potential is known as an *excitatory postsynaptic potential* (EPSP), and the membrane is said to be *hypopolarized.* There are two ways in which several EPSPs can combine to reach threshold and elicit an action potential: (1) In *spatial summation,* several presynaptic neurons simultaneously release neurotransmitter to a single postsynaptic neuron; (2) in *temporal summation,* the EPSPs result from the rapid successive discharge of neurotransmitter from the same axon terminal.

Inhibitory neurotransmitters are those that increase the postsynaptic membrane's permeability to potassium and chloride ions, resulting in a hyperpolarized membrane that exhibits an *inhibitory postsynaptic potential* (IPSP). During the time the membrane is hyperpolarized, the potential is farther below threshold, making it more difficult to generate an action potential.

9.16 What specific drugs can influence synaptic transmission?

Reserpine can inhibit uptake and storage of the neurotransmitter *norepinephrine* in synaptic vesicles. *Botulinum toxin* can inhibit the release of the neurotransmitter *acetylcholine* from synaptic vesicles.

Amphetamines can stimulate the release of *norepinephrine* from synaptic vesicles. *Atropine* can block receptors for *acetylcholine* on the postsynaptic membrane. *Cholinergic drugs* bind to receptors for acetylcholine, where they mimic the neurotransmitter. *Anticholinesterase drugs* inhibit the destruction or metabolism of acetylcholine.

Parkinson's disease is a progressive neurological disorder (dopamine deficiency in the extrapyramidal system) in which nerve cells within the basal nuclei of the brain are destroyed. Parkinson's disease occurs typically in middle or late life with a gradual progression over a prolonged course. Symptoms include tremor of the hands; weakness; rigidity of the large joints, which causes a stooped fixed posture, and a shuffling gait. The symptoms can be partially treated with exercise, heat, massage, the use of anticholinergic drugs, antihistamines, and L-dopa (a precursor of dopamine that can cross the blood-brain barrier).

Alzheimer's disease is the most common cause of *senile dementia.* The cause of Alzheimer's disease is still being investigated, but there is evidence that it is associated with the loss of neurons that terminate on the hippocampus and cerebral cortex (areas of the brain linked to memory storage) and that use acetylcholine as the neurotransmitter. Autopsies of people who have died of Alzheimer's disease show neurotic plaques (degenerated axons and deposits of amyloid protein) that are similar to plaques observed in patients with Down syndrome.

Review Exercises

Multiple Choice

1. Which kind of neuroglial cells are *not* found in the CNS? (*a*) astrocytes, (*b*) ependymal cells, (*c*) microglia, (*d*) satellite cells, (*e*) oligodendrocytes

2. The neuroglia that have functions similar to white blood cells are (*a*) oligodendrocytes, (*b*) astrocytes, (*c*) microglia, (*d*) ependymal cells, (*e*) lymphocytes.

3. The speed of a nerve impulse is independent of (*a*) the diameter of the nerve fiber, (*b*) the physiological condition of the nerve, (*c*) the presence of myelin, (*d*) the length of the nerve fiber, (*e*) the presence of neurolemmocytes.

4. The basic unit of the nervous system is (*a*) the axon, (*b*) the dendrite, (*c*) the neuron, (*d*) the cell body, (*e*) the synapse.

5. Depolarization of the membrane of a nerve cell occurs by the rapid influx of (*a*) potassium ions, (*b*) chloride ions, (*c*) organic anions, (*d*) sodium ions.

6. A transmitter substance released into the synaptic cleft is (*a*) cholinesterase, (*b*) acetylcholine, (*c*) ATP, (*d*) RNA, (*e*) all of the preceding.

7. At a synapse, impulse conduction normally (*a*) occurs in both directions, (*b*) occurs in only one direction, (*c*) depends on acetylcholine, (*d*) depends on epinephrine.

8. In a resting neuron, (*a*) the membrane is electrically permeable, (*b*) the outside of the membrane is positively charged, (*c*) the outside is negatively charged, (*d*) the potential difference across the membrane is zero.

9. Dendrites carry nerve impulses (*a*) toward the cell body, (*b*) away from the cell body, (*c*) across the body of the nerve cell, (*d*) from one nerve cell to another.

10. The enzyme that destroys acetylcholine is (*a*) ATPase, (*b*) epinephrine, (*c*) cholinesterase, (*d*) lipase, (*e*) acetylcholinase.

11. The transmitter substance in the presynaptic neuron is contained in (*a*) the synaptic cleft, (*b*) the neuron vesicle, (*c*) the synaptic gutter, (*d*) the mitochondria.

12. The interior surface of the membrane of a nonconducting neuron differs from the exterior surface in that the former is (*a*) negatively charged and contains less sodium, (*b*) positively charged and contains less sodium, (*c*) negatively charged and contains more sodium, (*d*) positively charged and contains more sodium.

13. The presence of myelin gives a nerve fiber its (*a*) gray color and degenerative abilities, (*b*) white color and increased rate of impulse transmission, (*c*) white color and decreased rate of impulse transmission, (*d*) gray color and increased rate of impulse transmission.

14. During repolarization of the neuronal membrane, (*a*) sodium ions rapidly move to the inside of the cell, (*b*) sodium ions rapidly move to the outside of the cell, (*c*) potassium ions rapidly move to the outside of the cell, (*d*) potassium ions rapidly move to the inside of the cell.

15. The arrival on a given neuron of a series of impulses from a series of terminal axons, thereupon producing an action potential, is an example of (*a*) temporal summation, (*b*) divergence, (*c*) generation potential, (*d*) spatial summation.

16. Neural regulation differs from endocrine regulation in that the former (*a*) is quick, precise, and localized, (*b*) is slower and more pervasive, (*c*) does not require conscious activity, (*d*) has longer lasting effects.

17. The gray matter of the brain consists mainly of neuron cell (*a*) axons, (*b*) dendrites, (*c*) secretions, (*d*) bodies.

18. The tightly packed coil of the neurolemmocyte membrane that encircles certain kinds of axons is called (*a*) a myelin sheath, (*b*) a neurolemma, (*c*) a node, (*d*) gray matter.

19. The interruptions occurring at regular intervals along a myelin-coated axon are (*a*) neurofibril nodes, (*b*) synapses, (*c*) synaptic clefts, (*d*) gap junctions.

20. The junction between two neurons is called (*a*) a neurospace, (*b*) an axon, (*c*) a synapse, (*d*) a neural junction.

21. An IPSP is mediated by (*a*) an increase in permeability to all cations; (*b*) selective permeability to calcium, sodium, and potassium; (*c*) an increase in permeability to all anions; (*d*) selective permeability to potassium and chloride ions.

22. The general depolarization toward threshold of a cell membrane when excitatory synaptic activities predominate is known as (*a*) facultation, (*b*) differentiation, (*c*) inhibition, (*d*) facilitation.

23. Examples of neurotransmitters are (*a*) adenine and guanine, (*b*) thymine and cytosine, (*c*) acetylcholine and norepinephrine, (*d*) none of the preceding.

24. Clusters of neuron cell bodies found in the CNS are termed (*a*) nerve clusters, (*b*) ganglia, (*c*) axons, (*d*) nuclei.

25. Which of the following occur within the peripheral nervous system? (*a*) oligodendrocytes, (*b*) ependymal cells, (*c*) microglia, (*d*) satellite cells

True or False

____ **1.** There are basically only two different types of cells in the nervous system.

____ **2.** The axon is the cytoplasmic neuronal extension that conducts impulses toward the cell body.

____ **3.** A polarized nerve fiber has an abundance of sodium ions on the outside of the axon membrane.

____ **4.** Glial cells sustain the CNS neurons metabolically, support them physically, and regulate ionic concentrations in the extracellular space.

____ **5.** Dendrites are usually longer than axons.

____ **6.** Every postsynaptic neuron has only one synaptic junction on the surface of its dendrites.

____ **7.** A single EPSP is sufficient to cause an action potential.

____ **8.** Only EPSPs show temporal and spatial summations.

____ **9.** Chemical synapses operate in only one direction.

____ **10.** All synapses are inhibitory.

____ **11.** A nerve impulse can travel along an axon for an indefinite distance without distortion or loss of strength.

____ **12.** The nerve impulse is all or nothing.

____ **13.** The resting potential in a nerve cell is caused by the high concentration of potassium outside the cell.

____ **14.** The permeability of the neuron's cell membrane to sodium decreases as the membrane is depolarized.

____ **15.** The sodium pump operates by diffusion, and thus requires no ATP for its operation.

____ **16.** Hyperpolarization of the postsynaptic membrane by an excitatory synapse produces an EPSP.

____ **17.** The myelin sheath surrounds the dendrites.

____ **18.** Transmission across the synaptic junction is diffusion of sodium.

____ **19.** Two transmitter substances in the nervous system are dopamine and acetylcholine.

____ **20.** Neuroglia have an action potential response.

____ **21.** Motor neurons convey information from receptors in the periphery to the CNS.

____ **22.** Somatic motor nerves innervate skeletal muscle, and autonomic nerves innervate smooth muscle, cardiac muscle, and glands.

Completion

1. The majority of specialized junctions that receive stimuli from other neurons are located on the ________________ and ________________ ________________ of the neuron.
2. Only 10% of the cells in the nervous system are ________________, and the remainder are ________________.
3. ________________ cells have one long axon and multiple short, highly branched dendrites extending from the cell body.
4. The velocity with which an action potential is transmitted down the membrane depends on the fiber ________________ and on whether or not the fiber is ________________.
5. On a myelinated neuron, the action potential appears to jump from one node to another. This method of propagation is called ________________ ________________.
6. Within the peripheral nervous system, myelin is formed by the ________________.
7. A junction between two neurons, where the electrical activity in the first influences the excitability of the second is called a ________________.
8. The transmitter substance is stored in small membrane–enclosed ________________ in the synaptic knob.
9. When an action potential depolarizes the synaptic knob, small quantities of transmitter substance are released into the ________________ ________________.
10. The adding together of two or more EPSPs that originate at different places, resulting in depolarization of the membrane, is called ________________ ________________.
11. The interval from the onset of an action potential until repolarization is about one-third complete, during which no stimulus can elicit another response, is referred to as the ________________ ________________ ________________.
12. With repetitive stimulation, there is a progressive decline in synaptic transmission due to depletion of the store of neurotransmitter in the axon terminal. This is referred to as ________________ ________________ ________________.
13. A chronic degenerative disease that progressively destroys the myelin sheaths of neurons is called ________________ ________________.
14. A cluster of nerve cell bodies in the peripheral nervous system is referred to as a ________________.

Labeling

Label the structures indicated on the figure to the right.

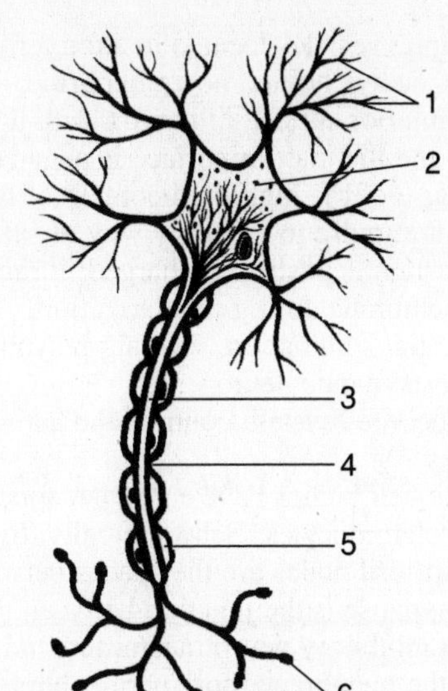

1. ______________________
2. ______________________
3. ______________________
4. ______________________
5. ______________________
6. ______________________

Matching

Match the kind of neuron with its description or function.

____	**1.** Multipolar neuron	(*a*) found only in the CNS
____	**2.** Sensory neuron	(*b*) one dendrite and one axon
____	**3.** Association neuron	(*c*) a single branch connected to a cell body
____	**4.** Unipolar neuron	(*d*) carries information toward the CNS
____	**5.** Bipolar neuron	(*e*) one long axon and many dendrites

Answers and Explanations for Review Questions

Multiple Choice

1. (*d*) Satellite cells are small flattened cells that support neuron cell bodies within the ganglia of the PNS.
2. (*c*) Microglia actively phagocytize pathogens and cellular debris within the CNS.
3. (*d*) The length of a nerve fiber has no bearing on the speed impulse conduction.
4. (*c*) The neuron, or nerve fiber, is the basic unit of the nervous system because it is at the neuron level that the activities of the system are carried out.
5. (*d*) When the membrane of a neuron is stimulated, there is an increase in the permeability of the membrane to sodium at that point. As sodium ions move inward, the membrane becomes depolarized.
6. (*b*) Acetylcholine is one of many transmitted substances that may be released into the synaptic cleft by the synaptic vesicles of a presynaptic neuron.
7. (*b*) Synaptic function occurs in only one direction because neurotransmitters are stored in synaptic vesicles in the presynaptic neurons.
8. (*b*) The exterior surface of a resting neuron is positively charged. There are more sodium ions outside the membrane.
9. (*a*) Dendrites transmit impulses toward the cell body of the neuron, and axons transmit impulses away from the cell body.
10. (*c*) Cholinesterase, or acetylcholinesterase, is the enzyme that chemically breaks down acetylcholine.

11. (*b*) Synaptic vesicles located in axon terminals contain neurotransmitter chemicals that include, for example, acetylcholine, norepinephrine, and glycine.
12. (*a*) The interior surface of neuron with a resting potential contains less sodium than the exterior surface, and the interior surface is negatively charged (–70 to –90 mV).
13. (*b*) Because of the high lipid content of the myelin sheath, it is white in color. The myelin sheath allows an impulse to travel by way of saltatory conduction (impulses jump from node to node).
14. (*c*) Potassium is positively charged; it rapidly moves from the interior surface to the exterior surface of the membrane during repolarization.
15. (*d*) In spatial summation, several presynaptic neurons simultaneously release neurotransmitters to a single postsynaptic neuron.
16. (*a*) The nervous system controls the activities of the body that are often very precise and localized responses.
17. (*d*) Nerve cell bodies have a grayish appearance whereas myelin sheaths are white.
18. (*a*) Neurolemmocytes (Schwann cells) form the myelin sheath in the PNS.
19. (*a*) Neurofibril nodes are the spaces between neurolemmocytes.
20. (*c*) The synapse is the junction between two neurons.
21. (*d*) Most inhibitory neurotransmitters induce hyperpolarization of the postsynaptic membrane by making the membrane more permeable to K^+, Cl^-, or both.
22. (*d*) A hypopolarized membrane is said to be facilitated.
23. (*c*) Acetylcholine and norepinephrine are two important neurotransmitters.
24. (*d*) Clusters of neuron cell bodies in the CNS form nuclei, but in the PNS they form ganglia.
25. (*d*) Satellite cells support ganglia within the PNS.

True or False

1. False; there are neurons and at least six types of neuroglia
2. False; the axon conducts impulses away from the cell body
3. True
4. True
5. False; dendrites are usually shorter than axons, although some dendrites are as long as axons
6. False; there may be many synaptic junctions on the surface of a single dendrite
7. False; by definition, an EPSP is subthreshold
8. True
9. True
10. False; many synapses are excitatory
11. True
12. True
13. False; potassium is in greater concentration on the inside of the cell
14. False; the membrane's permeability to sodium increases during depolarization
15. False; the sodium pump operates by active transport and thus requires energy (ATP)
16. False; an EPSP produces hypopolarization
17. False; the myelin sheath generally surrounds axons; some dendrites are myelinated
18. False; transmission across a synapse is the diffusion of a neurotransmitter
19. True
20. False; neuroglia function to support nerve fibers, not to transmit impulses
21. False; motor transmission is from the CNS to the periphery
22. True

Completion

1. dendrites, cell body
2. neurons, neuroglia
3. Multipolar
4. diameter, myelinated
5. saltatory conduction
6. neurolemmocytes (Schwann cells)
7. synapse
8. vesicles
9. synaptic cleft
10. spatial summation
11. absolute refractory period
12. synaptic fatigue
13. multiple sclerosis
14. ganglion

Labeling

1. Dendrites
2. Cell body of neuron
3. Axon
4. Neurofibril node
5. Neurolemmocyte
6. Axon terminals

Matching

1. (*e*)
2. (*d*)
3. (*a*)
4. (*c*)
5. (*b*)

Central Nervous System 10

Objective A To describe the structure and functions of the *central nervous system* in general terms.

The **central nervous system (CNS)** consists of the *brain* and *spinal cord.* The CNS is protected by a bony enclosure (the *cranium* and *vertebral column*) and the membranous *meninges* (see Objective H). The CNS is bathed in *cerebrospinal fluid* and contains gray and white matter. The functions of the CNS include body orientation and coordination, assimilation of experiences (learning), and programming of instinctual behavior.

10.1 What are gray matter and white matter composed of, and where are they located?

The *gray matter* consists of either nerve cell bodies and dendrites or of unmyelinated axons and neuroglia. It forms the outer convoluted cerebral cortex and cerebellar cortex. It also exists as special clusters of nerve cell bodies, called ***nuclei***, deep within the white matter. In the spinal cord, the gray matter is deep to the white matter. The *white matter*, consisting of aggregations of myelinated axons and associated, forms the ***tracts***, or bundled nerve fibers, within the CNS.

10.2 How large is the brain, how many neurons does it contain, and how are the neurons interconnected?

The brain of an adult weighs about 1.5 kg (3.3 lb) and is composed of an estimated 100 billion (10^{11}) neurons. Neurons communicate with one another by means of innumerable synapses between axons and dendrites. Neurotransmitter chemicals called *neuropeptides* (see table 10.3) transmit nerve impulses across synapses and act on postsynaptic neurons in the CNS. These specialized protein messengers account for specific mental functions.

Objective B To describe the *embryonic development* of the brain into the forebrain, midbrain, and hindbrain, and to explain how this correlates with the division of the brain into five mature regions derived from the three initial ones.

The brain begins its embryonic development as the front end of the *neural tube* starts to grow rapidly and to differentiate. By the fourth week after conception, three distinct swellings are evident: the **prosencephalon** (*forebrain*), the **mesencephalon** (*midbrain*), and the **rhombencephalon** (*hindbrain*). Further development, during the fifth week, results in the formation of five mature regions: the **telencephalon** and the **diencephalon** derive from the forebrain; the mesencephalon remains unchanged; and the **metencephalon** and **myelencephalon** form from the hindbrain (fig. 10.1).

Differential cell division and growth in specific areas of the brain cause some areas to become larger than others. It is uncertain what triggers proliferation of cells, growth, and specialization in one area as compared to another, but it is known that substances consumed by a pregnant mother (alcohol, for example) can significantly alter normal brain development.

10.3 List the principal structures in each of the five regions of the brain and indicate their general functions.

See table 10.1.

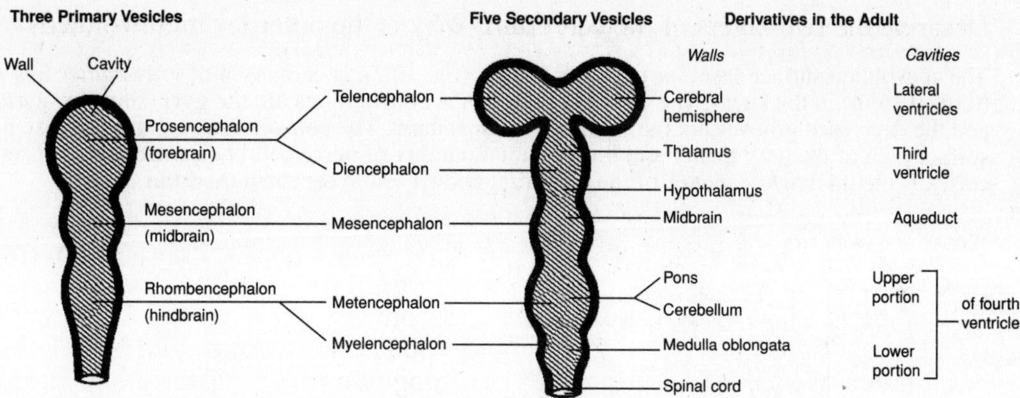

Figure 10.1 Developmental changes in the embryonic brain.

Table 10.1 Regions of the Brain and Their Principal Structures

Region	*Structure*	*Function*
Telencephalon	Cerebrum	Control of most sensory and motor activities; reasoning, memory, intelligence, etc.; instinctual and limbic (emotional) functions
Diencephalon	Thalamus	Relay center: All impulses (except olfactory) going into cerebrum synapse here
	Hypothalamus	Regulation of urine formation, body temperature, hunger, heartbeat, etc.; control of secretory activity in anterior pituitary; instinctual and limbic functions
	Pituitary gland	Regulation of other endocrine glands
Mesencephalon	Superior colliculus	Visual reflexes
	Inferior colliculus	Auditory reflexes
	Cerebral peduncles	Coordinating reflexes; contain many motor fibers
Metencephalon	Cerebellum	Balance and motor coordination
	Pons	Relay center; contains repiratory nuclei
Myelencephalon	Medulla oblongata	Relay center; contains many nuclei; visceral autonomic center (e.g., respiration, heart rate, vasoconstriction)

Objective C To describe the *cerebrum* and the functions of the cerebral lobes.

The **cerebrum** consists of five paired lobes within two convoluted **cerebral hemispheres**. The hemispheres are connected by the **corpus callosum**. The cerebrum accounts for about 80% of the brain's mass and is concerned with higher functions, including perception of sensory impulses, instigation of voluntary movement, memory, thought, and reasoning.

10.4 Describe the two layers of the cerebrum. Why is the outer layer convoluted?

The convoluted surface layer, or **cerebral cortex** (fig. 10.2), is composed of *gray matter* 2–4 mm (0.08–0.16 in) in thickness. The elevated folds of the convolutions are the **gyri** (singular, *gyrus*), and the depressed grooves are the **sulci** (singular, *sulcus*). The convolutions greatly increase the surface area of the gray matter and thus the total number of nerve cell bodies. Beneath the cerebral cortex is the thick *white matter* of the cerebrum known as the **cerebral medulla**.

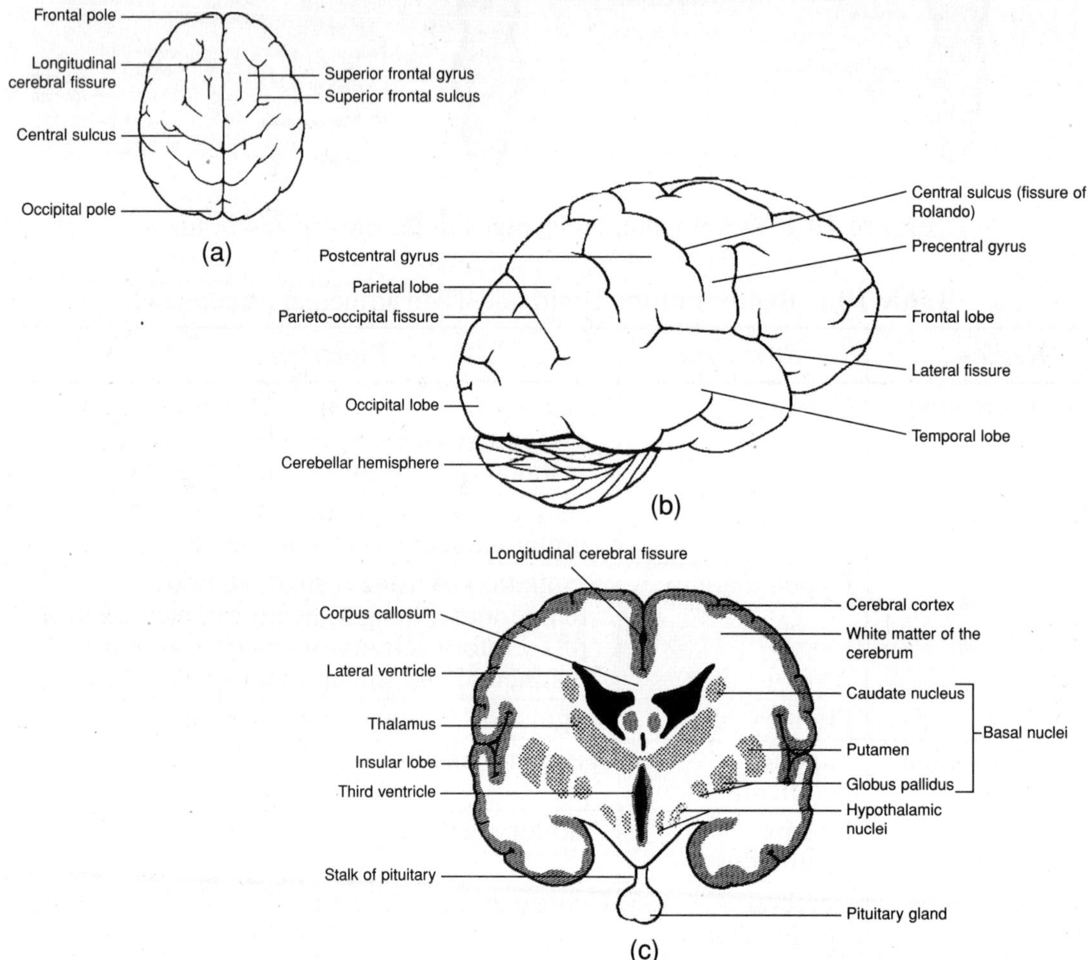

Figure 10.2 The cerebrum. *(a)* A superior view, *(b)* a lateral view, and *(c)* a coronal view.

10.5 What are the specific functions of the paired cerebral lobes?

See table 10.2.

10.6 Distinguish between a sulcus and a fissure.

A **sulcus** is a shallow depression or groove between the *gyri* of the convoluted cerebral cortex. Several of them are named as important landmarks of the brain. The most noted of these is the **central sulcus** between the *precentral gyrus* of the frontal lobe and the *postcentral gyrus* of the parietal lobe (see fig. 10.2).

A **fissure** is a deep groove between major structures of the cerebrum. The most obvious of these is the **longitudinal cerebral fissure** separating the cerebrum into *right* and *left cerebral hemispheres*. The **lateral fissure** separates the frontal lobe from the temporal lobe, and the **parieto-occipital fissure** separates the temporal lobe from the occipital lobe.

Table 10.2 The Cerebral Lobes and Their Functions

Cerebral lobes	*Functions*
Frontal lobe	Voluntary motor control of skeletal muscles; personality (with limbic system); intellectual process (e.g., concentration, planning, decision making); verbal communication
Parietal lobe	Somatesthetic interpretation (e.g., cutaneous and muscular sensations); understanding and utterance of speech
Temporal lobe	Interpretation of auditory sensations; auditory and visual memory
Occipital lobe	Integration of movements in focusing the eye; correlation of visual images with previous visual experiences and other sensory stimuli; conscious seeing
Insular	Memory; integration of other cerebral activities

10.7 Where is the motor speech area and why is it important?

The **motor speech area** (Broca's area) is located in the left inferior gyrus of the frontal lobe, immediately anterior to the lateral sulcus (fig. 10.3). Mental activity in the motor speech area causes selective stimulation of motor centers elsewhere in the frontal lobe, which, in turn, causes coordinated contraction of skeletal muscles in the pharynx and larynx. At the same time, motor impulses are sent to the respiratory centers (see problem 10. 19) to regulate air movement across the vocal cords. The combined muscular stimulation translates thought patterns into speech.

Speech and language disorders are broadly categorized as *aphasias*. These conditions vary in severity from moderate language problems to complete loss of the power of expression by speech, writing, or comprehension of spoken or written language. Certain types of aphasias are congenital. Others are acquired from trauma or disease that afflicts the language centers within the brain.

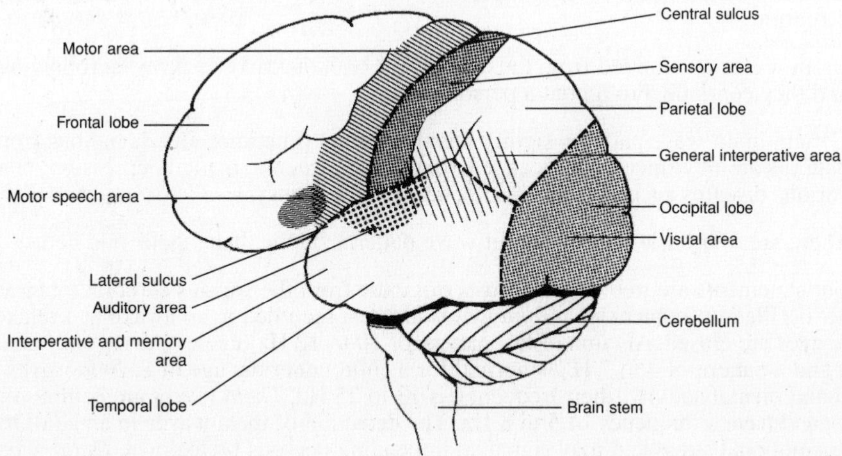

Figure 10.3 Principal motor and sensory areas of the cerebral cortex.

10.8 *True or false*: The cerebral hemispheres communicate one with the other by nerve impulses passing through fiber tracts.

True. Impulses travel not only between the lobes of a cerebral hemisphere, but also between the right and left cerebral hemispheres and to other regions of the brain.

There are three types of fiber tracts within the white matter. They are named on the basis of location and the direction in which they conduct impulses (fig 10.4). **Association fibers** are confined to a given hemisphere, where they conduct impulses between neurons in various lobes. **Commissural fibers** connect the neurons and gyri of one hemisphere with those of the other. The *corpus callosum* and *anterior commissure* (see fig. 10.4) are composed of commissural fibers. **Projection fibers** form *descending tracts,* which transmit impulses from the cerebrum to other parts of the brain and spinal cord, and *ascending tracts,* which transmit impulses from the spinal cord and other parts of the brain to the cerebrum. A *decussation* is where projection fibers cross from one side of the CNS to the other.

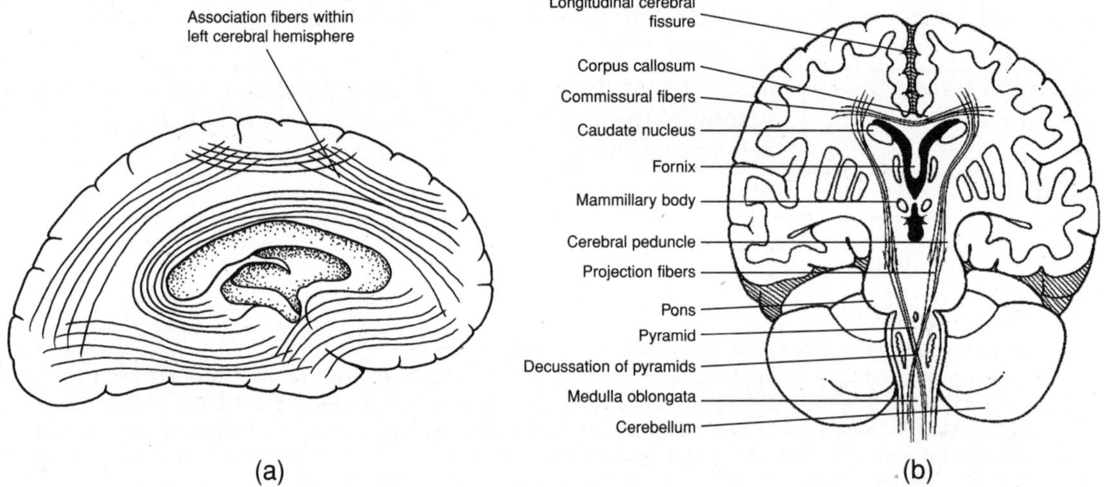

Figure 10.4 Fiber tracts within the brain. (*a*) A sagittal view of a cerebral hemisphere and (*b*) a coronal view of the cerebrum, midbrain, and brain stem.

10.9 Comment on the truth or falsity of the following statements concerning brain waves as recorded in an electroencephalogram (EEG).

(a) Brain waves are the collective expressions of millions of action potentials from neurons of the cerebrum.

(b) Brain waves are emitted from the developing brain as early as 8 weeks following conception, and they continue throughout a person's life.

(c) Certain brain-wave patterns signify healthy mental functions, and deviations from these patterns are of clinical significance in diagnosing trauma, mental depression, hematomas, and various diseases such as tumors, infections, and epilepsy.

(d) There are four basic kinds of brain-wave patterns: alpha, beta, theta, and delta.

All four statements are true. **Brain waves** originate from the various cerebral lobes and have distinct oscillation frequencies. *Alpha waves* are best recorded in an awake and relaxed person whose eyes are closed. An alpha EEG pattern of 10 to 12 Hz (cycles per second) is normal for an adult, and a pattern of 4 to 7 Hz is normal for a child under the age of 8. *Beta waves* accompany visual and mental activity; their frequency is 13 to 25 Hz. *Theta waves* are common in newborn infants and have a frequency of 5 to 8 Hz. The detection of theta waves in an adult may indicate severe emotional stress and may signal an impending nervous breakdown. *Delta waves* are common in a person who is asleep or in one who is awake but who has brain damage; they have a low frequency of 1 to 5 Hz.

10.10 What are the basal nuclei?

The **basal nuclei** (*basal ganglia*) are specialized paired masses of gray matter located deep within the white matter of the cerebrum. They are made up of the **corpus striatum** and other structures of the mesencephalon. The corpus striatum consists of the **caudate nucleus** and the **lentiform nucleus**. The lentiform nucleus, in turn, consists of the **putamen** and the **globus pallidus**.

Neural diseases, such as *Parkinson's disease*, or physical trauma to the basal nuclei generally cause a variety of motor dysfunctions, including rigidity, tremor, and rapid and aimless movements. Drug therapy may be somewhat effective in treating these disorders. Experimental treatments include brain tissue transplants.

Objective D To describe the location and structure of the *diencephalon* and to explain the autonomic functions of its chief components—the *thalamus*, *hypothalamus*, *epithalamus*, and *pituitary gland*.

The **diencephalon**, a major autonomic region of the forebrain, is almost completely surrounded by the cerebral hemispheres of the telencephalon. The *third ventricle* (problem 10.27) forms a midplane cavity within the diencephalon.

10.11 What is the structure of the thalamus and what are its functions?

The **thalamus** (fig. 10.5) is a large ovoid mass of gray matter. It is actually a paired organ, with each portion located immediately below the *lateral ventricle* (see problem 10.27) of its respective cerebral hemisphere. The thalamus is a relay center for all sensory impulses, except smell, to the cerebral cortex. It also is involved in the initial autonomic response of the body to intensely painful stimuli, and is therefore partially responsible for the physiological state of shock that frequently follows serious trauma.

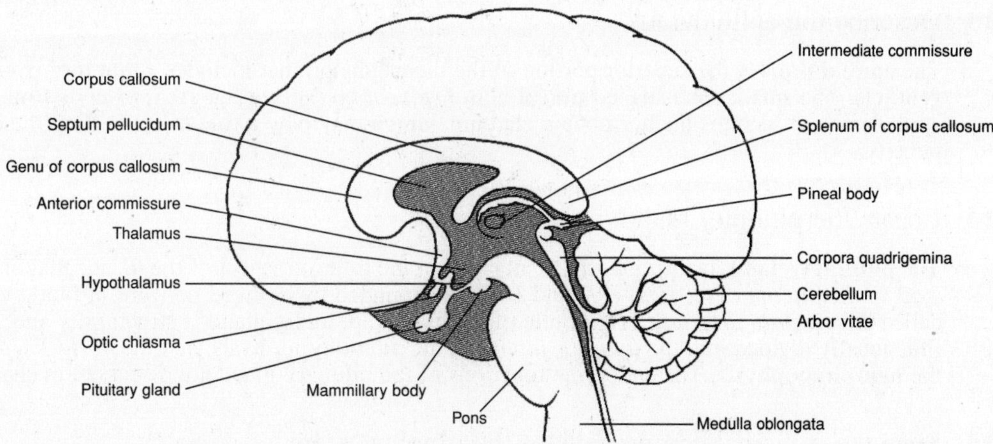

Figure 10.5 A sagittal section of the brain.

10.12 Which autonomic function is *not* performed by the hypothalamus? (*a*) heart rate, (*b*) respiration control, (*c*) body-temperature regulation, (*d*) regulation of hunger and thirst, (*e*) sexual response (*b*). The **hypothalamus** (see fig. 10.5) consists of several nuclei interconnected to other vital parts of the brain. Although most of its functions relate to regulation of visceral activities, the hypothalamus also performs emotional (limbic) and instinctual functions. Its principal functions are as follows:

Cardiovascular regulation. Impulses from the posterior hypothalamus produce autonomic acceleration of the heartbeat; impulses from the anterior portion produce autonomic deceleration.

Body-temperature regulation. Nuclei in the anterior portion of the hypothalamus monitor the temperature of the surrounding arterial blood. In response to above-normal temperatures, the hypothalamus initiates impulses that cause heat loss through sweating and dilation of cutaneous vessels. In response to below-normal temperatures, the hypothalamus relays impulses that cause contraction of cutaneous vessels and shivering.

Regulation of water and electrolyte balance. *Osmoreceptors* in the hypothalamus monitor the osmotic concentration of the blood. Viscosity of the blood due to lack of water causes antidiuretic hormone (ADH) to be produced and released from the posterior pituitary. At the same time, a thirst center within the hypothalamus causes the feeling of thirst.

Regulation of gastrointestinal activity and hunger. In response to sensory impulses from abdominal viscera, the hypothalamus regulates glandular secretions and peristalsis in the GI tract. Levels of glucose, fatty acids, and amino acids in the blood are monitored by a feeding center in the lateral hypothalamus. When sufficient amounts of food have been ingested, a satiety center in the midportion of the hypothalamus inhibits the feeding center.

Regulation of sleeping and wakefulness. The sleep center and the wakefulness center of the hypothalamus function with other parts of the brain to determine the level of conscious alertness.

Sexual response. Specialized *sexual-center nuclei* within the superior portion of the hypothalamus respond to sexual stimulation and are responsible for the feeling of sexual gratification.

Emotions. Specific nuclei within the hypothalamus interact with the rest of the limbic system (see problem 10.15) in causing such emotional responses as anger, fear, pain, and pleasure.

Control of endocrine functions. The hypothalamus produces neurosecretory chemicals that stimulate the anterior pituitary to release various hormones.

10.13 Describe the epithalamus.

The **epithalamus** is the superior portion of the diencephalon that includes a thin roof over the third ventricle. The small, cone-shaped **pineal gland** (*pineal body*) (see fig. 10.5) extends from the epithalamus; it secretes the hormone melatonin, which may play a role in controlling the onset of puberty.

10.14 Locate the pituitary gland.

The **pituitary gland**, or *hypophysis*, is attached to the inferior aspect of the diencephalon by the *stalk of the pituitary* (see figs. 10.2 and 10.5). Surrounded by a ringed network of blood vessels called the *cerebral arterial circle* (circle of Willis), the pituitary gland is structurally and functionally divided into the *anterior pituitary*, the **adenohypophysis**, and the *posterior pituitary*, the **neurohypophysis**. The endocrine functions of the pituitary gland are discussed in chapter 13.

10.15 State the principal components of the limbic system.

The **limbic system** is a roughly doughnut-shaped neuronal loop inside the brain, with the thalamic region in the "hole" and the cerebral cortex "outside" (fig. 10.6). Besides involving the hypothalamus, the limbic system includes three structures that are named after their shapes: the *amygdala* ("almond"), the *hippocampus* ("sea horse"), and the *fornix* ("arch"). The limbic system generates emotions. It also is involved in short-term memory through the hippocampus.

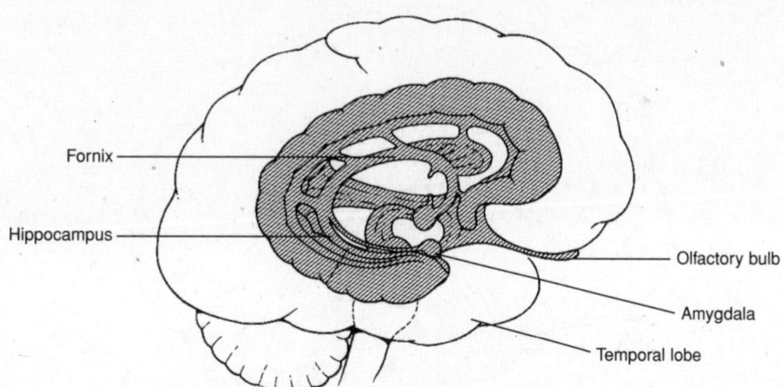

Figure 10.6 The limbic system.

Objective E To describe the location of the *mesencephalon* and the functions of its various structures.

The **mesencephalon**, or midbrain, is a short section of the brain stem between the diencephalon and the pons (see fig. 10.5). It contains the *corpora quadrigemina*, concerned with visual and auditory reflexes, and the *cerebral peduncles*, composed of fiber tracts. It also contains specialized nuclei that help to control posture and movement.

10.16 What are the functions of the superior and inferior colliculi?

The **corpora quadrigemina** are the four rounded elevations on the superior portion of the midbrain (see fig. 10.5). Of these, the two upper eminences, the **superior colliculi**, are concerned with visual reflexes; the two posterior eminences, the **inferior colliculi**, are responsible for auditory reflexes.

10.17 Do the cerebral peduncles contain only motor fibers?

No. The **cerebral peduncles** are composed of both motor and sensory fibers. They support the cerebrum and connect it to other regions of the brain.

10.18 What are the functions of the nuclei within the midbrain?

The **red nucleus** is gray matter that connects the cerebral hemispheres and the cerebellum. Its reddish color is due to a rich blood supply. It functions in reflexes concerned with motor coordination and maintaining posture. Another pigmented nucleus, the **substantia nigra**, is inferior to the red nucleus and is thought to inhibit involuntary movements. Its dark color is due to a high content of melanin.

Objective F To describe the *metencephalon*.

The **metencephalon** is the region of the brain stem that contains the *pons* and the *cerebellum* (see fig. 10.5). The pons consists of fiber tracts that relay impulses from one region of the brain to another. The cerebellum coordinates skeletal muscle contractions.

10.19 Besides serving as a relay center, what other functions has the pons?

Many of the cranial nerves originate from nuclei located within the **pons**. Other nuclei of the pons, in the *apneustic* and *pneumotaxic centers*, cooperate with nuclei in the rhythmicity area of the medulla oblongata to regulate the rate of breathing (fig. 10.7).

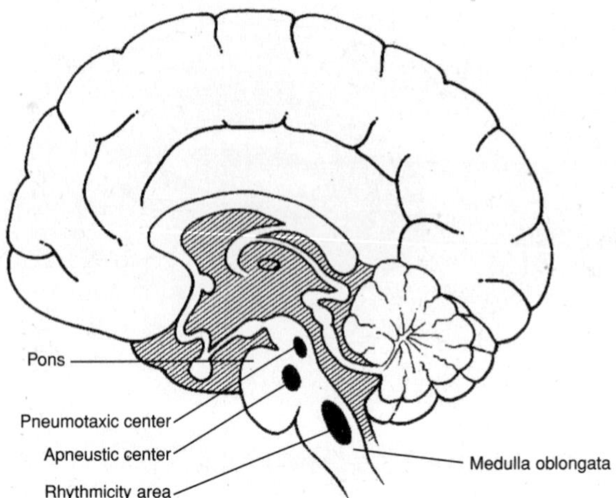

Figure 10.7 The respiratory centers in the pons and medulla oblongata.

10.20 Comment on the truth or falsity of the following statements concerning the cerebellum.

(a) It consists of two hemispheres that are convoluted at the surface.

(b) It functions totally at the subconscious (involuntary) level.

(c) It is the second-largest structure of the brain and is composed of a thin outer layer of gray matter and tracts of white matter, collectively called the *arbor vitae*.

(d) It coordinates skeletal-muscle contractions in response to incoming impulses from proprioceptors within muscles, tendons, joints, and sensory organs.

All four statements are true. The two principal functions of the cerebellum are to coordinate body movement and to maintain balance. In order to perform these functions, the cerebellum is in constant communication with other neurological structures through the **cerebellar peduncles**, which are fiber tracts that extend into and support the cerebellum.

Objective G To describe the location and structure of the *medulla oblongata* and to state its functions.

Connected to the spinal cord and composing much of the brain stem, the **medulla oblongata** is the principal structure within the *myelencephalon*. The medulla oblongata contains nuclei for cranial nerves and vital autonomic functions. The *reticular formation*, which arouses the cerebrum, is partially located in the myelencephalon.

10.21 Describe the decussation of projection fibers in the medulla oblongata.

The medulla oblongata consists primarily of white matter in the form of descending and ascending tracts that communicate between the spinal cord and various parts of the brain. Most of the projection fibers that form these tracts *decussate*, or cross to the other side, through the pyramidal region of the medulla oblongata (see fig. 10.4), permitting one side of the brain to receive information from and send information to the opposite side of the body.

10.22 State the functions of some of the nuclei of the medulla oblongata?

The gray matter of the medulla oblongata consists of a number of important nuclei (fig. 10.8) for cranial nerves (motor and sensory components), sensory relay to the thalamus, and motor relay from the cerebrum to the cerebellum.

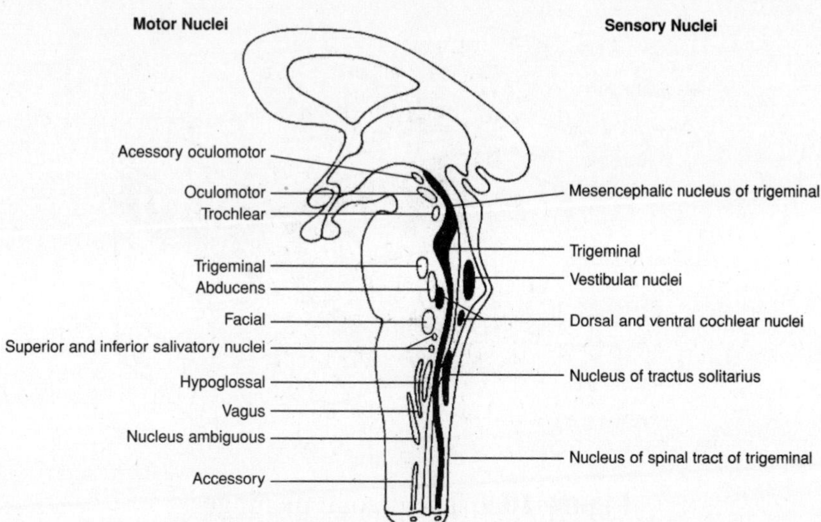

Figure 10.8 The brain-stem nuclei.

10.23 What are the autonomic functions of the medulla oblongata?

In addition to the nuclei of problem 10.22, three nuclei within the medulla oblongata function as autonomic centers for controlling visceral functions.

Cardiac center. Both inhibitory fibers (through the vagus nerves) and accelerator fibers (through spinal nerves T1–T5) arise from nuclei of the cardiac center.

Vasomotor center. Impulses from the vasomotor center cause the smooth muscles of arteriole walls to contract, thus raising the blood pressure.

Respiratory center (or *rhythmicity area*) The rate and depth of breathing is controlled by nuclei of this center, along with those of the pons (see problem 10.19).

10.24 The reticular formation is said to house the reticular activating system. Explain.

The **reticular formation** is a complex network of nuclei and ascending and descending nerve fibers within the brain stem. Functioning as the **reticular activating system (RAS)**, the reticular formation generates a continuous flow of impulses to rouse the cerebrum, unless inhibited by other parts of the brain. The RAS is sensitive to chemical changes within, or trauma to, the brain. Severe trauma to the reticular formation may cause a person to become comatose.

Objective H To describe the protective *meninges* of the CNS.

The entire CNS is protected by three connective tissue membranous coverings called **meninges** (singular, *meninx*). In order from the outside in, these are the **dura mater,** the **arachnoid,** and the **pia mater** (see fig. 10.9).

10.25 Are the meninges uniform throughout the CNS?

No. The *cranial dura mater* is divided into a thicker *periosteal layer* and a thinner *meningeal layer*. In certain areas of the brain, the two layers of the cranial dura mater are separated to form enclosed *dural sinuses* that collect venous blood and drain it to the internal jugular veins of the neck. The spinal dura mater is not double-layered.

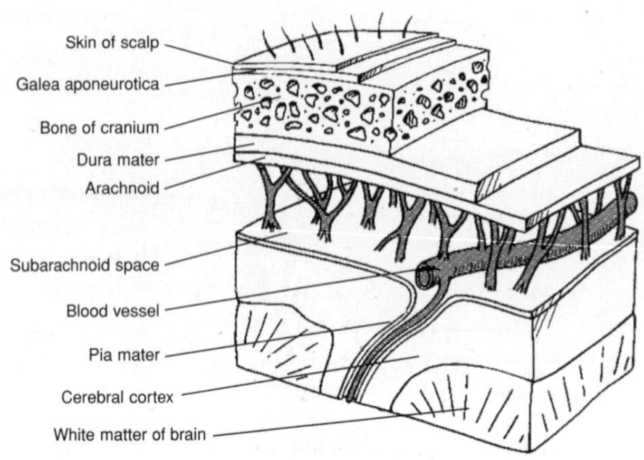

Figure 10.9 The cranial meninges.

10.26 Contrast the epidural and the subarachnoid spaces.

The *spinal dura mater* forms a tough tubular sheath around the spinal cord. The **epidural space** is a vascular area between the sheath and the vertebral canal. It contains loose fibrous and adipose connective tissues that form a protective pad around the spinal cord. The **subarachnoid space** is located between the arachnoid and the pia mater. It is maintained by delicate weblike strands (see fig. 10.9) and contains *cerebrospinal fluid* (see Objective I).

An *epidural block* is an injection of an anesthetic solution in the area where the spinal nerves pass through the epidural space. It is administered frequently in the lower lumbar area (between L3 and L4) to women in labor. As the name implies, the injection for an epidural block does not penetrate the dura mater. By contrast, a *spinal tap* (or lumbar puncture) administered in the same location punctures the dura mater. A spinal tap is performed to assess the condition of the cerebrospinal fluid and to look for signs of *spinal meningitis* or other neurological diseases.

Objective I To describe the properties and functions of cerebrospinal fluid.

Survey

Cerebrospinal fluid (CSF) is a clear, lymphlike fluid formed by active transport of substances from blood plasma in the *choroid plexuses* (see problem 10.28). CSF forms a protective cushion around and within the CNS; it also buoys the brain. CSF circulates through the ventricles of the brain, the central canal of the spinal cord, and the subarachnoid space around the CNS.

10.27 Describe the ventricles of the brain.

The **ventricles of the brain** (fig. 10.10) consist of a series of cavities that are connected to one another and to the **central canal of the spinal cord**. Each cerebral hemisphere contains one of the two **lateral ventricles** (combined *first* and *second ventricles*). The **third ventricle** is located in the diencephalon and is connected to the lateral ventricles by the two **interventricular foramina**. The fourth ventricle is located in the brain stem. It is connected to the third ventricle by the **mesencephalic aqueduct** (*cerebral aqueduct*) and meets the central canal inferiorly.

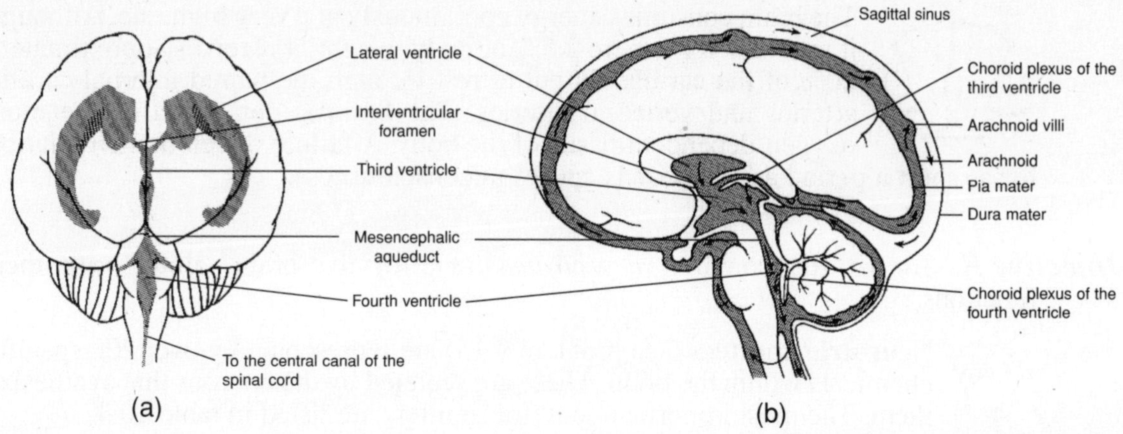

Figure 10.10 The ventricles of the brain. (*a*) An anterior view and (*b*) a sagittal view. The flow of CSF is indicated with arrows in (*b*).

10.28 What are the physical characteristics of CSF?

Cerebrospinal fluid has a specific gravity of 1.007, which is a density close to that of brain tissue. The CSF effectively reduces the weight of the brain by 97%. Thus, the 1500-gram brain, suspended in CSF, has a buoyed weight of approximately 45 grams. Since the CNS lacks lymphatic circulation, CSF moves cellular wastes into the venous return at its places of drainage into the arachnoid villi (see fig. 10.10*b*). CSF is continuously produced (about 800 mL/day) by masses of specialized capillaries called **choroid plexuses** that are located in the roofs of the ventricles (see fig. 10.10). A standing volume of 140 to 200 mL of CSF is maintained at a fluid pressure of about 10 mmHg.

Hydrocephalus is a condition in which CSF builds up within the ventricles of the brain. *Congenital*, or *primary*, *hydrocephalus* results from a developmental obstruction of CSF pathways. In a newborn, whose skull bones have not yet fused, this condition causes its head to enlarge. *Acquired hydrocephalus* results from diseases such as meningitis, or from trauma. After the cranial sutures have fused, hydrocephalus is more likely to result in brain damage.

Objective J To explain the importance of the *blood-brain barrier* in maintaining homeostasis within the brain.

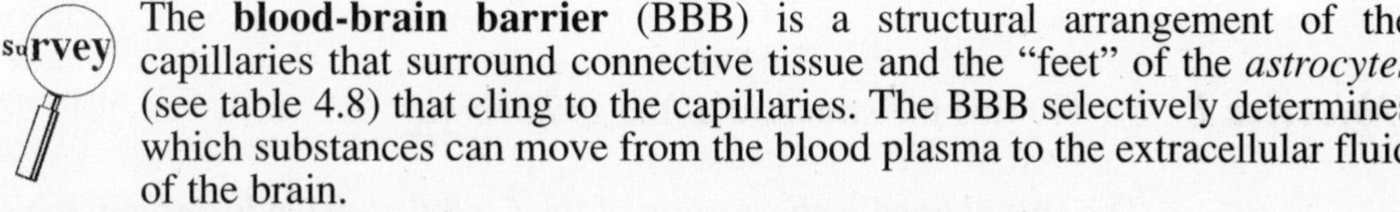

The **blood-brain barrier** (BBB) is a structural arrangement of the capillaries that surround connective tissue and the "feet" of the *astrocytes* (see table 4.8) that cling to the capillaries. The BBB selectively determines which substances can move from the blood plasma to the extracellular fluid of the brain.

10.29 *True or false*: Alcohol passes readily through the BBB because it is a lipid-soluble compound.

True. Fat-soluble compounds readily pass through the BBB, as do H_2O, O_2, CO_2, and glucose. The inorganic ions Na^+, K^+, and Cl^- pass more slowly, so that their concentrations are different in the brain than in the blood plasma. Other substances, such as macroproteins, lipids, creatinine, urea, inulin, certain toxins, and most antibiotics, are restricted in passage. The BBB is an important factor to consider when planning drug therapy for neurological disorders.

The brain consumes energy continuously at a very high rate. Although it accounts for a mere 2.5% of body weight, it receives approximately 20% of the cardiac output at rest through the paired internal carotid arteries and vertebral arteries. The brain is composed of the most oxygen-dependent tissue of the body. A failure of cerebral circulation for as short a period as 10 seconds causes unconsciousness.

Objective K To list the common *neurotransmitters* of the brain, along with their functions.

Neurotransmitters (see problem 9.15) are represented by over 200 specific chemicals within the brain. These are secreted by the neurons that synthesize them. The most important neurotransmitters are listed in table 10.3.

Table 10.3 Principal Neurotransmitters of the Brain

	Neurotransmitter	*Function*
Excitatory	Acetylcholine	Facilitates transmission of nerve impulses across synapses
	Epinephrine, norepinephrine	Arouse the brain and maintain alertness
	Serotonin	Temperature regulation, sensory perception, onset of sleep
	Dopamine	Motor control
Inhibitory	Gamma-aminobutyric acid (GABA)	Motor coordination through inhibition of certain neurons
	Glycine	Inhibits transmission along certain spinal cord tracts
Neuropeptides (short-chain amino acids)	Enkephalins, endorphins	Block transmission and perception of pain
	Substance P	Aids in transmission of impulses from pain receptors

Objective L To describe the structure of the *spinal cord*.

The **spinal cord** is the portion of the CNS that extends through the vertebral canal of the vertebral column to the level of the first lumbar vertebra (L1) (fig. 10.11). It is continuous with the brain through the foramen magnum of the skull (see table 6.3). The spinal cord consists of centrally located gray matter, involved in reflexes, and peripheral ascending and descending tracts of white matter that conduct nerve impulses to and from the brain. Thirty-one pairs of spinal nerves arise from the spinal cord (see Objective C in chapter 11).

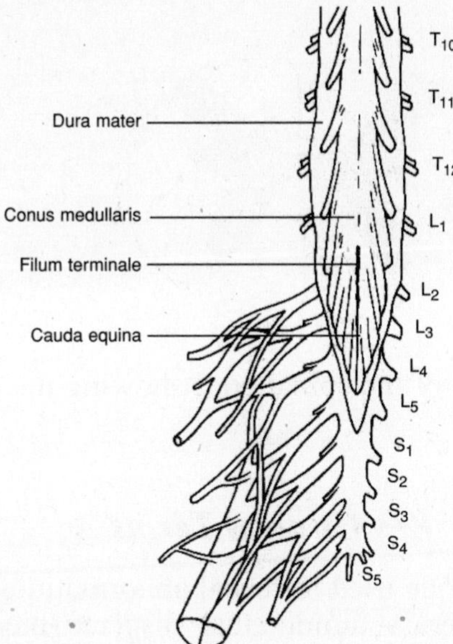

Figure 10.11 The inferior portion of the spinal cord.

10.30 Discuss the three neurological structures of the lumbar region indicated in figure 10.11.

At the level of L1, the **conus medullaris** is the terminal portion of the spinal cord. The **filum terminale** is a supportive fibrous strand of the pia mater that continues inferiorly from the spinal cord. The nerve trunks that radiate from the conus medullaris are collectively known as the **cauda equina** ("horse's tail").

10.31 Describe the geometry of the gray matter and the white matter within the spinal cord.

The deep *gray matter* has, in cross section, a four-horned or letter-H appearance (fig. 10.12). The **posterior** (*dorsal*) **horns** receive the axons of sensory fibers that enter the spinal cord from a spinal nerve; the **anterior** (*ventral*) **horns** contain the dendrites and cell bodies of motor neurons that leave the spinal cord to enter a spinal nerve. At the thoracic and lumbar levels, there are also **lateral horns.** These contain preganglionic sympathetic neurons whose axons leave via the anterior root.

The superficial *white matter* is composed primarily of myelinated fibers. Those of common origin, destination, and function form tracts. The tracts are separated by the horns of gray matter into three regions called the **posterior, lateral,** and **anterior funiculi** (fig. 10.12). Subdivisions of the funiculi, containing fibers from more than one tract, are called *fasciculi.*

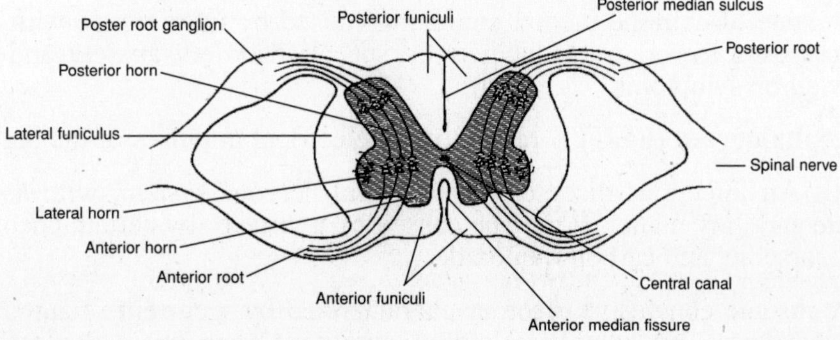

Figure 10.12 A cross section of the spinal cord and the roots of a paired spinal nerve.

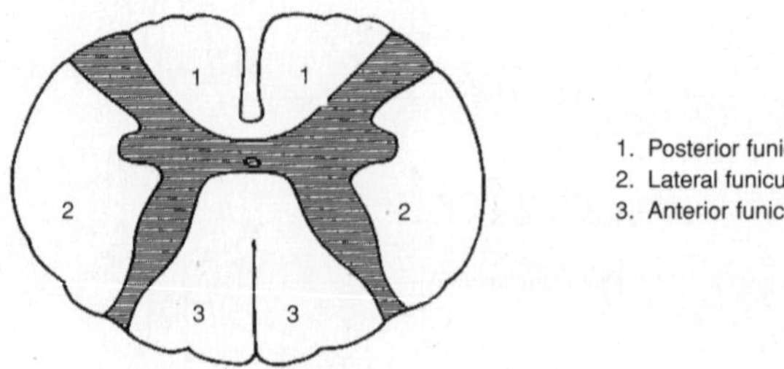

Figure 10.13 A cross section of the spinal cord showing the location of the funiculi.

Key Clinical Terms

Cerebral angiography A technique used to reveal abnormalities of cerebral blood vessels, such as aneurysms, or brain tumors that displace blood vessels. A radiopaque substance is injected into the carotid arteries; then X-ray films are taken of the blood vessels of the brain.

Cerebral concussion A transient state of unconsciousness following head injury and damage to the brain stem.

Cerebral palsy A motor nerve disorder caused by a permanent brain defect or an injury at birth or soon after. Symptoms may include paralysis, lack of coordination, and other dysfunctions of motor and sensory mechanism.

Cerebrovascular disease Any pathological change in cerebral blood vessels. Cerebrovascular diseases include aneurysms, atherosclerosis, embolism, infarction, thrombosis, stroke, and hemorrhage.

Chorea A nervous disorder characterized by bizarre, abrupt, involuntary movements. It may be hereditary or a result of rheumatic fever.

Coma Varying degrees of unconsciousness from any of a number of causes.

Convulsions Spasmodic contractions of muscles, associated with semiconsciousness or unconsciousness. Convulsions are the result of extreme irritability of the nervous system brought on, for example, by such things as brain damage, infection, or prolonged high fever.

Delirium A state of extreme mental confusion caused by interference with the metabolic processes of the brain. Hallucinations, speech disorders, anxiety, and disorientation may all be symptoms.

Electroencephalogram (EEG) A record of the electrical impulses of the brain.

Encephalitis An infectious disease of the central nervous system, with damage to both white and gray matter. It may be caused by a virus or by certain chemicals, such as lead, arsenic, and carbon monoxide.

Epilepsy A chronic convulsive disorder characterized by recurrent seizures and impaired consciousness. Epilepsy has a strong hereditary basis, but it also can be caused by head injuries, tumors, and childhood infectious diseases.

Multiple sclerosis (MS) A remitting and relapsing neurological disease that destroys the myelin of neurons. MS causes gradual paralysis and progressively severe disturbances in speech, vision, and mentation. Patients with advanced MS have difficulty in walking and suffer from body tremors, weakness, and exaggerated reflexes. The cause of MS is presently unknown, and treatment is limited.

Review Exercises

Multiple Choice

1. The white matter of the CNS is always (*a*) deep to the gray matter, (*b*) unmyelinated, (*c*) arranged into tracts, (*d*) composed of sensory fibers only.

2. Which of the following are the three initial developmental regions of the brain? (*a*) telencephalon, prosencephalon, rhombencephalon; (*b*) rhombencephalon, prosencephalon, mesencephalon; (*c*) metencephalon, myelencephalon, prosencephalon; (*d*) prosencephalon, diencephalon, mesencephalon

3. The third ventricle is located in (*a*) the cerebrum, (*b*) the forebrain, (*c*) the hindbrain, (*d*) the midbrain, (*e*) the cerebellum.

4. Neuropeptides are (*a*) neurotransmitter chemicals, (*b*) neuroglia, (*c*) products of the choroid plexuses, (*d*) nutrients for brain tissue, (*e*) both a and c.

5. The thalamus is located in (*a*) the telencephalon, (*b*) the mesencephalon, (*c*) the diencephalon, (*d*) the metencephalon, (*e*) the myelencephalon.

6. Regarding the cerebrum, which of the following is a *false* statement?
(*a*) It accounts for about 80% of the brain's mass.
(*b*) It consists of four paired lobes.
(*c*) It contains a thin superficial layer of convoluted gray matter.
(*d*) It is located within the telencephalonic region of the brain.

7. Which is *not* a lobe of the cerebrum? (*a*) parietal lobe, (*b*) insula, (*c*) occipital lobe, (*d*) temporal lobe, (*e*) sphenoidal lobe

8. Which lobe-function pairing is *incorrect*? (*a*) frontal lobe–sensory interpretation, (*b*) parietal lobe–speech patterns, (*c*) occipital lobe–vision, (*d*) temporal lobe–memory, (*e*) parietal lobe–somatesthetic interpretation

9. The basal nuclei form all of the following *except* (*a*) the putamen, (*b*) the caudate nucleus, (*c*) the globus pallidus, (*d*) the infundibulum.

10. Clusters of neuron cell bodies embedded in the white matter of the brain are referred to as (*a*) nuclei, (*b*) gyri, (*c*) sulci, (*d*) ganglia, (*e*) fasciculi.

11. Tracts of white matter that connect the right and left cerebral hemispheres are composed of (*a*) decussation fibers, (*b*) association fibers, (*c*) commissural fibers, (*d*) projection fibers.

12. Brain waves common to a healthy sleeping person and a brain-damaged awake person are called (*a*) alpha waves, (*b*) beta waves, (*c*) gamma waves, (*d*) theta waves, (*e*) delta waves.

13. Parkinson's disease and other motor disorders are attributed to dysfunction of, or trauma to (*a*) the pons, (*b*) the basal nuclei, (*c*) the parietal lobe, (*d*) the thalamus, (*e*) the corpus striatum.

14. The inability of a patient to perceive pain might be due to a tumor or trauma of (*a*) the insular lobe, (*b*) the hypothalamus, (*c*) the red nucleus, (*d*) the thalamus, (*e*) the pons.

15. Symptoms of fluctuating body temperature, intense thirst, and insomnia might indicate that a patient has dysfunction of (*a*) the hypothalamus, (*b*) the pons, (*c*) the medulla oblongata, (*d*) the pituitary gland, (*e*) the cerebellum.

16. Which property of blood is *not* monitored by the hypothalamus? (*a*) osmotic concentration, (*b*) PCO_2 content, (*c*) fatty acid content, (*d*) blood glucose levels, (*e*) amino acid levels

17. Which of the following is *not* involved with motor impulses or motor coordination? (*a*) red nucleus, (*b*) cerebellum, (*c*) basal nuclei, (*d*) precentral gyrus, (*e*) none of the preceding

18. The corpora quadrigemina, composed of the superior and inferior colliculi, is located in (*a*) the telencephalon, (*b*) the mesencephalon, (*c*) the diencephalon, (*d*) the metencephalon, (*e*) the constellation Aries.

19. The capillary network that develops in the roof of the third and fourth ventricles is called (*a*) the choroid plexus, (*b*) the sulcus limitans, (*c*) the hyperthalamic plexus, (*d*) the cerebral plexus, (*e*) the cerebral arterial circle.

20. Which brain structure–autonomic function pairing is *incorrect*? (*a*) pons–respiration, (*b*) corpus callosum–blood pressure, (*c*) medulla oblongata–respiration, (*d*) thalamus–intense pain, (*e*) hypothalamus–body temperature

21. An abnormal production of antidiuretic hormone (ADH) could result from a dysfunction of (*a*) the hypothalamus, (*b*) the choroid plexus, (*c*) the medulla oblongata, (*d*) the reticular activation system, (*e*) the pineal gland.

22. Regarding the medulla oblongata, which of the following is a *false* statement?
(*a*) It is the site of decussation of many sensory and motor fibers.
(*b*) It is located within the mesencephalon.
(*c*) It contains specialized nuclei for certain cranial nerves.
(*d*) It functions as cardiac, vasomotor, and respiratory centers.

23. The meninx in contact with the brain and spinal cord is (*a*) the pia mater, (*b*) the dura mater, (*c*) the perineural mater, (*d*) the arachnoid.

24. Cerebrospinal fluid (CSF) is found within (*a*) the epidural space, subarachnoid space, and dural sinuses; (*b*) the subarachnoid space, dural sinuses, and ventricles; (*c*) the central canal, epidural space, and subarachnoid space; (*d*) the ventricles, central canal, and subarachnoid space; (*e*) the central canal, epidural space, and ventricles.

25. Regarding cerebrospinal fluid, which of the following is a *false* statement?
(*a*) It has a specific gravity of 1.007 and buoys the brain.
(*b*) It maintains a volume of 140 to 200 mL and a fluid pressure of 10 mmHg.
(*c*) It moves metabolic wastes away from the cells of nervous tissue.
(*d*) It is produced in the choroid plexuses and drains into the cerebral arterial circle.

26. The presence of theta waves in an adult is an indication of (*a*) visual activity, (*b*) dreaming, (*c*) brain damage, (*d*) severe emotional stress, (*e*) none of the preceding.

27. The mesencephalic (cerebral) aqueduct links (*a*) the lateral ventricles, (*b*) the lateral ventricles and the third ventricle, (*c*) the third and fourth ventricles, (*d*) the lateral ventricles and the fourth ventricle, (*e*) the first and the second ventricles.

28. The spinal cord ends at the level of (*a*) the coccyx, (*b*) the first lumbar vertebra, (*c*) the sacrum, (*d*) the sciatic nerve.

29. The blood-brain barrier restricts passage of (*a*) lipids, (*b*) Na^+, (*c*) Cl^-, (*d*) H_2O, (*e*) lipid-soluble compounds.

30. Body temperature, sensory perception, and the onset of sleep are partially regulated by the neurotransmitter (*a*) glycine, (*b*) serotonin, (*c*) acetylcholine, (*d*) dopamine, (*e*) enkephalin.

31. The terminal portion of the spinal cord is known as (*a*) the cordis terminale, (*b*) the conus medullaris, (*c*) the cauda equina, (*d*) the bulbis caudis, (*e*) the filum terminale.

32. Which region of the brain is farthest from the spinal cord? (*a*) mesencephalon, (*b*) telencephalon, (*c*) myelencephalon, (*d*) metencephalon, (*e*) diencephalon

33. For substances within the blood to reach the neurons within the brain, they must first pass through a cellular membrane derived in part from (*a*) neurolemmocytes, (*b*) microglia, (*c*) astrocytes, (*d*) ganglia, (*e*) nuclei.

34. A patient with symptoms of tremor, halting speech, and an irregular gait may have experienced trauma to (*a*) the cerebrum, (*b*) the pons, (*c*) the cerebellum, (*d*) the thalamus, (*e*) the hypothalamus.

35. Blockage of the flow of cerebrospinal fluid may result in (*a*) meningitis, (*b*) hydrocephalus, (*c*) paraplegia, (*d*) encephalitis, (*e*) all of the preceding.

36. Two components of the basal nuclei are (*a*) the caudate nucleus and lentiform nucleus, (*b*) the globus pallidus and infundibulum, (*c*) the hypothalamic nucleus and red nucleus, (*d*) the insula and putamen.

37. Which is *not* involved in the transmission or perception of pain? (*a*) substance P, (*b*) thalamus, (*c*) enkephalins, (*d*) posterior horns, (*e*) none of the preceding

38. A disease of the nervous system in which the myelin sheaths of neurons are altered by the formation of plaques is (*a*) multiple sclerosis, (*b*) epilepsy, (*c*) cerebral palsy, (*d*) Parkinson's disease, (*e*) neurosyphilis.

39. Which two structures of the brain control respiration? (*a*) pons and hypothalamus, (*b*) cerebrum and hypothalamus, (*c*) pons and medulla oblongata, (*d*) hypothalamus and pituitary gland

40. Trauma to the superior colliculi would most likely affect (*a*) speech, (*b*) auditory perception, (*c*) coordination and balance, (*d*) vision, (*e*) perception of pain.

True or False

____ **1.** The thalamus is an important relay center in that all sensory impulses (except olfaction) going to the cerebrum synapse there.

____ **2.** The cerebral longitudinal fissure separates the two cerebral hemispheres, and the central sulcus separates the precentral gyrus from the postcentral gyrus.

____ **3.** The convoluted cerebral cortex and the convoluted surface of the cerebellum are the only parts of the brain that contain gray matter.

____ **4.** All ventricles of the brain are paired, except for the fourth.

____ **5.** The posterior horns of the spinal cord contain motor neurons only.

____ **6.** The motor speech (Broca's) area of the brain is generally within the left cerebral hemisphere.

____ **7.** The gyri and sulci form the convolutions of the cerebral cortex that greatly increase the surface area of the white matter.

____ **8.** Both the hypothalamus and the medulla oblongata mediate vasoconstriction and vasodilation in regulating blood pressure.

____ **9.** Association fibers are confined to a single hemisphere and serve to relay impulses to the various cerebral lobes.

____ **10.** An alpha brain-wave pattern is a healthy sign in a person who is awake but relaxed, and a beta brain-wave pattern is a healthy sign in a person who is awake and mentally alert.

____ **11.** The hypothalamus is a component of the limbic system that helps determine one's emotions.

____ **12.** The pineal gland, the hypothalamus, and the pituitary gland all have neuroendocrine functions.

____ **13.** The cerebral arterial circle constitutes the blood-brain barrier, which selectively determines which components of the blood can enter the CNS.

____ **14.** Cerebrospinal fluid is produced in the choroid plexuses; flows through the cavities, spaces, and canals of the CNS; and drains through the arachnoid villi into the venous blood draining the head.

____ **15.** The reticular activating system of the brain generates emotions.

Completion

1. The ____________________ ____________________ is the meninx closest to the brain.
2. The ____________________ gyrus is the principal motor area of the cerebrum.
3. ____________________ are neuroglial cells that participate in the blood–brain barrier.
4. The ____________________ ____________________ are tracts of white matter within the cerebellum.
5. The ____________________ lobe of the cerebrum is deep to the others.
6. ____________________ fibers connect the right and left cerebral hemispheres.
7. Cerebrospinal fluid flows through the ____________________ ____________________ of the spinal cord.

8. The spinal cord ends at the ____________________ ____________________ at the level of L1.

9. Collectively, the first and second ventricle constitute the ____________________ ventricle of the brain.

10. Brain waves are recorded as an ____________________.

Labeling

Label the structures indicated on the figure to the right.

1. ____________________

2. ____________________

3. ____________________

4. ____________________

5. ____________________

6. ____________________

7. ____________________

8. ____________________

9. ____________________

10. ____________________

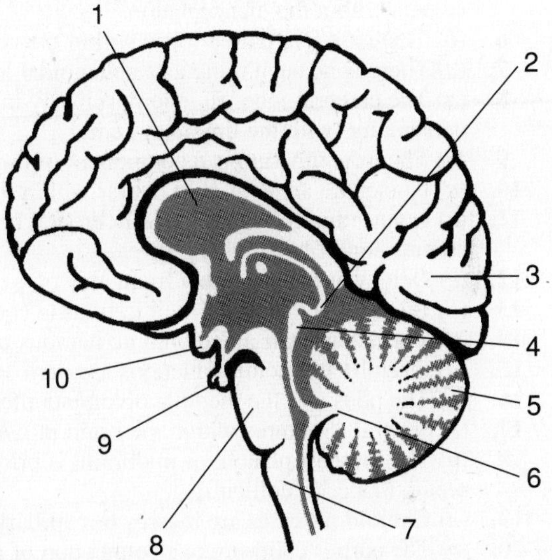

1
2
3
4
10
5
9
6
7
8

Matching

Match the structure with its description or function.

____	**1.** Thalamus	(*a*) area of decussation
____	**2.** Postcentral gyrus	(*b*) arouses the cerebrum
____	**3.** Medulla oblongata	(*c*) somatesthetic area
____	**4.** Reticular formation	(*d*) auditory reflexes
____	**5.** Choroid plexus	(*e*) drain cerebrospinal fluid
____	**6.** Arachnoid villi	(*f*) responds to intense pain
____	**7.** Inferior colliculi	(*g*) secretes melatonin
____	**8.** Pineal gland	(*h*) monitors osmotic concentration of blood
____	**9.** Pons	(*i*) produces cerebrospinal fluid
____	**10.** Hypothalamus	(*j*) apneustic center

Answers and Explanations for Review Exercises

Multiple Choice

1. (*c*) The white matter in the CNS consists of tracts that convey sensations from one structure or region to another.
2. (*b*) The rhombencephalon differentiates into the myelencephalon and the metencephalon, and the prosencephalon differentiates into the diencephalon and the telencephalon.
3. (*d*) The third and fourth ventricles are unpaired along the midline within the midbrain and hindbrain respectively.
4. (*a*) Neuropeptides are protein molecules produced within the brain.
5. (*c*) The thalamus, epithalamus, hypothalamus, and pituitary gland are autonomic nervous system centers within the diencephalon.
6. (*b*) There are five paired lobes within the cerebrum.
7. (*e*) There is no such thing as a sphenoidal lobe in a cerebral hemisphere, but there is a frontal lobe.
8. (*a*) The cerebral lobes function primarily in voluntary movement, higher intellectual processes, and personality (with the limbic system).
9. (*d*) The infundibulum is a component of the stalk of the pituitary gland.
10. (*a*) Nuclei are areas of gray matter within the white matter, where nerve impulses are processed.
11. (*c*) Connecting the right and left cerebral hemispheres, the corpus callosum is composed of commissural fibers.
12. (*e*) Delta waves have a low frequency of 1 to 5 Hz and are normal during sleep.
13. (*b*) Basal nuclei are important in muscle coordination during body movement.
14. (*d*) The thalamus is an autonomic nervous center that responds to intense pain.
15. (*a*) Over 10 autonomic functions are performed by the hypothalamus in maintaining homeostasis.
16. (*b*) The pons and the medulla oblongata monitor respiratory gases and control respiratory rates.
17. (*e*) Several structures within the brain influence motor coordination and balance.
18. (*b*) The mesencephalon, or midbrain, is primarily concerned with hearing (inferior colliculi) and seeing (superior colliculi).
19. (*a*) Choroid plexuses are masses of capillary networks that produce cerebrospinal fluid.
20. (*b*) The corpus callosum is a connection of nerve fibers between the two cerebral hemispheres.
21. (*a*) The hypothalamus influences the production of ADH by the posterior pituitary.
22. (*b*) The medulla oblongata is located within the myelencephalon.
23. (*a*) The pia mater adheres to the surface of the CNS, actually following the contours of the sulci and gyri.
24. (*d*) All of the spaces, canals, and subarachnoid space of the CNS contain CSF.
25. (*d*) CSF drains from the CNS through the arachnoid villi into the venous return from the head.
26. (*d*) The presence of theta waves may even presage a nervous breakdown.
27. (*c*) The mesencephalic aqueduct traverses the midbrain (mesencephalon) connecting the unpaired third and fourth ventricles.
28. (*b*) Because the spinal cord ends at L1, a spinal tap can be performed below this level without risk of spinal cord puncture.
29. (*a*) Lipids cannot traverse the blood-brain barrier ("fathead" is a misnomer).
30. (*b*) Although it is also produced elsewhere in the body, the serotonin produced in the brain has the specific functions of influencing body temperature, meditating sensory perception, and regulating sleep.
31. (*b*) The conus medullaris is the point of spinal cord termination at the level of L1.
32. (*b*) The telencephalon is the superomost region of the brain. The cerebrum is located within the telencephalon.
33. (*c*) Astrocytes are specialized glial cells that help form the blood-brain barrier.
34. (*c*) The cerebrum controls all skeletal muscle contraction, voluntary and involuntary.
35. (*b*) Hydrocephalus is bulging of the brain due to inadequate drainage of CSF.
36. (*a*) Consisting of the caudate nucleus, lentiform nucleus, and others, the basal nuclei influence motor control.

37. (*e*) Substance P mediates pain perception; the thalamus responds to intense pain; enkephalins dampen pain perception; and the posterior (dorsal) horns are composed of sensory neurons that transmit pain sensations.
38. (*a*) Multiple sclerosis refers to the multiple scarlike growths (scleroses) on neurological tissues.
39. (*c*) Apneustic and pneumotaxic centers are located within the pons, and the rhythmicity area is located in the medulla oblongata.
40. (*d*) The superior colliculi function in eye-hand coordination.

True or False

1. True
2. True
3. False; nuclei are clusters of gray matter located within the white matter.
4. False; both the third and fourth ventricles are unpaired
5. False; the posterior horns contain sensory neurons only
6. True
7. False; gyri and sulci form convolutions of gray matter
8. False; vasoconstriction and vasodilation to maintain blood pressure are functions of the medulla oblongata
9. True
10. True
11. True
12. True
13. False; the cerebral arterial circle provides a rich blood supply to the brain, especially the pituitary gland
14. True
15. False; the limbic system generates emotions; the reticular activating system stimulates (alerts) the brain

Completion

1. pia mater
2. precentral
3. Astrocytes
4. arbor vitae
5. insular
6. Commissural
7. central canal
8. filum terminale
9. lateral
10. electroencephalogram

Labeling

1. Corpus callosum
2. Pineal body
3. Occipital lobe
4. Corpora quadrigemina
5. Cerebellum
6. Arbor vitae
7. Medulla oblongata
8. Pons
9. Pituitary gland
10. Optic chiasma

Matching

1. (*f*)
2. (*c*)
3. (*a*)
4. (*b*)
5. (*i*)
6. (*e*)
7. (*d*)
8. (*g*)
9. (*j*)
10. (*h*)

Peripheral and Autonomic Nervous Systems 11

Objective A To review the *organization of the nervous system* and to distinguish between the structural and functional divisions.

Anatomically, the nervous system is divided into the **central nervous system (CNS)** and the **peripheral nervous system (PNS)**. The CNS includes the *brain* and the *spinal cord* (see chapter 10). The PNS (described in this chapter) includes the *cranial nerves*, arising from the inferior aspect of the brain, and the *spinal nerves*, arising from the spinal cord.

The **autonomic nervous system (ANS)** is a functional division of the nervous system. It consists of components within the CNS and specific nerves. The ANS is subdivided into *sympathetic* and *parasympathetic divisions* (see problem 11.22) that provide innervation to smooth and cardiac muscles, as well as glands. The ANS functions autonomically to maintain homeostasis and carry out many involuntary functions in the body.

Reference is frequently made to a **somatic nervous system** in connection with the innervation of skeletal muscles, which have both voluntary and involuntary contraction. The designation **visceral nervous system** refers to the autonomic innervation of visceral organs (those organs within the thoracic and abdominopelvic cavities).

11.1 Are the nerves of the PNS sensory only, motor only, or mixed?

Most peripheral nerves are composed of both motor and sensory neurons; they are thus **mixed nerves**. Some cranial nerves, however, are composed either of sensory neurons only or of motor neurons only. **Sensory nerves** serve the special senses (see chapter 12) of taste, smell, sight, hearing, and balance. **Motor nerves** conduct impulses to muscles, causing them to contract, or to glands, causing them to secrete.

11.2 What function is served by ganglia in the PNS?

In the PNS, the cell bodies of neurons are clumped together as **ganglia**. Ganglia are sites for possible synapses of neurons between organs and the spinal cord.

11.3 What are dermatomes and why are they clinically important?

A **dermatome** is the area of the skin innervated by all the cutaneous neurons of a given spinal or cranial nerve (fig. 11.1). The pattern of dermatome innervation is of clinical importance in anesthetizing a particular portion of the body. Abnormally functioning dermatomes also provide clues about injury to the spinal cord or to specific spinal nerves.

Objective B To identify the 12 pairs of *cranial nerves* and their functions.

Cranial nerves connect the brain to structures of the head, neck, and trunk. Most are mixed nerves, some are totally sensory nerves, and others are primarily motor nerves. The names of the cranial nerves indicate their primary functions or their general distribution. The cranial nerves are also identified by Roman numerals in order of appearance from front to back (see table 11.1 and fig. 11.2).

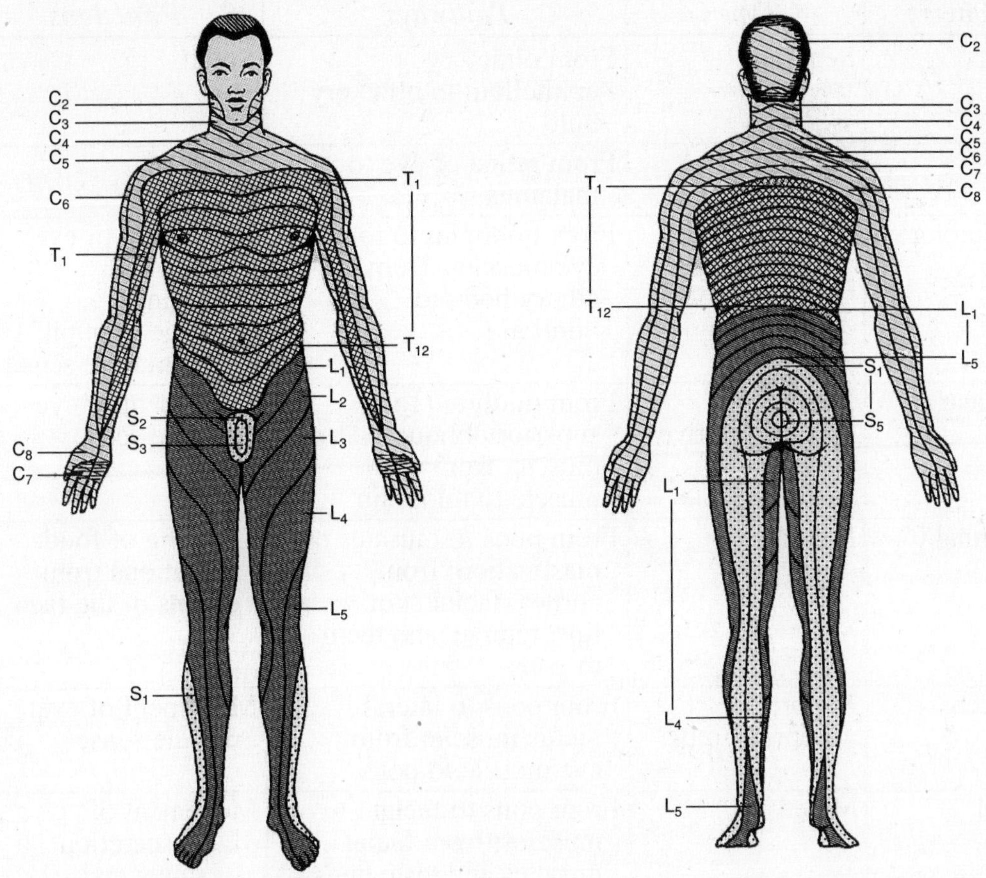

Figure 11.1 Pattern of dermatomes and the spinal nerves involved.

11.4 Where do the cranial nerves attach to the brain?

The cranial nerves emerge from the inferior surface of the brain and pass through foramina of the skull (see fig. 6.10 and table 6.3). The first two pairs of cranial nerves are attached to the forebrain; the remaining 10 pairs are attached to the brain stem. Sensory nerves originate in nerve trunks and sensory organs, and terminate at brain nuclei; motor nerves originate at brain nuclei.

11.5 What do the olfactory, optic, and vestibulocochlear nerves have in common?

They are the purely *sensory cranial nerves* (see chapter 12). The **olfactory nerve** consists of bipolar neurons that function as *chemoreceptors* and relay sensory impulses of smell from mucous membranes of the nasal cavity. The **optic nerve** conducts sensory impulses from the *photoreceptors* (rods and cones) in the retina of the eye. The **vestibulocochlear nerve** consists of a *vestibular branch*, arising from the vestibular organs of equilibrium and balance, and a *cochlear branch*, arising from the spiral organ of hearing.

Table 11.1 The Cranial Nerves

Cranial nerve	*Type*	*Pathways*	*Functions*
I Olfactory	Sensory	From olfactory epithelium to olfactory bulb	Smell
II Optic	Sensory	From retina of eye to thalamus	Sight
III Oculomotor	Motor; proprioceptive	From midbrain to four eye muscles; from ciliary body to midbrain	Movement of eye and eyelid; focusing; change in pupil size; muscle sense
IV Trochlear	Motor; proprioceptive	From midbrain to superior oblique muscle; from eye muscle to midbrain	Movement of eye muscle sense
V Trigeminal	Mixed	From pons to muscles of mastication; from cornea, facial skin, lips, tounge, and teeth to pons	Chewing of food; sensations from organs of the face
VI Abducens	Motor; proprioceptive	From pons to lateral rectus muscle; from eye muscle to pons	Movement of eye; muscle sense
VII Facial	Mixed	From pons to facial muscles; from facial muscles and taste buds to pons	Movement of face; secretion of saliva and tears; muscle sense; taste
VIII Vestibulo-cochlear	Sensory	From organs of hearing and balance to pons	Hearing; balance and posture
IX Glosso-pharyngeal	Mixed	From medulla oblongata to pharyngeal muscles; from pharyngeal muscles and taste buds to medulla oblongata	Swallowing, secretion of saliva; muscle sense; taste
X Vagus	Mixed	From medulla oblongata to viscera; from viscera to medulla	Visceral muscle movement; visceral sensations
XI Accessory	Motor; proprioceptive	From medulla oblongata to pharynx and neck muscles; from neck muscles to medulla	Swallowing and head movement; muscle sense
XII Hypoglossal	Motor; proprioceptive	From medulla oblongata to muscles of the tongue; from tongue muscles to medulla	Speech and swallowing; muscle sense

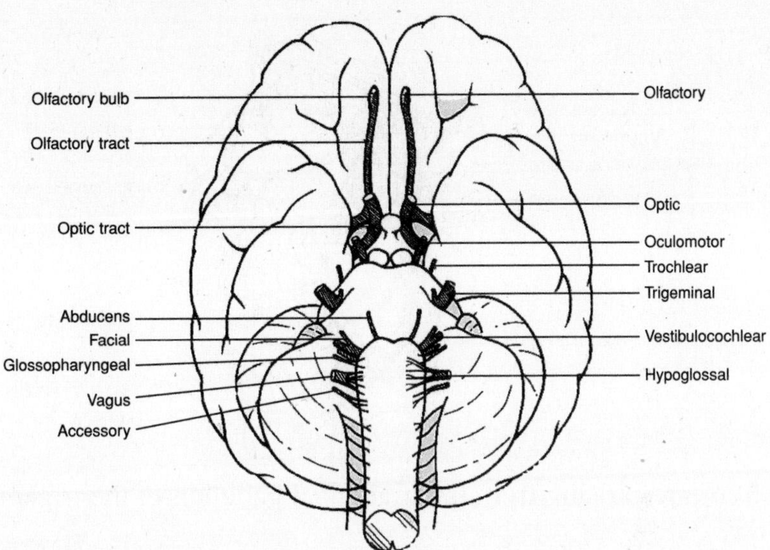

Figure 11.2 The cranial nerves as seen in an inferior view of the brain.

11.6 Which cranial nerves innervate the muscles that move the eyeball?

Movements of the eyeball are controlled by six extrinsic eye (ocular) muscles. The **oculomotor nerve** innervates the *superior, inferior,* and *medial recti muscles* and the *inferior oblique muscle* (see fig. 12.5). The **abducens nerve** innervates the *lateral rectus muscle*, and the **trochlear nerve** innervates the *superior oblique muscle*.

In the event of a concussion or other head injury, part of a quick neurological assessment for cranial nerve damage is to have the patient follow finger movements with the eyes. An inability to look cross-eyed may signal damage to the *oculomotor nerve*; problems with lateral eye movements, damage to the *abducens nerve*; and trouble looking downward, away from the midline, damage to the *trochlear nerve*.

11.7 Of all the cranial nerves, which is most important to a dentist?

A knowledge of the **trigeminal nerve** (fig. 11.3) is essential in the practice of dentistry. This paired cranial nerve conveys sensory information from the face, nasal area, tongue, teeth, and jaws; it supplies motor innervation to the muscles of mastication (see fig. 8.2). The trigeminal nerve gives rise to three separate nerves that branch from the trigeminal ganglion (fig. 11.3). The **ophthalmic nerve** conveys sensory innervation to the anterior scalp, skin of the forehead, upper eyelid, surface of the eyeball, lacrimal (tear) gland, side of the nose, and upper mucosa of the nasal cavity. The **maxillary nerve** conveys sensory innervation to the lower eyelid, lateral and inferior mucosa of the nasal cavity, palate and portions of the pharynx, teeth and gums of the upper jaw, upper lip, and skin of the cheek. The **mandibular nerve** conveys sensory innervation to the teeth and gums of the lower jaw, anterior two-thirds of the tongue, mucosa of the mouth, auricle of the ear, and lower part of the face. It is the motor portion of the mandibular nerve that serves the muscles of mastication.

Surface and bony landmarks of the oral cavity are invaluable to a dentist in administering an anesthetic prior to filling or extracting a particular tooth. Alveolar nerves can be desensitized by injections near the roots of specific teeth. An anesthetic injection near the mental nerve desensitizes the lower incisors. A maxillary nerve block, performed by injecting near the sphenopalatine ganglion, desensitizes the teeth in the upper jaw.

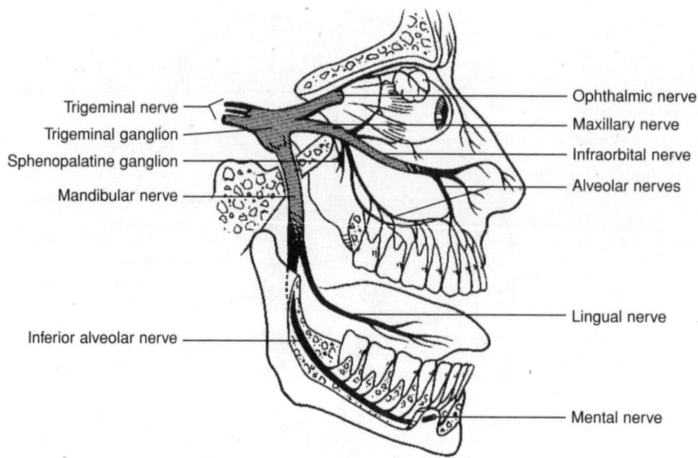

Figure 11.3 Nerves arising from the trigeminal ganglion of the trigeminal nerve.

Tic douloureux, also called *trigeminal neuralgia*, is a disorder of the trigeminal nerve characterized by severe recurring pain in one side of the face. Because the pain cannot be treated with drugs on a long-term basis, denervation eventually may be required for the patient. Caution must then be exercised while eating, however, so as not to unknowingly chew the cheek.

11.8 What are the functions of the facial nerve?

The **facial nerve** (fig. 11.4) provides motor innervation to the facial muscles and salivary glands. It also conducts sensory impulses from taste buds on the anterior two-thirds of the tongue (see problem 12.2).

Bell's palsy is a temporary functional disorder of a facial nerve, usually of sudden onset. The facial muscles on the affected side lose tonus, causing them to sag. Bell's palsy is thought to be virally caused. There is no treatment for the condition, and usually recovery is complete.

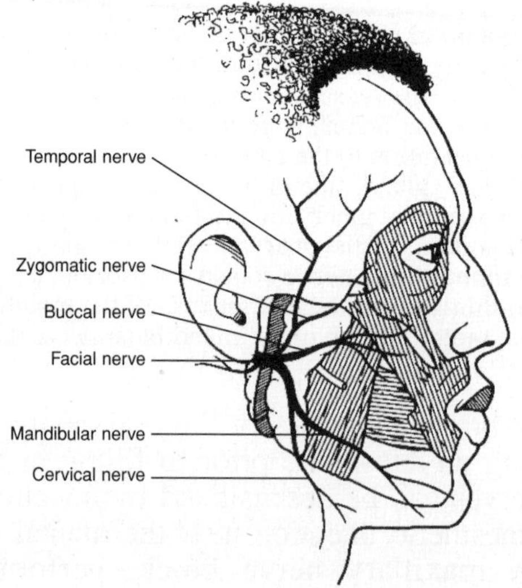

Figure 11.4 Nerves arising from the facial nerve.

11.9 Describe the distribution of the vagus nerve.

The paired **vagus nerves** are the principal autonomic nerves that provide visceral innervation (fig. 11.5). Autonomic impulses through the vagus nerves regulate digestive activities, including glandular secretions and peristalsis. Sensory fibers of the vagus nerves convey sensations of hunger (hunger pangs), abdominal distension, intestinal discomfort, and laryngeal movement.

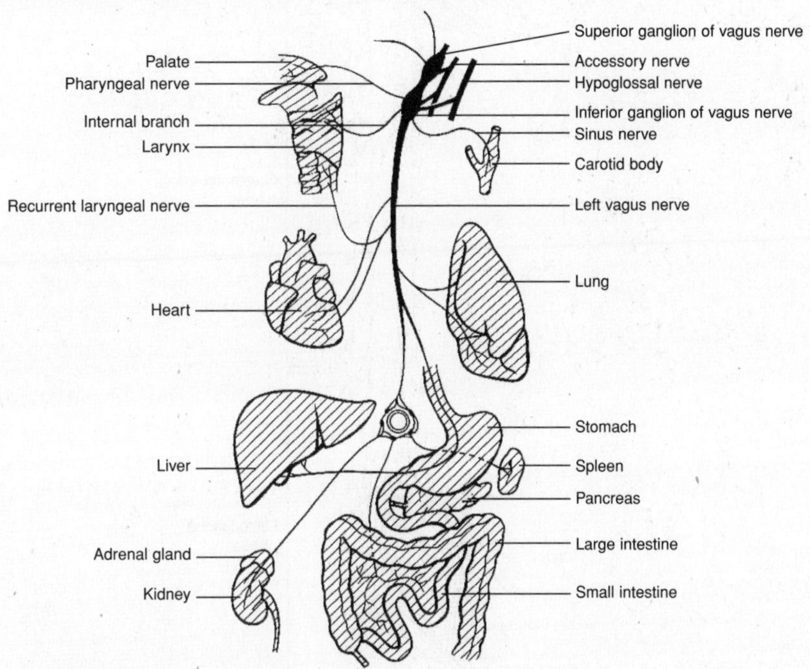

Figure 11.5 Innervation pattern of the left vagus nerve.

11.10 *True or false*: Because the cranial nerves emerge from the inferior surface of the brain, they are well protected from trauma.

False. A blow to the head not only may cause trauma at the point of impact, but also at the opposite side of the skull, where the brain rebounds off the cranium. A blow to the top of the head, for example (as in an automobile accident), may damage the cranial nerves from the rebound of the brain off the floor of the cranium. Neurological examinations following traumatic head injuries routinely involve testing for dysfunctions of cranial nerves.

Objective C To locate and describe the spinal nerves.

Survey

The 31 pairs of **spinal nerves** are grouped as follows: 8 *cervical nerves*, 12 *thoracic nerves*, 5 *lumbar nerves*, 5 *sacral nerves*, and 1 *coccygeal nerve* (fig. 11.6). The first pair of cervical nerves (C1) emerges between the occipital bone of the skull and the first cervical vertebra (the atlas). The rest of the spinal nerves exit the spinal cord and vertebral canal through *intervertebral foramina* (see problem 6.26). Each spinal nerve is a *mixed nerve*, attached to the spinal cord by a *posterior* (*dorsal*) *root* of sensory fibers and an *anterior* (*ventral*) *root* of motor fibers.

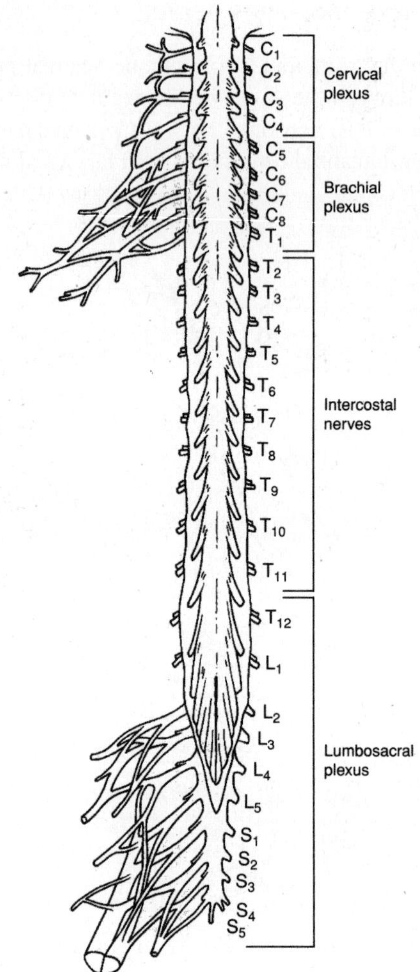

Figure 11.6 The spinal cord, spinal nerves, and plexuses.

11.11 Trace the branching of the spinal nerves.

Upon emergence through the intervertebral foramina, the anterior roots (immediately) and the posterior roots (after swelling into *posterior* (*dorsal*) *root ganglia*, where the cell bodies of the sensory neurons are located) become, respectively, *anterior* and *posterior rami* (fig. 11.7). These rami further divide, or ramify. Except in the thoracic nerves T2–T12, the anterior rami of different spinal nerves combine and then split again, forming a network known as a *plexus*. There are four plexuses of spinal nerves: the *cervical plexus, brachial plexus, lumbar plexus,* and *sacral plexus* (see fig. 11.6). The last two may be referred to jointly as the *lumbosacral plexus*. Nerves that emerge from a plexus no longer carry a spinal designation, but instead are named according to the structure or region they innervate.

11.12 Identify some common nerves and their sites of origin.

Of the hundreds of nerves in the body, several paired nerves stand out because of their size and broad area of innervation. The paired **phrenic nerves** arise from the cervical plexuses (right and left), travel through the thorax, and innervate the diaphragm. Impulses through these nerves cause contraction of the diaphragm and inspiration of air.

The **axillary, radial, musculocutaneous, ulnar,** and **median nerves** arise from the brachial plexus and innervate the shoulder and upper extremity. When you hit your "funny bone" at the elbow, it is the ulnar nerve that is traumatized.

The **femoral, obturator,** and **saphenous nerves** arise from the lumbar plexus and innervate portions of the hip and lower extremity.

The large **sciatic nerve** (which consists of **tibial** and **common fibular nerves**) arises from L4–S3 of the sacral plexus, passes through the pelvis, and extends down the posterior aspect of the thigh within the sciatic sheath. It is the largest nerve in the body. A posterior dislocation of the hip joint will generally injure the sciatic nerve. A herniated disc, pressure from the uterus during pregnancy, or an improperly administered injection into the buttock may damage the roots leading to the sciatic nerve or the nerve itself.

Compression of a nerve may have serious consequences, including paralysis. Even a temporary compression of the sciatic nerve, for example, as you sit on a hard surface for a period of time may result in the perception of tingling in the limb as you stand up. The limb is said to have "gone to sleep."

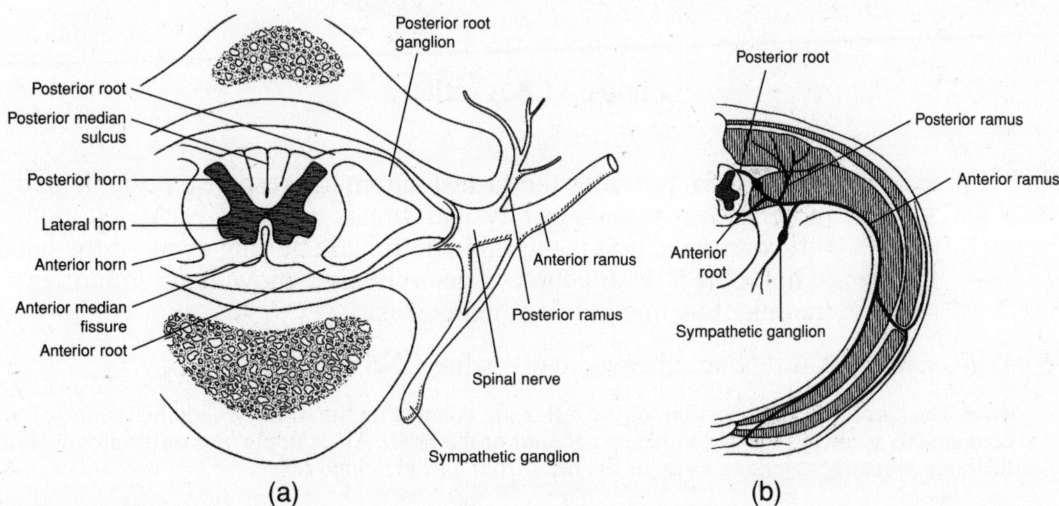

Figure 11.7 (*a*) A cross section of the spinal cord, a spinal nerve, and the rami. (*b*) A detailed pattern of spinal nerve innervation of the body wall and sympathetic innervation to visceral organs.

Objective D To be able to trace a spinal *reflex arc*.

There are five components in a typical **reflex arc** (fig. 11.8).

Receptor. Located within the skin, a tendon, a joint, or some other peripheral organ, a receptor consists of *dendritic endings of a sensory neuron* that responds to specific stimuli, such as sudden pressure or pain.

Sensory neuron. Extending from the receptor through the *posterior root*, the sensory (afferent) neuron conveys stimuli to the *posterior horn* of the spinal cord. The cell bodies of sensory neurons are located in *posterior root ganglia*.

Center. The axon of a sensory neuron synapses with an *association neuron* (also called an *interneuron* or *internuncial neuron*) within the center. The center in the spinal cord appears as an H of gray matter.

Motor neuron. Beginning at a synapse with the association neuron, the motor neuron conveys impulses from the *anterior horn* of the spinal cord, through the *anterior root*, to the effector organ.

Effector. The effector is a *muscle* or *gland* that responds to a motor impulse by contracting or secreting, respectively.

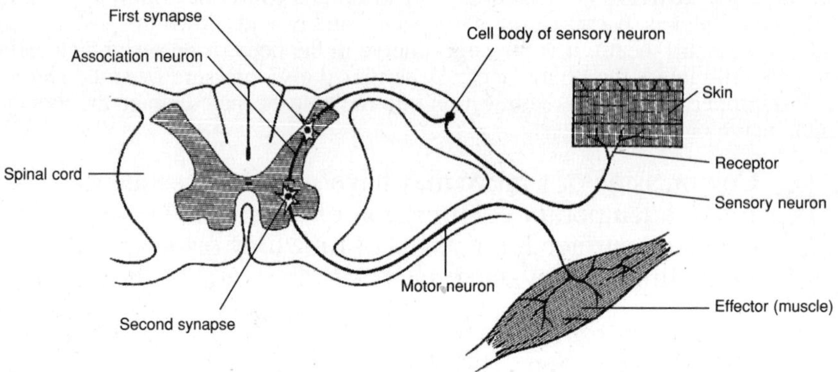

Figure 11.8 A reflex arc.

A *reflex arc* provides the fastest automatic response possible to avoid more serious trauma or physical threats to the body. An example of a reflex arc in action is the rapid automatic pulling away of the hand as a hot object is touched. This reflexive movement minimizes the trauma, thus maintaining homeostasis.

11.13 *True or false.* A reflex arc always involves the CNS.

True. The "arc," or center portion, of the reflex arc connecting the sensory with the motor components is always located in the spinal cord or the brain. An example of a reflex arc involving the brain is the rapid jerking away of the head from a sudden loud noise.

11.14 Which parts of a reflex arc constitute a motor unit?

Recall from chapter 7 (Objective F) that a motor unit consists of a motor neuron coupled with the specific skeletal muscle fibers that it innervates. This means that the motor unit is represented by the motor neuron and a specific cluster of skeletal muscle fibers, as shown in fig. 11.8.

Checking a patient's reflexes (*reflex testing*) is a common part of a routine physical examination. Deep tendon reflex testing provides information about the functioning of receptors, sensory nerves, synapses, and the spinal cord. It also checks for motor reflex problems. The functioning of these structures may be altered by developmental problems, drugs, or certain diseases.

11.15 What happens to make a person aware of a reflex?

Synapses on both sides of an association neuron (see fig. 11.8) permit communication with tracts up and down the spinal cord. For example, when a person steps on a piece of broken glass with a bare foot, the injured foot is reflexively pulled away from the harmful object. As the foot is pulled away, and in a near simultaneous movement, the arms are extended to maintain balance on one foot. Within milliseconds, a pain sensation is conveyed to the brain and the person is aware of what has happened, and even the nature of the reflexive response.

11.16 Give an example of a *monosynaptic reflex* (a reflex arc without an association neuron), and a *polysynaptic reflex* (a reflex arc that involves more than one association neuron).

Monosynaptic reflex. An example is the *knee-jerk reflex*. Tapping the patellar ligament with a rubber mallet causes the quadriceps femoris muscle to stretch, which provokes impulses from intrafusal spindle receptors at the tendinous attachment of the muscle. The impulses are conducted along the sensory neuron to the spinal cord, where the sensory neuron synapses directly with the motor neuron. This stimulates the contraction of extrafusal fibers, and thus the whole muscle. As the quadriceps femoris muscle contracts, the knee joint extends.

Polysynaptic reflex. An example is the *withdrawal reflex.* When a painful stimulus contacts the skin—for example, a sharp or hot object—a sensory receptor is activated. Sensory impulses are transmitted through a sensory neuron to the spinal cord, where two or more association neurons are stimulated. One association neuron generates impulses to a motor neuron, which initiates a response such as foot or hand withdrawal; the other association neurons conduct impulses to the brain, so that the person becomes aware of the painful event.

Objective E To distinguish further between the *ANS* and the *somatic system.*

The autonomic and somatic components of the nervous system are compared in table 11.2.

Table 11.2 A Comparison of the Autonomic and Somatic Nervous Systems

Autonomic	*Somatic*
Functions automatically, generally without conscious awareness	Conscious or voluntary regulation
Fibers synapse once (at a ganglion) after they leave the CNS: Cell body in CNS Synapse Heart	Fibers do not synapse after they leave the CNS: Cell body in CNS Skeletal muscle
Effector cells can be either stimulated or inhibited	Effects on skeletal muscle fibers always stimulatory

11.17 What specific physiological activities are regulated by the ANS?

The ANS is instrumental in maintaining homeostasis. Autonomic responses include regulating the following: the diameter of blood vessels (and thus blood pressure), GI secretion, the diameter of the pupils, micturition (see Objective J in chapter 21), sweating, the glomerular filtration rate in the kidneys, diameter of the bronchioles, erection of the penis, basal metabolism, liver glycogenolysis, body temperature, and adrenal medulla secretion. (Not a complete list.)

11.18 How do the ANS and CNS interact?

Sensory impulses from visceral organs are carried via sensory nerves of the ANS to the CNS, where they mainly influence centers within the hypothalamus, brain stem, and spinal cord. These centers integrate the sensory visceral input with input from higher brain centers (the cerebral cortex and limbic system). The appropriate responses are then sent back to the visceral organs through motor nerves of the ANS.

Objective F To compare the *sympathetic* and *parasympathetic divisions of the ANS* as to origin of preganglionic fibers, location of ganglia, and neurotransmitter substances.

The sympathetic and parasympathetic divisions of the autonomic nervous system are compared in table 11.3.

Table 11.3 A Comparison of the Sympathetic and Parasympathetic Divisions of the ANS

Feature	*Sympathetic division*	*Parasympathetic division*
Orgin of preganglionic fibers	Thoracolumbar nerves	Craniosacral nerves
Location of ganglia	Far from visceral effector organs (see problem 11.19)	Near or within visceral effector organs
Neurotransmitter substances	In ganglia, acetylcholine; in effector organs, norepinephrine	In ganglia, acetylcholine; in effector organs, acetylcholine

11.19 Describe the ganglia in both the sympathetic and parasympathetic divisions of the ANS.

Sympathetic division. There are two types of sympathetic ganglia: *sympathetic chain ganglia* and *collateral ganglia*. **Sympathetic chain ganglia**, or *paravertebral ganglia*, are interconnected by neuron fibers to form two chains lateral to the spinal cord. There are 22 ganglia in each chain (3 cervical, 11 thoracic, 4 lumbar, and 4 sacral). As diagrammed in the upper part of fig. 11.9*a*, preganglionic neurons leave the spinal cord, pass through anterior roots into spinal nerves, and then pass from the spinal nerves via *white rami communicantes* into the sympathetic chains. There, most of them synapse (on postganglionic neurons) in the chain ganglia. Some of the postganglionic neurons travel back into spinal nerves via *gray rami communicantes*, while the rest pass directly to the viscera. **Collateral**, or *prevertebral*, **ganglia** are found outside the sympathetic chain, in the vicinity of the viscera and arteries. As diagrammed in the lower part of figure 11.9*a*, some (preganglionic) neurons synapse (on postganglionic neurons) in collateral ganglia (*celiac, superior mesenteric,* and *inferior mesenteric*).

Parasympathetic division. All parasympathetic ganglia are called *terminal ganglia* because they are located close to or in the target organ. Two examples of parasympathetic innervation are schematized in fig. 11.9*b*.

11.20 Explain why parasympathetic fibers and sympathetic fibers are referred to as *cholinergic* and *adrenergic*, respectively.

The reason is that acetylcholine and norepinephrine are the respective chemical transmitters released at effector organs in the parasympathetic and sympathetic divisions.

In three exceptions—those of sweat glands, some blood vessels within skeletal muscles and the external genitalia, and the adrenal medulla—the innervating sympathetic fibers are cholinergic.

11.21 What are the types of acetylcholine receptors (cholinergic) in the ANS?

Muscarinic receptors are located on effector cells innervated by postganglionic neurons of the parasympathetic division and on those effector cells innervated by postganglionic cholinergic neurons of the sympathetic division (see problem 11.19). **Nicotinic receptors** are located at the ganglia in both the sympathetic and parasympathetic divisions.

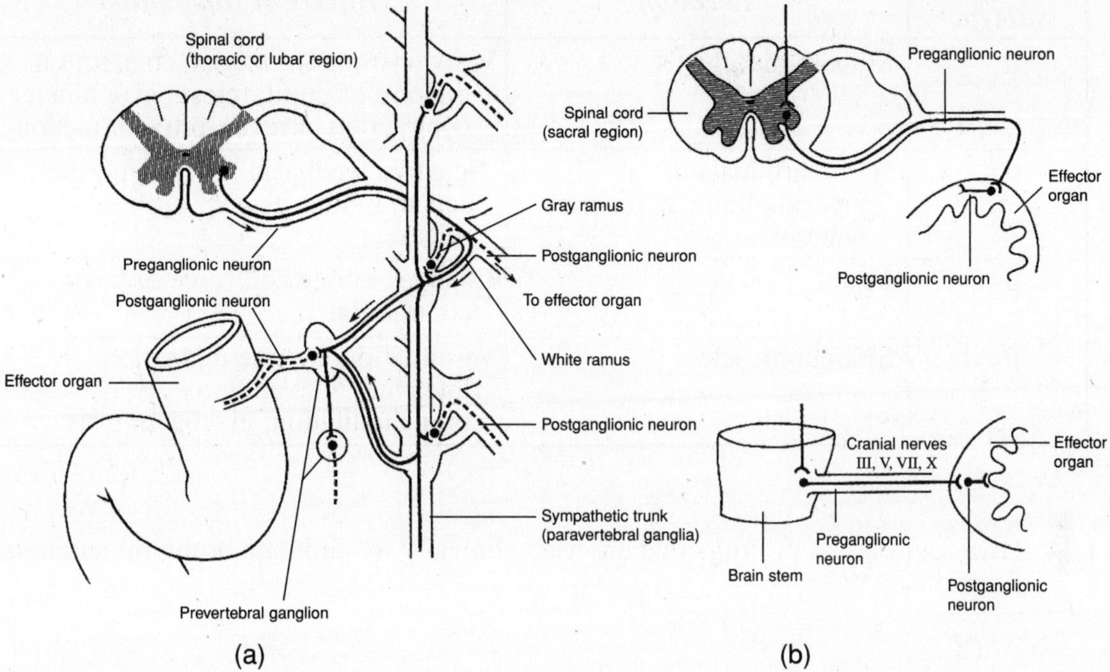

Figure 11.9 Innervation of (*a*) sympathetic ganglia and (*b*) parasympathetic ganglia.

11.22 What are the types of norepinephrine receptors (adrenergic) in the ANS?

There are two main types, called *alpha* (α) *receptors* and *beta* (β) *receptors*, each divided into two subtypes (table 11.4). Norepinephrine stimulates mainly alpha receptors; *epinephrine* stimulates both alpha and beta receptors approximately equally. *Isoproterenol*, a synthetic catecholamine, stimulates mainly beta receptors.

Objective G To be able to predict the effects of *sympathetic* versus *parasympathetic stimulation* on specific organs.

The heart, as well as most smooth muscles and visceral organs of the body, is innervated by both sympathetic and parasympathetic fibers. One division stimulates, while the other one inhibits. The two divisions are usually activated reciprocally; that is, as the activity of one is enhanced, the activity of the other is diminished. To predict the effects of each division on a specific organ, use the following rule of thumb:

Sympathetic stimulation activates the body in states of stress, fear, and rage (the "fight-or-flight" reaction), and during strenuous physical activity.

Parasympathetic stimulation maintains body functions under quiet, day-to-day living conditions; it decreases heart rate and promotes digestion and absorption of food.

11.23 List the organs that are innervated by the ANS and indicate the effects of sympathetic and parasympathetic stimulation on each organ (see tables 11.4 and 11.5).

Table 11.4 Types of Norepinephrine Receptors

Receptor subtype	*Location*	*Effects of stimulation*
α_1	Smooth muscle	Vasoconstriction, uterine contraction, dilation of pupil, intestinal sphincter contraction, arrector pili contraction
α_2	Axon terminals of postganglionic adrenergic neurons	Negative feedback: norepinephrine acts to inhibit its own further release
β_1	Heart	Changes in rate and force of heart contraction
β_2	Smooth muscle	Vasodilation, uterine relaxation, intestinal relaxation, bronchodilation, glycogenolysis

11.24 Give four classes of drugs that are used clinically to stimulate or inhibit autonomic functions.

Adrenergic receptor stimulants. These include *epinephrine, norepinephrine, isoproteranol, ephedrine,* and *amphetamine.* Prescribed to dilate bronchial tubes, treat cardiac arrest, dilate pupils, delay absorption of local anesthetics, elevate mood of patient.

Adrenergic receptor antagonists. These include *phentolamine, phenoxybenzamine, prazosin* (alpha blockers); *propranolol, timolol, nadolol* (beta blockers). Prescribed to lower blood pressure in cases of *pheochromocytoma* (alpha blockers); lower blood pressure, reduce frequency of anginal episodes, treat heart arrhythmias, reduce intraocular pressure in cases of glaucoma (beta blockers).

Cholinergic receptor stimulants. These include *acetylcholine* and its mimics—*methacholine, carbachol, bethanecol.* Prescribed to stimulate GI tract and urinary bladder postoperatively, lower intraocular pressure in glaucoma, dilate peripheral blood vessels, terminate curarization, treat myasthenia gravis.

Cholinergic receptor antagonists. These include *atropine, scopolamine,* and *dicyclomine* (antimuscarinic agents). Prescribed to treat Parkinson's disease, dilate pupil, control motion sickness, treat peptic ulcers and hypermobility of the GI tract, decrease salivary and bronchial secretion (preoperative use of atropine).

Table 11.5 A Comparison of Sympathetic and Parasympathetic Activity

Organ or gland	*Sympathetic (adrenergic or cholinergic) stimulation*	*Parasympathetic (cholinergic) stimulation*
Heart	Increased rate and strength of contraction	Decreased rate and strength of contraction
Skin	Vasoconstriction (adrenergic); vasodilation, blushing (cholinergic)	None
Skeletal muscles	Vasoconstriction (adrenergic); vasodilation (cholinergic)	None
Blood vessels	Mostly constriction	Dilation in a few organs (e.g., penis)
Viscera	Vasoconstriction (adrenergic to abdominal viscera)	Vasodilatation (abdominal viscera)
Reproductive organs	Vasodilation (cholinergic to external genitalia)	Vasodilation (external genitalia)
Hair (arrector pili muscle)	Contraction and erection of hair, "goose bumps"	None
Bronchioles	Dilation	Constriction
GI tract	Decreased activity and tone	Increased activity (peristalsis) and tone
Gallbladder and ducts	Inhibition	Stimulation
Anal sphincter	Closing stimulated	Closing inhibited
Urinary bladder	Muscle tone aided	Contraction
Ciliary muscle of eye	Relaxation (for far vision)	Contraction (for near vision)
Iris of eye	Dilation of pupil	Constriction of pupil
Sweat glands	Stimulation of secretion (cholinergic)	None
Nasal, lacrimal, salivary, gastric, intestinal, and pancreatic glands	Vasoconstriction and inhibited secretion	Vasodilation and stimulated secretion
Pancreatic islets	Decreased secretion of insulin	Increased secretion
Liver	Stimulation of glycogen hydrolysis with release of glucose into blood	None
Adrenal medulla	Increased secretion of norepinephrine and epinephrine (which increase heart rate, blood pressure, blood sugar)	None

Review Exercises

Multiple Choice

1. The chemical transmitter between sympathetic postganglionic fibers and the effector organs is (*a*) norepinephrine, (*b*) acetylcholine, (*c*) adrenaline, (*d*) epinephrine.

2. Most body organs are innervated by (*a*) the parasympathetic division of the ANS, (*b*) the sympathetic division of the ANS, (*c*) both divisions of the ANS, (*d*) the CNS.

3. Parasympathetic fibers arise from which set of cranial nerves? (*a*) III, V, IIX, and X; (*b*) IV, V, IX, and X; (*c*) III, VII, IX, and X; (*d*) V, IX, X, and XII

4. A preganglionic fiber entering the sympathetic chain *cannot* (*a*) synapse with postganglionic neurons at the first ganglion it meets, (*b*) travel down the sympathetic chain before synapsing with postganglionic neurons, (*c*) end in the sympathetic chain without having synapsed, (*d*) pass through the sympathetic chain without having synapsed.

5. The cell bodies of the preganglionic neurons of the sympathetic division are located within (*a*) the cervical and sacral regions of the spinal cord, (*b*) the white matter of the spinal cord, (*c*) the lateral horns of the spinal cord gray matter, (*d*) the brain and sacral region.

6. The autonomic nervous system is responsible for which function(s)? (*a*) motor, (*b*) sensory, (*c*) motor and sensory, (*d*) none of the preceding

7. The white ramus of each spinal nerve has attached to it (*a*) a prevertebral ganglion, (*b*) a chain ganglion, (*c*) a posterior root ganglion, (*d*) the celiac ganglion.

8. Which pair of actions describes the effect of the sympathetic division of the ANS on the pupil of the eye and the GI tract? (*a*) dilates/inhibits, (*b*) dilates/stimulates, (*c*) constricts/inhibits, (*d*) constricts/stimulates

9. Which of the following would *not* result from sympathetic stimulation? (*a*) glucogenolysis, (*b*) contraction of the spleen, (*c*) secretion of catecholamines from the adrenal medulla, (*d*) profuse secretion of the salivary glands

10. One reason for the division of the ANS is that (*a*) sympathetic signals are transmitted from the spinal cord to the periphery through two successive neurons, in contrast to one neuron for parasympathetic signals; (*b*) sympathetic fibers alone innervate organs in the abdominal cavity; (*c*) sympathetic fibers alone arise from the spinal cord; (*d*) the effects of the two divisions on the organs are usually antagonistic.

11. The sympathetic division of the ANS does *not* (*a*) arise from thoracolumbar levels, (*b*) summon energy during an emergency, (*c*) stimulate bile secretion from the gallbladder, (*d*) dilate the bronchial tubes.

12. The lacrimal gland is innervated by (*a*) the facial cranial nerve, (*b*) the optic cranial nerve, (*c*) the ophthalmic nerve, (*d*) the oculomotor cranial nerve, (*e*) the maxillary nerve.

13. Consider the following statements about the parasympathetic division of the ANS:

(i) All its neurons release acetylcholine as their primary neurotransmitter substance.

(ii) The cell bodies of its postganglionic neurons lie in or near the organ innervated.

(iii) The cell bodies of its preganglionic neurons lie in the cervical and sacral spinal cord.

Of these statements: (*a*) all are true, (*b*) none are true, (*c*) i and ii are true, (*d*) ii and iii are true, (*e*) iii is true.

14. Consider the following statements about the sympathetic division of the ANS:

(i) All of its neurons release norepinephrine as their primary neurotransmitter substance.

(ii) All the cell bodies of its postganglionic neurons lie in or near the organ innervated.

(iii) The cell bodies of its preganglionic neurons lie in the thoracic and lumbar spinal cord.

Of these statements: (*a*) i is true, (*b*) ii is true, (*c*) iii is true, (*d*) i and iii are true, (*e*) all are true.

15. Autoreceptors of the sympathetic division of the ANS that are involved in negative feedback are (*a*) the α_1, (*b*) the α_2, (*c*) the β_1, (*d*) the β_2.

16. Beta receptors are stimulated by (*a*) methoxamine, (*b*) acetylcholine, (*c*) isoproteranol, (*d*) atropine.

17. The receptors for acetylcholine at the ganglia of both the sympathetic and parasympathetic divisions are (*a*) muscarinic receptors, (*b*) blocked by atropine, (*c*) nicotinic receptors, (*d*) stimulated by isoproteranol.

18. Which type of receptor is found in the heart? (*a*) alpha, (*b*) beta, (*c*) nicotinic, (*d*) GABA

19. Which class of drugs may be used to treat bronchial asthma? (*a*) cholinergic, (*b*) anticholinesterase, (*c*) adrenergic, (*d*) adrenergic blockers

20. Cholinergic blockers have as an unwanted side effect (*a*) increased gastric secretion, (*b*) spasms in the GI tract, (*c*) diarrhea, (*d*) dry mouth.

21. A patient scheduled for surgery confides in his nurse the night before that he is "terribly scared." Which of the following indicate(s) increased sympathetic activity in this patient? (*a*) Patient complains that his mouth feels dry, (*b*) patient's gown is moist with perspiration, (*c*) patient appears pale, (*d*) patient's pupils are widely dilated, (*e*) all of the preceding

22. Which of the following is *not* a function of the ANS? (*a*) innervation of all visceral organs, (*b*) transmission of sensory and motor impulses, (*c*) regulation and control of vital activities, (*d*) conscious control of motor activities

23. Concerning (1) the heart, (2) glands, (3) smooth muscle, and (4) certain skeletal muscles, we can say that the ANS innervates (*a*) 1, 2, and 3; (*b*) 1, 3, and 4; (*c*) 2 and 4; (*d*) 1, 2, 3, and 4.

24. Atropine (which blocks muscarinic receptors) is liable to cause (*a*) weakness of cardiac muscles, (*b*) an increase in the resting heart rate, (*c*) an excessive flow of saliva, (*d*) overactivity of the small intestine.

25. A ganglion is an aggregate of nerve cell bodies (*a*) inside the brain or spinal cord, (*b*) outside the brain and spinal cord, (*c*) in the spinal cord only, (*d*) in the brain only.

26. A cranial nerve that affects eye movement is (*a*) the optic nerve, (*b*) the trigeminal nerve, (*c*) the trochlear nerve, (*d*) the hypoglossal nerve.

27. The cranial nerve with the greatest distribution is (*a*) the trigeminal nerve, (*b*) the vagus nerve, (*c*) the abducens nerve, (*d*) the accessory nerve.

28. Taste sensation is mediated by which cranial nerves? (*a*) trigeminal and facial, (*b*) trochlear and abducens, (*c*) facial and glossopharyngeal, (*d*) trigeminal and glossopharyngeal

29. In a patient with a contusion over the parotid region, the facial muscles on one side of the face are paralyzed, one eye can't be shut, and the corner of the mouth droops. Which cranial nerve is damaged? (*a*) the abducens nerve, (*b*) the facial nerve, (*c*) the glossopharyngeal nerve, (*d*) the accessory nerve, (*e*) the hypoglossal nerve

30. The knee-jerk reflex in response to a mallet tap over the patellar ligament (*a*) is a conditioned reflex, (*b*) is a polysynaptic reflex, (*c*) has its reflex center in the spinal cord, (*d*) is mediated by a three-neuron reflex arc.

31. Which pairing of nerve and organ innervation is *incorrect*? (*a*) phrenic nerve–diaphragm, (*b*) vagus nerve–abdominal viscera, (*c*) glossopharyngeal nerve–taste buds, (*d*) abducens nerve–facial muscles, (*e*) sciatic nerve–lower extremity

32. An inability to walk a straight line may indicate damage to which cranial nerve? (*a*) the vestibulocochlear nerve, (*b*) the trochlear nerve, (*c*) the facial nerve, (*d*) the hypoglossal nerve, (*e*) the accessory nerve

33. The rectus eye muscle capable of causing the eyeball to turn laterally in a horizontal plane is innervated by which cranial nerve? (*a*) the optic nerve, (*b*) the abducens nerve, (*c*) the facial nerve, (*d*) the oculomotor nerve, (*e*) the trochlear nerve

34. Which of the following is *not* a plexus of the spinal nerves? (*a*) the cervical plexus, (*b*) the sacral plexus, (*c*) the choroid plexus, (*d*) the brachial plexus, (*e*) the lumbar plexus

35. Which of the following cranial nerves is *not* a mixed nerve? (*a*) the abducens nerve, (*b*) the glossopharyngeal nerve, (*c*) the trigeminal nerve, (*d*) the vagus nerve, (*e*) the vestibulocochlear nerve

True or False

____ **1.** Cranial nerves innervate only structures of the head and neck.

____ **2.** The extrinsic ocular muscles are innervated by three different cranial nerves.

____ **3.** All spinal nerves are mixed nerves.

____ **4.** The parasympathetic division of the ANS functions in meeting stressful and emergency conditions.

____ **5.** An inability to shrug the shoulders may indicate a dysfunction of the facial nerves.

____ **6.** Erection of the penis is primarily a parasympathetic response.

____ **7.** The posterior (dorsal) root of a spinal nerve consists of sensory neurons only.

____ **8.** Compression of the brachial plexus could result in paralysis of the hand.

____ **9.** The optic nerve controls movement of the eye.

____ **10.** Included in the PNS are 12 pairs of cranial nerves, 31 pairs of spinal nerves, and 4 plexuses of the spinal column.

____ **11.** The olfactory, optic, and vestibulocochlear nerves are the only cranial nerves that are purely sensory.

____ **12.** Bell's palsy is a temporary functional disorder of the facial (seventh cranial) nerve.

____ **13.** All reflex arcs involve the CNS.

____ **14.** The sympathetic division of the ANS is craniosacral in its origin.

____ **15.** There are seven cervical vertebrae and eight cervical nerves.

Completion

1. A __________________ is the area of the skin innervated by all the cutaneous neurons of a given spinal or cranial nerve.

2. The __________________ cranial nerve innervates the lateral rectus (ocular) muscle.

3. __________________ __________________ is a disorder of the trigeminal (fifth cranial) nerve characterized by severe recurring pain in one side of the face.

4. The __________________ nerve is the branch of the trigeminal (fifth cranial) nerve that innervates the lower jaw and teeth, skin over the lower jaw, and the tongue.

5. There are __________________ cervical nerves, __________________ thoracic nerves, __________________ lumbar nerves, __________________ sacral nerves, and __________________ coccygeal nerve.

6. The autonomic nervous system is divided into the __________________, or adrenergic division, and the __________________, or cholinergic division.

7. __________________ receptors are located at the ganglia in both sympathetic and parasympathetic divisions of the ANS.

8. __________________ fibers do not synapse after they leave the CNS.

9. The portion of the ANS that is thoracolumbar in its origin is the __________________ division.

10. The __________________ cranial nerve conveys sensations from the retina of the eye to the thalamus.

Labeling

Label the structures indicated on the figure to the right.

1. ______________________________
2. ______________________________
3. ______________________________
4. ______________________________
5. ______________________________
6. ______________________________

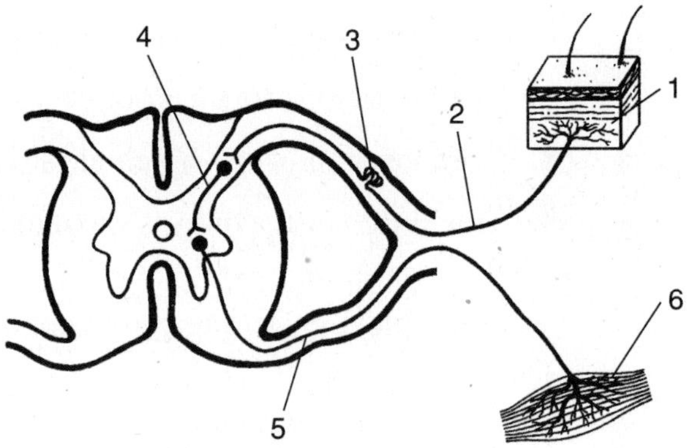

Answers and Explanations for Review Exercises

Multiple Choice

1. (*a*) Norepinephrine is the neurotransmitter at the effector organs of the sympathetic division of the ANS, with three exceptions–those of sweat glands, some blood vessels within skeletal muscles and the external genitalia, and the adrenal medulla. Sympathetic postganglionic fibers innervating these effectors secrete acetylcholine (are cholinergic).
2. (*c*) Most organs have both sympathetic and parasympathetic innervation, with one division stimulating and the other division inhibiting.
3. (*c*) The oculomotor (cranial nerve III), facial (cranial nerve VII), glossopharyngeal (cranial nerve IX), and vagus (cranial nerve X) conduct parasympathetic impulses.
4. (*c*) None of the preganglionic neurons end in the sympathetic chain without having synapsed or having left the chain to synapse at a more distant ganglion.
5. (*c*) Cell bodies of the preganglionic neurons in the sympathetic division originate in the lateral horns of the spinal cord gray matter in the thoracolumbar region.
6. (*c*) The autonomic nervous system has both sensory and motor components.
7. (*b*) The white ramus is a branch between a spinal nerve and a chain ganglion.
8. (*a*) Sympathetic stimulation dilates the pupil while it inhibits activity in the GI tract.
9. (*d*) All digestive processes are inhibited by sympathetic stimulation, including salivary gland secretion.
10. (*d*) Because the sympathetic and parasympathetic divisions are antagonistic, they continuously regulate the activity of effector organs.
11. (*c*) Bile secretion is a digestive function inhibited by sympathetic stimulation.
12. (*c*) The ophthalmic nerve is the superior branch of the trigeminal cranial nerve that innervates the anterior scalp, upper eyelid, surface of the eye, and lacrimal gland.
13. (*c*) The anatomical origin is cranial-sacral, not cervical-sacral.
14. (*c*) Norepinephrine is not always the neurotransmitter (there are three exceptions) and postganglionic neurons may be very long.

15. (*b*) Norepinephrine acts to inhibit its own further release.
16. (*c*) Isoproterenol is a synthetic catecholamine that stimulates mainly beta receptors.
17. (*c*) Nicotinic receptors are at the ganglia in both ANS divisions.
18. (*b*) Only the beta receptors are found in the heart.
19. (*c*) Adrenergic, or sympathetic, activation causes bronchial dilation via beta receptors.
20. (*d*) Inhibition of the salivary glands causes the mouth to be dry.
21. (*e*) All of these symptoms result from sympathetic stimulation.
22. (*d*) As the name implies, the autonomic nervous system functions without conscious awareness.
23. (*a*) The ANS does not innervate any skeletal muscles.
24. (*b*) Muscarinic receptors are associated with the parasympathetic division (they slow the heart). Therefore, blocking the muscarinic receptors would cause an increase in the resting heart rate.
25. (*b*) A ganglion is an aggregate of nerve cell bodies in the PNS; an aggregate in the CNS is called a nucleus.
26. (*c*) The trochlear nerve is one of three that functions in eye movement.
27. (*b*) The vagus nerve innervates effector organs in the thoracic and abdominal cavities.
28. (*c*) Both the facial and the glossopharyngeal nerves provide sensory innervation to the tongue.
29. (*b*) Damage to the facial nerve causes the entire side of the face to sag because muscle tonus is lost.
30. (*c*) The reflex center is within the gray matter of the spinal cord.
31. (*d*) The abducens nerve innervates the eye, but not facial muscles.
32. (*a*) The paired vestibulocochlear nerve innervates the organs associated with balance and equilibrium.
33. (*b*) The abducens nerve innervates the lateral rectus eye muscle.
34. (*c*) The choroid plexus produces cerebrospinal fluid.
35. (*e*) The vestibulocochlear nerve is sensory only.

True or False

1. False; the vagus nerves innervate thoracic and abdominal viscera
2. True
3. True
4. False; the sympathetic division functions under stressful conditions
5. False; the accessory nerves regulate muscles that move the shoulders
6. True
7. True
8. True
9. False; the abducens, oculomotor, and trochlear cranial nerves innervate muscles that move the eye
10. True
11. True
12. True
13. True
14. False; the sympathetic division of the ANS is thoracolumbar in its origin
15. True

Completion

1. dermatome
2. abducens
3. Trigeminal neuralgia
4. mandibular
5. 8, 12, 5, 5, 1
6. sympathetic, parasympathetic
7. Nicotinic
8. Somatic
9. sympathetic
10. optic

Labeling

1. Receptor organ
2. Sensory neuron
3. Cell body of sensory neuron
4. Association neuron
5. Motor neuron
6. Effector organ

Sensory Organs 12

Objective A To explain what is meant by *sensory organs* and to list the six special senses.

Survey

Sensory organs are specialized extensions of the nervous system that contain sensory (afferent) neurons adapted to respond to specific stimuli and conduct nerve impulses to the brain. Because sensory organs are very specific as to the stimuli to which they respond, they act as energy filters that allow perception of only a narrow range of energy. For example, the rods and cones within the eye respond to a precise range of light waves and normally do not respond to X rays, radio waves, or to ultraviolet and infrared light.

The senses of the body are classified as **general senses** or **special senses** according to the complexity of the receptors and the neural pathways (nerves and tracts) involved. General senses include the cutaneous receptors (touch, pressure, heat, cold, and pain) within the skin. Collectively, the cutaneous receptors are said to provide the *sense of touch* (see problem 5.19 and table 5.2). Special senses are localized in complex receptor organs and have extensive neural pathways. The special senses are the *senses of taste, smell, sight, hearing,* and *balance.*

12.1 Which of the senses depend on the sensory neurons termed *chemoreceptors*, and which involve *photoreceptors*?

Chemoreceptors are specialized neurons that respond to chemical stimuli. Moist environments are necessary for these receptors to function. The senses of smell and taste rely on chemoreceptors.

Photoreceptors are specialized neurons that respond to light waves. The rods and cones within the eyes (see problem 12.12) are photoreceptors.

Objective B To describe the receptors and the neural pathway for the sense of *taste.*

Survey

Receptors for the **sense of taste** (*gustation*) are located in taste buds on the surface of the tongue. The taste buds are associated with peglike projections of the tongue mucosa called **lingual papillae** (fig. 12.1). A few taste buds are also located in the mucous membranes of the palate and pharynx. A taste bud contains a cluster of 40 to 60 **gustatory cells**, as well as many more **supporting cells** (fig. 12.2). Each gustatory cell is innervated by a sensory neuron.

The four primary taste sensations are *sweet* (evoked by sugars, glycols, and aldehydes); *sour* (evoked by H^+, which is why all acids taste sour); *bitter* (evoked by alkaloids); and *salty* (evoked by anions of ionizable salts).

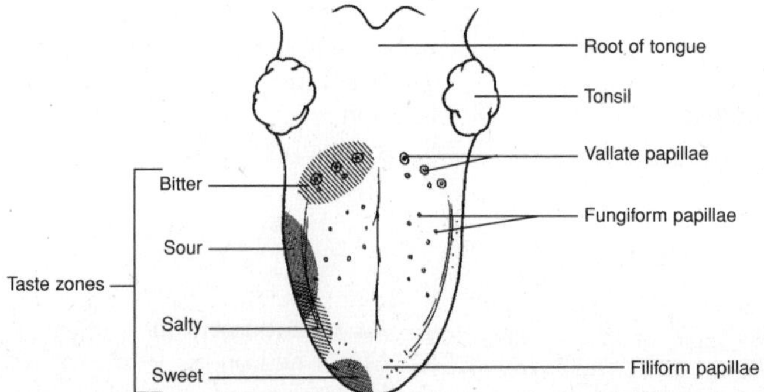

Figure 12.1 The surface of the tongue showing the locations of the taste buds and the taste zones.

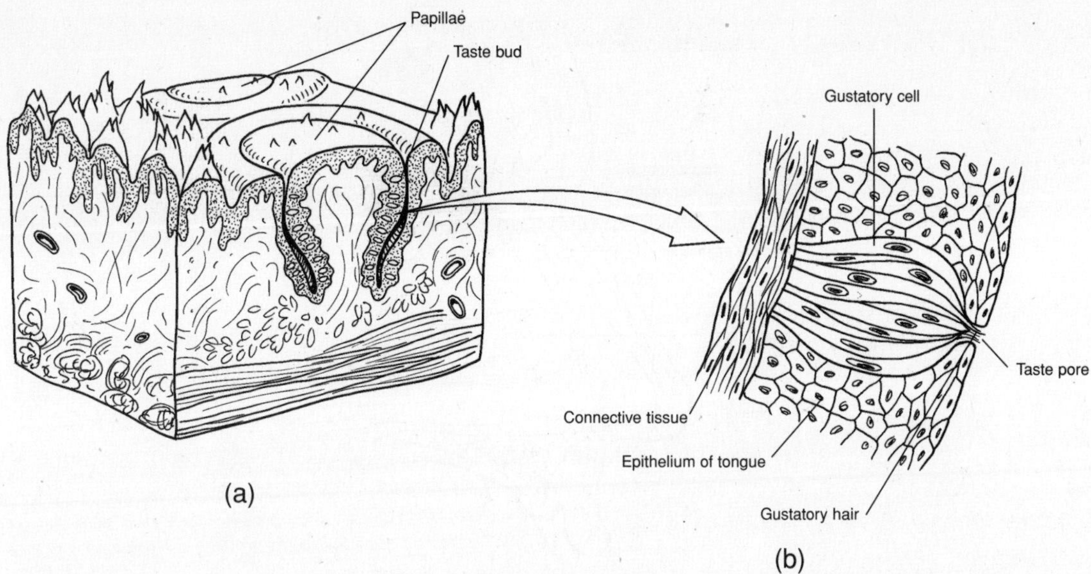

Figure 12.2 (*a*) A lingual papilla and (*b*) a taste bud containing gustatory cells.

12.2 What are the three kinds of lingual papillae and where are they located?

Vallate papillae. The largest but fewest in number, they are arranged in an inverted V-shape pattern on the back of the tongue (see fig. 12.1).

Fungiform papillae. Knoblike in appearance, they are found on the tip and sides of the tongue.

Filiform papillae. Short and thickened in appearance, they are found on the anterior two-thirds of the tongue.

12.3 Do taste receptors undergo adaptation?

Yes. With continuous exposure to a taste stimulus, there is a decrease in sensory neuron transmission.

12.4 Which cranial nerves conduct taste sensations to the brain?

Sensory innervation of the tongue and pharynx is by the chorda tympani branch of the *facial nerve* from the anterior two-thirds of the tongue, the *glossopharyngeal nerve* from the posterior third of the tongue, and the *vagus nerve* from the pharyngeal region (see table 11.1).

12.5 Which areas of the brain receive impulses from the taste receptors?

Taste sensations are transmitted to the *brain stem* (*nucleus solitarius*), then to the *thalamus* (*nucleus ventralis posteromedialis*), and finally to the sensory *cerebral cortex* (*postcentral gyrus* on the lateral convexity), where taste perception occurs.

Objective C To identify the receptors and neural pathway for the sense of *smell*.

Receptors for the **sense of smell** (*olfaction*) are located in each lateral side of the nasal cavity, in the nasal mucosa of the superior nasal concha (fig. 12.3). Like taste receptors, smell receptors are chemoreceptors. For smell, however, the chemicals are originally airborne and become dissolved in the mucous layer lining the superolateral part of the nasal cavity.

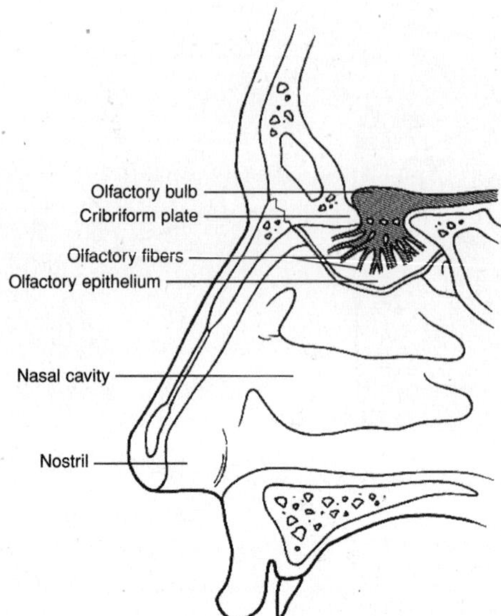

Figure 12.3 Olfactory receptors within the olfactory epithelium of the nasal cavity.

12.6 What are the characteristics of an odorant—a chemical molecule that stimulates a smell receptor?

The odorant must be volatile (to reach the smell receptor), water-soluble (to penetrate the moist mucous membrane covering the receptor), and lipid-soluble (to penetrate the cell membrane of the olfactory receptor cell).

12.7 Does adaptation of smell receptors occur?

Yes. Smell receptors adapt very rapidly to continued exposure to odorants (50% adaptation within the first second). As compared to other mammals, humans have a poor sense of smell. Seemingly, detecting the presence of an odor is more important for us than determining its intensity.

12.8 Do all of the volatile chemicals in the nose stimulate smell receptors?

No. Only about 2% or 3% of the air that is inhaled comes in contact with the olfactory receptors because of their location above the main airstream. Olfaction can be greatly increased by forceful sniffing, which draws the volatile chemicals into contact with olfactory receptors.

12.9 Which of the cranial nerves innervate the olfactory mucosa?

The *olfactory nerve* transmits most impulses related to smell (see table 11.1). However, some irritating chemicals (such as pepper) stimulate the *trigeminal nerve* as well. Irritating chemicals generally initiate a protective and reflexive sneeze and/or cough. Olfactory sensory sensations are conveyed along each *olfactory tract* to the olfactory portions of the *cerebral cortex* (*prepiriform cortex, subcallosal gyrus,* and *olfactory tubercle*), where olfactory perception occurs.

Objective D To describe the accessory structures of the eye.

Accessory structures of the eye either protect the eye or enable eye movement. Each eye is protected by a **bony orbit** (see problem 6.13) composed of facial and cranial bones. Other accessory structures include the following:

Eyebrow. The eyebrow (fig. 12.4a) consists of short, thick hairs above the eye that help to prevent perspiration and airborne particles from entering the eye. They also shade the eyes from the sun.

Eyelids. The two eyelids (*palpebrae*) cover and protect the eyes from desiccation, foreign matter, and sunlight. Each eyelid is covered with skin and contains muscle fibers, a **tarsal plate** (of dense fibrous connective tissue), **tarsal glands** (specialized sebaceous glands), and **ciliary glands** (sweat glands). The numerous **eyelashes** attached to the eyelids protect the eye from airborne particles.

Lacrimal apparatus. The lacrimal apparatus consists of the **lacrimal gland**, (fig. 12.4b) which secretes *lacrimal fluid* (tears), and the **lacrimal canals**, which drain the fluid into the **lacrimal sac**. Lacrimal fluid lubricates the anterior surface of the eye, in contact with the eyelids. This fluid also contains the *lysozyme*, a bactericidal polysaccharide.

Eye muscles. Six *extrinsic eye muscles* (*ocular muscles*) attached from the bony orbit to the eyeball are responsible for the various eye movements (fig. 12.5). The *superior rectus* rotates the eye superiorly; the *medial rectus* rotates the eye medially; the *lateral rectus* rotates the eye laterally, the *superior oblique* rotates the eye inferolaterally; and the *inferior oblique* rotates the eye superolaterally. In addition, the *levator palpebrae superioris* (see fig. 12.4) elevates the superior eyelid and the *orbicularis oculi* (see fig. 8.1) constricts the eyelids.

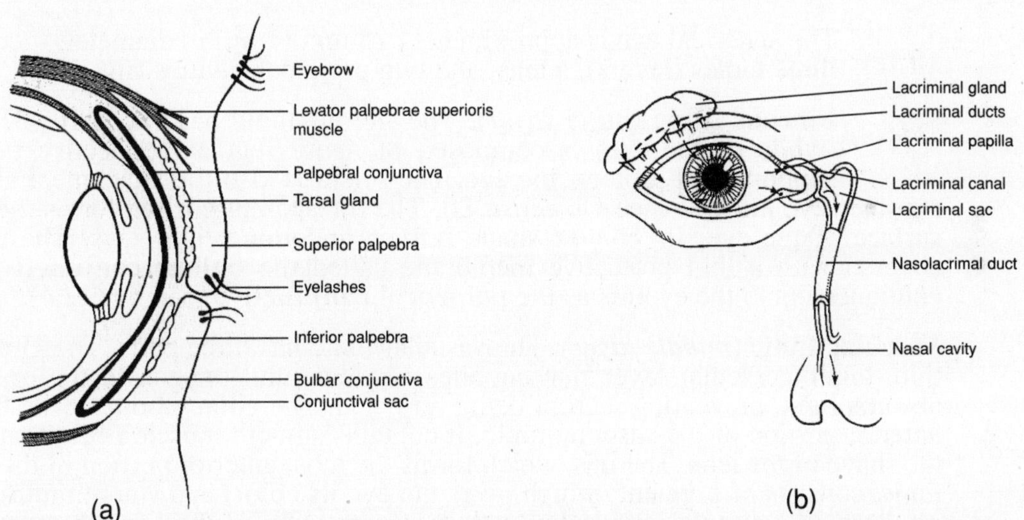

Figure 12.4 Accessory structures of the eye. (*a*) A sagittal view of the anterior eye and the eyelids; (*b*) the lacrimal gland and the pathway of drainage (see arrows) for lacrimal fluid.

12.10 Why do you get a "runny nose" when you cry?

Lacrimal fluid drains across the anterior surface of the eyes into the lacrimal canals, through the lacrimal sacs, and through the nasolacrimal ducts, which empty into the nasal cavity. Normally, the lacrimal fluid (tears) will flow posteriorly through the nasal cavity and into the pharynx. When a person cries, however, the tears are so copious that drainage may spill from the eyes onto the cheeks as well as out the nostrils. Shedding emotional tears is a behavior particular to humans.

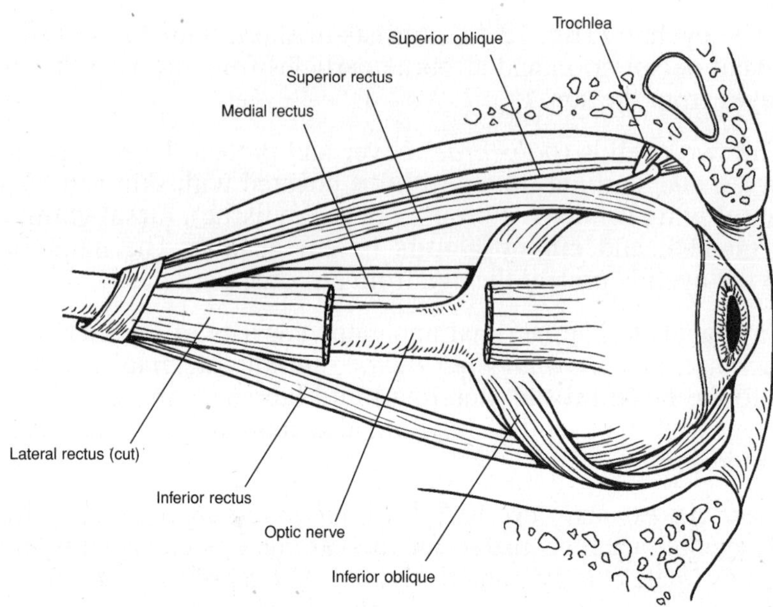

Figure 12.5 The extrinsic eye muscles.

Objective E To describe the structure of the *eye*.

The spherical **eye** is approximately 25 mm (1 in.) in diameter. It consists of three tunics (layers), a lens, and two principal cavities (fig. 12.6).

Fibrous tunic (*outer layer*). The fibrous tunic has two parts. The **sclera** (white of the eye) is composed of dense regular connective tissue that supports and protects the eye. The sclera is also the attachment site of the extrinsic eye muscles (see Objective D). The transparent **cornea** forms the anterior surface of the eye. Its convex shape refracts incoming light rays. The cornea is covered with a thin protective membrane called the **bulbar conjunctiva** that is continuous onto the eyelids as the **palpebral conjunctiva** (see fig. 12.4).

Vascular tunic (*middle layer*). The vascular tunic has three parts. The **choroid** is a thin, highly vascular layer that supplies nutrients and oxygen to the eye. It also absorbs light, preventing it from being reflected. The **ciliary body** is the thickened anterior portion of the vascular tunic. It contains smooth muscle fibers that regulate the shape of the lens. The **iris**, which forms the most anterior portion of the vascular tunic, consists of pigment (which gives the eye its color) and smooth muscle fibers arranged in a circular and radial pattern. Contraction of the smooth muscle fibers regulates the diameter of the **pupil**, which is the opening in the center of the iris.

Internal tunic (*inner layer, or retina*). This receptor component of the eye contains two types of photoreceptors. **Cones** function at high light intensities and are responsible for daytime color vision and acuity (sharpness). **Rods** function at low light intensities and are responsible for night (black-and-white) vision. In addition, the *retina* contains *bipolar cells*, which synapse with the rods and cones, and *ganglion cells*, which synapse with the bipolar cells (see problem 12.12). The axons of the ganglion cells course along the retina to the **optic disc** and form the **optic nerve**. The **fovea centralis** is a shallow pit at the back of the retina that contains only cones. It is the area of keenest vision. Surrounding the fovea centralis is the **macula lutea**, which also has an abundance of cones.

Lens. The lens is a transparent, biconvex structure composed of tightly arranged proteins. It is enclosed in a **lens capsule** and held in place by the **suspensory ligament** (composed of *zonular fibers*) that attaches to the ciliary body. The lens focuses light rays for near and far vision.

Cavities of the eye. The interior of the eye is separated by the lens into an **anterior cavity** and a **posterior cavity** (*vitreous chamber*). The anterior cavity is partially subdivided by the iris into an *anterior chamber* (between the cornea and the iris) and a *posterior chamber* (between the iris and the lens). The anterior cavity contains a watery fluid called *aqueous humor*. The posterior cavity contains a transparent jellylike substance called *vitreous humor*.

Aqueous humor is continuously produced by the ciliary body. It flows from the posterior chamber through the pupil and into the anterior chamber. From there, it drains into a vascular network at the base of the lens called the **scleral venous sinus** (*canal of Schlemm*). Vitreous humor is produced prenatally. Additional small amounts are produced as the eye increases in size, but it is not continuously produced as is aqueous humor.

A *cataract* is the loss of transparency of the lens. It is a chemical change in the protein of the lens caused by injury, poisons, infections, or age degeneration. Untreated cataracts in both lenses is a common cause of blindness. Cataracted lenses can be surgically removed and replaced with prosthetic lenses, thus restoring sight.

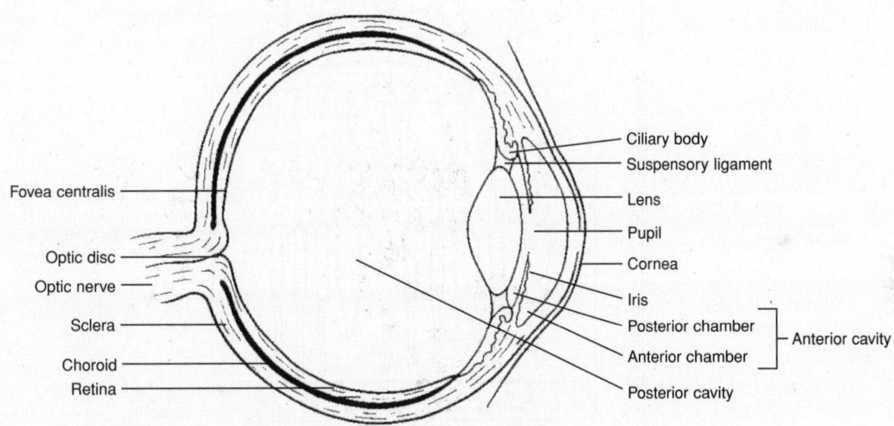

Figure 12.6 The internal anatomy of the eye.

Glaucoma is an abnormal increase in the intraocular pressure of the eye. Aqueous humor does not drain through the scleral venous sinus as quickly as it is produced. Accumulation of fluid causes compression of the blood vessels in the choroid and compression of the optic nerve. Blindness results as retinal cells die and the optic nerve atrophies. With early detection, glaucoma can be effectively treated with medications.

12.11 What are the refractory structures (media) of the eye?

Incoming light rays are refracted (bent) such that a sharp, inverted (upside-down) image is focused on the fovea centralis. In the order through which the light rays pass, the refractive structures are the cornea, aqueous humor, lens, and vitreous humor. The greatest degree of refraction is provided by the cornea, but the most important refractive structure is the lens. The curvature of the highly elastic lens can be actively changed so as to maintain sharply focused images as the eye moves.

Astigmatism is a condition in which an irregular curvature of the cornea or lens of the eye distorts the refraction of light rays. This condition is suspected if there are blurred areas in a person's field of vision. Correction for astigmatism requires a careful assessment of the irregularities and a prescription of specially ground lenses.

12.12 What are the layers of the retina?

Functionally, the retina is composed of two layers (fig. 12.7). The thin *pigmented layer* is in contact with the choroid, and the thick *nervous layer* is the visual portion. The nervous layer contains three distinct groupings of cells. In the order in which they conduct impulses, they are the **rod** and **cone cells, bipolar neurons**, and **ganglion neurons**. The *optic nerve* consists of a convergence of the axons from the ganglion neurons. It is interesting to note that the incoming light rays must first pass by the ganglion neurons and the bipolar neurons before they stimulate the rods and cones to conduct nerve impulses in the opposite direction.

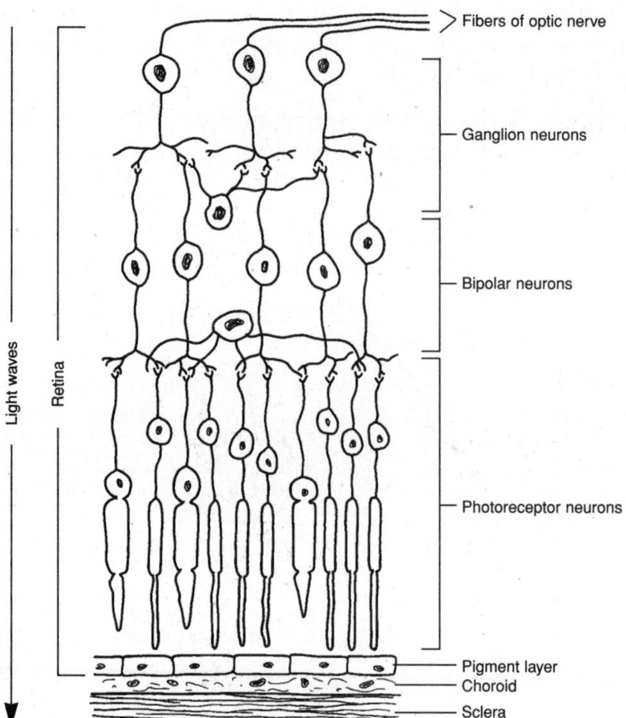

Figure 12.7 The neurons of the retina.

12.13 Which are more numerous, rods or cones?

Rods number over 100 million per eye and are thinner and more elongated than cones. They are more numerous toward the periphery of the retina. Cones number about 7 million per eye and are concentrated in the fovea centralis and the surrounding macula lutea.

12.14 Do all cones respond to the entire visible spectrum?

No. The cones fall into three classes, with absorption peaks corresponding to the three primary colors—blue, green, and orange-red (fig. 12.8).

Color blindness is the inability to distinguish colors, particularly reds and greens. Red-green color blindness affects about 5% of the U.S. population. True color blindness, or *monochomatism*, is extremely rare. In this condition, only shades of black and white are seen. Most vertebrate animals lack color vision.

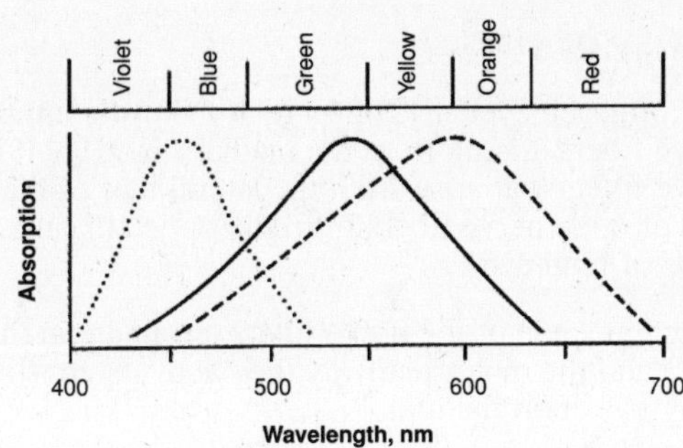

Figure 12.8 The visual spectrum.

Objective F To describe the field of vision and the visual pathway.

The *field of vision* is what is what a person visually perceives. With good eyesight, the focal point in the field of vision is sharp and clear. Away from the focal point, the image is less clear, and is actually hazy at the periphery of the field of vision.

A person does not see with the eyes. Rather light rays striking the photoreceptors in the retina cause the transmission of visual sensations (nerve impulses) to the occipital cerebral lobes, where visual perception occurs.

12.15 What are the three visual fields within the field of vision?

The anterior position of the human eyes permits an overall field of vision of about 180°. There are three visual fields within the field of vision (fig. 12.9). The **macular field** provides the area of keenest vision because the light rays from this focal point activate the photoreceptors in the fovea centralis and macula lutea in both eyes. The **binocular field** is that portion of the field of vision that is viewed from both eyes, but not keenly focused upon. It provides a clear image, but not as sharp an image as the macular field. The **monocular field** is the portion of the field of vision that is viewed by one eye, and not shared by the other eye. It is the hazy peripheral vision. Depth perception requires that both eyes work together to accurately focus on the object; hence, the monocular field is not viewed as three-dimensional.

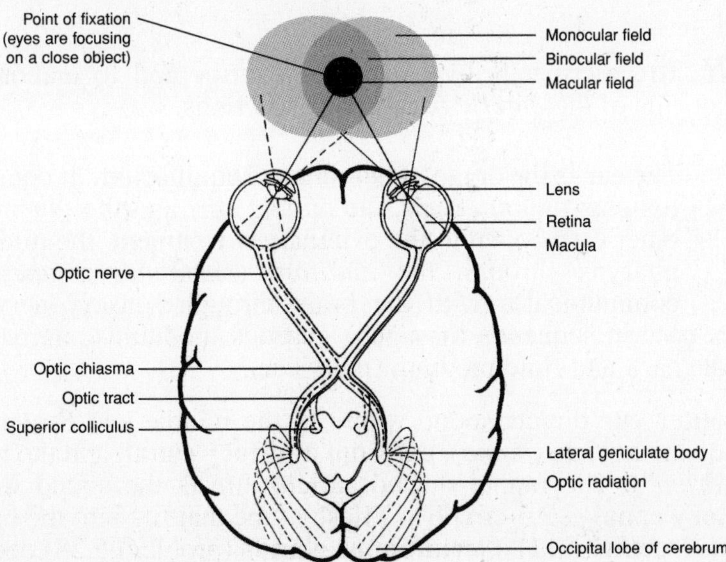

Figure 12.9 Visual fields of the eyes and neural pathways.

12.16 What is the neural pathway of vision?

The two **optic nerves** (one from each eyeball) converge at the **optic chiasma** (see fig. 12.9). However, only the optic nerve fibers arising from the medial (nasal) half of each retina cross to the opposite side. The optic nerve fibers that arise from the lateral half of the retina do not cross. The **optic tract** is a continuation of optic nerve fibers from the optic chiasma, and is composed of nerve fibers arising from the retinas of both eyes.

As an optic tract enters the brain, some of the nerve fibers terminate in the *superior colliculus*. These fibers (from both eyes) and the motor pathway they activate constitute the **tectal system**, which is responsible for body-eye coordination.

Approximately 75% of the fibers in an optic tract pass to the **lateral geniculate body** of the *thalamus*, where they synapse with neurons whose axons constitute the **optic radiation** pathway. Visual information is then transmitted through the optic radiation to the **striate cortex** area of the occipital cerebral lobe. This entire system involving the neural pathway from the lateral geniculate body to the striate cortex is known as the **genicolostriate system** and is responsible for perception of the visual field.

Objective G To describe how the eye focuses on objects at various distances.

For an image to be focused on the retina, the more distant the object, the flatter must be the lens. Adjustments in lens shape, accomplished by the ciliary muscles in the ciliary body (see fig. 12.6), are called *accommodation*. When these smooth muscles contract, the zonular fibers within the suspensory ligament slacken, causing the lens to thicken and become more convex.

In *myopia* (*nearsightedness*), the eye is too long for the refractive power of the lens, and far objects are focused at a point in front of the retina. The eye can focus on very near objects. Myopia is treated with concave lenses. In *hyperopia* (*farsightedness*), the eyeball is too short for the lens, and near objects are focused behind the retina. Distant objects are focused correctly. To treat hyperopia, convex lenses are used.

Objective H To describe the *ear* in general terms and to elaborate on the structural components of the *outer ear* and their functions.

The **ear** is the organ of hearing and equilibrium. It consists of three principal regions: the *outer ear*, the *middle ear*, and the *inner ear* (fig. 12.10). The outer ear is open to the external environment, the middle ear is open to the pharynx through the *auditory* (*eustachian*) *tube*, and the inner ear communicates with the brain through sensory nerves. Incoming sound waves pass in sequence through a gaseous medium (external ear), solid medium (middle ear), and fluid medium (inner ear).

The **outer ear** directs sound waves to the middle ear. Structures of the outer ear include the **auricle** (*pinna*), **external auditory canal**, and the **tympanic membrane** ("*eardrum*"). The funnel-shaped auricle directs the sound waves to the **external auditory canal**, a 2.5-cm (1-in.) fleshy tube that fits into the bony *external acoustic meatus* (see fig. 6.11). **Ceruminous glands** (problem 5.28) deep within the external auditory canal secrete protective *cerumen* (ear wax). The thin **tympanic membrane** conducts sound waves to the middle ear.

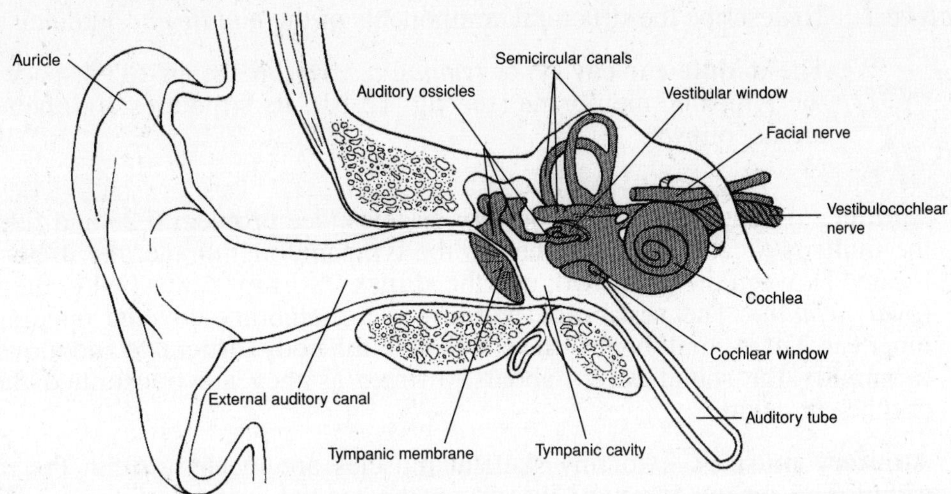

Figure 12.10 The ear.

A *ruptured tympanic membrane* ("*broken eardrum*") may occur as the result of infections or trauma. A middle-ear infection (*acute purulent otitis media*) in children is common following a cold or tonsillitis. The pathogens gain entry into the middle ear through the auditory tube. An intense *earache* is a common symptom of a middle-ear infection. The pressure from the inflammation may eventually rupture the tympanic membrane permitting drainage of pus. Spontaneous perforation of the tympanic membrane from an infection or a loud noise usually heals rapidly, but scar tissue may form and lessen sensitivity to sound vibrations.

12.17 What are the common physical parameters used to describe a sound wave?

There are two: *amplitude* and *frequency* (fig. 12.11). The **amplitude** is the "height" of the wave; the *power* or *intensity* of the wave is proportional to the square of its amplitude. Intensity translates psychologically into *loudness*, and is measured (on a logarithmic scale) in *decibels* (dB).

Frequency is the number of oscillations ("back-and-forth," in the case of sound) the wave makes in a unit of time. Frequency translates into *pitch*, and is measured in hertz (Hz), where 1 Hz = 1 cycle per second.

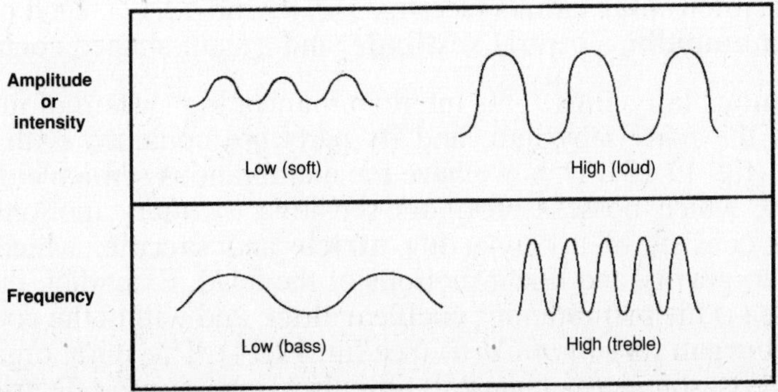

Figure 12.11 Profiles of sound-wave amplitude and frequency.

Objective I To describe the structural components of the *middle ear* and their functions.

The **middle-ear cavity**, or *tympanic cavity*, is the air-filled space medial to the tympanic membrane (see fig. 12.10). Its structures and their functions are as follows:

Auditory ossicles. The three auditory ossicles (see problem 6.23 and fig. 6.15) are the **malleus** ("hammer"), attached to the tympanic membrane; the **incus** ("anvil"), located between the other two; and the **stapes** ("stirrup"), attached to the *vestibular (oval) window*. The **vestibular window** is a membrane-covered opening into the inner ear. These small bones (the smallest in the body) articulate and move as levers to amplify the sound waves about 20 times as they are transmitted through the middle-ear cavity.

Auditory muscles. Two tiny skeletal muscles are located within the middle-ear cavity. The **tensor tympani** inserts on the medial surface of the malleus and is innervated by the trigeminal nerve. The **stapedius** inserts on the neck of the stapes and is innervated by the facial nerve. These two muscles function reflexively to reduce the pressure of loud sounds before it can injure the inner ear.

Auditory (eustachian) tube. The auditory tube connects the middle-ear cavity to the pharynx. With this connection, the air pressure is equalized on both sides of the tympanic membrane. The auditory tube also permits moisture to drain from the middle-ear cavity.

A *myringotomy* is a surgical opening of the tympanic membrane to relieve pressure or release pus from the middle ear. A tiny tube may be implanted to help keep the auditory tube open. A myringotomy may be performed in a child who is subject to repeated middle-ear infections and accompanying earaches. The tube, which is eventually sloughed out of the ear, prohibits further infections by allowing drainage through the auditory tube.

Objective J To describe the structural components of the *inner ear* and their functions.

The **inner ear** contains not only the organs of hearing, but also those of equilibrium and balance. Its structures and their functions are as follows:

Bony labyrinth. This is a network of cavities in the petrous part of the temporal bone (see fig. 6.15). The cavities consist of three bony **semicircular canals** (see figs. 12.10 and 12.12), each of which swells into a globular **ampulla**, a central **vestibule**, and a snail-shaped **cochlea**.

Membranous labyrinth. This intercommunicating system of membranous ducts is seated in the bony labyrinth, and its parts are conamed with those of the bony labyrinth (fig. 12.13). Thus we have the membranous **semicircular canals** and their **ampullae**, which possess receptors sensitive to rotary motions of the head. The **vestibule** consists of a connecting **utricle** and **saccule**, which possess receptors sensitive to gravity and linear motions of the head. Extending through the center of the cochlea is the membranous **cochlear duct**, and within the cochlear duct is found the **spiral organ** (*organ of Corti*) (see fig. 12.14). The spiral organ is a "transducer" that converts sound (mechanical) impulses into nerve (electrical) impulses. The membranous labyrinth is filled with a fluid called *endolymph*, and to the outside of the membranous labyrinth is a fluid called *perilymph*.

The **vestibular window** (*oval window*) is located at the footplate of the stapes, where it transfers sound waves from the solid medium of the auditory ossicles to the fluid medium of the cochlea. The **cochlear window** (*round window*) is positioned directly below the vestibular window, where it reverberates in response to loud sounds.

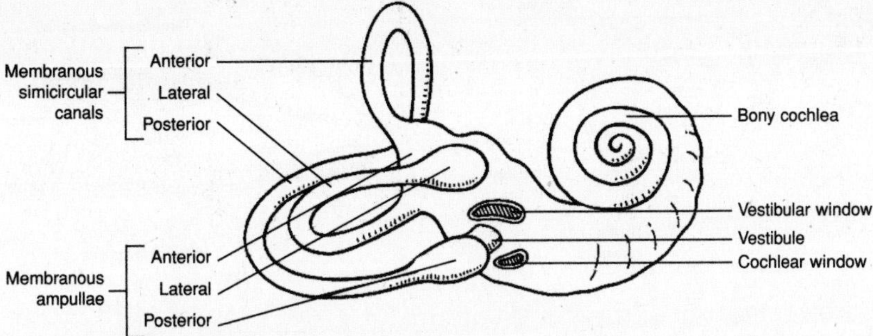

Figure 12.12 The bony labyrinth of the inner ear.

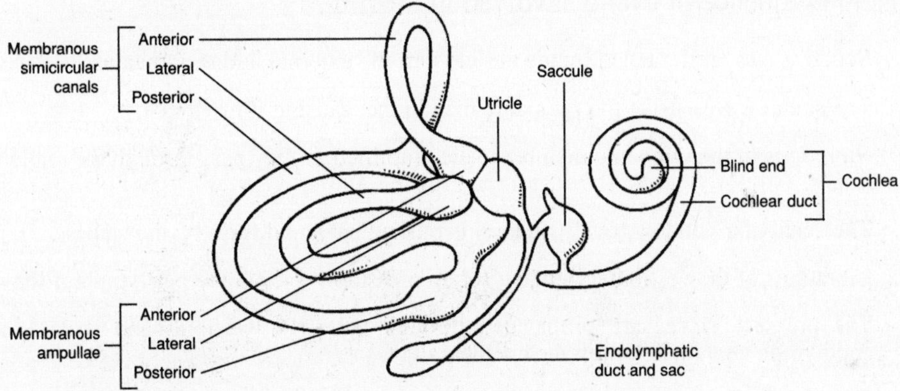

Figure 12.13 The membranous labyrinth of the inner ear.

12.18 Describe the cochlea in detail.

The **cochlea** has three chambers: an upper **scala tympani**, a lower **scala vestibuli**, and a middle **cochlear duct** (fig. 12.14). The scala tympani is continuous with the scala vestibuli, and both contain perilymph. The cochlear duct is bordered by the **vestibular membrane** and the **basilar membrane**. It contains endolymph. It also contains the **hair cells** that are embedded in the basilar membrane and that contact the **tectorial membrane**. The cochlear duct and the structures it contains constitute the **spiral organ** (*organ of Corti*). The spiral organ is considered the functional unit of hearing because it is here that the fluid vibrations of the mechanical sound waves stimulate the hair cells (dendritic endings of neurons) causing nerve impulses (sound sensations) to be conveyed through the cochlear nerve to the brain for perception.

High-frequency sound waves activate hair cells closer to the vestibular window at the base of the cochlea. Low-frequency sound waves activate hair cells farther away from the vestibular window, toward the top of cochlea.

Tinnitus is a ringing perception in one or both ears when no sound is present. It is caused by abnormal stimulation of either the spiral organ or the cochlear nerve. Tinnitus accompanies most ear disorders as well as other diseases including cardiovascular disease and anemia. Loud noises, nicotine, caffeine, and alcohol may aggravate the condition.

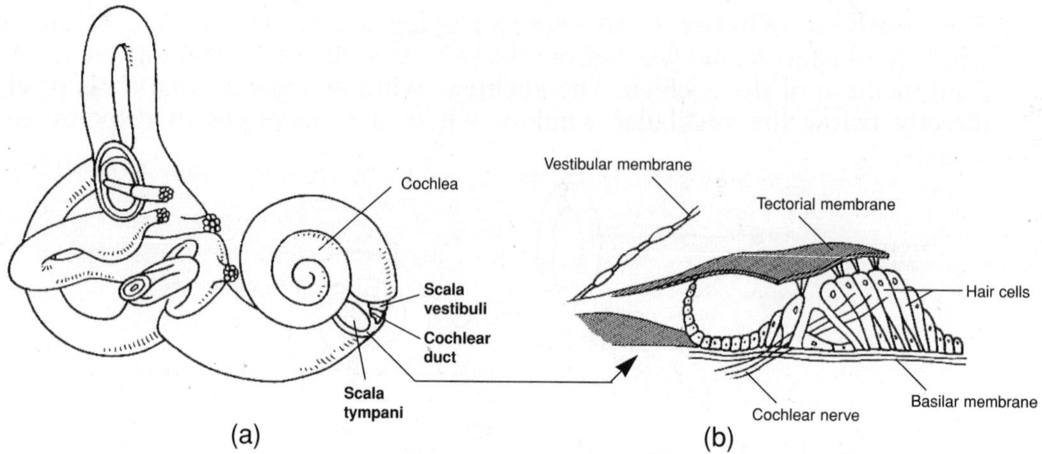

Figure 12.14 The cochlea is shaped like a snail shell. (*a*) The three chambers of the cochlea (in bold) and (*b*) the spiral organ.

12.19 List the sequence of events involved in hearing.

1. Sound waves are funneled by the auricle into the external auditory meatus.
2. The sound waves strike the tympanic membrane, causing it to vibrate.
3. Vibrations of the tympanic membrane are amplified as they pass through the malleus, incus, and stapes.
4. The vestibular window (oval window) is pushed back and forth by the stapes.
5. Vibrations of the vestibular window set up pressure waves in the perilymph of the cochlea.
6. The pressure waves are propagated through the scala vestibuli and scala tympani to the endolymph contained within the cochlear duct.
7. Stimulation of the hair cells within the spiral organ of the cochlea causes the generation of nerve impulses in the cochlear nerve (a portion of the vestibular [eighth cranial] nerve), which pass into the pons of the brain.

Deafness, which refers to any hearing loss, is of two types. *Conduction deafness* is caused by a defect of the outer or middle ear that inhibits sound transmission. An example would be an immovable stapes that would interfere with the transmission of sound through the middle-ear chamber. *Perception deafness* is caused by defects of structures of the cochlea or defects of the cochlear nerve. Conduction deafness usually can be corrected, while perception deafness can be corrected only rarely.

12.20 Explain how changes in body motion (that involve the head) are monitored by hair cell receptors within the vestibular organs (the three semicircular canals, the utricle, and the saccule).

Whenever the head is moved—or, more precisely, *accelerated*—in a certain direction, the hair cells of the vestibular organs move with the head. However, because of inertia, the endolymph within the vestibular organs tends to keep its original position in space; thus, it pushes in the opposite direction, against the hair-cell receptors, thereby stimulating them. The information generated by the receptors, in the form of nerve impulses, is transmitted to the CNS, where it helps to regulate postural reflexes and equilibrium.

Receptors in the roughly spherical utricle and saccule detect linear acceleration in any given direction. Receptors in the semicircular canals detect rotational acceleration—also in any direction, since the semicircular canals are disposed in perpendicular planes.

Review Exercises

Multiple Choice

1. The structure that is in direct contact with the tympanic membrane is (*a*) the stapes, (*b*) the incus, (*c*) the malleus, (*d*) the semicircular canals.

2. Which of the following is *not* a structure of the eye? (*a*) bulbular conjunctiva, (*b*) suspensory ligament, (*c*) basilar membrane, (*d*) macula lutea, (*e*) ciliary body

3. Movement of the eye superolaterally is the function of which muscle? (*a*) superior rectus, (*b*) lateral rectus, (*c*) inferior oblique, (*d*) superior oblique

4. Which of the following terms does *not* apply to how light rays are processed in the eyes? (*a*) refraction, (*b*) accommodation, (*c*) inversion, (*d*) conversion, (*e*) dispersion

5. The first structure of the eye contacted by incoming light rays is (*a*) the bulbular conjunctiva, (*b*) the cornea, (*c*) the anterior chamber, (*d*) the iris, (*e*) the pupil.

6. In the central region of the retina there is a yellowish spot, the macula lutea, with a depression in its center that produces the sharpest vision. This depression is called (*a*) the optic disc, (*b*) the rods and cones, (*c*) the vitreous body, (*d*) the fovea centralis, (*e*) the ganglion cells.

7. Which of the following is *not* a refractive medium of the eye? (*a*) lens, (*b*) vitreous humor, (*c*) pupil, (*d*) cornea, (*e*) aqueous humor

8. The modality of taste that is sensed over the tip of the tongue is (*a*) sweet, (*b*) sour, (*c*) bitter, (*d*) salty.

9. Which structure separates the external auditory canal from the middle-ear chamber? (*a*) auditory membrane, (*b*) vestibular membrane, (*c*) tympanic membrane, (*d*) acoustic membrane

10. Aqueous humor produced by the ciliary body is secreted into the posterior chamber and enters the anterior chamber through (*a*) the pupil, (*b*) the scleral venous sinus, (*c*) the vitreous body, (*d*) the suspensory ligament, (*e*) the lens capsule.

11. The basic functional unit of hearing is (*a*) the utricle, (*b*) the auricle, (*c*) the spiral organ, (*d*) the semicircular canals.

12. Transmission of sound waves through the inner ear occurs through (*a*) nerve fibers, (*b*) a gaseous medium, (*c*) auditory ossicles, (*d*) a fluid medium, (*e*) a solid medium.

13. Which is the proper sequence of visual sensory transmission from stimulation of photoreceptors located on the medial side of the retina?
    (*a*) optic nerve, lateral geniculate body, optic radiation, optic tract, cerebral cortex
    (*b*) optic nerve, optic chiasma, lateral geniculate body, optic tract, cerebral cortex, optic radiation
    (*c*) optic nerve, optic chiasma, optic tract, lateral geniculate body, optic radiation, cerebral cortex
    (*d*) optic nerve, optic tract, lateral geniculate body, optic radiation, cerebral cortex

14. When the eyeball is too long and an image is focused in front of the retina, the condition is termed (*a*) presbyopia, (*b*) hyperopia, (*c*) myopia, (*d*) astigmatism.

15. Which of the following is *not* a type of lingual papilla? (*a*) vallate papilla, (*b*) glossal papilla, (*c*) fungiform papilla, (*d*) filiform papilla

16. The suspensory ligament extends from (*a*) the ciliary body to the lens capsule, (*b*) the fovea centralis to the optic disc, (*c*) the retina to the vitreous humor, (*d*) the conjunctiva to the inner surfaces of the eyelids, (*e*) the orbit to the sclera of the eye.

17. Identify the organ/innervation mismatch. (*a*) glossopharyngeal nerve–tongue, (*b*) optic nerve–eye, (*c*) facial nerve–olfactory epithelium, (*d*) cochlear nerve–spiral organ, (*e*) vestibular nerve–semicircular canals.

18. Which portion of the cochlea responds to low-frequency sound waves? (*a*) the portion closest to the vestibular window, (*b*) the middle portion, (*c*) the portion closest to the cochlear nerve, (*d*) the end portion

19. The hair cells in the spiral organ are supported by (*a*) the basilar membrane, (*b*) the vestibule, (*c*) the tectorial membrane, (*d*) the utricle, (*e*) the cochlear plate.

20. Aqueous humor is drained from the anterior cavity of the eye through (*a*) the tarsal duct, (*b*) the scleral venous sinus, (*c*) the nasolacrimal duct, (*d*) the optic canal.

21. The fleshy outer portion of the ear is referred to as (*a*) the auricle, (*b*) the external auditory canal, (*c*) the acoustic apparatus, (*d*) the otic fold.

22. Which of the following is the correct sequence for passage of sensory impulses through the cells of the retina? (*a*) ganglion neurons, rods and cones, bipolar neurons; (*b*) rods and cones, bipolar neurons, ganglion neurons; (*c*) rods and cones, ganglion neurons, bipolar neurons; (*d*) ganglion neurons, bipolar neurons, rods and cones

23. Ceruminous glands secrete (*a*) lacrimal fluid, (*b*) mucus into the middle-ear chamber, (*c*) aqueous humor, (*d*) cerumen, (*e*) endolymph.

24. Conduction deafness involves structures in (*a*) the outer ear and middle ear, (*b*) the cochlea, (*c*) the inner ear, (*d*) the auditory pathway to the brain.

25. Which of the following structures would be directly involved in glaucoma? (*a*) vitreous humor, (*b*) sclera, (*c*) lens, (*d*) scleral venous sinus, (*e*) nasolacrimal duct

True or False

____ **1.** The special senses are localized in complex receptor organs and have extensive neural pathways.

____ **2.** Taste buds occur on the surface of the tongue, but are also found in smaller number in the mucosa of the palate and pharynx.

____ **3.** The pitch of a sound is directly related to the wave frequency.

____ **4.** Lacrimal fluid (tears) contains the enzyme amylase.

____ **5.** Contraction of the lateral rectus muscle rotates the eye laterally, away from the midline.

____ **6.** The anterior chamber is located between the cornea and the iris and is filled with vitreous humor.

____ **7.** The malleus is the bone in the middle ear that is attached to the vestibular window.

____ **8.** Vibrations of the vestibular window set up compressional waves in the perilymph of the cochlea.

____ **9.** The saccule, semicircular canals, and cochlea constitute the vestibular organs.

____ **10.** The auditory canal equalizes the pressure on the inside of the tympanic membrane to that on the outside of the membrane.

____ **11.** The photoreceptive rods and cones are sensitive to color and to black and white, respectively.

____ **12.** The vitreous humor is a permanent refractive medium in the posterior cavity of the eye, while the aqueous humor is a constantly replaced refractive medium in the anterior cavity of the eye.

____ **13.** Foramina within the cribriform plate are associated with olfaction.

____ **14.** An awareness of the position of the head as it relates to gravity is due to stimulation of hair cells in the utricle.

____ **15.** A dysfunction of the facial (seventh cranial) nerve would inhibit a person's ability to detect sweet taste.

Completion

1. Alkaloids elicit the ____________________ taste sensation.

2. The anterior two-thirds of the tongue is innervated by the ____________________ cranial nerve.

3. The posterior eye cavity contains a transparent, jellylike substance called ____________________ ____________________.

4. A ____________________ causes the lens to lose its transparency.

5. ____________________ is the condition resulting from an irregular curvature of the cornea.

6. True color blindness is referred to as ____________________.

7. Movement of the thin ____________________ ____________________ transmits sound waves from the outer to the middle ear.

8. The ____________________ (oval) window is located at the footplate of the stapes, and the ____________________ (round) window is located at the end of the scala vestibuli.

9. The ____________________ organ (organ of Corti) is the functional unit of hearing.

10. The ____________________ organs are the functional units of balance and equilibrium.

Labeling

Label the structures indicated on the figure to the right.

1. ______________________
2. ______________________
3. ______________________
4. ______________________
5. ______________________
6. ______________________
7. ______________________
8. ______________________
9. ______________________
10. ______________________

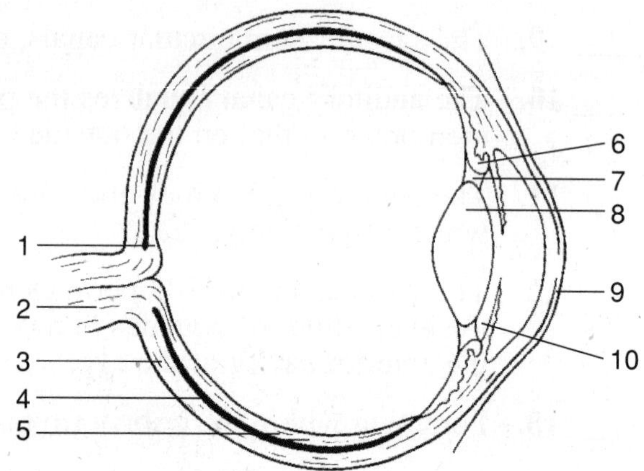

Matching

Match the structure to its function.

____	**1.** Cornea	(*a*)	provides a sharp visual image
____	**2.** Tarsal gland	(*b*)	secretes lacrimal fluid (tears)
____	**3.** Fovea centralis	(*c*)	vibrates in response to sound waves
____	**4.** Optic radiation	(*d*)	attaches to lens capsule
____	**5.** Auditory tube	(*e*)	refracts light rays
____	**6.** Lacrimal gland	(*f*)	secretes ear wax
____	**7.** Suspensory ligament	(*g*)	equalizes air pressure
____	**8.** Ceruminous gland	(*h*)	secretes an oily substance
____	**9.** Ciliary body	(*i*)	transmits sensory impulses
____	**10.** Basilar membrane	(*j*)	secretes aqueous humor

Answers and Explanations for Review Exercises

Multiple Choice

1. (*c*) In the order through which sound waves pass through the auditory ossicles in the middle-ear chamber, these bones are the malleus (contacting the tympanic membrane), incus (in the middle), and stapes (contacting the vestibular window).
2. (*c*) The basilar membrane is in the spiral organ within the cochlea of the ear.
3. (*c*) Because of its point of attachment on the sclera, the inferior oblique is the muscle that acts to turn

the eye superolaterally.

4. (*e*) Dispersion is the opposite of refraction. Refraction is an important function of the eye because it converges the light rays onto a focal point.
5. (*a*) The membranous bulbular conjunctiva is a thin protective covering over the anterior surface of the eye.
6. (*d*) Containing only cones, the fovea centralis is the region of the retina that provides the keenest vision.
7. (*c*) The pupil is not an anatomical structure; it is an opening within the lens for the passage of light rays.
8. (*a*) Taste buds that respond to molecules that elicit a sweet taste are located on the tip of the tongue.
9. (*c*) The tympanic membrane vibrates in response to sound waves, which activates the auditory ossicles in the middle-ear chamber.
10. (*a*) The pupil permits the passage of light waves and the passage of aqueous humor.
11. (*c*) Contained within the cochlea, the spiral organ is the basic functional unit of hearing because it transforms fluid vibrations from sound waves (mechanical energy) into a nerve impulse (electrical energy).
12. (*d*) The fluid medium of perilymph surrounds the cochlear duct within the cochlea.
13. (*c*) Only the nerve fibers that originate from the medial side of the retina (responding to the lateral field of vision) cross the optic chiasma to the opposite side of the brain. Nerve fibers of the optic nerve that arise from the lateral side of the retina (responding to the medial field of vision) do not cross at the optic chiasma to the opposite side.
14. (*c*) Myopia, or nearsightedness, may be corrected with a biconcave lens or sometimes by surgical procedures (radial keratotomy or photorefractive keratectomy).
15. (*b*) Vallate papillae are located at the back of the tongue, fungiform papillae are located on the tip and sides of the tongue, and filiform papillae are located on the anterior two-thirds of the tongue.
16. (*a*) The degree of tension in the suspensory ligament extending from the ciliary body to the lens capsule determines the shape of the lens.
17. (*c*) The olfactory epithelium lining the superior border of the nasal cavity is innervated by the olfactory (first cranial) nerve.
18. (*d*) High-frequency sounds activate sensory receptors near the vestibular window, whereas low-frequency sounds activate sensory receptors distant from the vestibular window. A gradation of sound frequencies is elicited in the area between.
19. (*a*) Hair cells are supported within the basilar membrane and contact the tectorial membrane, where they are stimulated.
20. (*b*) Produced by the ciliary body, aqueous humor flows into the posterior chamber through the pupil and into the anterior chamber. From there, it drains out of the eye at the scleral venous sinus.
21. (*a*) The auricle, or pinna, is the fleshy appendage on the side of the head that is commonly referred to as the ear.
22. (*b*) Incoming light rays that pass through the neural layer of the retina first activate the rods and cones, then the bipolar neurons, and finally the ganglion cells.
23. (*d*) Ear wax, or cerumen, is a protective waterproofing substance secreted by the cerumenous gland in the external auditory canal.
24. (*a*) Involving structures of the outer and middle ear, conduction deafness may result from impacted cerumen, a ruptured tympanic membrane, or immovable auditory ossicles.
25. (*d*) Poor drainage of aqueous humor results in excessive intraocular pressure that may cause deterioration of the retina and/or the optic nerve.

True or False

1. True
2. True
3. True

4. False; lacrimal fluid contains lysozyme
5. True
6. False; the anterior chamber is filled with aqueous humor
7. False; the malleus is attached to the tympanic membrane, and the stapes is attached to the vestibular window
8. True
9. False; the saccule, utricle, and semicircular canals constitute the vestibular organs
10. True
11. False; cones respond to colors and rods respond to black and white
12. True
13. True
14. True
15. True

Completion

1. bitter
2. facial (VII)
3. vitreous humor
4. cataract
5. Astigmatism
6. monochromatism
7. tympanic membrane
8. vestibular, cochlear
9. spiral
10. vestibular

Labeling

1. Fovea centralis
2. Optic nerve
3. Sclera
4. Choroid
5. Retina
6. Ciliary body
7. Suspensory ligament
8. Posterior cavity
9. Cornea
10. Lens

Matching

1. (*e*)
2. (*h*)
3. (*a*)
4. (*i*)
5. (*g*)
6. (*b*)
7. (*d*)
8. (*f*)
9. (*j*)
10. (*c*)

Endocrine System 13

Objective A To describe the *endocrine system* in general terms and to compare endocrine responses and neural responses with respect to speed and duration.

The **endocrine system** consists of *endocrine glands* that secrete specific chemicals called *hormones* into the blood or surrounding interstitial fluid. The endocrine system functions closely with the nervous system in regulating and integrating body processes. More specifically, hormones cause changes in the metabolic activities in specific cells, and nerve impulses cause muscles to contract or glands to secrete. In general, the action of hormones is relatively slow and the effects are prolonged, whereas the action of nerve impulses is fast and the effects are of short duration.

Endocrinology is the study of endocrine glands, the hormones they secrete, and the effects they have on their *target cells,* or *target tissues.*

Objective B To define a *hormone* and to describe the various classes of hormones.

A **hormone** is a chemical messenger secreted by an endocrine gland. Its chemical composition is such that it has its effect on specific receptor sites on target cells. Hormones are classified according to chemical structure and the location of the cell membrane receptors on their target cells.

13.1 Distinguish between the classes of hormones on the basis of their chemical structure.

Amines or amino acid derivatives (catecholamines). Contain atoms of carbon, hydrogen, and nitrogen and are characterized by an amine (NH_2) group. *Examples*: epinephrine, norepinephrine, thyroxine (T_4), triiodothyronine (T_3).

Polypeptides. Composed of long chains of amino acids. *Examples:* adrenocorticotropic hormone (ACTH), calcitonin, cholecystokinin, gastrin, glucagon, human growth hormone (HGH), insulin, melanocyte-stimulating hormone (MSH), oxytocin, parathyroid hormone (PTH), prolactin (PRL), secretin, somatostatin, and vasopressin (antidiuretic hormone, ADH).

Glycoproteins. Consist of large proteins combined with carbohydrates. *Examples*: follicle-stimulating hormone (FSH), human chorionic gonadotropin (hCG), luteinizing hormone (LH), and thyroid-stimulating hormone (TSH).

Steroids. Consists of lipids synthesized from cholesterol. *Examples*: aldosterone, cortisol, estradiol, progesterone, and testosterone.

Fatty acid derivatives. Composed of long hydrocarbon acid chains. *Examples*: prostaglandins, leukotrienes, and thromboxane.

13.2 Distinguish between the classes of hormones based on the location of the cell membrane receptors on their target cells.

Group I hormones. Bind to intracellular receptors and are lipophilic (lipid-soluble, enabling them to cross cell membranes). Group I hormones include the steroid hormones, iodothyronines, and calcitrol.

Group II hormones. Bind to cell surface receptors and are hydrophilic (water-soluble, enabling them to remain in the extracellular fluid). Group II hormones include the polypeptide, protein, glycoprotein, and catecholamine hormones.

13.3 Where do hormones have their effect?

Hormones are specific as to which cells they affect and the cellular changes they elicit. The arrival of a hormone at a target site triggers a sequential series of biochemical events that leads to a specific response (action). The hormone binds to specific, high-affinity protein receptors located either on the cell surface, in the cytoplasm (intracellular), or in the nucleus. Steroids and thyroid hormones are lipophilic and readily enter the cell. Receptors for catecholamines (epinephrine and norepinephrine), polypeptides, and glycoproteins are located on or in cellular membranes. They are generally insoluble in lipids and cannot passively cross the cell membrane.

Objective C To define *negative feedback* and *positive feedback* and to explain their importance in regulating the secretion of hormones.

Negative feedback involves a cascade, or chain, of biochemical or physiological events. Generally, an increased amount of end product inhibits the production, mechanism, or action of a starting substance to prohibit further synthesis of the end product.

Example: A › B › C › D

As A progresses through B and C to D, the amount of D increases. Substance D, however, is an inhibitor of substance A. As levels of substance D increase, substance A receives "negative feedback" to prevent the process from continuing to produce more D.

By contrast, in the case of **positive feedback**, D would stimulate A to further produce increased amounts of B, and so on to D. There are few positive feedback mechanisms in the body (see problem 13.5).

Homeostasis is maintained by the continual adjustments of endocrine function in response to changes in our environment. Negative feedback occurs when the product or result of activity in the endocrine system inhibits the factors that produced the product or result so as to maintain a normal range of values. Positive feedback increases the deviation from normal values, and thus is not homeostatic.

13.4 Give a model for the negative feedback mechanism that often regulates the production or secretion of many hormones.

Shown in fig. 13.1 is an outline of the negative feedback mechanism associated with cortisol and the hypothalamic-pituitary-adrenal axis.

13.5 Give an example of a positive feedback mechanism.

The secretion of oxytocin during labor accompanying childbirth is a *positive feedback mechanism.* As the baby is forced toward the vagina (birth canal) through uterine contractions, the increased pressure on the mother's cervix stimulates pressure receptor cells in the wall of the cervix. Nerve impulses are sent to the pituitary gland causing the release of oxytocin. Oxytocin is then carried by the blood to the uterus, causing the uterine muscles to contract even more vigorously and frequently. These contractions force the baby farther into the vagina, and childbirth is accomplished. Once the baby is born, the pressure stimulus for oxytocin release ends and the positive feedback mechanism is stopped.

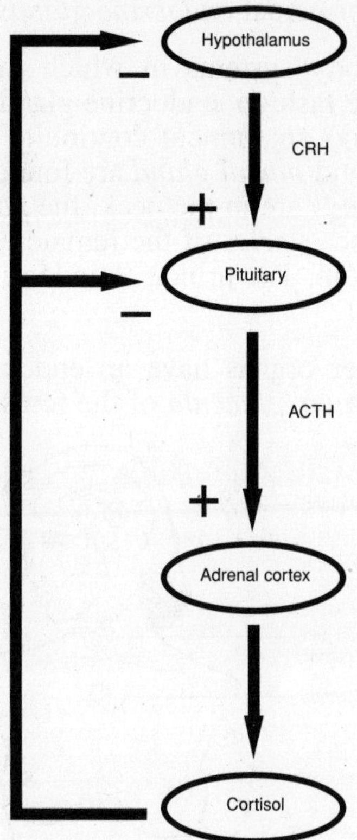

Figure 13.1 Negative feedback within the hypothalamic-pituitary-adrenal axis.

13.6 Describe the characteristics of the body's endogenous (inherent) endocrine rhythms.

Temporal oscillations ranging from a few minutes to 24 hours (circadian) to a year occur in endocrine function (fig. 13.2). The nature of circadian rhythms has been studied in humans isolated in a soundproof chamber with no time cues. Under such conditions, individuals exhibit certain 24-hour rhythms in hormonal secretion.

Circadian rhythms in hormonal secretion have important medical implications. A blood plasma cortisol level of 150 mg/mL at 8 A.M. is normal, whereas the same level observed at 8 P.M. may indicate *hypercortisolism* (Cushing's syndrome). Differentiation between pathologic and normal blood plasma levels of certain hormones is improved by carefully selecting the time the sample is taken (time of day, month, or year).

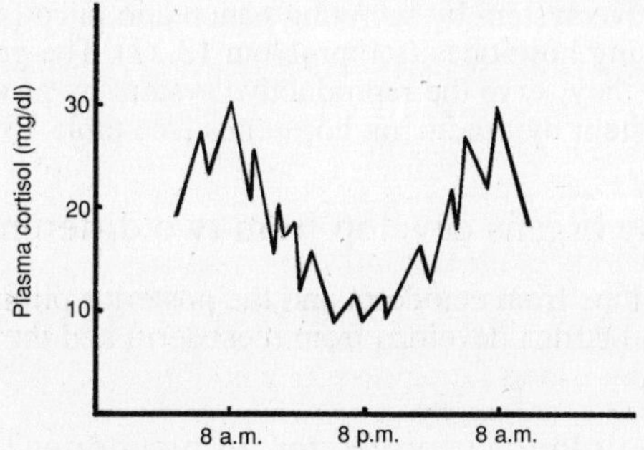

Figure 13.2 Plasma cortisol levels on a daily cycle.

Objective D To identify the principal *endocrine glands* and to list their secretions.

survey

Unlike other body systems in which the organs are physically hooked together in some fashion, endocrine glands are widely scattered throughout the body with no anatomical continuity (fig. 13.3). The *pituitary gland, hypothalamus,* and *pineal gland* are found within the skull; the *thyroid* and *parathyroid glands* are in the neck; the *pancreas* and *adrenal glands* are in the abdominal region; the *ovaries* of the female are in the pelvis; and the *testes* of the male are in the scrotum. The principal endocrine glands and their secretions are summarized in table 13.1.

In addition, several other organs have an endocrine function. These include the *thymus, stomach, duodenum, placenta* of the fetus, and even the *heart.*

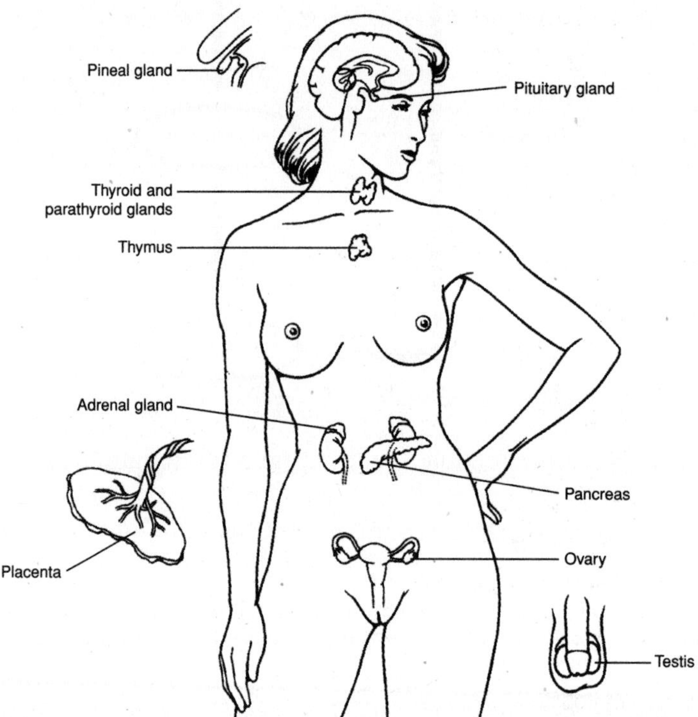

Figure 13.3 The principal endocrine glands.

13.7 What is a mixed gland?

A *mixed gland* is one that directly serves two or more body systems. The pancreas is a mixed gland because it serves the digestive system by secreting pancreatic juice (see problem 19.36) and the endocrine system by releasing hormones (see problem 13.27). The gonads (testes and ovaries) are also mixed glands because they serve the reproductive system by producing gametes (see problem 23.1) and the endocrine system by producing hormones (see table 13.1).

13.8 Which of the endocrine organs develop from two different germ layers?

The anterior pituitary develops from ectoderm and the posterior pituitary develops from the neuroectoderm. The adrenal cortex develops from mesoderm and the adrenal medulla develops from neuroectoderm.

13.9 Which of the endocrine glands secrete steroid hormones?

The testes, the ovaries, and the adrenal glands all secrete steroid hormones.

Table 13.1 The Principal Endocrine Glands and Their Secretions

Gland		*Hormones*
Pituitary gland	Adenohypophysis (anterior pituitary)	Human growth hormone (HGH or GH) Thyroid-stimulating hormone (TSH) Adrenocorticotropic hormone (ACTH) Prolactin (PRL) Follicle-stimulating hormone (FSH) Luteinizing hormone (LH)
	Neurohypophysis (posterior pituitary)	Antidiuretic hormone (ADH) Oxytocin
Thyroid gland		Thyroxine (T_4) Triiodothyronine (T_3) Calcitonin
Parathyroid glands		Parathyroid hormone (PTH)
Adrenal gland	Adrenal cortex	Cortisol Corticosterone (*glucocorticoids*) Aldosterone Deoxycorticosterone (*mineralocorticoids*)
	Adrenal medulla	Epinephrine and norepinephrine
Pancreas		Insulin Glucagon
Testes		Testosterone (an *androgen*)
Ovaries		Estradiol (an *estrogen*) Progesterone

Objective E To describe the structure of the *pituitary gland* and to identify the secretory cells of the anterior pituitary.

The small, pea-shaped **pituitary gland** (cerebral hypophysis) is located on the inferior side of the brain. It is positioned in the sella turcica of the sphenoid bone. The pituitary gland is attached to the brain by the *pituitary stalk* (fig. 13.4). The *infundibulum* is the portion of the pituitary stalk that connects the hypothalamus to the posterior lobe of the pituitary gland.

The pituitary gland is divided into an *anterior lobe,* or **adenohypophysis,** and a *posterior lobe*, or **neurohypophysis**. The adenohypophysis consists of a *pars distalis, pars tuberalis,* and *pars intermedia*. The *pars intermedia* makes up the anterior part of the pituitary stalk.

The anterior lobe forms from an invagination (Rathke's pouch) of the pharyngeal epithelium–thus, the epithelial nature of its cells. The posterior lobe forms from and outgrowth of the hypothalamus and contains axons from the neurosecretory cells of the hypothalamus, along with neuroglia-like cells (pituicytes).

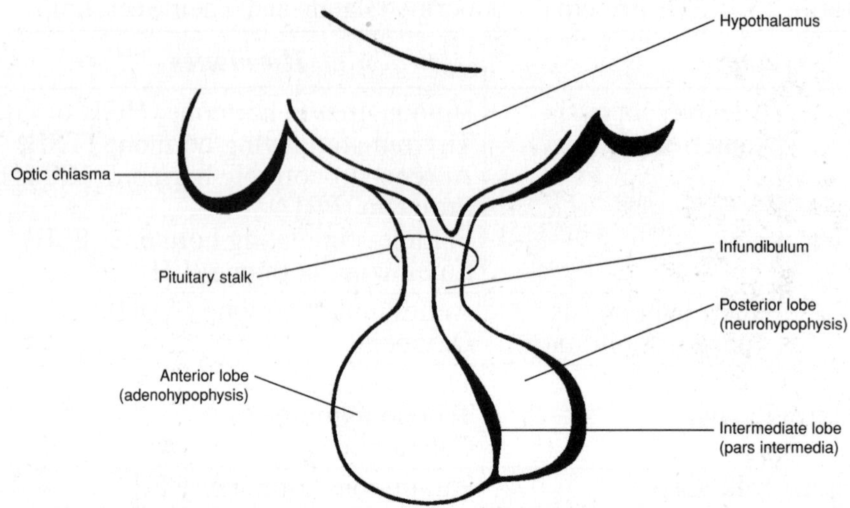

Figure 13.4 The structure of the pituitary gland.

The pituitary gland is a relatively common site for the development of a tumor. As the tumor begins to grow, it frequently causes hypersecretion of pituitary hormones. Growth of soft body tissues and altered reproductive cycles are common symptoms. Surgical removal of a tumor of the pituitary gland is called a *hypophysectomy*. The surgical approach is usually through the nasal cavity and the sphenoidal sinus to the sella turcica. Over 70% of the pituitary gland can be surgically removed without loss of normal hormonal function.

13.10 Secretory cells of the anterior pituitary are categorized into three groups, according to their staining properties (table 13.2). Identify these secretory cells.

Table 13.2 Secretory Cells of the Anterior Pituitary

Category	*Stain*	*Hormone*
Acidophils	Acidic stain	Human growth hormone (HGH) and prolactin (PRL)
Basophils	Basic stain	Thyroid-stimulating hormone (TSH), follicle-stimulating hormone (FSH), and luteinizing hormone (LH)
Chromophobes	Stain-resistant	Adrenocorticotropic hormone (ACTH)

Objective F To state the origin and effect of each of the pituitary hormones.

The pituitary hormones and their effects on the target tissues are summarized in table 13.3.

Table 13.3 Summary of the Pituitary Hormones

Source cell	*Hormone*	*Target tissue*	*Effect*
Somatotrophs	HGH	Bones; soft tissue	Accelerates rate of body growth, stimulates uptake of amino acids into cells and protein synthesis; promotes carbohydrate and fat breakdown
Thryrotrophs	TSH	Thyroid gland	Promotes growth and development of thyroid gland; stimulates synthesis and release of thyroid hormones
Corticotrophs	ACTH	Adrenal cortex	Promotes growth and development of adrenal cortex; stimulates secretion of glucocorticoids
Lactotrophs	PRL	Mammary glands	Promotes development of mammary glands; stimulates milk production
Gonadotrophs	FSH	Ovaries and testes	Female: stimulates growth of ovarian follicles Male: stimulates spermatogenesis
Luteotrophs	LH	Ovaries and testes	Female: stimulates maturation of follicle cells, promotes ovulation and development of corpus luteum, and stimulates corpus luteum to secrete estrogens and progesterone Male: stimulates interstitial cells to secrete testosterone
Supraoptic and paraventricular nuclei of the hypothalamus	ADH	Kidney tubules	Facilitates water reabsorption in the distal convoluted tubules and collecting ducts
	Oxytocin	Mammary glands and uterus	Stimulates contraction of uterine muscles; stimulates secretion of milk from the breast

13.11 What regulates the secretion of prolactin?

Under the influence of an increased production of somatomammotropin from the placenta, prolactin levels increase progressively during pregnancy. During lactation, stimulation of the nipple by a nursing infant initiates a neuroendocrine reflex that results in increased prolactin secretion. This stimulates milk production for the next episode of nursing.

Prolactin enhances breast development and milk production in females. In nonlactating, premenopausal women who are not pregnant, prolactin levels average 10–20 ng/mL. During pregnancy and lactation, prolactin levels may reach 500 ng/mL. The function of prolactin in males is uncertain, although some evidence indicates that it may increase testicular LH receptors. Prolactin levels in males average 5 ng/mL.

13.12 What are the mechanisms by which growth hormone stimulates the growth of body cells?

Protein synthesis is a major prerequisite for tissue growth because proteins are largely responsible for cellular structure and (as enzymes) regulate all cellular function. GH promotes protein synthesis by (1) stimulating amino acid uptake by cells; (2) increasing synthesis of tRNA, the limiting factor in protein synthesis; and (3) increasing the number and aggregation of ribosomes.

13.13 What are some of the factors that stimulate GH secretion?

Hypoglycemia. A 50% reduction in blood glucose will result in a fivefold increase in GH secretion.

Muscular activity. Walking 30 minutes will cause GH levels to rise.

Amino acids. Increased amounts stimulate GH secretion.

Stress (catecholamines). Increased amounts also stimulate GH secretion.

13.14 Give examples of GH secretion disorders.

Dwarfism. Decreased GH secretion before normal height has been reached. *Symptoms*: Small body, but normally proportioned; mild obesity with lack of appetite; tender, thin skin. *Treatment*: GH injections.

Gigantism. Excess GH before closure of the epiphyseal growth plates in long bones. *Symptoms*: Pathological acceleration of growth; if a tumor is involved, vision may be impaired. *Treatment*: Surgical removal of the tumor or the pituitary gland (hypophysectomy).

Acromegaly. Excess GH after closure of the epiphyseal plates. *Symptoms*: Large jaw; thickened and puffy nose; large ears, tongue, and head; increased basal metabolic rate (BMR); loss of visual fields. *Treatment*: Irradiation, radioisotope implantation, or surgical removal of the tumor or pituitary gland.

Growth hormone abuse. With the development of recombinant DNA techniques, commercially produced growth hormone became available in 1985. Although expensive, growth hormone is used to treat pituitary dwarfs. Not infrequently, parents seek GH treatment for their normally sized children in the hope of increasing the children's chances for athletic success. Growth hormone is also used by body builders in place of anabolic steroids because it, too, has the ability to increase muscle mass and strength. Moreover, the presence of this hormone in the body is difficult to detect because it is rapidly broken down. The potential long-term effects of growth hormone treatment are unknown.

13.15 What are the mechanisms that stimulate oxytocin and ADH release?

Oxytocin. Stretching of the uterus late in pregnancy initiates impulses to the hypothalamus that signal the posterior pituitary to release oxytocin. Through a positive feedback mechanism, the oxytocin then stimulates strong uterine contractions that accompany the period of labor during parturition. In addition, oxytocin has an important function in lactation. Mechanical stimulation of the nipple during nursing initiates impulses via the hypothalamus that signal the posterior pituitary to release oxytocin. The oxytocin then stimulates contractions in the myoepithelial cells surrounding the lactiferous alveoli of the mammary gland, thus causing the "let-down" or secretion of milk.

ADH. Both a decrease in body water (dehydration) and an increase in plasma osmolarity stimulate ADH secretion. ADH causes increased water reabsorption in the kidney tubules. Through a negative feedback mechanism, water is returned to the body fluids and the plasma osmotic pressure is decreased to normal levels.

A dysfunction of the posterior pituitary results in deficient ADH secretion, causing a condition called *diabetes insipidus*. Symptoms of this renal disorder include polyuria (excessive urination), polydipsia (excessive thirst), and severe electrolyte imbalance. Diabetes insipidus is treated by injections of ADH.

13.16 Why is oxytocin sometimes administered to a woman following parturition?

Oxytocin causes the uterus to shrink and constricts the uterine vessels, thus minimizing the possible danger of hemorrhage.

Objective G To describe the anatomy and physiology of the *thyroid gland.*

An anterior view of the thyroid gland is shown in fig. 13.5; a posterior view is shown in fig. 13.7. Biosynthesis of the thyroid hormones (under the stimulation of TSH) is diagrammed in fig. 13.6.

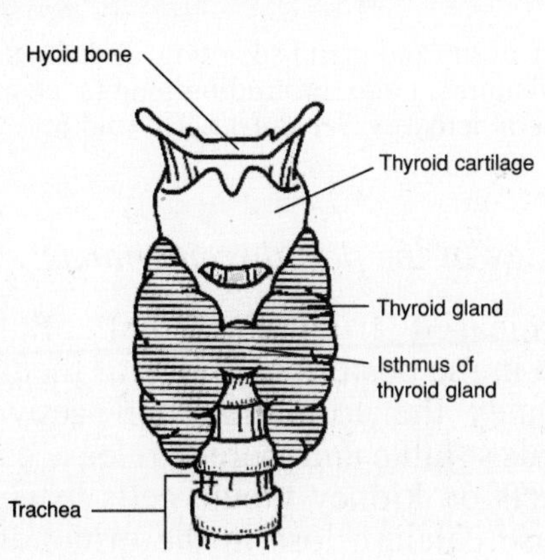

Figure 13.5 An anterior view of the thyroid gland.

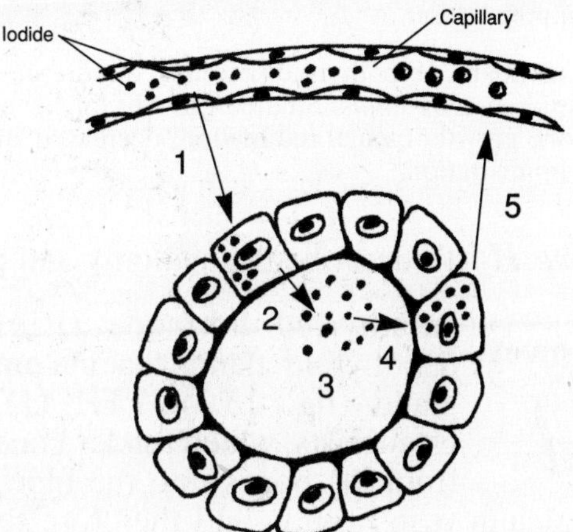

Figure 13.6 Thyroid hormone synthesis and secretion.

13.17 Describe events in the thyroid follicles that result in the synthesis and secretion of thyroid hormones.

1. Iodide is actively transported from the blood plasma into thyroid follicle cells (see fig. 13.6).
2. Iodide and thyroglobulin are secreted into the lumen.
3. Iodide is oxidized to iodine and attached to tyrosines in the thyroglobulin, forming mono- and diiodotyrosines (MIT, DIT). Coupling of MIT and DIT forms triiodothyronine (T_3, or triiodothyronine); coupling of two DITs forms tetraiodotyrosine (T_4, or thyroxine).
4. Under the influence of TSH, the colloid is taken up by endocytosis into the thyroid follicle cells.
5. T_3 and T_4 are removed from thyroglobulin and secreted.
6. T_3 and T_4 are transported in the blood in association with plasma proteins: thyroid-binding globulin (TBG), thyroxine-binding prealbumin (TBPA), and albumin.

13.18 Describe the actions of thyroid hormones T_3 and T_4.

The thyroid hormones (1) accelerate metabolic rate and oxygen consumption in all body tissues, (2) increase body temperature, (3) affect growth and development in early life, (4) accelerate glucose absorption, and (5) enhance the effects of the sympathetic division of the autonomic nervous system.

13.19 What are some common disorders associated with thyroid dysfunction?

Goiter. When dietary intake of iodine is low (below 10 μg/day), T_3 and T_4 synthesis becomes inadequate, and secretion declines. As blood plasma levels of the two hormones fall, the negative feedback mechanism causes an increased release of TSH from the anterior pituitary. The excessive TSH causes the thyroid to hypertrophy, producing a goiter that may become very large. Exposure to cold can also bring about increased secretion of TSH.

Graves' disease (thyrotoxicosis). Hyperthyroid secretion (excessive secretion by the thyroid gland). *Symptoms*: Loss of weight; rapid pulse; warm, moist skin; increased appetite; increased BMR; tremor; goiter; exophthalmos (bulging eyes); muscular weakness. *Treatment*: Removal of a portion of the thyroid gland, radioiodine, and antithyroid drugs.

Myxedema. Hypothyroid secretion (insufficient secretion of the thyroid gland) in adults. *Symptoms*: Weight gain; slow pulse; dry, brittle hair; decreased BMR; lack of energy; sensation of coldness; diminished perspiration; weakness. *Treatment*: Thyroid hormone (T_3 and T_4) administration.

Cretinism. Hypothyroid secretion (severe insufficiency of thyroid gland secretion) in infants and children. *Symptoms*: Stunted growth; thickened facial features; large, protruding tongue; abnormal bone growth; mental retardation; decreased BMR; general lethargy. *Treatment*: Thyroid hormone administration.

Objective H To describe the anatomy and physiology of the *parathyroid glands*.

Survey

Parathyroid hormone (PTH) is released from the small, flattened parathyroid glands that are embedded in the posterior surface of the thyroid gland (fig. 13.7). PTH (1) stimulates the formation and activity of osteoclasts, which render bone minerals soluble and thereby release calcium from the bones into the blood; (2) acts on kidney tubule cells to increase calcium reabsorption and therefore to decrease calcium loss in the urine; and (3) increases the synthesis of 1/25-dihydroxycholecalciferol, which increases calcium absorption from the GI tract. As all three lead to increased plasma calcium levels, it follows that secretion of PTH will be evoked by a drop in plasma calcium (or magnesium) concentration.

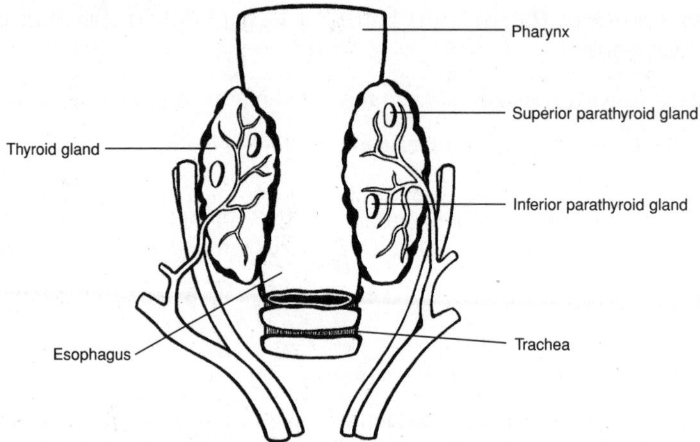

Figure 13.7 A posterior view of the thyroid gland showing the parathyroid glands.

13.20 What two cell types are found in the parathyroid glands?

The cells that secrete PTH are called *principal cells* (chief cells). They have a clear cytoplasm. *Oxyntic* ("acid-secreting") *cells*, which have granules in their cytoplasm, are dispersed throughout the parathyroid glands. The function of oxyntic cells is not known.

13.21 Why is it important that adequate plasma calcium levels be maintained?

Calcium participates in essentially all known biological functions. Among these are transmission of nerve impulses, muscle contraction, cell division, coagulation of blood, release of neurotransmitters, secretory processes of endocrine and exocrine glands, and enzyme function. The following disorders are associated with imbalances in plasma calcium levels:

Hypoparathyroidism. Deficient secretion of PTH. *Symptoms*: Hypocalcemia (low plasma calcium levels); neuromuscular hyperactivity; paresthesia (numbness and tingling around the mouth, in the tips of the fingers, and sometimes in the feet); convulsions. *Treatment*: Ergocalciferol with oral calcium.

Primary hyperparathyroidism. Excessive secretion of PTH. *Symptoms*: Hypercalcemia (high levels of plasma calcium), although most patients are relatively asymptomatic. At the present time there is no satisfactory protocol for treatment of hyperparathyroidism. Some forms of managing this condition include use of (1) glucocorticoids—when malignant neoplasms are involved, (2) mithramycin—a toxic antibiotic that inhibits bone reabsorption, (3) oral phosphate, (4) estrogen, and (5) calcitonin.

Objective I To describe the anatomy and physiology of the *adrenal gland.*

Survey

The **adrenal** (*suprarenal*) **glands** are embedded in adipose tissue at the superior borders of the kidneys. Each adrenal gland is triangular in shape (see fig. 13.8) and consists of an outer adrenal cortex and an inner adrenal medulla. The adrenal cortex is composed of three layers, or zones, as indicated in figure 13.8*c*. The actions of the steroid hormones (table 13.1) secreted by the adrenal cortical zones and the adrenal medulla are as follows:

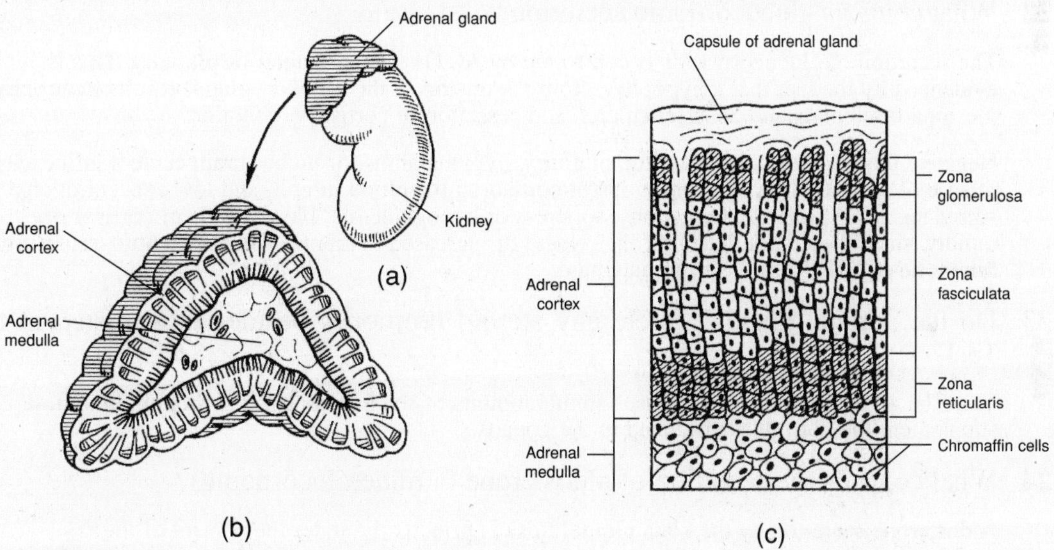

Figure 13.8 The adrenal gland. (*a*) The position of the adrenal gland at the superior border of a kidney; (*b*) the adrenal cortex and the adrenal medulla as seen in a sectioned adrenal gland; and (*c*) the three histological layers, or zones, of the adrenal cortex.

Glucocorticoids. (1) Regulate carbohydrate and lipid metabolism, stimulate synthesis of glucose from noncarbohydrates (gluconeogenesis), increase blood glucose and liver glycogen storage, accelerate the breakdown of proteins; (2) in large doses, inhibit inflammatory responses (capillaries fail to dilate, less edema occurs, fewer white blood cells migrate into the inflamed area); (3) promote vasoconstriction; and (4) help the body to resist stress.

Mineralocorticoids. Regulate the concentration of extracellular electrolytes (cations), especially sodium and potassium.

The effects of the **amine hormones** from the adrenal medulla are listed in table 13.4.

Table 13.4 Functions of the Amine Hormones Epinephrine and Norepinephrine

Epinephrine	*Norepinephrine*
Elevates blood pressure by increased cardiac output and peripheral vasoconstriction	Elevates blood pressure through generalized vasoconstriction
Accelerates respiratory rate and dilates respiratory passageways	Similar effect, but less marked
Increases efficiency of muscular contraction	Similar effect, but less marked
Increases rate of glycogen breakdown into glucose, so level of blood glucose rises	Similar effect, but less marked
Increases conversion of fats to fatty acids, so level of blood fatty acids rises	Similar effect, but less marked
Increases release of ACTH and TSH from the adenohypophysis	No effect

13.22 What controls glucocorticoid secretion?

The secretion of glucocorticoids is controlled by ACTH from the anterior pituitary. This is evidenced by the fact that a hypophysectomy (excision of the pituitary gland) results in atrophy of the zona fasciculata and zona reticularis, and cessation of cortisol production.

Negative feedback of cortisol on the pituitary, hypothalamus, or higher brain centers influences the release of ACTH. High concentrations of cortisol in the blood inhibit, and low concentrations stimulate, ACTH release. In response to stress or hypoglycemia, blood levels of cortisol rise rapidly, since these stimuli trigger the release of increased amounts of **corticotropin-releasing hormone (CRH)** from the hypothalamus.

13.23 Do the adrenal glands secrete any steroid hormones besides those listed in table 13.1?

Yes. The adrenal cortex also releases small amounts of sex hormones. It is thought that these supplement the hormones produced in the gonads.

13.24 What controls the secretion of aldosterone (a mineralocorticoid)?

Aldosterone secretion by the zona glomerulosa is principally under the control of the reninangiotensin system, the plasma potassium concentration, and, to a limited extent, ACTH. A "flowchart" for aldosterone production is shown in fig. 13.9.

13.25 What are some common disorders associated with adrenal gland dysfunction?

Cushing's disease (syndrome). Excess glucocorticoids (cortisol), with mineralocorticoid levels usually normal. *Symptoms*: Thick arms, legs, and skin; red cheeks; poor wound healing; round "moon" face; high blood pressure; decreased antibody formation; hyperglycemia (excessive blood sugar); muscle weakness. *Treatment*: Surgical removal of portions of the pituitary gland or adrenal glands; irradiation; hormone replacement therapy.

Addison's disease. Insufficient glucocorticoids and mineralocorticoids. *Symptoms*: Loss of electrolytes and body fluids; low blood pressure; hypoglycemia (insufficient blood sugar); weakness; loss of appetite; inability to withstand stress; increased pigmentation. *Treatment*: Administration of glucocorticoids and mineralocorticoids.

Adrenogenital syndrome. Excessive secretion of androgens from the adrenal cortex. *Symptoms*: In young children, premature puberty and enlarged genitalia; in mature women, development of masculine traits. *Treatment*: Surgical removal, if tumor is causing hypersecretion.

Pheochromocytoma. Tumor of the chromaffin cells of the adrenal medulla, with hypersecretion of epinephrine and norepinephrine. *Symptoms*: High blood pressure, increased BMR; hyperglycemia; nervousness; sweating. *Treatment*: Surgical removal of tumor.

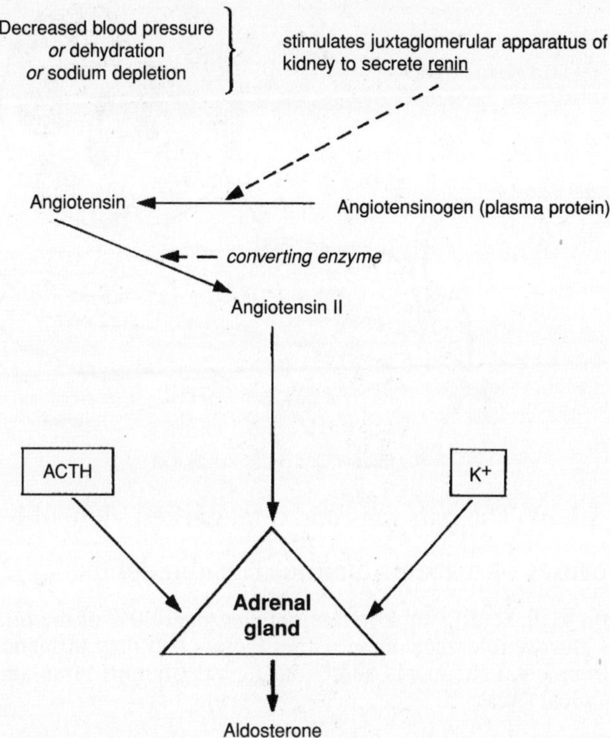

Figure 13.9 The sequence of events in aldosterone production.

13.26 What factors stimulate the adrenal medulla to secrete epinephrine (adrenaline) and norepinephrine (noradrenaline)?

Adrenal medullary secretion is prompted by sympathetic impulses during stress and in emergency situations in which the body is prepared for "fight or flight."

Objective J To identify the *pancreatic hormones* and to explain their physiological effects.

The endocrine portion of the pancreas (fig. 13.10) consists of scattered clusters of cells called **pancreatic islets** (*islets of Langerhans*). **Glucagon** is secreted by alpha cells, which constitute 20% of each pancreatic islet. Alpha cells are located mainly on the periphery of islets and are innervated by cholinergic fibers. **Insulin** is secreted by beta cells, constituting 75% of each pancreatic islet. Beta cells are located mainly at the center of the islet and are innervated by adrenergic fibers. **Somatostatin** is secreted by delta cells, which constitute 5% of each pancreatic islet. The delta cells are scattered throughout the islets.

13.27 What are the physiological effects of the pancreatic hormones?

Insulin stimulates movement of blood glucose across the cell membrane, stimulates glycolysis, and lowers blood glucose levels. Glucagon stimulates glycogenolysis and maintains blood glucose levels during fasting or starvation. Somatostatin, which has insulin-like properties, stimulates incorporation of sulfur into cartilage and stimulates collagen formation.

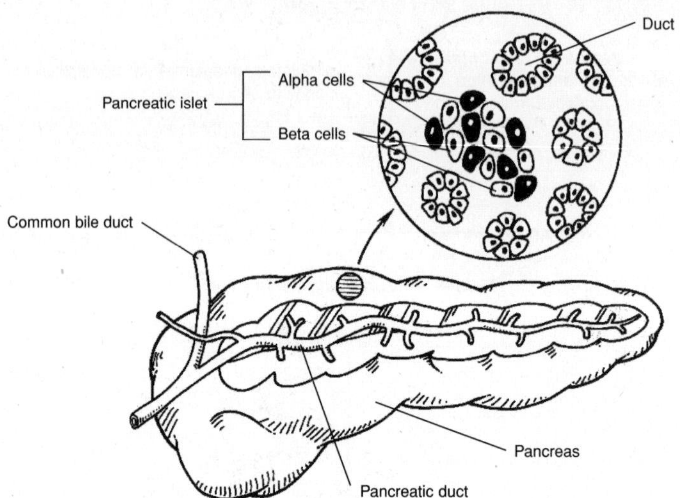

Figure 13.10 The pancreas and a magnified view of a pancreatic islet.

13.28 What are causes of diabetes mellitus (insulin deficiency)?

Predisposition to diabetes is inherited; moreover, over 20% of the relatives of diabetic patients have an abnormal glucose tolerance curve. Other factors that may influence the development of diabetes include environmental chemicals, infectious agents (mumps virus), autoimmune events, nutrition, and psychological stress.

13.29 What are the two types of diabetes mellitus?

Insulin-dependent, or *juvenile*, *diabetes* requires insulin injections. It is often severe and complicated by ketoacidosis (acetone breath). Insulin-dependent diabetes is usually contracted in youth, but may occur at any age. Non-insulin-dependent, or *maturity-onset*, *diabetes* does not require insulin injections. It is mild, and ketoacidosis is rare. This type is often associated with obesity and usually improves with weight loss. It is often treated with oral hypoglycemic drugs to stimulate insulin release from beta cells.

13.30 State the symptoms of diabetes mellitus.

(1) Glycosuria, or glucose in the urine; (2) polyuria, or increased urine volume; (3) polydipsia, or increased fluid intake; (4) hyperglycemia, or high blood glucose levels: (5) weakness; (6) loss of weight; (7) ketoacidosis; (8) vascular abnormalities.

13.31 What test is used to determine whether or not a patient has diabetes mellitus?

The *oral glucose tolerance test* (fig. 13.11) is useful in diagnosing diabetes mellitus. A glucose dose (2 g glucose/kg body weight) is given to the fasting patient. Diabetes is present if, just before the dose, the blood glucose level exceeded 115 mg/dL blood, or if the levels 1, 1.5, and 2 hours after the dose exceed 185, 165, and 140 mg/dL blood, respectively.

Insulin shock. Insulin shock may occur in a diabetic patient when too much insulin is injected for the patient's caloric intake and exercise level. The symptoms associated with insulin excess are mainly associated with brain function. The brain uses glucose as its major source of energy. With insulin excess, more glucose than is necessary is transported into the cells of the body. The result is a lowering of the blood glucose level so that the brain cannot function properly. Symptoms of decreased brain function may include confusion, fainting, unconsciousness, and possible death.

Chronic Complications of Diabetes Mellitus

Ophthalmologic complications: Microaneurysms, dot hemorrhages, exudates, retinal edema, and growth of vascular and fibrous tissue within the retina.

Renal complications: Thickening of the basement membrane of the capillaries of the glomeruli, proteinuria, hypoalbuiminemia, hypertension, and edema.

Neuralgic complications: Sensory loss, pains in the chest and abdominal area, motor neuropathy, and autonomic neuropathy (tachycardia, hypotension, nausea, vomiting, dysphasia, constipation, diarrhea, impotence).

Cardiovascular complications: Atrophic brown spots and necrobiosis.

Infections: Bacteriuria, candidal esophagitis, and candidal vaginitis (yeast infections).

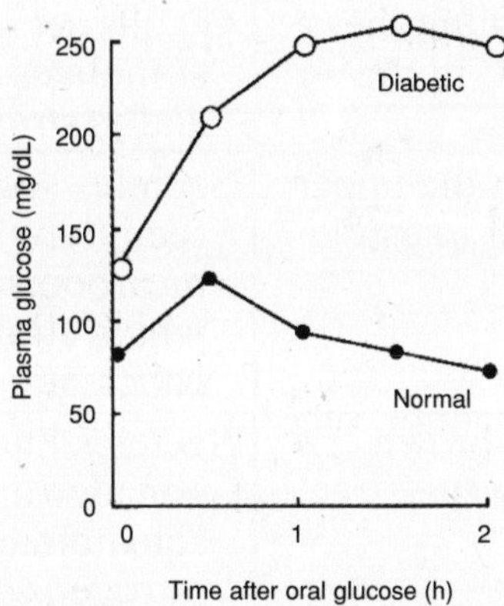

Figure 13.11 Diabetic and normal oral glucose tolerance curves.

Objective K To examine other organs and glands that have an endocrine function: the *thymus*, *pineal gland*, *gastric* and *duodenal mucosae*, and *placenta*.

See table 13.5.

13.32 What are the endocrine functions of the pineal gland?

The **pineal gland** is the major source of plasma melatonin in humans. Melatonin is synthesized from serotonin (5-hydroxytryptamine). At the present time, the exact role of melatonin in humans is not known; however, clinical observations indicate that precocious puberty may occur in males whose pineal gland has been destroyed by tumors. Therefore, it has been suggested that the pineal exerts an antigonadotropic effect. (In birds and rodents, melatonin has been implicated in the regulation of reproductive functions in relationship to diurnal light cycles).

13.33 What are the functions of human chorionic gonadotropin?

The trophoblastic tissue of the placenta begins to secrete human chorionic gonadotropin (hCG) shortly after implantation of the fertilized ovum. Secretion increases up to about the seventh week of pregnancy and then declines to a comparatively low value at about the sixteenth week. The major function of hCG is to maintain the corpus luteum, and thus the secretion of estrogen and progesterone, so as to prevent menstruation. Between the second and third month, the placenta assumes the role of estrogen and progesterone production, and the corpus luteum is no longer needed. In the male fetus, hCG stimulates the production of testosterone, which is essential to male sexual differentiation and development.

Table 13.5 Other Endocrine Organs

Organ	*Description/Location*	*Endocrine function*
Thymus	Bilobed organ positioned in the upper mediastinum, in front of the aorta and behind the manubrium of the sternum	Secretes the hormone thymosin, which stimulates T-lymphocyte activity
Pineal gland	Small, cone-shaped gland located in the roof of the third ventricle, near the corpora quadrigemina	Secretes the hormone melatonin, which affects the secretion of gonadotropins and ACTH from the anterior pituitary
Gastric mucosa	Epithelial cells lining the stomach; G cells in the glandular walls	G cells secrete gastrin, which stimulates gastric juice secretion and gastric motility
Duodenal mucosa	Epithelial cells in the upper part of the small intestine	Secretes secretin, which stimulates secretion of pancreatic juice rich in bicarbonate, and cholecystokinin, which stimulates secretion of pancreatic juice rich in enzymes
Placenta	Vascular reddish-brown oval structure in the pregnant uterus	Secretes human chorionic gonadotropin (hCG), human somatomammatropin (hCS), estrogens, and progesterone

Review Questions

Multiple Choice

1. A hormone is best described as (*a*) an internal secretion that is transported through ducts, (*b*) an internal secretion with many effects, (*c*) a chemical secreted by a gland, (*d*) a chemical produced in one part of the body that is transported in the blood to another place, where it acts in a regulatory capacity.

2. Which of the following is *not* a steroid hormone? (*a*) estrogen, (*b*) cortisone, (*c*) adrenaline, (*d*) testosterone, (*e*) none of the preceding

3. The portion of the pituitary gland that arises from the roof of the primitive oral cavity is (*a*) the adenohypophysis, (*b*) the pars nervosa, (*c*) the neurohypophysis, (*d*) the infundibulum, (*e*) the hypothalamus.

4. The endocrine gland that is formed from two different germ layers is (*a*) the ovary, (*b*) the thyroid gland, (*c*) the pancreas, (*d*) the adrenal gland.

5. The alpha cells of the pancreas secrete (*a*) insulin, (*b*) enzymes, (*c*) glucagon, (*d*) none of the preceding.

6. The group of adrenocortical hormones concerned with electrolyte balance is (*a*) the glucocorticoids, (*b*) the mineralocorticoids, (*c*) the androgens, (*d*) epinephrine and norepinephrine.

7. The adrenal medulla secretes (*a*) cortisone, (*b*) cortisol, (*c*) epinephrine, (*d*) acetylcholine.

8. Which hormone stimulates testosterone secretion? (*a*) LH, (*b*) progesterone, (*c*) FSH, (*d*) ACTH

9. The secretion of ACTH from the pituitary stimulates the release of (*a*) aldosterone from the adrenal medulla, (*b*) cortisol from the adrenal cortex, (*c*) epinephrine from the adrenal medulla, (*d*) renin from the kidney.

10. Oxytocin and ADH are stored in (*a*) the adenohypophysis, (*b*) the anterior pituitary, (*c*) the posterior pituitary, (*d*) the kidneys.

11. Hypersecretion of growth hormone after closure of the epiphyseal plates causes (*a*) acromegaly, (*b*) myxedema, (*c*) Addison's disease, (*d*) gigantism, (*c*) none of the preceding.

12. A marked deficiency of hormone secretion by the thyroid gland in a young child causes (*a*) acromegaly, (*b*) repressed mental and physical growth, (*c*) bulging eyes, (*d*) high basal metabolic rate, (*e*) all of the preceding.

13. Which of the following is *not* a pituitary hormone? (*a*) hGH, (*b*) LH, (*c*) PRL, (*d*) testosterone, (*e*) oxytocin

14. Calcium levels in the blood are increased by (*a*) calcitonin, (*b*) heparin, (*c*) dicumarol, (*d*) parathyroid hormone, (*e*) vitamin E.

15. Milk ejection from the mammary gland is assisted by (*a*) oxygen, (*b*) prolactin, (*c*) oxytocin, (*d*) prostate hormone, (*e*) ADH.

16. Releasing hormones are synthesized in (*a*) the hypothalamus, (*b*) the hypophysis, (*c*) the pancreas, (*d*) the posterior pituitary, (*e*) the ovary.

17. Which of the following hormones originates from the supraoptic and paraventricular nuclei of the hypothalamus? (*a*) prolactin, (*b*) estrogen, (*c*) antidiuretic hormone, (*d*) luteinizing hormone, (*e*) growth hormone

18. Which target tissue will receive the hormone produced by the corticotrophs? (*a*) thyroid gland, (*b*) pancreas, (*c*) prostate, (*d*) adrenal cortex, (*e*) adrenal medulla

19. Which of the following is (are) *not* influenced by parathyroid hormone? (*a*) kidneys, (*b*) bones, (*c*) small intestine, (*d*) muscles, (*e*) none of the preceding

20. The hormone whose action resembles stimulation through the sympathetic division of the autonomic nervous system is (*a*) epinephrine, (*b*) cortisol, (*c*) androgens, (*d*) aldosterone, (*e*) melatonin.

21. Secretion of which hormone would be increased in the case of an iodine-deficiency goiter? (*a*) TSH, (*b*) thyroxine, (*c*) T_3, (*d*) all of preceding

22. The hormone released from the anterior pituitary that stimulates the development of the seminiferous tubules of the testes is called (*a*) PRL, (*b*) ACTH, (*c*) FSH, (*d*) LH.

23. Which of the following statements about glucocorticoids is *true*?
(*a*) The major glucocorticoid in humans is cortisol.
(*b*) They are secreted by the zona fasciculata of the adrenal cortex.
(*c*) Secretion of these hormones is decreased in Addison's disease.
(*d*) All of the above are true.

24. The basal metabolic rate can reflect dysfunction of (*a*) the pituitary gland, (*b*) the parathyroid glands, (*c*) the adrenal gland, (*d*) the thyroid gland, (*e*) the pancreas.

25. What is the proper sequence of adrenal cortex zones, from the outside in? (*a*) zona glomerulosa, zona fasciculata, zona reticularis; (*b*) zona glomerulosa, zona reticularis, zona fasciculata; (*c*) zona reticularis, zona fasciculata, zona glomerulosa; (*d*) zona fasciculata, zona reticularis, zona glomerulosa

26. A symptom of diabetes mellitus is: (*a*) glyconemia, (*b*) polydipsia, (*c*) weight gain, (*d*) hypoglycemia.

27. Which of the following is a mixed gland? (*a*) adrenal gland, (*b*) pituitary gland, (*c*) thyroid gland, (*d*) pancreas

28. Through negative feedback, a hormone may shut off the secretion of an anterior pituitary hormone by: (*a*) stimulating the release of a (hypothalamic) releasing hormone, (*b*) inhibiting the release of a (hypothalamic) inhibiting hormone, (*c*) inhibiting the release of a (hypothalamic) releasing hormone, (*d*) all of the preceding.

29. Stimulation of the mother's nipples by a nursing baby initiates sensory impulses which pass into the central nervous system and eventually reach the hypothalamus. These impulses result in (*a*) synthesis and release of prolactin from the posterior pituitary, (*b*) release of lactogenic hormone from the anterior pituitary, (*c*) release of oxytocin from the posterior pituitary, (*d*) release of prolactin-inhibiting factor.

30. Choose the *true* statement about a person with type I (insulin-dependent) diabetes mellitus.
(*a*) There is little or no insulin secretion.
(*b*) Dietary treatment may not suffice.
(*c*) There is hyperglycemia.
(*d*) Ketoacidosis and dehydration may develop.
(*e*) All of the above are true.

True or False

____ **1.** Inhibition or stimulation of transport across the cell membrane is one of the major hormonal actions.

____ **2.** The major mode of action of steroid hormones is to increase protein synthesis in specific target-organ cells.

____ **3.** Two hormones are never present in the blood at the same time.

____ **4.** The adrenal medulla secretes adrenaline and noradrenaline.

____ **5.** An enlarged thyroid gland is referred to as a goiter.

____ **6.** The cells of a parathyroid gland respond directly to the glucose concentration in the blood.

____ **7.** Aldosterone, secreted from the posterior pituitary, is involved in the regulation of sodium and potassium.

____ **8.** The posterior pituitary is not composed of true glandular tissue.

____ **9.** All hormones are steroids, amino acid derivatives, peptides, or proteins.

____ **10.** Blood glucose levels, muscular activity, and stress all influence hGH release.

Completion

1. Antidiuretic hormone, ADH, is also known as __________________.
2. Hormones that cross the cell membrane are said to be __________________, while those that cannot are __________________.
3. The __________________ gland and the __________________ function together as an integrated unit.
4. The technical name of the posterior pituitary is __________________ and the technical name of the anterior pituitary is __________________.
5. Developmentally, the anterior pituitary is formed from an invagination of the pharyngeal epithelium known as __________________ __________________.
6. __________________ are stain-resistant and secrete corticotropes.
7. __________________ enhances breast development and milk production in females, while __________________ allows for the let-down of milk and causes uterine contractions.
8. Hyperthyroid secretion in infants and children is known as __________________.
9. Sex hormones, in addition to being produced in the ovaries and testes, are also produced in minimal amounts in the __________________ __________________.
10. A tumor of the chromaffin cells of the adrenal medulla is known as a __________________.

Labeling

Label the structures indicated on the figure to the right.

1. __________________________
2. __________________________
3. __________________________
4. __________________________
5. __________________________

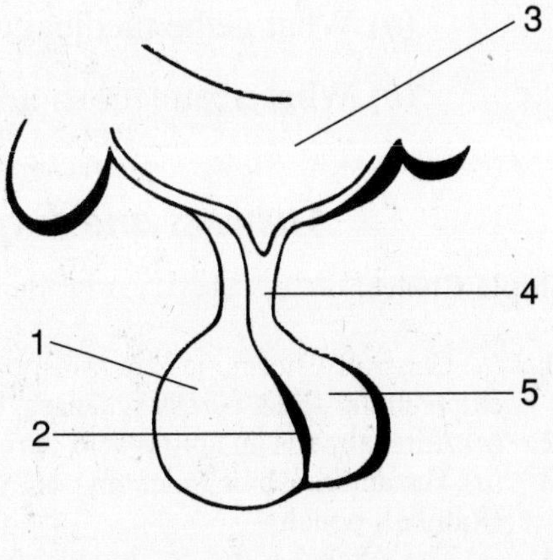

Matching

Match the disease or condition with its description.

____	**1.** Dwarfism	(*a*) hyposecretion of thyroxin
____	**2.** Graves disease	(*b*) hypersecretion of thyroxin
____	**3.** Precocious puberty in males	(*c*) hyposecretion of somatotropin
____	**4.** Cretinism	(*d*) hypersecretion of somatotropin
____	**5.** Tetany	(*e*) hyposecretion of parathyroid hormone
____	**6.** Diabetes mellitus	(*f*) hyposecretion of ADH
____	**7.** Diabetes insipidus	(*g*) hyposecretion of insulin
____	**8.** Acromegaly	(*h*) hypersecretion of testosterone

Clinical Cases

1. A 40-year-old man complained to his physician of polyuria, nocturia, and polydipsia. He was found to produce 7 to 10 liters of urine per day. Blood sugar level was 97 mg% (or 97 mg/dL serum), and PBI (protein-bound iodine) was 6 μg/dL serum.

(*a*) What is the medical diagnosis?

(*b*) What treatment should be used?

2. A 50-year-old man visited a clinic complaining of dry skin and hair, constipation, intolerance to cold, and diminished vigor. In addition, he said that he had gained weight and he had a puffy look. His pulse rate was 55 beats/min and his blood pressure 110/70 mmHg.

(*a*) What is the medical diagnosis?

(*b*) What treatment should be used?

Answers and Explanations for Review Questions

Multiple Choice

1. (*d*) Generally, hormones are transported in the blood. However, local hormones may be transported in extracellular fluid, across synapses, or in external excretions (pheromones).
2. (*c*) Adrenaline is an amino acid derivative.
3. (*a*) The anterior lobe (adenohypophysis) is formed from an invagination of the pharyngeal epithelium (Rathke's pouch).
4. (*e*) The adrenal cortex is formed from mesoderm and the adrenal medulla from neuroectoderm.
5. (*c*) Glucagon is secreted from alpha cells and insulin is secreted from beta cells in the pancreatic islets.
6. (*b*) Mineralocorticoids (aldosterone—most importantly) regulate extracellular electrolytes such as sodium and potassium.
7. (*c*) Epinephrine and norepinephrine are secreted from the adrenal medulla.
8. (*a*) LH stimulates testosterone secretion in males; in females, it stimulates maturation of the follicle,

ovulation, and development of the corpus luteum.

9. (*b*) ACTH stimulates the secretion of glucocorticoids (cortisol the most important in humans) from the adrenal cortex.
10. (*c*) Oxytocin and ADH are stored in and released from the posterior pituitary, but they are produced in the hypothalamus.
11. (*a*) Acromegaly is caused by excess hGH in adults; symptoms may include a large jaw, nose, ears, tongue and head; increased basal metabolic rate; and loss of visual fields.
12. (*b*) Also referred to as cretinism, hypothyroidism in infants and children is characterized by retarded mental and physical development.
13. (*d*) Testosterone is secreted by the testes.
14. (*d*) Parathyroid hormone increases blood calcium and calcitonin decreases blood calcium levels.
15. (*c*) Oxytocin stimulates milk secretion and stimulates strong contractions in the pregnant uterus.
16. (*a*) Releasing hormones are produced by neurosecretory neurons in the hypothalamus.
17. (*c*) Antidiuretic hormone (ADH) is produced in the hypothalamus but released from the posterior pituitary.
18. (*d*) Corticotrophs secrete ACTH, which stimulates the adrenal cortex to secrete cortisol.
19. (*e*) All the organs or tissues listed are influenced by PTH: kidneys–reabsorption of calcium; bones–release of calcium; small intestine–absorption of calcium; muscle–calcium required for proper contraction.
20. (*a*) Epinephrine causes the fight-or-flight actions similar to stimulation through the sympathetic division of the autonomic nervous system.
21. (*a*) TSH from the pituitary would be increased because, with an iodine-deficiency goiter, there would be a reduction of T_3 and T_4. Because of the low T_3 and T_4, there would be no negative feedback to inhibit the release of TSH.
22. (*c*) FSH (follicle stimulating hormone) stimulates spermatogenesis in the seminiferous tubules of the testes.
23. (*d*) All of the listed statements are true.
24. (*d*) A major function of the thyroid hormones is to regulate basal metabolic rate and body temperature.
25. (*a*) Refer to fig. 13.8.
26. (*b*) Because diabetics urinate more, they are thirstier and drink more water (polydipsia).
27. (*d*) The pancreas is both an endocrine (insulin and glucagon) and exocrine (pancreatic juice) gland.
28. (*c*) Inhibited release of a releasing hormone will lead to a reduction in the secretion of a specific anterior pituitary hormone.
29. (*c*) Oxytocin released from the posterior pituitary stimulates the secretion of milk during nursing.
30. (*e*) All of the listed statements are true.

True or False

1. True
2. True
3. False; many hormones are present in the blood
4. True
5. True
6. False; parathyroid cells respond to calcium blood levels
7. False; aldosterone is secreted from the adrenal cortex, not the posterior pituitary
8. False; the posterior pituitary is composed of neural axons
9. False; some hormones, such as *leukotrienes*, are fatty acid derivatives
10. True

Completion

1. vasopressin
2. lipophilic, hydrophilic
3. pituitary, hypothalamus
4. neurohypophysis, adenohypophysis
6. Chromophobes
7. Prolactin, oxytocin
8. cretinism
9. adrenal cortex

5. Rathke's pouch

10. pheochromocytoma

Labeling

1. Anterior lobe (adenohypophysis)
2. Intermediate lobe (pars intermedia)
3. Hypothalamus
4. Infundibulum
5. Posterior lobe (neurohypophysis)

Matching

1. (*c*)
2. (*b*)
3. (*h*)
4. (*a*)
5. (*e*)
6. (*f*)
7. (*g*)
8. (*d*)

Clinical Cases

1. (*a*) diabetes insipidus; (*b*) antidiuretic hormone (ADH)
2. (*a*) myxedema; (*b*) thyroid hormone treatment

Cardiovascular System: Blood 14

Objective A To describe the nature of *blood* as a part of the cardiovascular system and to explain its functions.

Blood is a fluid connective tissue that is pumped by the heart through the vessels (arteries, arterioles, capillaries, venules, and veins) of the cardiovascular system.

14.1 What are the principal functions of blood?

Transport. Blood transports oxygen and nutrients to the body tissues and carbon dioxide and waste materials from the tissues to the organs of excretion. It also transports hormones from endocrine glands to their target tissues.

Acid-base regulation. Blood functions to control respiratory acidosis (low pH) or alkalosis (high pH) through the bicarbonate buffer system. High levels of hydrogen ions combine with bicarbonate to form carbonic acid, which dissociates immediately to form carbon dioxide and water; as carbon dioxide is exhaled, blood becomes less acidic and pH levels stabilize.

Thermoregulation. Under conditions of hyperthermia, the blood carries excess heat to the body surface for temperature regulation.

Immunity. Leukocytes (white blood cells) are transported in the blood to sites of injury or invasion by disease-causing agents.

Hemostasis. Thrombocytes (platelets) and clotting proteins minimize blood loss when a blood vessel is damaged.

14.2 What is the blood volume of an average person?

The volume of whole blood is about 4.5 liters in women and 5.5 liters in men. To demonstrate that the average is indeed about 5 liters, recall that blood weight is about 7% of total body weight, and that 150 lb is a reasonable average body weight. The average person will thus have

$$(0.07)\ (150\text{ lb}) = 10.5\text{ lb of blood}$$

Now, 1 lb of blood occupies approximately 1 pint or 500 mL; therefore, the average blood volume will be

$$(10.5\text{ lb})\ (500\text{ mL/lb}) = 5250\text{ mL} = 5.25\text{ L}$$

Objective B To describe the composition of blood.

Blood is composed of a liquid matrix (blood plasma) and several types of formed elements (red blood cells, white blood cells, and platelets) (see figs. 14.1 and 14.2). The blood plasma contains a variety of proteins and many other small molecules and ions. Blood minus the formed elements and the clotting proteins is called **serum**.

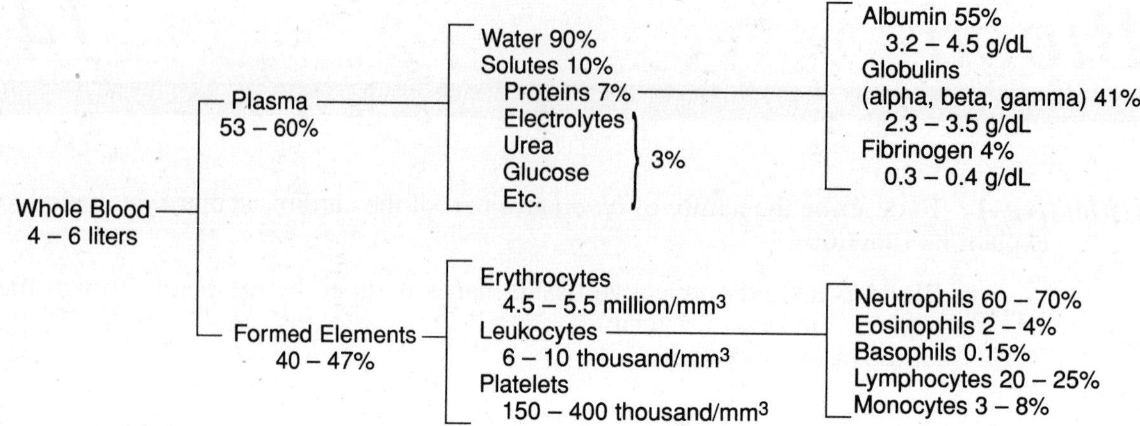

Figure 14.1 The composition of blood.

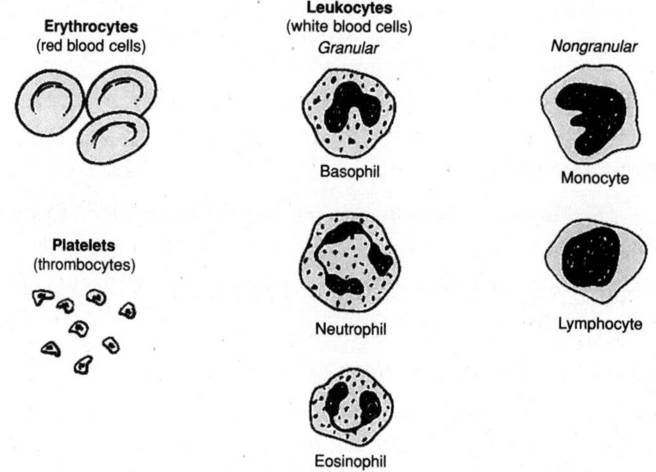

Figure 14.2 The appearance of blood cells.

Objective C To describe *erythrocytes* (red blood cells) in terms of origin, structure, and function.

Survey

An **erythrocyte**, or red blood cell (RBC), is a flexible, biconcave, anucleated cell. RBCs are manufactured at several sites in the body. During embryonic development, *erythropoiesis* (the manufacture of red blood cells) occurs first in the definitive yolk sac. Production then moves to the liver, spleen, and bone marrow. In children, RBCs are produced in the bone marrow of long bones of the arms and legs. In adults, RBCs are produced in the bone marrow of the ribs, sternum, vertebrae, and pelvis. The main constituent (about one-third by weight) of the RBCs is **hemoglobin**, and the essential function of these cells is to carry oxygen, reversibly trapped by hemoglobin, to all parts of the body.

14.3 What is the hematocrit and how is it measured?

The **hematocrit** is the percentage of total blood volume occupied by the erythrocytes. It ranges from 40% to 54% in men and from 38% to 47% in women. It is measured by centrifuging a blood sample in a capillary tube. For example, if the tube were 100 mm long and if the packed red blood cells occupied the distal 45 mm, the hematocrit would be 45%.

Hematology is a branch of biology and a clinical discipline that studies the morphology and composition of the blood and blood-forming tissues. Clinicians who work in hematology units of hospitals and clinics analyze blood to detect infection and disease.

14.4 What conditions would cause a change in the hematocrit?

Anemia (*low hematocrit*) may be caused by a decreased rate of red blood cell production or excessive loss of red blood cells (see table 14.1).

Polycythemia (*high hematocrit*) may be caused by excessive red blood cell production.

Table 14.1 A Summary of the Various Anemias

Type	*Cause*	*Symptoms*	*Treatment*
Hemorrhagic	Blood loss	Shock	Transfusion
Aplastic	Bone marrow destruction by drugs, chemicals, or radiation	Fatigue and susceptibility to infection (WBCs also affected)	Transfusion; removal of chemical or irradiator
Nutritional	Deficiency in folic acid, vitamin B_{12}, or iron	If any, fatigue; neurological deficits	Folic acid, vitamin B_{12}, or iron administration
Hemolytic	Increased destruction of RBCs	If any, fatigue and jaundice	Various

Pernicious anemia is a type of nutritional anemia. It occurs when the parietal cells of the stomach fail to manufacture a substance (*intrinsic factor*) that is required for the eventual absorption of vitamin B_{12} in the small intestine. In the absence of intrinsic factor (owing to autoimmune destruction of parietal cells), B_{12} is not absorbed and the result is pernicious anemia.

Objective D To outline the process of *erythropoiesis* and to describe the structure and function of *hemoglobin*.

Erythropoiesis (Gk. *erythros*, red; *poiesis*, making) is the manufacture of red blood cells. The sequence of cellular differentiation in erythropoiesis is as follows:

$$\text{hemocytoblast} \rightarrow \text{proerythroblast} \rightarrow \text{erythroblast} \rightarrow \text{normoblast} \rightarrow \text{reticulocyte}^{*} \rightarrow \text{erythrocyte}$$

*The nucleus is lost at the reticulocyte stage.

14.5 What substances are required for erythrocyte production?

Table 14.2 Substances Needed for Erythrocyte Production

Substance	***Function***
Protein	Cell membrane structure
Lipid	Cell membrane structure
Amino Acid	Globin portion of hemoglobin
Iron	Incorporated into hemoglobin
Vitamin B_{12}	DNA synthesis
Folic acid	DNA synthesis
Copper	Catalyst for hemoglobin synthesis
Cobalt	Aids in hemoglobin synthesis

14.6 If about 2.5 million RBCs are formed each second in the bone marrow, and if RBCs are destroyed at the same rate in the liver and spleen, calculate the average lifetime, *T*, of a red blood cell.

The concentration of RBCs is roughly 5 million per mm^3, and the volume of blood is about 5L = 5 million mm^3. Thus, the standing population of RBCs is approximately

$$(5 \times 10^6/mm^3)\ (5 \times 10^6\ mm^3) = 2.5 \times 10^{13}$$

Under hemostasis, this population will "turn over" once during the lifetime of the average RBC; that is,

$$(2.5 \times 10^6)T = 2.5 \times 10^{13} \text{ or } I = 107 \text{ seconds} = \text{approximately 120 days}$$

14.7 What factors cause fluctuations in erythrocyte number?

Any condition that decreases oxygen in the body tissues will, by a negative feedback mechanism, increase erythropoiesis, e.g., high altitude (30% greater hematocrit at 14,000 ft than at sea level), muscle exercise, anemia, or chronic emphysema. *Temperature*: Increased body temperature increases the number of RBCs. *Sex*: After puberty, males have a higher hematocrit than females. Age: Infants have a relatively high hematocrit. *Time of day*: The RBC count is highest in early evening.

14.8 Describe the feedback mechanism mentioned previously (in problem 14.7).

In response to low oxygen concentration, the kidneys secrete the hormone *erythropoietin*. Erythropoietin travels in the blood to the bone marrow, where it stimulates erythropoiesis. The increased number of erythrocytes transport more oxygen to the tissues.

Blood doping is a technique sometimes used by athletes to increase the oxygen-carrying capacity of their blood, and thus their endurance. It involves withdrawing some of the athlete's RBCs and then reinjecting them a few days before a competitive event. After the blood is withdrawn, the erythrocytes are quickly replaced. Then, when the stored blood is reinfused, a temporary polycythemia results. The intended effect may be achieved—there may be up to a 10% increase in aerobic capacity. Blood doping is illegal, however, and not without risk. It can impair the flow of blood, as well as cause flulike symptoms. Injecting synthetic erythropoietin to stimulate RBC production is another technique used to increase athletes' endurance.

14.9 What is the chemical makeup of hemoglobin?

Hemoglobin (Hb) consists of *globin* (four polypeptide chains; fig. 14.3) and *heme* (four Fe^{2+} porphyrin molecules; fig. 14.4). Each erythrocyte contains approximately 280 million hemoglobin molecules. Each iron portion of heme is able to combine with four molecules of oxygen. This means that a single erythrocyte can transport over a billion molecules of oxygen.

Figure 14.3 A heme molecule.

Figure 14.4 A porphyrin ring.

14.10 Can hemoglobin bind other gas molecules besides oxygen?

Yes. Carbon dioxide (CO_2) and carbon monoxide (CO) also bind to hemoglobin. Hemoglobin, when saturated with oxygen, is called *oxyhemoglobin*. It is cherry red in color. When oxyhemoglobin loses its oxygen, it becomes bluish purple. Hemoglobin in combination with carbon dioxide is called *carbaminohemoglobin*. Oxygen and carbon dioxide have distinct carry sites on the Hb molecule. Carbon monoxide combined with Hb is called *carboxyhemoglobin*. CO binds to a heme and has 200 times the affinity for the heme that oxygen has. It is this exclusion of oxygen that makes carbon monoxide so dangerous a gas.

14.11 When disintegrating erythrocytes are phagocytosed in the spleen and liver, how is the hemoglobin molecule broken down?

1. Hemoglobin > heme + globin
2. Heme > Fe^{2+} + porphyrin
3. Globin > protein > amino acids

Porphyrin is changed from a ring structure (fig. 14.3) to a straight-chain structure called *biliverdin* ("green of bile"), which in turn is reduced to the straight-chain *bilirubin* ("red of bile"). Bilirubin, carried from the liver in the bile, may be excreted in the feces as *stercobilin* or in the urine as *urobilin*. Feces and urine owe their brown or yellowish color to these bilirubin products. When yellowish bilirubin accumulates in the blood to an abnormally high degree, it yellows the skin (*jaundice*). Causes of jaundice include liver disease, excess red blood cell destruction, or bile duct obstruction (feces will be gray).

Objective E To describe the origin of *platelets* and to explain how they function.

Platelets, or *thrombocytes*, are small cellular fragments that originate in the bone marrow from a giant cell known as a **megakaryocyte.** The megakaryocytes form platelets by pinching off bits of cytoplasm and extruding them into the blood. Platelets contain several clotting factors, calcium ions, ADP, serotonin, and various enzymes; they play an important role in **hemostasis** (the arrest of bleeding).

14.12 How do platelets function?

In the event of a vessel defect or injury, platelets aggregate to form a plug. As they aggregate, they release adenosine diphosphate (ADP). The ADP makes the surface of platelets sticky, so that they adhere to the growing layers of aggregated platelets. In addition, thromboxane A_2 is released from the surface membranes of aggregating platelets. This prostaglandin derivative further enhances platelet aggregation. The platelet plug aids in reducing blood loss at the site of damage by three mechanisms: (1) physically sealing the vessel defect, (2) releasing chemicals that cause vasoconstriction, and (3) releasing other chemicals that stimulate blood clotting (serotonin, epinephrine, thromboxane A_2).

Objective F To explain the mechanism of *hemostasis*.

The major events are (1) *constriction of the blood vessels*; (2) *plugging of the wound by aggregated platelets*; and (3) *clotting of the blood into a mass of fibrin*, which augments the plug in sealing the wound and providing a framework for repair.

14.13 List the chemicals, or factors, involved in the clotting process.

The *factors*, nearly all of which are produced in the liver, are designated by Roman numerals according to their order of discovery. The numerical order, therefore, does not reflect the reaction sequence.

I = fibrinogen
II = prothrombin
III = thromboplastin
IV = calcium
V = labile factor
VII = SPCA (*serum prothrombin conversion accelerator*)
VIII = AHF (*antihemophilic factor*)
IX = PTC (*plasma thromboplastic component*), also called Christmas factor
X = Stuart-Prower factor
XI = PTA (*plasma thromboplastin antecedent*)
XII = Hageman factor
XIII = fibrin stabilizing factor

Note: Factor VI is no longer considered a separate entity.

14.14 Describe the two pathways that initiate clotting.

Refer to fig. 14.5. The *intrinsic pathway* is activated when blood is exposed to a negatively charged surface, such as that provided by collagen at the site of a wound or by the glass of a test tube. All factors that bring about clotting by means of the intrinsic pathway are present in the blood. The *extrinsic pathway* is activated by tissue thromboplastin, which is released when vascular walls or other tissue are damaged. The final steps in both pathways are identical.

14.15 Give examples of the actions of anticoagulants.

Citrates and *oxalates* (organic biochemical molecules) bind calcium, which is essential at several steps in the clotting process. *Heparin*, a protein released from the liver, prevents the activation of factor IX and interferes with thrombin action. *Dicoumarol* and *Coumadin* block the formation of prothrombin and factors VII, IX, and X by interfering with vitamin K, which acts as a catalyst in the synthesis of these chemicals in the liver.

14.16 Cite some disorders in which there is excessive bleeding.

Hemophilia is a hereditary lack, by altered biosynthesis, of a single clotting factor. Lack of VIII causes hemophilia A (*classical hemophilia*); lack of IX causes hemophilia B (*Christmas disease*). In *vitamin K deficiency*, clotting factors are not properly synthesized in the liver. In *thrombocytopenia*, the concentration of thrombocytes is too low, and the patient may develop hundreds of small hemorrhages (which appear as small purplish blotches on the skin) throughout the body tissues.

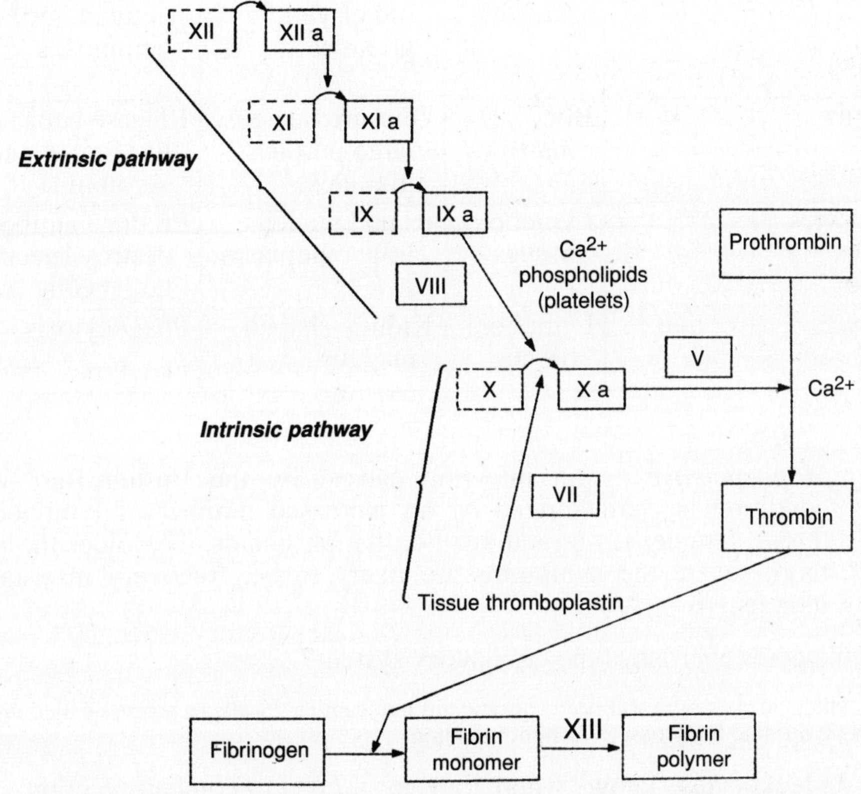

Figure 14.5 The intrinsic and extrinsic pathways of hemostasis.

Objective G To distinguish between the five types of *leukocytes* (white blood cells).

The various leukocyte types are compared in table 14.3.

14.17 List some diseases that cause increases in the various leukocytes.

Neutrophils: Appendicitis, pneumonia, tonsillitis
Eosinophils: Hay fever, asthma, parasitic infestations
Basophils: Smallpox, nephrisis, myxedema
Lymphocytes: Whooping cough, mumps, mononucleosis
Monocytes: Tuberculosis, typhus

Table 14.3 A Comparison of the Five Types of Leukocytes

Type	*Avg. No./ mm³*	*Origin*	*Description*	*Function*
Neutrophils	5400	Bone marrow	Lobed nucleus, fine granules	Phagocytosis
Eosinophils	275	Bone marrow	Lobed nucleus, red or yellow granules	May phagocytize antigen-antibody complexes
Basophils	35	Bone marrow	Obscure nucleus, large purple granules	Release heparin, histamine, and serotonin
Lymphocytes (B cells, T cells)	2750	Lymphoid tissues	Round nucleus, little cytoplasm	Produce antibodies, destroy specific target cells
Monocytes	540	Lymphoid tissues	Kidney-shaped nucleus	Phagocytosis

Mononucleosis is an infection caused by the Epstein-Barr virus. This condition is characterized by an increased number of lymphocytes, sore throat, fatigue, fever, and swollen lymph glands. The older the person, the more severe the symptoms are likely to be. Recovery may take several months.

14.18 Do leukocytes ever leave the circulatory system?

Yes. Leukocytes have the ability to squeeze through capillary walls (a process called *diapedesis*) and move out into body tissues to fight infection.

14.19 How do leukocytes "know" where they are needed to combat infection?

Infected tissues release certain chemicals (e.g., *leukotaxine*) that locally increase the permeability of capillary walls. Circulating leukocytes are recruited to the infected area by the chemical attractants—a process termed *chemotaxis*.

Objective H To list the major components of *blood plasma* and to describe the functions of the *albumins*, *globulins*, and *electrolytes*.

Blood plasma consists of the following:

1. Water
2. Proteins (albumins, globulins, fibrinogens)
3. Electrolytes (Na^+, K^+, Ca^{2+}, Mg^{2+}, Cl^-, HCO_3^-, HPO_4^{2-}, SO_4^{2-})
4. Nutrients (glucose, amino acids, lipids, cholesterol, vitamins, trace elements)
5. Hormones
6. Dissolved gasses (carbon dioxide, oxygen, nitrogen)
7. Waste products (urea, uric acid, creatinine, bilirubin)

14.20 What are the characteristics and functions of albumins?

Albumins (MW = 69,000) are the smallest and most abundant proteins in blood plasma. They are produced in the liver and play an important role in maintaining the osmotic pressure of the blood. They also act as important blood buffers and are partly responsible for the viscosity of blood.

14.21 State the major functions of globulins and identify the four types.

The **globulin** factions of the blood protein contain numerous substances that serve a variety of functions, including transport (thyroid hormone, cholesterol, and iron), enzymatic action; clotting; and immunity. They may be separated by electrophoresis into four types: ***alpha 1*** (e.g., fetoprotein, antitrypsin, lipoproteins); ***alpha 2*** (e.g. antithrombin, cholinesterase); ***beta*** (e.g., transferrin, plasminogen, prothymbin); and ***gamma*** (e.g., IgG, IgA, IgM, IgD, IgE; Ig = immunoglobulin).

14.22 What purpose do the electrolytes serve?

Many of the ions transported within the blood are necessary in membrane transport, blood osmolarity, and neurological function.

Review Questions

Multiple Choice

1. Granules are not visible in (*a*) neutrophils, (*b*) lymphocytes, (*c*) eosinophils, (*d*) basophils.

2. Which of the following four components of the blood are necessary for clotting? (*a*) calcium, vitamin K, albumin, globulin; (*b*) calcium, heparin, prothrombin, fibrinogen; (*c*) calcium, prothrombin, fibrinogen, platelets; (*d*) calcium, prothrombin, platelets, vitamin A

3. In the adult, the majority of leukocytes are (*a*) basophils, (*b*) eosinophils, (*c*) lymphocytes, (*d*) neutrophils.

4. The chief function of the serum albumin in the blood is to (*a*) produce antibodies, (*b*) form fibrinogen, (*c*) maintain colloidal osmotic pressure, (*d*) remove waste products.

5. Calcium ions are necessary for the formation of (*a*) fibrinogen, (*b*) thromboplastin, (*c*) thrombin, (*d*) prothrombin.

6. A differential blood count (*a*) gives the number of red blood cells per cubic millimeter, (*b*) determines the percentage of erythrocytes per cubic millimeter, (*c*) gives the number and variety of leukocytes in each 200 counted, (*d*) determines the platelet count.

7. The intrinsic factor necessary for the complete maturation of red blood cells is derived from (*a*) bone marrow, (*b*) vitamin B_6, (*c*) the liver, (*d*) the mucosa of the stomach.

8. A hemoglobin measurement of 15 g per 100 mL (or 1 dL) of blood is (*a*) within the normal limits, (*b*) subnormal, (*c*) above normal, (*d*) low, but satisfactory.

9. Which of the following is most consistent with a diagnosis of appendicitis? (*a*) an increase in monocytes, (*b*) an increase in erythrocytes, (*c*) leukopenia, (*d*) an increase in neutrophils

10. Which concentration would be an indication of anemia? (*a*) thrombocytes-300,000/mm^3, (*b*) hematocrit - 43%, (*c*) hemoglobin-17 g/dL, (*d*) erythrocytes-3.8 million/mm^3

11. The leukocyte that is *not* involved in phagocytosis but that does secrete the anticoagulant heparin is (*a*) the basophil, (*b*) the monocyte, (*c*) the eosinophil, (*d*) the lymphocyte.

12. Iron-deficiency anemia (*a*) is more common in men than in women, (*b*) is characterized by increased numbers of leukocytes, (*c*) should generally be treated by intramuscular injections of iron, (*d*) is the form of anemia typically accompanying chronic blood loss from the body.

13. Erythrocyte production (*a*) is stimulated by high estrogen levels in the blood, (*b*) falls if the stomach loses its ability to produce intrinsic factor, (*c*) occurs in the spleen in normal adults at sea levels, (*d*) is stimulated by a rise in the concentration of amino acids in the arterial blood.

14. For blood clotting to occur normally, (*a*) heparin must be inactive, (*b*) there must be a sufficient dietary intake of vitamin C, (*c*) tissue damage outside of the vessel must occur, (*d*) the liver must have an adequate supply of vitamin K.

15. Which of the following is *not* a plasma protein? (*a*) lamellated corpuscle, (*b*) globulin, (*c*) fibrinogen, (*d*) platelet

16. Which of the following does *not* stimulate erythropoietin production? (*a*) hemorrhage, (*b*) chronic emphysema, (*c*) stress-induced release of epinephrine into the system, (*d*) decreased oxygen delivery to the tissues

17. Insufficient vitamin B_{12} in the body may result in (*a*) hemolytic anemia, (*b*) pernicious anemia, (*c*) aplastic anemia, (*d*) an embolus.

18. The percent volume of whole blood occupied by packed red blood cells is referred to as (*a*) the hematocrit, (*b*) the formed elements, (*c*) the erythrocytic fraction, (*d*) the sedimentation index.

19. Production of red blood cells in a mature adult occurs in all the following areas *except* (*a*) the sternum, (*b*) the ribs, (*c*) the skull bones, (*d*) the vertebrae, (*e*) the os coxae.

20. Plasma proteins constitute what percentage of the blood plasma volume? (*a*) 17–19%, (*b*) 7–9%, (*c*) 25–27%, (*d*) 52–55%

21. The general term for reactions that prevent or minimize loss of blood from the vessels if they are injured or ruptured is (*a*) stabilization energy, (*b*) homeostasis, (*c*) syneresis, (*d*) hemostasis.

22. Hemostasis does *not* involve (*a*) contraction of smooth muscles in blood vessel walls, (*b*) adherence of platelets to damaged tissue, (*c*) clot retraction, (*d*) increased renin-angiotensin activity.

23. What is the correct order of these events?
(1) conversion of fibrinogen to fibrin
(2) clot retraction and leakage of serum
(3) thromboplastin production
(4) conversion of prothrombin to thrombin
(*a*) 3, 2, 1, 4; (*b*) 3, 4, 1, 2; (*c*) 3, 4, 2, 1; (*d*) 4, 1, 3, 2

24. Which factor is *not* synthesized in hemophilia A? (*a*) VIII, (*b*) VII, (*c*) IX, (*d*) XIII

25. Blood minus the formed elements and clotting proteins is called (*a*) plasma, (*b*) serum, (*c*) albumen, (*d*) globulin.

True or False

____ **1.** Blood functions in transport, pH balance, thermoregulation, and immunity mechanisms.

____ **2.** Thrombocytes contain clotting factors that include calcium, iron, thiamin, and oxalic acid.

____ **3.** Polycythemia is an unusually high hematocrit.

____ **4.** Erythrocyte production requires folic acid, copper, protein, polysaccharides, and biliverdin.

____ **5.** A heme molecule consists of a nitrogen-containing organic ring called porphyrin and one atom of iron.

____ **6.** The major mechanisms of hemostasis are plugging, clotting, and constriction.

____ **7.** Thromboplastin is released when vascular walls or other tissue is damaged.

____ **8.** Calcium and phospholipids are required for the conversion of prothrombin to thrombin.

____ **9.** Hemoglobin, when saturated with carbon dioxide, is termed *carboxyhemoglobin*.

Completion

1. An excessive number of red blood cells is referred to as ________________.

2. ________________ is the manufacture of red blood cells.

3. Hemoglobin, when saturated with oxygen, is cherry red in color and is called ________________.

4. Thrombocytes are formed from giant cells called ________________.

5. The ________________ ________________ is activated when blood is exposed to a foreign surface.

6. In ________________, the concentration of thrombocytes is too low.

7. ________________ refers to the ability of leukocytes to squeeze through capillary walls.

8. Mononucleosis is an infection caused by the ________________ virus.

Labeling

Label the structures indicated on the figure to the right.

1. ______________________________
2. ______________________________
3. ______________________________
4. ______________________________
5. ______________________________

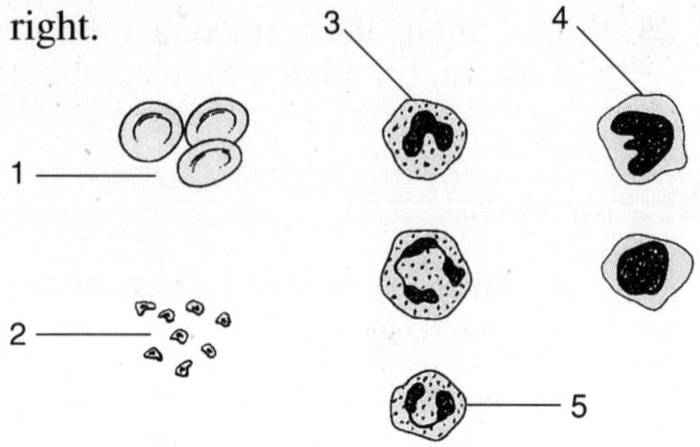

Matching

Match the cell with its description or function.

____	**1.** Thrombus	(*a*)	obscure nucleus; stains with large purple granules
____	**2.** Phagocytosis	(*b*)	formation of clots
____	**3.** Hematoma	(*c*)	granules that take up the red dye eosin
____	**4.** Eosinophil	(*d*)	enzymatically decomposes fibrin
____	**5.** Plasmin	(*e*)	lobed nucleus and fine granules; stains with neutral dyes
____	**6.** Neutrophil	(*f*)	accumulation of blood
____	**7.** Leukocyte	(*g*)	white blood cell
____	**8.** Lymphocyte	(*h*)	selective defender against invaders
____	**9.** Basophil	(*i*)	ingestion and digestion of particulate matter

Answers and Explanations for Review Questions

Multiple Choice

1. (*b*) Neutrophils, eosinophils and basophils are grouped together as granulocytes due to the presence of granules in their cytoplasm; lymphocytes and monocytes are classified as agranulocytes because they lack visible granules.
2. (*c*) Each is involved in an essential step in clotting.
3. (*d*) Neutrophils account for 65–70% of the white blood cells.
4. (*c*) Albumin is the smallest and most abundant of the plasma proteins; by virtue of its presence in the blood plasma and its absence in the interstitial fluid, it establishes an osmotic gradient between blood and interstitial fluid.
5. (*c*) Calcium is required for several of the steps in both the extrinsic and intrinsic pathways, and is also essential in converting prothrombin to thrombin.
6. (*c*) A differential white blood cell count gives the percent distribution of types of leukocytes.
7. (*d*) The intrinsic factor is produced in the mucosa of the stomach and is necessary for vitamin B_{12} absorption. This vitamin is essential for mitosis, and thus the formation of RBCs.
8. (*a*) A hemoglobin measurement of 15g/100 mL blood is the normal average value.
9. (*d*) A patient with appendicitis will show an increase in the number of neutrophils.
10. (*d*) An erythrocyte count of 3.8 million/mm^3 is abnormally low and would be an indication of anemia.

11. (*d*) Lymphocytes do not produce heparin, but they are involved in the production of antibodies and the distribution of specific target cells.
12. (*a*) Iron-deficiency anemia may occur more frequently in females because of lower levels of RBC production and loss of blood occasional to menstruation.
13. (*b*) Erythrocyte production will decrease with a lack of gastric secretion due to a decrease in the intrinsic factor produced by the parietal cells in the stomach.
14. (*d*) The liver requires adequate amounts of vitamin K in order to synthesize several of the clotting factors.
15. (*d*) Platelets (thrombocytes) are cellular fragments, not plasma proteins.
16. (*c*) Epinephrine does not affect erythropoietin production.
17. (*b*) In the absence of intrinsic factor, B_{12} is not absorbed; the result is pernicious anemia.
18. (*a*) The hematocrit is an indication of the oxygen-carrying capacity of the blood.
19. (*c*) The skull bones are not active in the production or RBCs.
20. (*b*) Plasma proteins (albumins, globulins, fibrinogen) account for 7% to 9% of the plasma volume.
21. (*d*) Hemostasis includes all mechanisms that prevent blood loss due to blood vessel injury.
22. (*d*) Renin–angiotensin activity is not related to hemostasis.
23. (*b*) When a blood vessel is severely damaged, a clot usually forms in the damaged area within 20 seconds.
24. (*a*) Hemophilia A (classical hemophilia) is caused by a lack of factor VIII.
25. (*b*) Plasma is blood minus the formed elements. Serum is blood minus the formed elements and the clotting proteins.

True or False

1. True
2. False; none of the listed substances are produced by thrombocytes in the clotting process
3. True
4. False; polysaccharides and biliverdin are not required
5. True
6. True
7. True
8. False; phospholipids are not required for the conversion of prothrombin to thrombin
9. False; hemoglobin saturated with carbon dioxide is called carbaminohemoglobin

Completion

1. polycythemia
2. Erythropoiesis
3. oxyhemoglobin
4. megakaryocytes
5. intrinsic pathway
6. thrombocytopenia
7. Diapedesis
8. Epstein-Barr

Labeling

1. Erythrocytes(red blood cells)
2. Platelets (thrombocytes)
3. Basophil
4. Monocyte
5. Eosinophil

Matching

1. (*b*)
2. (*i*)
3. (*f*)
4. (*c*)
5. (*d*)
6. (*e*)
7. (*g*)
8. (*h*)
9. (*a*)

Cardiovascular System: The Heart 15

Objective A To describe the *heart* and to locate it within the thorax.

Survey The **heart** is a hollow, four-chambered muscular organ that is specialized for pumping blood through the vessels of the body (fig. 15.1). It weighs about 255 grams in the female and 310 grams in the male, accounting for about 5% of the body weight. The heart is located in the *mediastinum* (see problem 1.21), where it is surrounded by a tough fibrous membrane called the **pericardium.** The **pericardial sac** is the actual compartment formed by the pericardium that encloses the heart.

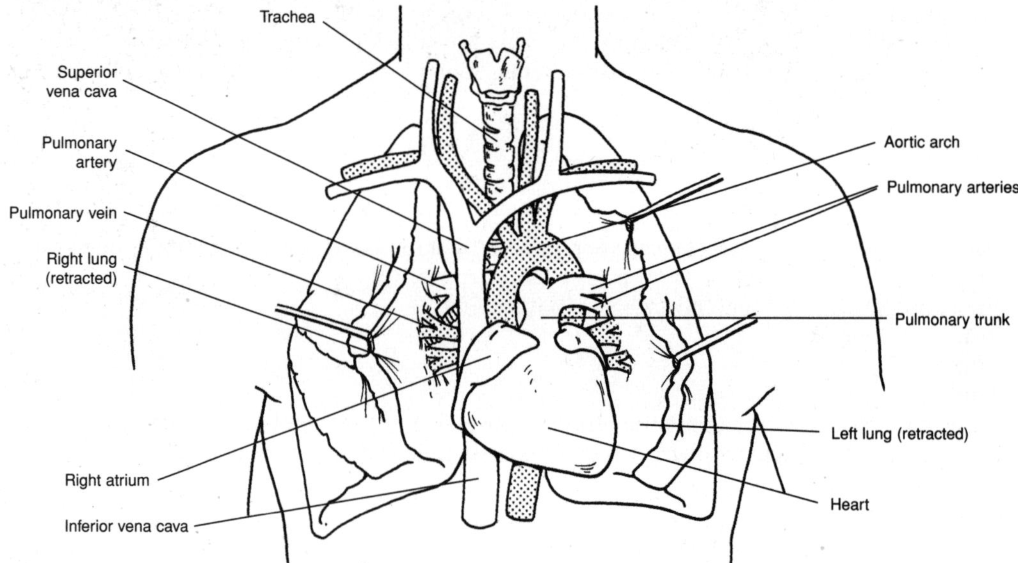

Figure 15.1 The heart, lungs, and associated vessels.

15.1 Which portion of the heart is the base and which is the apex?

About two-thirds of the heart is to the left (subject's left) of the midsagittal plane, with its *apex of the heart*, or cone-shaped end, pointing downward, in contact with the diaphragm. The *base of the heart* is the broad superior end, where the large vessels attach.

15.2 What is the relationship between the heart and lungs?

Ventilation of the lungs brings oxygen in contact with blood from the heart. The pumping action of the heart then circulates the oxygenated blood through the body and returns deoxygenated blood to the lungs for removal of carbon dioxide. The vessels that connect the heart and lungs are called *pulmonary vessels.*

15.3 What is the function of the pericardium?

The inner *serous layer of the pericardium* secretes *pericardial fluid* that lubricates the surface of the heart. The outer *fibrous layer* has the protective and separative function.

Pericarditis is an inflammation of the parietal pericardium that increases the secretion of pericardial fluid into the pericardial cavity. Because the fibrous layer of the pericardium is inelastic, the increase in fluid pressure impairs ventricular contraction and blood flow through the heart.

Objective B To trace the *development of the embryonic heart* from day 18 through day 25.

Development of the heart from undifferentiated mesoderm requires only 7 or 8 days. On day 19 after conception, specialized **heart cords** begin to migrate toward each other medially from the two longitudinal bands of splanchnic (visceral) mesoderm. By day 21, a hollow center has developed in each heart cord, and the structure is called a **heart tube** (fig. 15.2). By day 23, the heart tubes have fused into a single medial endocardial heart tube. By day 25, fusion is complete, dilations are occurring, and blood is being pumped.

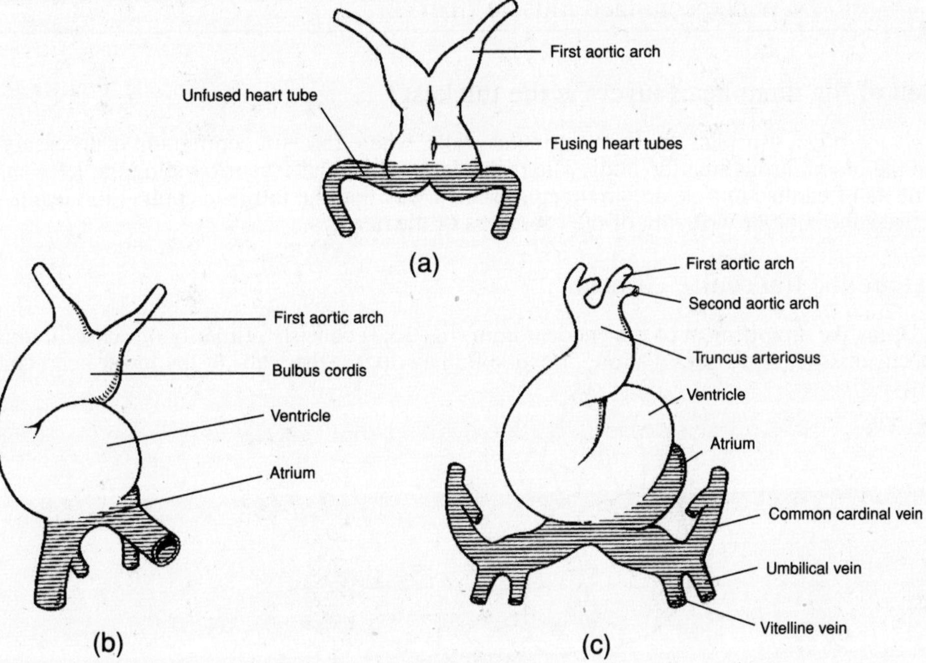

Figure 15.2 Development of the embryonic heart. Anterior views (*a*) at day 21, (*b*) at day 23, and (*c*) at day 25.

Partitioning of the heart chambers begins during the middle of the fourth week and is complete by the end of the fifth week. It is during this crucial period that congenital conditions such as heart murmurs, septal defects, patent foramen ovale, and stenosis may develop.

Objective C To contrast the three *layers of the heart wall* with respect to structure and function.

Refer to table 15.1.

Table 15.1 Layers of the Heart

Layer	*Structure*	*Function*
Epicardium (visceral pericardium)	Serous membrane of connective tissue, covered with epithelium and including blood capillaries, lymph capillaries, and nerve fibers	Lubricative outer covering
Myocardium	Cardiac muscle tissue, separated by connective tissues and including blood capillaries, lymph capillaries, and nerve fibers	Contractile layer to eject blood from heart chambers
Endocardium	Epithelial membrane and connective tissues, including elastic and collagenous fibers, blood vessels, and specialized muscle fibers	Strenghened protective inner lining of the chambers and valves

15.4 Which of the three heart layers is the thickest?

The *myocardium*, especially in the ventricular walls, where forceful contraction is necessary to pump the blood throughout the body. The muscular wall is thickest surrounding the left ventricle. The fibers of cardiac muscle are arranged in such a way that the intrinsic contraction results in an effective squeezing or wringing of the chambers of the heart.

15.5 What are the trabeculae carneae?

This latticelike arrangement of the endocardium (fig. 15.3) consists primarily of dense fibrous connective tissue; it provides a strong, flexible framework for the walls of the lower heart chambers.

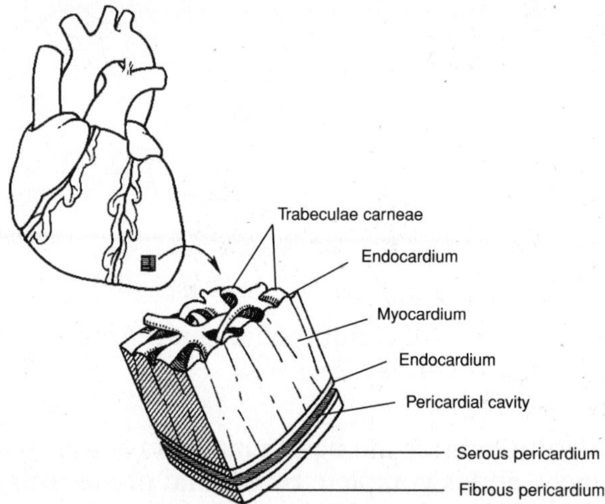

Figure 15.3 The heart wall, pericardial cavity, and the pericardium.

Objective D To describe the *chambers* and *valves* of the heart.

The heart is a four-chambered double pump (fig. 15.4). It consists of upper right and left **atria** that pulse together, and lower right and left **ventricles** that also contract together. The atria are separated by the thin, muscular **interatrial septum**, while the ventricles are separated by the thick, muscular **interventricular septum**. Two atrioventricular valves (AV valves), the **bicuspid** and **tricuspid valves**, are located between the chambers of the heart, and **semilunar valves** are present at the bases of the two large vessels (the pulmonary trunk and the aorta) that leave the heart.

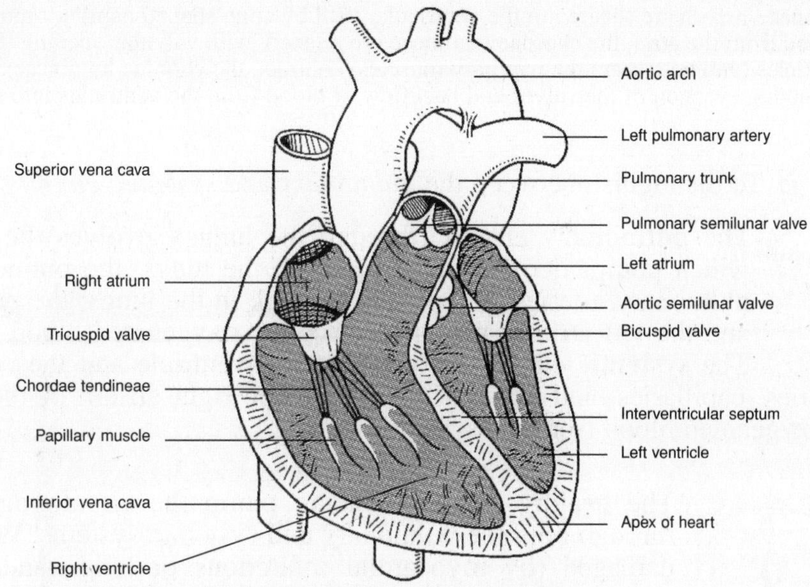

Figure 15.4 Internal anatomy of the heart.

15.6 Describe the workings of each of the heart valves.

See table 15.2.

Table 15.2 Valves of the Heart

Valve	*Location*	*Structure and function*
Tricuspid valve	Between right atrium and right ventricle	Composed of three cusps that prevent a backflow of blood from the right ventricle into the right atrium during ventricular contraction
Pulmonary semilunar valve	Between right ventricle and pulmonary trunk	Composed of three half-moon shaped flaps that prevent a backflow of blood from the pulmonary trunk into the right ventricle during ventricular relaxation
Bicuspid (mitral) valve	Between left atrium and left ventricle	Composed of two cusps that prevent a backflow of blood from the left ventricle to the left atrium during ventricular contraction
Aortic semilunar valve	Between left ventricle and ascending aorta	Composed of three half-moon-shaped flaps that prevent a backflow of blood from the aorta into the left ventricle during ventricular contraction

15.7 Describe the structure and functions of the papillary muscles and chordae tendineae?

Each cusp of the atrioventricular valves is held in position by strong tendinous cords, the chordae tendineae, which are secured to the ventricular wall by cone-shaped papillary muscles. As blood is ejected from the atria, the chordae tendineae are relaxed, with valvular opening. But as the ventricles (and with them the papillary muscles) contract, the chordae tendineae are pulled taut, preventing eversion of the valves and backflow of blood from the ventricles into the atria.

Objective E To distinguish between the *pulmonary* and *systemic circuits* of blood flow.

The **pulmonary circuit** (through the lungs) involves the right ventricle, which pumps deoxygenated blood to the lungs; the pulmonary trunk and pulmonary arteries; a capillary network in the lungs; the pulmonary veins; and the left atrium, which receives the oxygenated blood from the lungs. The **systemic circuit** involves the left ventricle and the remainder of the arteries, capillaries, and veins of the body. The right atrium of the heart receives deoxygenated blood from the systemic circuit.

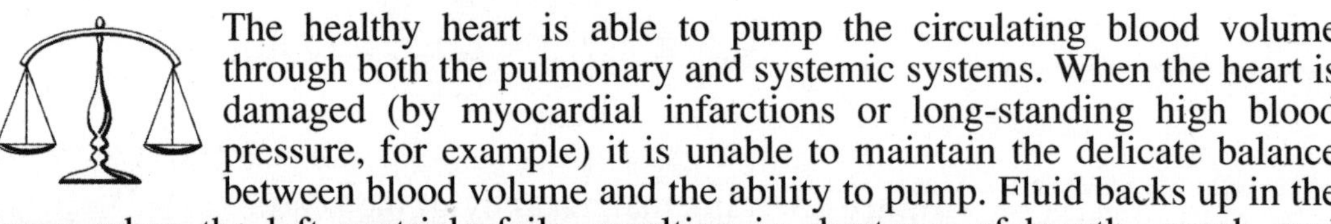

The healthy heart is able to pump the circulating blood volume through both the pulmonary and systemic systems. When the heart is damaged (by myocardial infarctions or long-standing high blood pressure, for example) it is unable to maintain the delicate balance between blood volume and the ability to pump. Fluid backs up in the lungs when the left ventricle fails, resulting in shortness of breath, cough, and respiratory distress. When the right ventricle weakens, fluid builds up in the peripheral tissues, leading to *edema* (swelling in the extremities) and liver engorgement.

15.8 Describe the flow of blood through the heart.

Blood fills both atria and begins to flow into both ventricles (fig. 15.5*a*). Next, the atria contract, emptying the remaining blood into the ventricles (fig. 15.5*b*). The ventricles then contract, forcing blood into the ascending aorta and pulmonary trunk (fig. 15.5*c*).

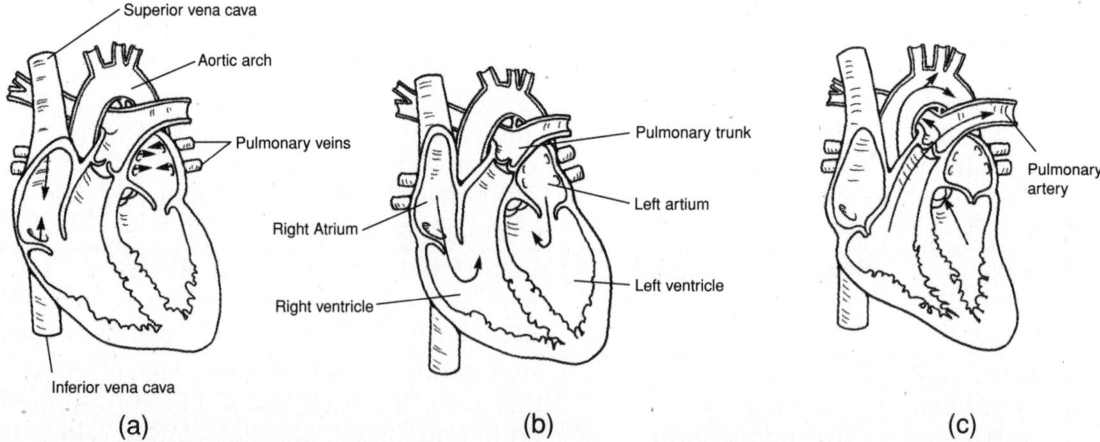

Figure 15.5 Blood flow through the heart. (*a*) Atria are filling, (*b*) atria are contracting, and (*c*) blood is being ejected from the heart by ventricular contraction.

15.9 Correlate the contractions of the heart chambers and the opening and closing of the heart valves and explain what causes the characteristic "lub-dub" sounds.

During atrial contraction, the atrioventricular valves are open and the semilunar valves are closed; during ventricular contraction, the reverse is true. The louder "lub," or *first sound*, is produced by the closing of the AV valves. The softer "dub," or *second sound*, is produced by the closing of the semilunar valves.

Objective F To explain how the fetal circulation differs from the circulation of a newborn.

Survey

The circulatory system of a fetus (fig. 15.6) is adapted to the fact that the fetal lungs are nonfunctional and that oxygen and nutrients are obtained from the placenta. Fetal circulation involves an umbilical cord that connects the placenta and the fetal umbilicus. The umbilical cord consists of an **umbilical vein** that transports oxygenated blood toward the heart and two **umbilical arteries** that return deoxygenated blood to the placenta. A **ductus venosus** allows blood to bypass the fetal liver; a **foramen ovale** permits blood to flow directly from the right atrium to the left; and a **ductus arteriosis** shunts blood from the pulmonary trunk to the aortic arch.

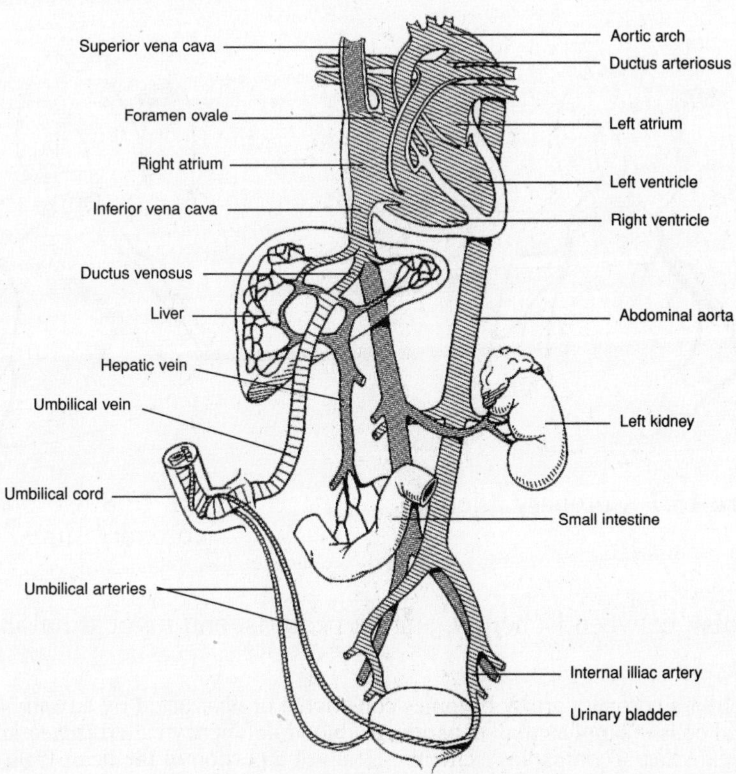

Figure 15.6 Fetal circulation.

The cardiovascular structures of the fetus undergo gradual transformations following birth to become other structures that persist throughout life. The umbilical vein forms the *round ligament* of the liver; the umbilical arteries atrophy to become the *lateral umbilical ligaments*; the ductus venosus forms the *ligamentum venosum*, a fibrous cord in the liver; the foramen ovale closes at birth and becomes the *fossa ovalis*, a depression in the interatrial septum; and the ductus arteriosis closes shortly after birth, atrophies, and becomes the *ligamentum arteriosum.*

Many newborn babies with congenital heart defects have insufficient oxygenated blood in the systemic circulation. One common congenital problem is a *patent foramen ovale*, in which the interatrial opening fails to close. The result of this and the other congenital heart defects is *cyanosis*, a bluish discoloration, and the infant is commonly called a "blue baby."

Objective G To describe *coronary circulation* to the myocardium of the heart.

Blood supply to the myocardium is provided by the **right** and **left coronary arteries**, which exit the ascending aorta just beyond the aortic semilunar valve (fig. 15.7). The left coronary artery gives rise to its major branches, the **anterior interventricular** and **circumflex arteries**, and the right coronary artery gives rise to the **posterior interventricular** and **marginal arteries**. The **great cardiac vein** and the **middle cardiac vein** return blood from the myocardial capillaries to the coronary sinus, and from there to the right atrium (fig. 15.8).

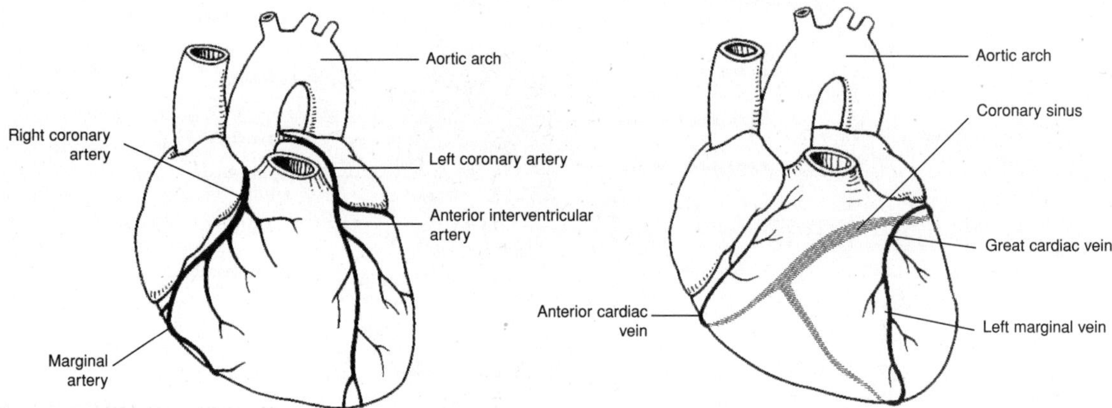

Figure 15.7 Coronary arteries.

Figure 15.8 Coronary veins and coronary sinus.

15.10 Distinguish between ischemia, angina pectoris, and myocardial infarction (or infarct).

If a branch of a coronary artery becomes constricted or obstructed by an *embolus* (clot), the myocardial cells it supplies may experience a blood deficiency called *ischemia. Angina pectoris* is the chest pain that accompanies ischemia. Death of a portion of the heart from ischemia is called *myocardial infarction* (heart attack).

Objective H To describe the *conduction system* of the heart.

The **conduction system** consists of *nodal tissues* (specialized cardiac muscle fibers) that initiate the conduction of depolarization waves through the myocardium. Depolarization waves cause the coordinated contractions that empty the heart chambers.

15.11 What is the "pacemaker" of the heart, where is it located, and what is its basic frequency?

The *pacemaker* of the heart is the **sinoatrial node** (SA node), located in the posterior wall of the right atrium (fig. 15.9). It typically depolarizes spontaneously at the rate of 70 to 80 times per minute, causing the atria to contract. Impulses from the SA node pass to the **atrioventricular node** (AV node), **atrioventricular bundle** (AV bundle, or *bundle of His*), and finally to the **conduction myofibers** (*Purkinje fibers*) within the ventricular walls. Stimulation of the conduction myofibers causes the ventricles to contract simultaneously.

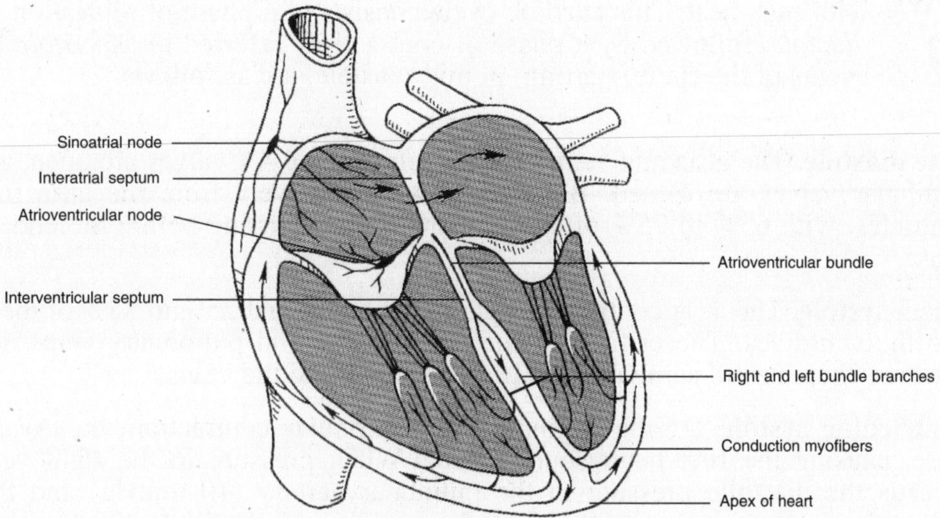

Figure 15.9 The conduction system of the heart.

Objective I To describe the *innervation of the heart*.

The SA and AV nodes are innervated by sympathetic and parasympathetic nerve fibers (fig. 15.10). Sympathetic impulses through the cardiac accelerator nerves accelerate heart action; parasympathetic impulses through the vagus (tenth cranial) nerves decelerate heart action. These autonomic impulses are regulated by the cardiac centers in the hypothalamus and medulla oblongata.

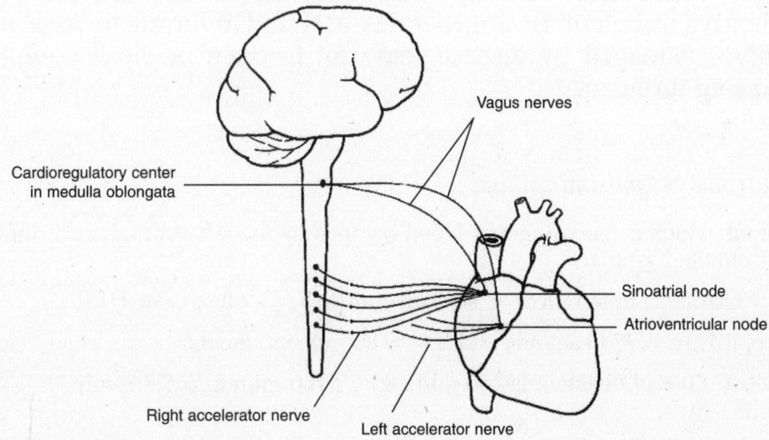

Figure 15.10 Autonomic innervation of the heart.

15.12 *True or False*. Norepinephrine and acetylcholine have a synergistic (cooperative) rate-changing action on the heart.

False. The effects of the two neurotransmitters—the former secreted by sympathetic postganglionic neurons, the latter by parasympathetic postganglionic neurons—are antagonistic.

Objective J To describe the *cardiac cycle*.

The atria and ventricles go through a sequence of events that is repeated with each beat. This **cardiac cycle** consists of a phase of relaxation, called *diastole*, followed by a phase of contraction, referred to as *systole*. Major events of the cycle, starting in mid-diastole, are as follows.

Late diastole. The atria and ventricles are relaxed, the AV valves are open, and the semilunar valves are closed. Blood is flowing passively from the atria into the ventricles, with 65% to 75% of ventricular filling occurring before the end of this phase.

Atrial systole. The atria contract and pump the additional 25% to 35% of the blood into the ventricles. The orifices of the venae cavae and pulmonary veins narrow; however, there is still some regurgitation of blood into the veins.

Ventricular systole. At the beginning of ventricular contraction, the AV valves close, causing the first heart sound, "lub." When pressure in the right ventricle exceeds the diastolic pressure in the pulmonary artery (10 mmHg) and the left ventricular pressure exceeds the diastolic pressure in the aorta (80 mmHg), the semilunar valves open and ventricular ejection begins. Under normal resting conditions, the pressure reaches 25 mmHg on the right side and 120 mmHg on the left side. The stroke volume, or volume of blood ejected from either ventricle, is 70 to 90 mL. At the end of systole, there are about 50 ml of blood remaining in each ventricle.

Early diastole. As the ventricles begin to relax, the pressure drops rapidly. The semilunar valves close, preventing backflow into the ventricles from the arteries and causing the second heart sound, "dub." Also, the AV valves open, and blood begins to flow from the atria into the ventricles.

When blood flows smoothly through heart valves and blood vessels, it flows silently; turbulent flow generates a sound referred to as a *heart murmur*. Valves damaged by disease may fail to open or close completely, thereby causing turbulence.

15.13 How is cardiac output calculated?

Cardiac output, which is the volume of blood pumped by the left ventricle in 1 minute, may be calculated from the formula:

$$\text{Cardiac output (C.O.)} = \text{stroke volume (S.V.)} \times \text{heart rate (H.R.)}$$

For instance, if H.R. = 72 beats/min and S.V. = 80 ml/beat, then:

$$\text{C.O.} = (72\ \text{beats/min})(80\ \text{ml/beat}) = 5760\ \text{ml/min} = 5.8\ \text{L/min}$$

15.14 The formula of problem 15.13 involves the stroke volume, the value of which is often unknown. Give an alternative procedure for determining cardiac output.

By the *Fick principle*, the amount (α) of a substance taken up by an organ (or the whole body) per unit time is equal to the arterial level (A.L.) of the substance minus the venous level (V.L.), times the blood flow. Since the blood flow equals the C.O., the following formula is derived:

$$\text{C.O.} = \frac{\alpha}{\text{A.L.}-\text{V.L.}}$$

For instance, the body's O_2 consumption is typically given by α = 250 ml/min, and typical blood levels of O_2 are A.L. = 190 ml/L blood. Thus, a typical cardiac output is

$$\text{C.O.} = \frac{250\ \text{ml/min}}{50\ \text{ml/L blood}} = 5\text{L blood/min}$$

15.15 Which of the following factors influence(s) cardiac output? (i) increased activity of the sympathetic nervous system, (ii) increased end-diastolic volume, (iii) decreased venous return to the heart, (iv) various forms of anemia.

All of the above affect cardiac output.

(i) Sympathetic stimulation increases the heart rate and the strength of heart contraction. It also initiates the release of epinephrine and norepinephrine from the adrenal medulla, which increase the cardiac output.
(ii) As the end-diastolic volume increases, the myocardium is increasingly stretched; as a result, the muscles contract with greater force, which leads to a greater stroke volume and cardiac output. This mechanism is known as *Starling's law of the heart* (or the *Frank-Starling law*).
(iii) In the case of decreased venous return (hemorrhage, etc.), the heart does not fill properly, which causes a decreased stroke volume and a decreased cardiac output.
(iv) Under most conditions of anemia, there is a reduction in blood viscosity, as well as localized vasodilation, due to diminished oxygen transport to the tissues. Both conditions produce a decrease in the total peripheral resistance, and therefore an increased cardiac output.

The normal heart rate (55–90 beats per minute [bpm]) depends on a balance between sympathetic and parasympathetic control. Beta adrenergic blocking drugs such as *propranolol* inhibit sympathetic input to the heart and reduce the heart rate. Conversely, exercise, stress, and low blood volume cause the release of catecholamines that stimulate an increase in rate. *Atropine*, an anticholinergic agent, has just the opposite effect of propranolol and is used clinically when a patient's heart rate becomes dangerously low.

15.16 Where is the stethoscope positioned on the chest wall to hear the heart sounds?

See table 15.3 and fig. 15.11.

Table 15.3 Placement of Stethoscope to Hear Sounds of Heart Valves

Heart valve	***Stethoscope position***
Tricuspid valve	5th intercostal space at sternum
Bicuspid (mitral) valve	5th intercostal space inferior to left nipple
Pulmonary semilunar valve	2nd intercostal space left of sternum
Aortic semilunar valve	2nd intercostal space right of sternum

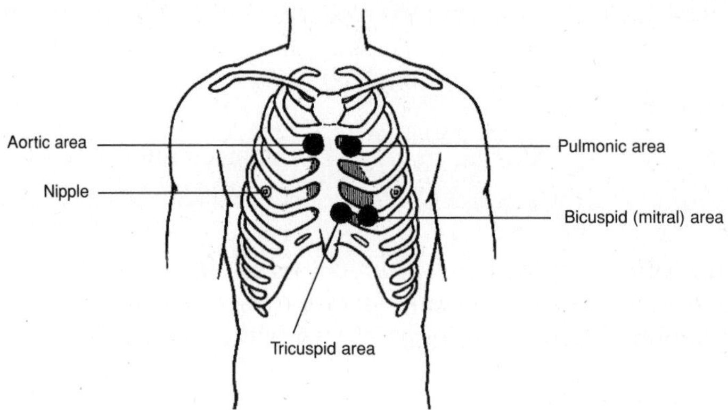

Figure 15.11 Sites of cardiac auscultation.

Objective K To identify the normal features of an *electrocardiogram.*

Survey

Because the body is a good conductor of electricity, potential differences generated by the depolarization and repolarization of the myocardium can be detected on the surface of the body and recorded as an **electrocardiogram** (**ECG** or **EKG**) (fig. 15.12).The *P wave* indicates depolarization of the atria. The *QRS complex* is the record of ventricular depolarization; the *T wave*, of ventricular repolarization. The short flat segment between S and T represents the refractory state of the ventricular myocardium; that between P and Q, a nonconductive phase of the AV node, during which atrial systole can be completed.

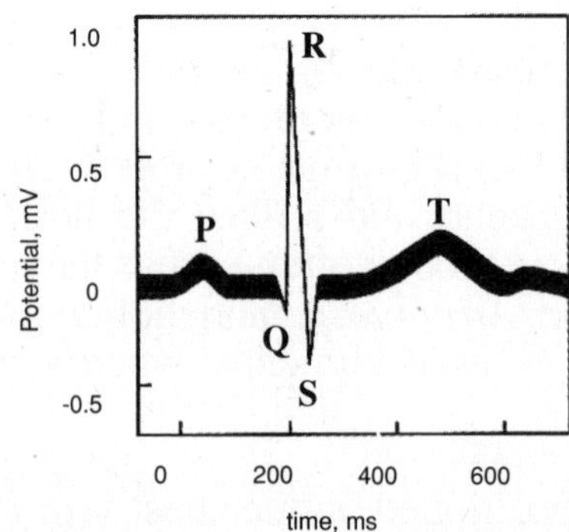

Figure 15.12 A normal electrocardiogram (ECG).

15.17 Describe the three conventional types of electrocardiographic leads.

Standard limb leads. Each lead joins two electrodes of opposite polarities (fig. 15.13).

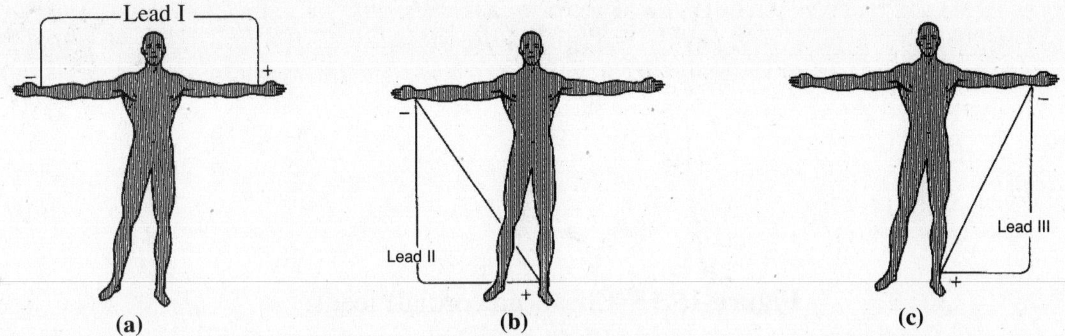

Figure 15.13 Standard limb leads.

Augmented unipolar limb leads. Each lead has one positive and two negative electrodes (fig. 15.14). Signal aVR is inverted relative to the other two.

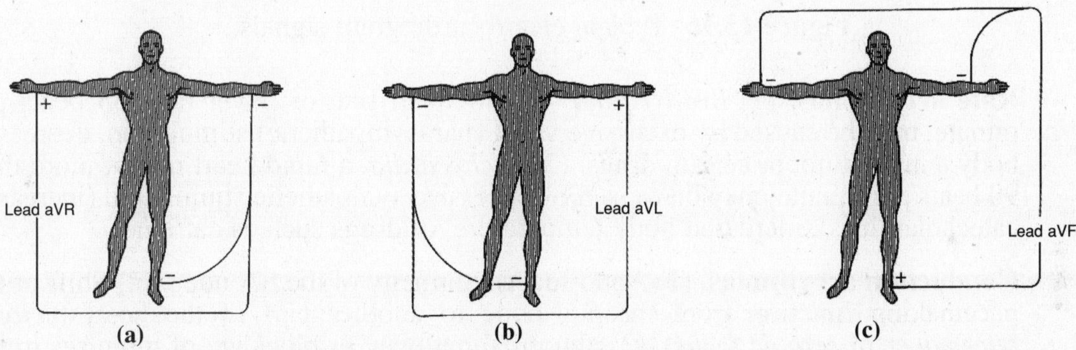

Figure 15.14 Augmented unipolar limb leads.

Chest (precordial) leads. Each lead joins a positive electrode, attached at one of the six sites shown in fig. 15.15, with three negative electrodes (arms and leg). The 12 limb and precordial leads all measure the same electrical activity of the heart, but each one "shows" the heart from a different viewpoint—much as 12 cameras surrounding an object would all show different angles of the same object. A skilled observer can use the 12 leads to construct a composite of the heart's electrical activity. Typical signals are shown in fig. 15.16.

Objective L To become familiar with *arrhythmias* detected by the electrocardiogram.

Deviations from normal heart rate or from normal electrical activity of the conduction system are referred to as **cardiac arrhythmias**.

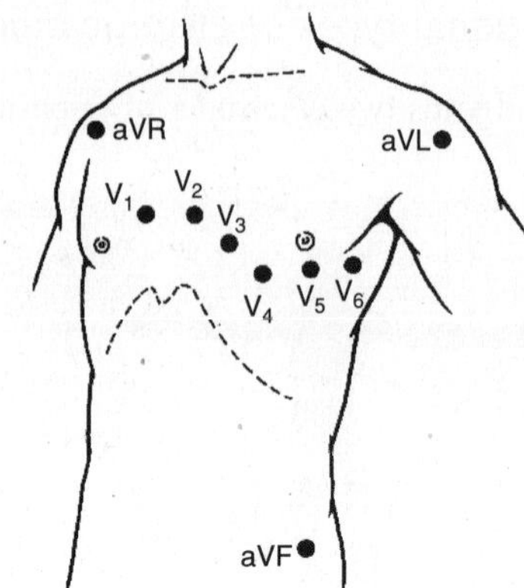

Figure 15.15 Chest (precordial) leads.

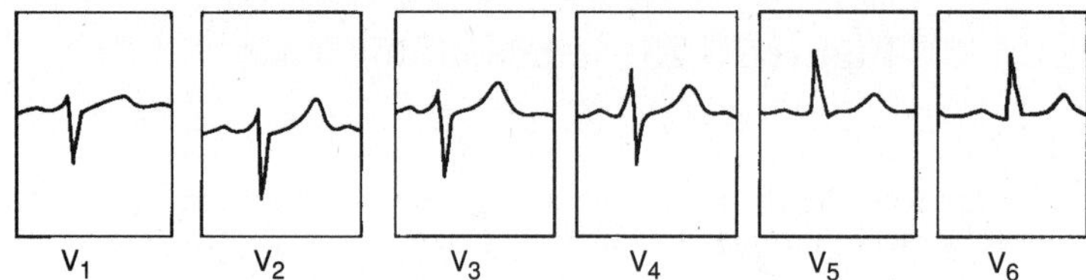

Figure 15.16 Typical electrocardiogram signals.

Rate arrythmias. (1) *Bradycardia*, a slow heart rate of fewer than 55 beats per minute, may be caused by excessive vagal (parasympathetic) stimulation, decreased body temperature, or certain drugs. (2) *Tachycardia*, a rapid heart rate of more than 90 beats per minute, may be caused by excessive sympathetic stimulation (increased catecholamines), increased body temperature, or drugs such as caffeine.

Conduction arrythmias. (1) Abnormal rhythmicity of the SA node. (2) Shift of the pacemaking function from the SA node to another part of the heart (*ectopic pacemaker* or *ectopic focus*). (3) abnormal pathway or blockage of impulses in the conduction system.

15.18 What are some causes of ectopic pacemaker activity.

Causes include ischemia or other localized heart damage; dilation of the atria due to hypertension; toxic irritants (e. g., nicotine, caffeine, alcohol); lack of sleep; anxiety; extremes in body temperature; departures from normal body pH.

15.19 What may cause heart block?

Impulses through the heart are sometimes blocked at critical points in the conduction system. Heart blocks may be due to (1) localized destruction of the conduction system as a result of infarct (see problem 15.10), (2) excessive stimulation of the vagus nerves, or (3) infection in the conduction system.

15.20 What are the characteristics of ECGs from patients who experience premature beats?

Premature beats are caused when an ectopic pacemaker fires so as to create waves that appear earlier in the cycle than they would normally.

Atrial premature depolarization or complex (fig. 15.17) is due to premature depolarization of an ectopic pacemaker in the atria. It may precede an interval of atrial flutter or fibrillation (see problem 15.21). This kind of premature beat is usually considered innocuous.

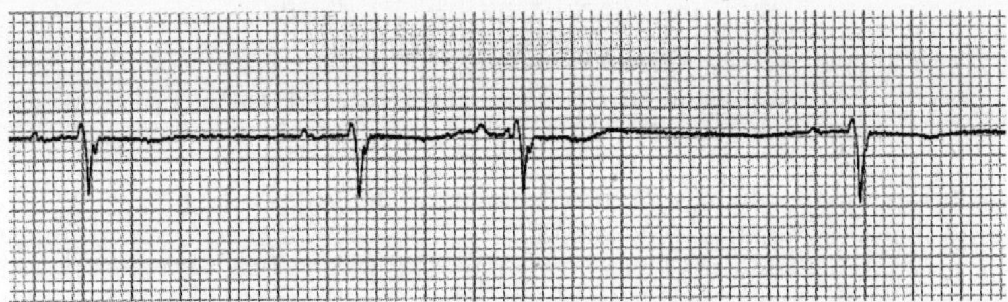

Figure 15.17 Atrial premature depolarization.

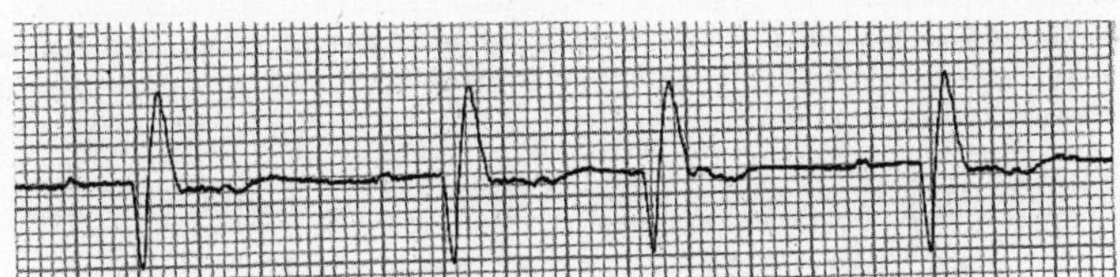

Figure 15.18 AV-nodal premature depolarization.

AV-nodal premature depolarization. Originates from an ectopic discharge in the AV node (fig. 15.18). The ECG shows a normal QRS complex that generally is not preceded by a P wave.

Premature ventricular depolarization (PVD). Originates from an ectopic pacemaker in the ventricles (fig. 15.19). The P wave is lacking in the ECG. The QRS complex is wide (because conduction is mainly through the muscle cells of the ventricle rather than through the conduction myofibers) and tall (one ventricle depolarizes slightly before the other). The T wave is often inverted (altered repolarization). One PVD may be coupled with one or several normal beats.

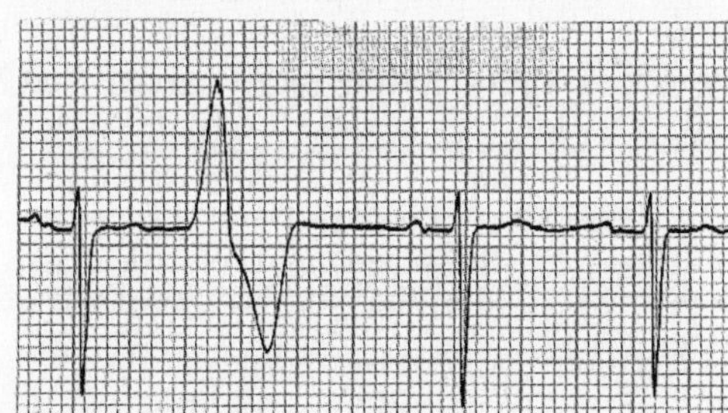

Figure 15.19 Premature ventricular depolarization (PVD).

15.21 What are the characteristics of ECGs from patients who experience flutter or fibrillation?

Rapid (300 beats/min) and regular atrial depolarizations due to a circus pathway in the atrial tissue are called atrial flutter (fig. 15.20). During atrial flutter, the ventricles are unable to respond to each atrial impulse, so that a partial block usually is present. It is marked by a 2:1 or 4:1 rhythm (the atria depolarize between 2 to 4 times for each ventricular contraction) and packed, regular P waves with a "sawtooth appearance."

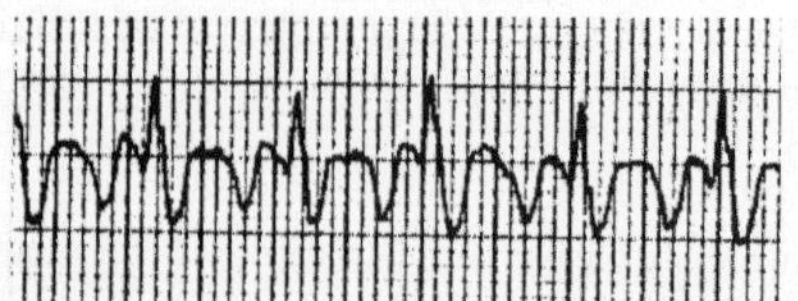

Figure 15.20 Atrial flutter ECG.

Atrial fibrillation. Caused by disorganized electrical activity in the atria (fig. 15.21). The P waves are absent and the baseline atrial electrical activity is irregular. The QRS complexes and T waves usually look normal, but their rhythm is irregular. This is because the ventricles respond to conducted impulses as soon as conduction myofibers repolarize.

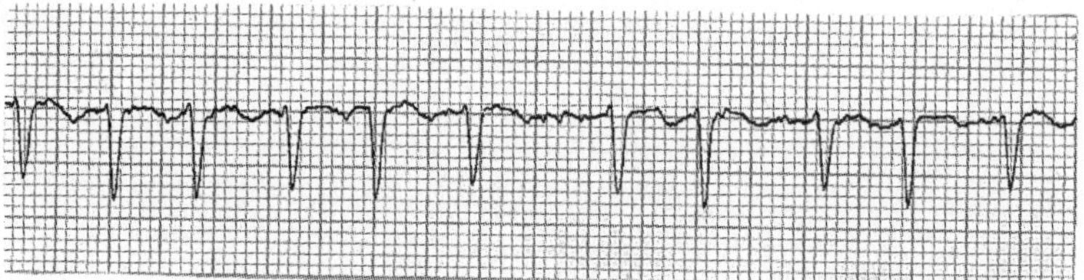

Figure 15.21 Atrial fibrillation ECG.

Ventricular tachycardia. Usually due to a single ectopic pacemaker in the ventricles (fig. 15.22). The ECG usually resembles a smooth sine wave. This condition is dangerous because the heart does not fill properly and cardiac output is decreased. Moreover, it may develop into ventricular fibrillation.

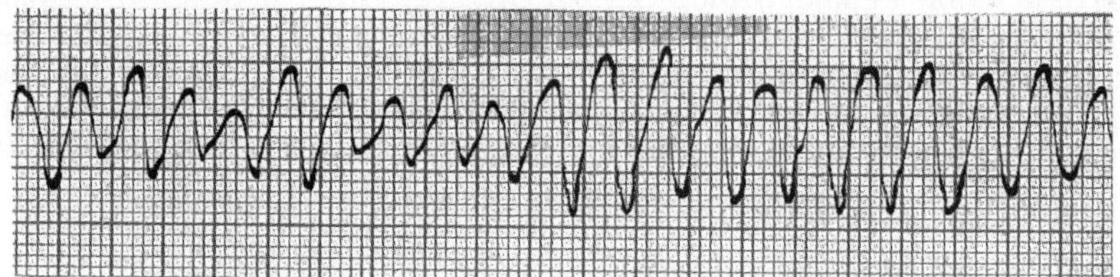

Figure 15.22 Ventricular tachycardia ECG.

Ventricular fibrillation. Caused by disorganized circus electrical activity in the ventricles. The ECG is chaotic-like "random noise." In this, the most serious of the spasmodic conditions, the blood pressure drops rapidly.

15.22 Characterize an ECG from a patient who has suffered myocardial infarction (heart attack).

A *myocardial infarction* is caused by a lack of blood flow to an area of the heart as a result of coronary vascular narrowing (spasmodic or from atherosclerosis) or vascular blockage (embolism). The QRS and T complexes change as cardiac muscle tissue progresses from early to late infarction. Ischemia is first reflected in ST segment depression (fig. 15.23). ST segment elevation heralds an early infarction. Late infarction is reflected in the T wave inversion. The deep Q wave is evidence of an old resolved infarction.

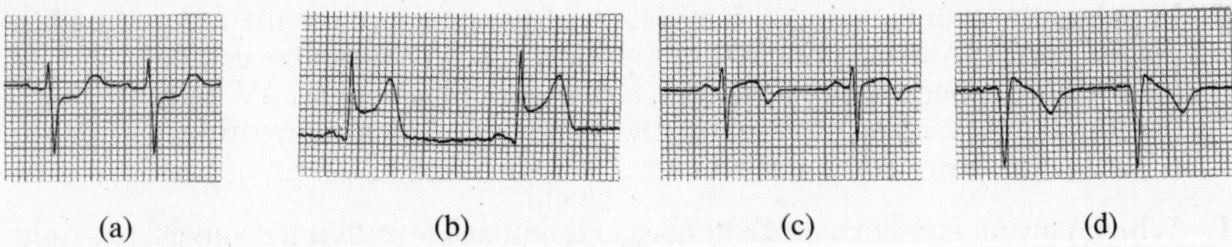

Figure 15.23 Typical ECGs of (*a*) ischemia, (*b*) early infarct, (*c*) late infarct, and (*d*) completed or old infarct.

Review Exercises

Multiple Choice

1. The prenatal heart begins to pump blood during (*a*) the fourth week, (*b*) the fifth week, (*c*) the sixth week, (*d*) the seventh week.

2. Which is a correct pairing for fetal circulation? (*a*) foramen ovale—right ventricle to left ventricle, (*b*) ductus venosus—umbilical vein to inferior vena cava, (*c*) foramen ovale—right atrium to pulmonary trunk, (*d*) ductus arteriosus—pulmonary artery to pulmonary vein

3. The valve that is located on the same side of the heart as the pulmonary semilunar valve is (*a*) the tricuspid valve, (*b*) the mitral valve, (*c*) the bicuspid valve, (*d*) the aortic semilunar valve.

4. A stenosis of the bicuspid heart valve might cause blood to back up into (*a*) the coronary circulation, (*b*) the venae cavae, (*c*) the pulmonary circuit, (*d*) the left ventricle.

5. In the fetus, fully oxygenated blood is carried by (*a*) the ductus arteriosis, (*b*) the umbilical artery, (*c*) the placental vein, (*d*) the umbilical vein.

6. After birth, the ductus arteriosus develops into (*a*) the fossa ovalis, (*b*) the ligamentum arteriosum, (*c*) the lateral umbilical ligament, (*d*) the ligamentum venosum, (*e*) the round ligament of the liver.

7. The outermost of the three layers of the heart is (*a*) the epicardium, (*b*) the supracardium, (*c*) the pericardium, (*d*) the endocardium.

8. The correct sequence for blood entering the heart through the venae cavae and leaving through the aorta is
(*a*) right atrium, left atrium, left ventricle, right ventricle
(*b*) left ventricle, left atrium, right ventricle, right atrium
(*c*) right atrium, right ventricle, left atrium, left ventricle
(*d*) left atrium, left ventricle, right atrium, right ventricle

9. The sinoatrial node (SA node) is situated in the wall of (*a*) the right atrium, (*b*) the interventricular septum, (*c*) the pulmonary trunk, (*d*) the superior vena cava, (*e*) the left ventricle.

10. Impulses through the conduction system of the heart follow the ordered path:
    (*a*) AV node, SA node, conduction myofibers, atrioventricular bundle
    (*b*) SA node, conduction myofibers, atrioventricular bundle, AV node
    (*c*) SA node, AV node, atrioventricular bundle, conduction myofibers
    (*d*) AV node, atrioventricular bundle, SA node, conduction myofibers

11. Which pairing is *incorrect*? (*a*) chordae tendineae—semilunar valves, (*b*) right ventricle—papillary muscles, (*c*) left ventricle—trabeculae carneae, (*d*) right atrium—coronary sinus, (*e*) left atrium—pulmonary veins

12. The heart is covered by (*a*) the pericardium, (*b*) the epicardium, (*c*) the supracardium, (*d*) the endocardium.

13. An increase in cardiac output follows all of the following *except* (*a*) physical exercise, (*b*) fever, (*c*) digestion, (*d*) parasympathetic stimulation through the vagus (tenth cranial) nerves.

14. The "lub" sound of the heart is caused by (*a*) closing of the AV valves, (*b*) closing of the semilunar valves, (*c*) blood rushing out of the ventricles, (*d*) filling of the ventricles, (*e*) a depolarization of the SA node.

15. Which occurs during systole? (*a*) ventricular filling, (*b*) atrial filling, (*c*) ventricular contraction, (*d*) atrial relaxation

16. To clearly hear the sound of the bicuspid valve, a stethoscope should be placed (*a*) to the right of the sternum at the second intercostal space, (*b*) to the left of the sternum at the second intercostal space, (*c*) to the left of the sternum at the fifth intercostal space inferior to the nipple, (*d*) to the right of the sternum at the fifth intercostal space.

17. When the atrioventricular bundle is completely interrupted, (*a*) the atria beat at an irregular rate, (*b*) the ventricles typically contract at 30 to 40 beats/min, (*c*) the PR intervals in the ECG are longer than normal but remain constant from beat to beat, (*d*) the QRS complex varies in shape from beat to beat.

18. Which of the following is *not* a condition of late diastole? (*a*) the atria and ventricles are relaxed, (*b*) the AV valves are open, (*c*) the aortic semilunar valve is open, (*d*) the ventricles receive blood from the atria, (*e*) both a and c are correct

19. During ventricular contraction, (*a*) all the blood is forced out of the ventricles, (*b*) some of the blood remains in the ventricles, (*c*) no blood is forced out of the ventricles, (*d*) some blood backflows into the atria.

20. Which of the following is *not* part of the pulmonary circuit? (*a*) the left atrium, (*b*) the pulmonary trunk, (*c*) the aortic semilunar valve, (*d*) the pulmonary veins, (*e*) the pulmonary semilunar valves

True or False

____ 1. Initiation of the heartbeat, at 8 weeks after conception, marks the transition from embryo to fetus.

____ 2. The pericardial sac secretes fluids that lubricate the surface of the heart.

____ **3.** Cutting the vagus (tenth cranial) nerves where they innervate the heart would increase the heart rate.

____ **4.** A patent (open) ductus arteriosus in an adult permits blood flow from the pulmonary trunk to the aortic arch.

____ **5.** The right atrium of the fetal heart receives relatively well oxygenated blood.

____ **6.** Epinephrine increases the rate, but not the strength, of heart contraction.

____ **7.** The mediastinum, pericardial cavity, and two pleural cavities are compartments of the thoracic cavity.

____ **8.** The heart is totally derived from embryonic mesoderm.

____ **9.** Chordae tendineae, papillary muscles, and trabeculae carneae are structural features unique to the ventricles of the heart.

____ **10.** *Angina pectoris* is the comprehensive term for heart attack.

Completion

1. The ______________________________ is the space between the lungs in the thoracic cavity where the heart is positioned.

2. The first sound of the heart, or "lub," is caused by closure of the ____________________ valves.

3. A patent foramen ovale is located within the __________________________ septum of the heart.

4. Depolarization of the _________________________ ______________________ causes ventricular contraction or systole.

5. A heart ___________________________ is cause by turbulent blood flow or backflow of blood across a valve.

6. ________________________ _________________________ are the roughened ridges of connective tissue lining the ventricles of the heart

7. ____________________________ is a heart rate or fewer than 60 beats/min.

8. A "pacemaker" at a site other than the SA node is referred to as an ____________________ pacemaker.

9. Systole is indicated by the ________________________ deflection of the ECG.

10. Stroke volume x heart rate = _________________________ ______________________.

Labeling

Label the structures indicated on the figure to the right.

1. ______________________
2. ______________________
3. ______________________
4. ______________________
5. ______________________
6. ______________________
7. ______________________
8. ______________________
9. ______________________
10. ______________________

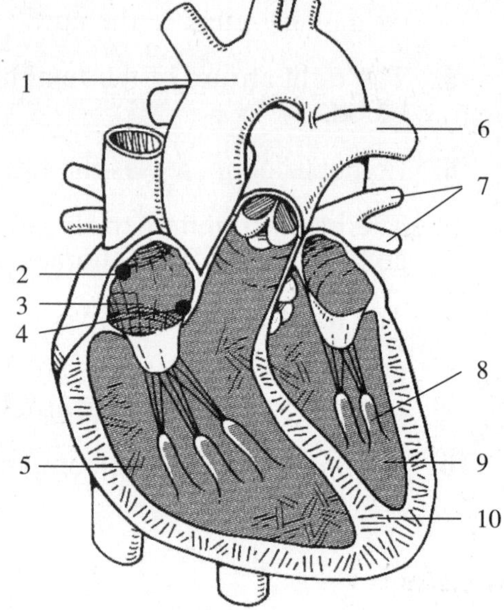

Matching

Match the cardiac event with its description.

____	**1.** P wave	(*a*)	atrial depolarization
____	**2.** First heart sound	(*b*)	cardiac output
____	**3.** Second heart sound	(*c*)	ventricular depolarization
____	**4.** QRS complex	(*d*)	ventricular repolarization
____	**5.** S.V. x H.R.	(*e*)	closure of the AV valves at the onset of systole
____	**6.** T wave	(*f*)	closure of the semilunar valves at the onset of diastole

Answers and Explanations for Review Exercises

Multiple Choice

1. (*a*) By day 25, the embryonic heart is sufficiently developed to pump blood.
2. (*b*) The ductus venosus ensures the rapid flow of oxygenated blood from the umbilical vein to the heart.
3. (*a*) Both the pulmonary semilunar valve and the right atrioventricular, or tricuspid, valve are located in the right side of the heart.
4. (*c*) With bicuspid stenosis (narrowing), the blood backs into the left atrium and pulmonary veins. This condition can cause pulmonary capillary congestion.
5. (*d*) The umbilical vein transports oxygenated blood from the placenta toward the fetal heart.
6. (*b*) The ligamentum arteriosum is a small connective tissue cord extending from the pulmonary trunk to the aortic arch.
7. (*a*) The epicardium, or visceral pericardium, is a thin protective serous membrane that adheres to the myocardium of the heart.
8. (*c*) The right atrium and right ventricle transport deoxygenated blood that arrives to the heart through systemic veins, and the left atrium and left ventricle transport oxygenated blood that arrives to the heart through the pulmonary veins.

9. (*a*) The SA node (pacemaker) is located in the posterior wall of the right atrium near the opening of the superior vena cava.
10. (*c*) Depolarization of the SA node causes atrial contraction and the conduction of impulses through the AV node and the AV bundle. Depolarization of the conduction myofibers causes ventricular contraction and the ejection of blood from the heart.
11. (*a*) The chordae tendineae extend from the papillary muscles to the cusps of the AV valves. Chordae tendineae are found only in the ventricles.
12. (*a*) The heart is enclosed in a loose-fitting serous sac called the pericardium or pericardial sac.
13. (*d*) Parasympathetic stimulation through the vagus nerves autonomically slows the heart rate, thus lowering the cardiac output.
14. (*a*) "Lub" is the first heart sound that occurs at the beginning of ventricular contraction as a result of closure of the AV valves. "Dub" is the second heart sound, immediately following "lub," that results from closure of the semilunar valves.
15. (*c*) Systole refers to ventricular contraction, and diastole refers to ventricular relaxation.
16. (*c*) To most clearly auscultate (hear with a stethoscope) the bicuspid (mitral) valve, the stethoscope is placed at the fifth intercostal space, inferior to the left nipple.
17. (*b*) When all impulses between the atria and ventricles are blocked, the ventricles will pace themselves at a rate of about 30 to 40 beats/min.
18. (*e*) During diastole, both semilunar valves are closed, preventing the backflow of blood from the ascending aorta and the pulmonary trunk. The atria are also relaxed in preparation for the arrival of blood from the vena cavae and the pulmonary veins.
19. (*b*) After ventricular contraction (systole), some blood (about 50 ml) remains in each ventricle and is known as the end systolic volume.
20. (*c*) Located at the base of the ascending aorta, the aortic semilunar valve is part of the systemic circuit.

True or False

1. False; the embryonic heart begins pumping blood at about day 25
2. True
3. True
4. True
5. True
6. False; epinephrine (adrenergic) increases both the heart rate and the force of contraction
7. True
8. True
9. True
10. False; angina pectoris is chest pain that is associated with ischemia (insufficient blood to the heart muscle), whereas heart damage is associated with a heart attack (myocardial infarction)

Completion

1. mediastinum
2. atrioventricular (AV)
3. interatrial
4. conduction myofibers
5. murmur
6. Trabeculae carnae
7. Bradycardia
8. ectopic
9. QRS
10. cardiac output

Labeling

1. Aortic arch
2. Sinoatrial node
3. Right atrium
4. AV node
5. Right ventricle
6. Pulmonary artery
7. Pulmonary veins
8. Papillary muscle
9. Left ventricle
10. Interventricular septum

Matching

1. (*a*)
2. (*e*)
3. (*f*)
4. (*c*)
5. (*b*)
6. (*d*)

Cardiovascular System: Vessels and Blood Circulation 16

Objective A To describe the functions of the cardiovascular system in general terms.

Transport: Nutrients and oxygen are carried to all body cells, waste products and carbon dioxide are carried from the cells to the organs of excretion, and hormones are carried from the endocrine glands to target tissues. *Thermoregulation*: The amount of heat lost from the body is regulated by the degree of blood flow through the skin. *Acid-base balance*: In cooperation with the respiratory and urinary systems, the cardiovascular system regulates (through buffer substances in the blood) the body pH. *Protection against disease*: The leukocytes are adapted to defend against foreign microbes and toxins.

Objective B To compare *arteries*, *capillaries*, and *veins* as to structure and function.

Blood is carried away from the heart in large vessels called **arteries**. These divide into smaller arteries, and the smaller arteries divide into **arterioles**. Arterioles divide into microscopic **capillaries** (the exchange area of the system). The capillaries converge to form vessels called **venules**, which join to form still larger vessels called **veins**. Veins return the blood to the heart.

The walls of blood vessels are composed of the following *tunics* (layers): the **tunica interna**, an inner layer of squamous epithelium, called *endothelium*, resting on a layer of connective tissue; the **tunica media**, a middle layer of smooth muscle fibers mixed with elastic fibers; and the **tunica externa**, an outer layer of connective tissue containing elastic and collagenous fibers (fig. 16.1 and table 16.1). The tunica adventitia of the larger vessels is infiltrated with a system of tiny blood vessels called the *vasa vasorum* ("vessels of the vessels") that nourish the more external tissues of the blood vessel wall.

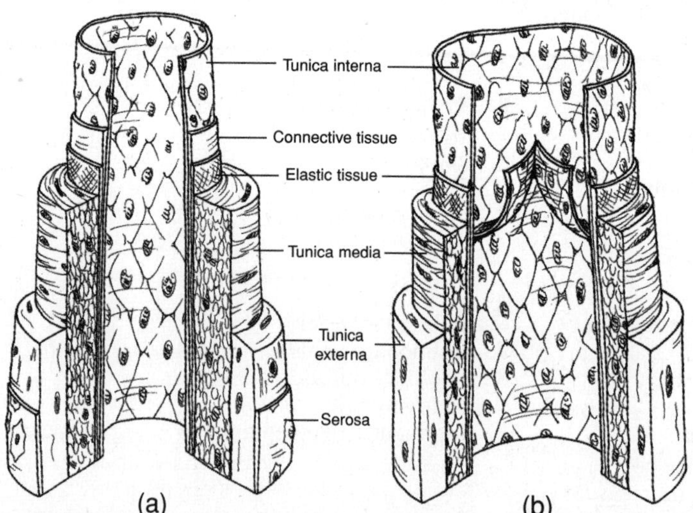

Figure 16.1 Structure of (*a*) an artery and (*b*) a vein.

Table 16.1 A Comparison of Vessels in the Cardiovascular System

Vessel	*Structure*	*Function*
Artery	Strong, elastic vessel consisting of three tunics; lumen diameter large relative to wall thickness	Distributive channel to body tissues; blood carried under high pressure (muscular wall and large lumen minimize pressure drop)
Arteriole	Thick layer of smooth muscle in tunica media; relatively narrow lumen	Alters diameter to control blood flow, dampens pulsate flow to a steady flow
Capillary	Wall composed of a single layer of *endothelium* (tunica interna); smooth muscle cuff (precapillary sphincter) at its origin regulates blood flow	Allows exchange of fluids, nutrients, and gases between the blood and the interstitial fluids
Vein	Thin, distensible vessel consisting of three tunics; lumen diameter very large; valves present	Carries blood from tissues to heart; serves as fluid reservoir (veins hold 60% to 75% of circulating blood volume); constricts in response to sympathetic stimulation; valves ensure unidirectional blood flow

16.1 Do capillaries exchange substances between the blood and the interstitial fluid in the same way throughout the body?

No. Among other things (see problem 16.3), the size and number of *fenestrations* (openings or pores) in capillaries vary with the function of the organ or tissue. Large fenestrations, and therefore increased exchange, are characteristic of endothelial cells of capillaries in the GI tract, renal glomeruli, and some glands. In the brain, capillaries have small fenestrations, or none at all, and the exchange of many substances is retarded (the blood-brain barrier; see chapter 10, Objective J).

16.2 Compare arterial blood pressure and venous blood pressure.

The most important variables affecting blood pressure are *cardiac rate, blood volume,* and total *peripheral resistance.* Arterial blood pressure is much greater than venous blood pressure due to ventricular contractions of the heart pulsating the blood into arteries and the recoiling of the arterial walls. Blood pressure decreases rapidly within the capillaries and is near zero where venous blood enters the heart.

16.3 List some factors that influence exchange between the blood and the interstitial fluid.

(1) A large surface area (about 700 m^2) for exchange because of the large number of capillaries in the body. (2) Fenestrations. (3) Diffusion—the principal mechanism of exchange. (4) Capillary hydrostatic pressure—the force that pushes fluids into the interstitial space; it ranges from 10 to 45 mmHg in most tissues. (5) Interstitial pressure, which varies depending on physiological conditions. (6) Capillary osmotic pressure, which is mainly due to plasma proteins (albumin). Normal osmotic pressure (23 to 28 mmHg), which causes reabsorption of fluid into the capillaries. (7) Interstitial osmotic pressure—movement of some proteins out of capillaries induces outward filtration of fluid into the interstitial space.

Objective C To identify the principal *systemic arteries.*

Survey

See fig. 16.2.

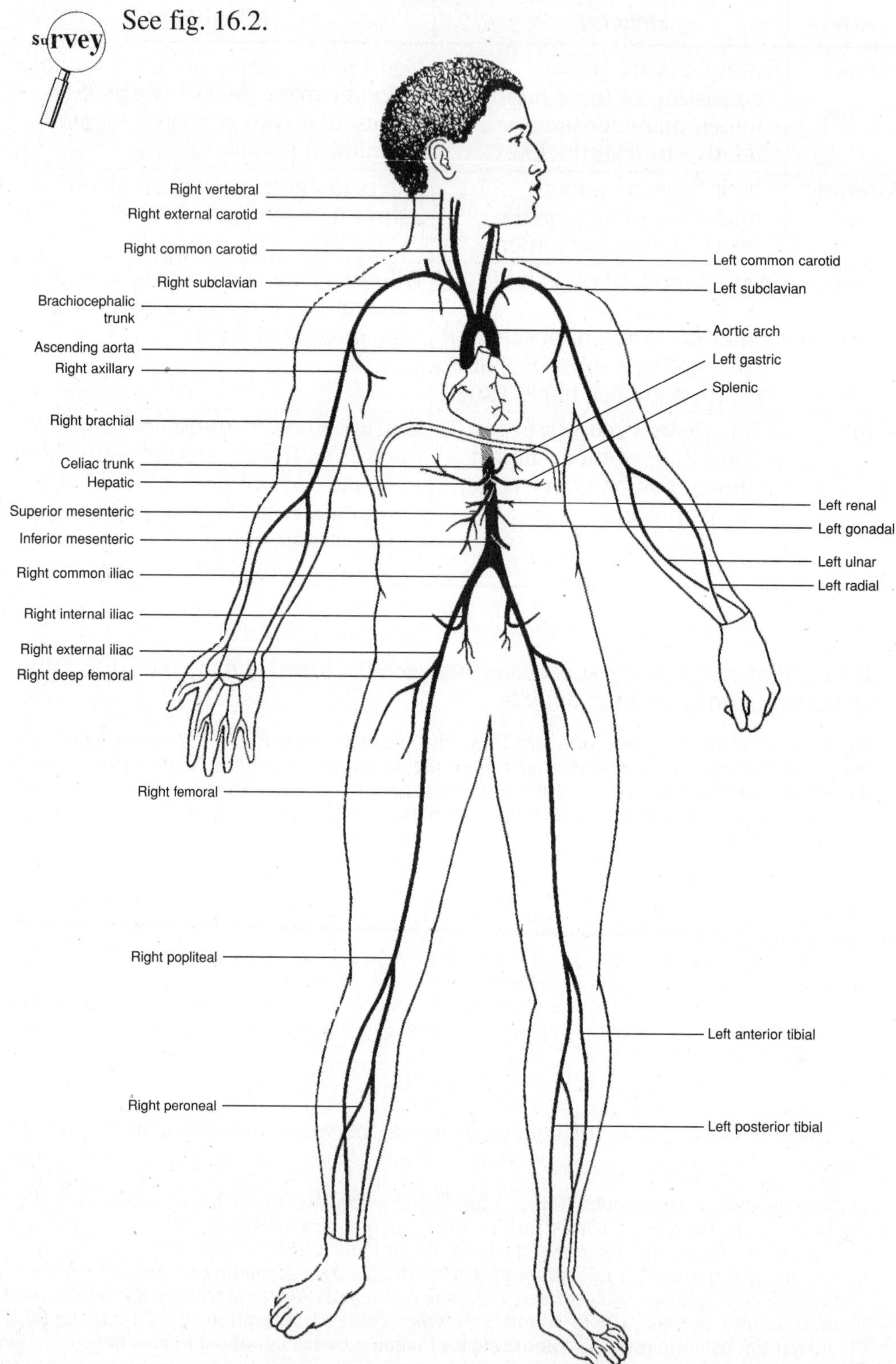

Figure 16.2 Principal arteries of the body.

16.4 Specify the arteries that branch from the aortic arch.

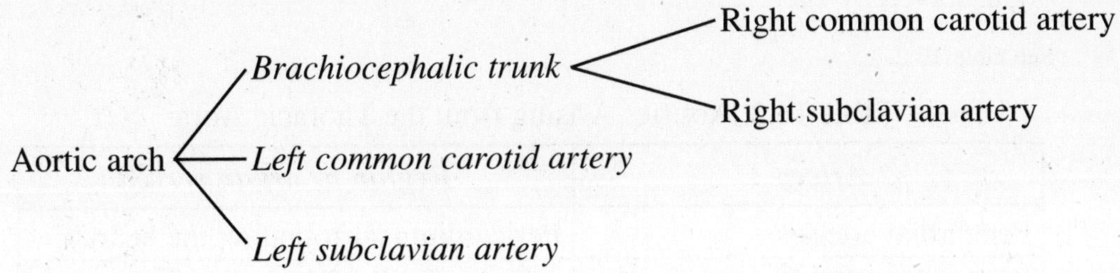

16.5 Supply the missing labels in fig. 16.3.

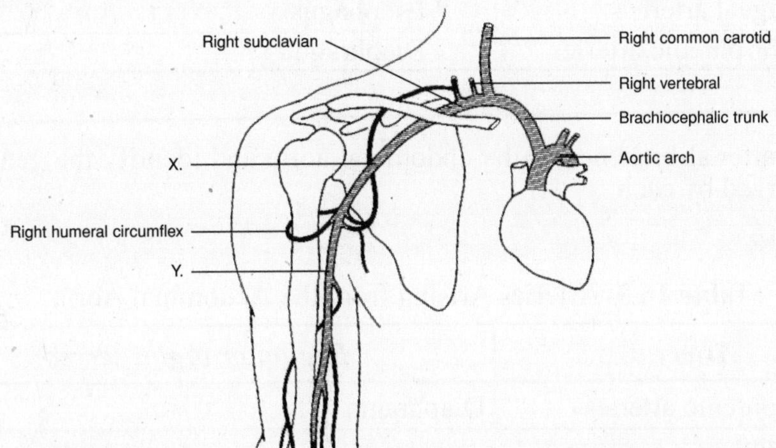

Figure 16.3 Arteries of the right neck and shoulder regions.

X = right axillary artery; Y = right brachial artery.

16.6 What are the four vessels that supply blood to the brain?

The paired internal carotid arteries and the paired vertebral arteries (see fig. 16.4).

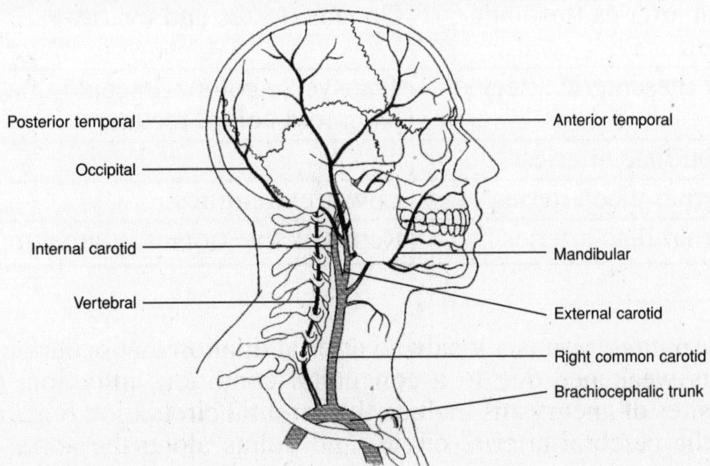

Figure 16.4 Arteries of the neck and head.

16.7 List the arterial branches of the thoracic aorta and identify the general region or organ served by each.

See table 16.2.

Table 16.2 Arteries Arising from the Thoracic Aorta

Artery	*Region or organ served*
Pericardial arteries	Pericardium surrounding the heart
Intercostal arteries	Thoracic wall (muscles of rib cage)
Bronchial arteries	Right and left bronchus
Esophageal arteries	Esophagus
Superior phrenic arteries	Diaphragm

16.8 List the arterial branches of the abdominal aorta and identify the general region or organ served by each.

See table 16.3.

Table 16.3 Arteries Arising from the Abdominal Aorta

Artery	*Region or organ served*
Inferior phrenic arteries	Diaphragm
Celiac trunk	
Hepatic artery	Liver, upper pancreas, duodenum
Splenic artery	Spleen, pancreas, stomach
Left gastric artery	Stomach, esophagus
Superior mesenteric artery	Small intestine, pancreas, cecum, appendix, ascending colon, transverse colon
Suprarenal arteries	Adrenal (suprarenal) glands
Renal arteries	Kidneys
Gonadal arteries (testicular; ovarian)	Gonads (testes and ovaries)
Inferior mesenteric artery	Transverse colon, descending colon, sigmoid colon, rectum
Common iliac arteries	
External iliac arteries	Lower extremities
Internal iliac arteries	Reproductive organs, gluteal muscles

An *aneurysm* is a localized arterial dilation that occurs where an arterial wall is weakened due to a congenital condition, infection, or trauma. Frequent sites of aneurysms include the cerebral circulation (e.g., a *berry aneurysm* on the cerebral arterial circle) and points along the aorta. Aneurysms may be detected through an *angiogram* and then surgically treated. A ruptured cerebral aneurysm is known as a *stroke*.

Objective D To identify the principal *systemic veins.*

Survey

See fig. 16.5.

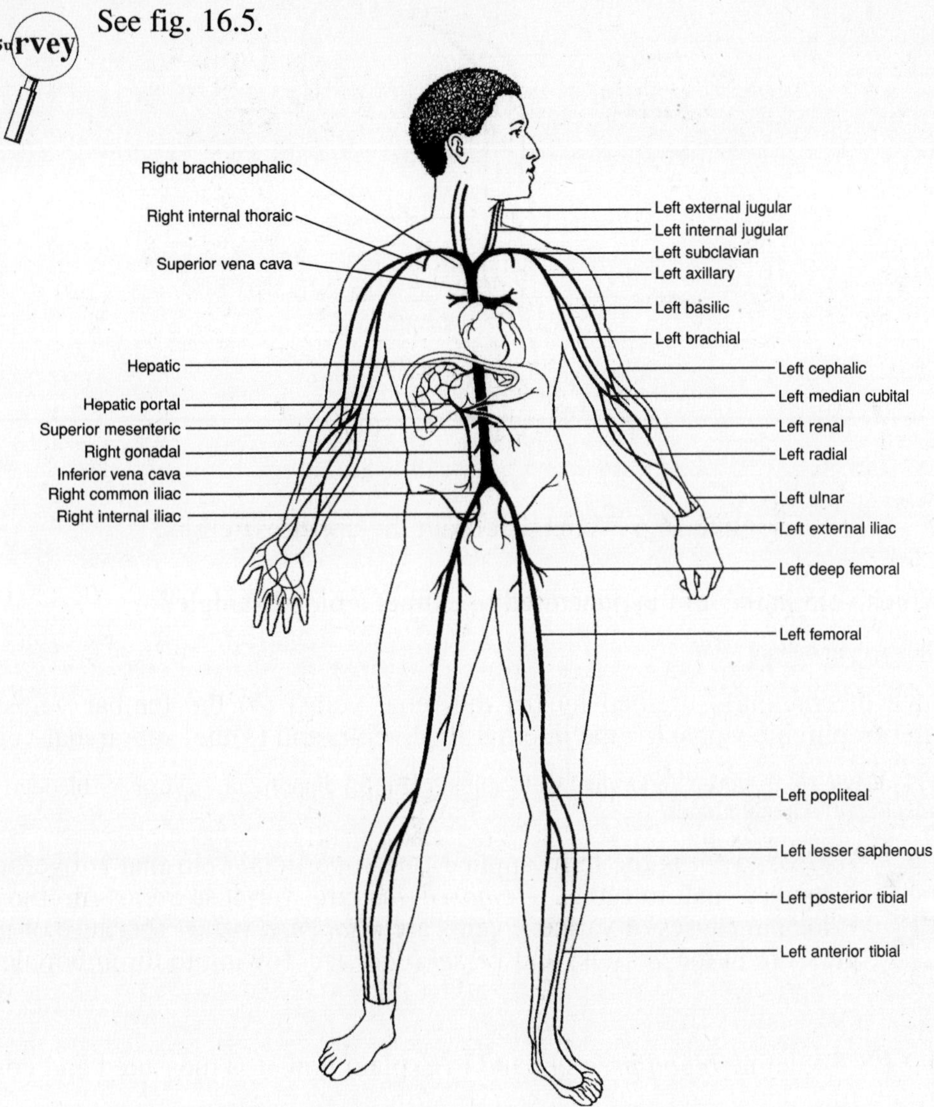

Figure 16.5 Principal veins of the body.

16.9 Identify the major vein that returns blood to the heart from the head, neck, and upper extremities, and the one that returns blood from the abdomen and lower extremities.

The superior vena cava and inferior vena cava, respectively.

16.10 Identify the paired vein that drains blood from the brain, meninges, and cranial nervous sinuses, and that passes down the neck adjacent to the common carotid artery and vagus nerve.

The internal jugular vein.

16.11 Classify the veins that drain the upper extremity as deep or superficial.

Deep: brachial, axillary, and subclavian veins (see fig. 16.5). *Superficial*: median antebrachial, median cubital, basilic, and cephalic veins (see fig. 16.6).

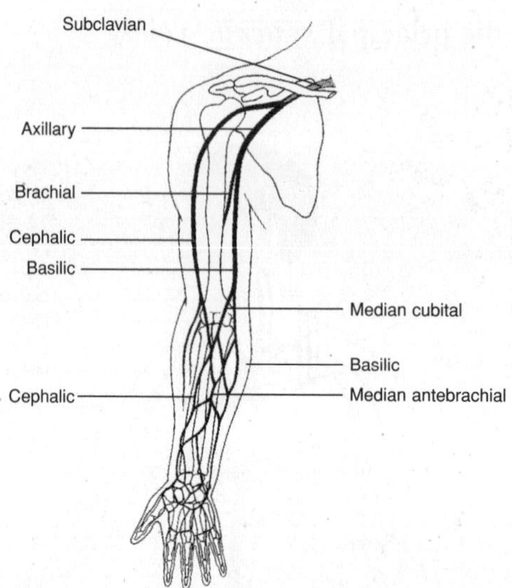

Figure 16.6 Veins that drain the upper extremity.

16.12 Which vein in the arm is punctured to extract a blood sample?

The median cubital vein.

16.13 State the region(s) drained by (*a*) the renal veins, (*b*) the lumbar veins, (*c*) the inferior phrenic veins, (*d*) the internal iliac veins, and (*e*)the suprarenal veins.

(*a*) kidneys; (*b*) posterior abdominal wall, spinal cord; (*c*) diaphragm; (*d*) urinary bladder, rectum, prostate, (*e*) adrenal glands.

Varicose vein is the term applied to a superficial vein that is overdistended, irregular, and tortuous. *Hemorrhoids* are varicose veins in the rectum. Principal causes of varicose veins are *weakened valves* (because of increased pressure in the vessels) and *vessel blockage* (owing to thrombophlebitis).

Objective E To define *blood pressure* and to explain how it is measured and controlled.

Blood pressure is the force per unit area exerted by the blood against the inner walls of blood vessels; it is due primarily to the action of the heart. The body adjusts blood pressure by altering the heart rate (increased heart rate increases pressure), blood volume (increased volume increases pressure), and peripheral resistance (decreased vessel diameter increases resistance, and with it pressure). Normal blood pressure is about 120/80:

Systolic pressure	120 mmHg
Diastolic pressure	– 80 mmHg
Pulse pressure	40 mmHg

16.14 Compare the blood pressures in the different types of blood vessels.

The systolic blood and diastolic blood pressures are much greater in systemic arteries than in pulmonary arteries (fig. 16.7). Blood pressure decreases in arteries proportionate to their distance from the heart. Blood pressure is very low in capillaries and only slight in veins.

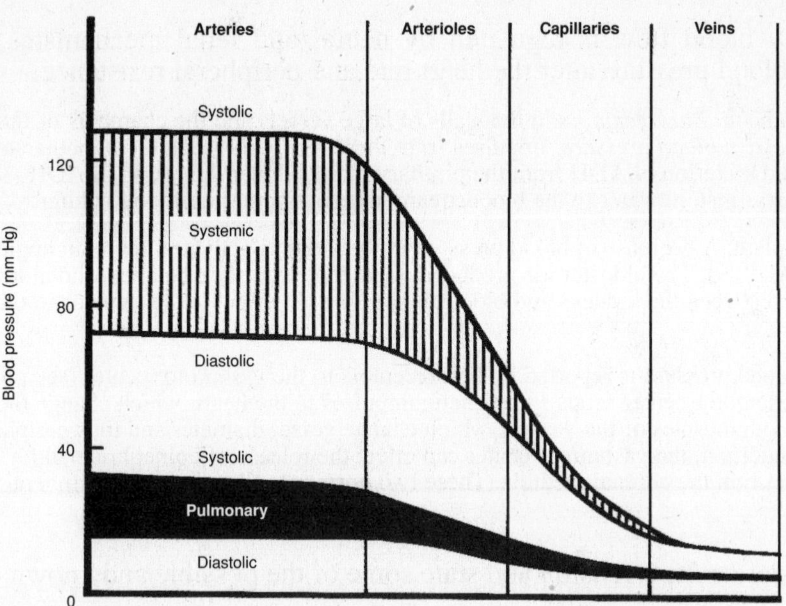

Figure 16.7 Relative blood pressures in systemic and pulmonary vessels.

16.15 Describe the use of the sphygmomanometer.

A cuff is wrapped around the arm and a stethoscope is placed over the brachial artery near the elbow. The cuff is inflated with air(the pump being a hand bulb) until the pressure is greater than the *systolic pressure*; this occludes (closes up) the artery, preventing blood flow to the lower arm. The pressure in the cuff is then lowered slowly. When it falls just below the systolic pressure, a turbulent flow of blood through the constricted area causes a sound to be heard in the stethoscope. A tapping sound is heard with each successive heartbeat as blood passes through the artery. The cuff pressure at which the first sound is heard is the *systolic pressure*. As the pressure in the cuff is further lowered, the tapping sounds get louder, then become softer and muffled, and finally disappear. These sounds are called *Korotkoff sounds*. The cuff pressure at the last sound is the *diastolic pressure*.

16.16 Make a sketch of the body, showing the pressure points where arterial pulsations can best be detected.

See fig. 16.8.

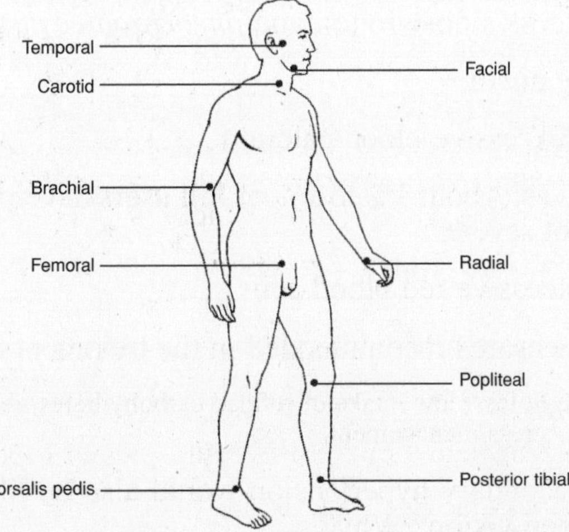

Figure 16.8 Pressure points for detection of arterial pulsations.

16.17 Explain how blood flow is regulated by neural and renal mechanisms, and how changes in blood pressure alter the heart rate and peripheral resistance.

Neural mechanism. *Baroreceptors* in the walls of large vessels and the chambers of the heart detect a *decrease in blood pressure*. Impulses from these receptors reach the hypothalamus, which elicits increased secretion of ADH from the pituitary gland. Under the action of ADH (see table 13.1), the kidneys restore water to the bloodstream, thereby increasing blood volume.

Renal mechanism. A decrease in blood pressure in the kidneys activates the renin angiotensin system (see fig. 13.9). The aldosterone produced alters the electrolyte balance, and along with it the water balance, between the kidneys and bloodstream. The net effect is the same as in the neural mechanism.

A change in blood pressure is reported by baroreceptors to the vasomotor center (see problem 10.23). The vasomotor center sends sympathetic impulses to the heart, which change the heart rate, and to the smooth muscles of the vessels, which change vessel diameter and thus peripheral resistance. In addition, the vasomotor center can effect the release of epinephrine and norepinephrine from the adrenal medulla. These two hormones likewise alter peripheral resistance and heart rate.

Objective F To define *hypertension* and state some of the possible and known causes for this condition.

Hypertension is a sustained elevation of the systemic arterial pressure. It is generally characterized by a systolic pressure that exceeds 160 mmHg and a diastolic pressure that exceeds 95 mmHg. Hypertension is classified as either of two types. **Primary** (essential) **hypertension** accounts for 85% to 90% of all cases and occurs without any known cause. It is found more often in females than in males, more often in black people than in white people, and runs in some families. Excessive salt intake, obesity, an unusually large volume of body fluids, psychoemotional stress, faulty responses of baroreceptors, and increased sensitivity of vessels to catecholamines are associated with primary hypertension but are not known to be causative factors. **Secondary hypertension**, which accounts for the remaining 10% to 15% of cases, is due to identifiable disorders:

Renal diseases. These include *renal ischemic disease* (narrowing of the renal arteries), *glomerulonephritis,* and *pyelonephritis.*

Adrenal diseases. These include Cushing's syndrome (see problem 13.25), primary *aldosteronism* (excess aldosterone), and *pheochromocytoma* (see problem 13.25).

Narrowing of the aorta.

Hypercalcemia. Excessive blood calcium.

Oral contraceptives. About 1% to 5% of Pill users develop elevated blood pressure, but it is usually not severe.

Polycythemia. Excessive red blood cells.

16.18 List the general measures recommended in the treatment of hypertension.

Regular exercise, weight loss, low intake of refined carbohydrates, restriction of salt intake, cessation of smoking, stress management.

16.19 Which cause of secondary hypertension would also be likely to produce headache, vertigo, and dimmed vision? Why?

Pheochromocytoma because of the elevated epinephrine secretion and excretion.

Objective G To define *arteriosclerosis* and to explain why this condition is considered a serious health problem.

Arteriosclerosis is a generalized degenerative vascular disorder that results in a thickening and hardening of the vessel wall (hence the common name, "hardening of the arteries"). Soft masses of fatty materials accumulate on the inside of the arterial wall (*atherosclerosis*) and later undergo calcification and hardening. The altered wall presents a rough surface that attracts platelets and macromolecules and leads to proliferation of the smooth muscle cells of the tunica media. These changes in the tunica intima and the tunica media result in a narrowed lumen and decreased blood flow.

16.20 Are only the large arteries affected by arteriosclerosis?

No. Although arteriosclerotic lesions often occur in large arteries, such as the aorta, they also occur in medium and small arteries, such as the coronary, renal, mesenteric, and iliac arteries.

16.21 Although the cause of arteriosclerosis is not yet well understood, the disease does appear to be positively correlated with six of the following conditions and negatively correlated with one of them: (*a*) high intake of saturated fats, (*b*) high intake of refined carbohydrates, (*c*) elevated blood pressure, (*d*) regular sustained exercise, (*e*) cigarette smoking, (*f*) obesity, (*g*) family history of heart disease. Which is the odd condition?

(*d*) Vigorous exercise for at least 30 minutes three times a week is recommended to sustain a healthy cardiovascular system.

Cerebrovascular disease is one of the most common neurological disorders in adults. It is usually the result of atherosclerosis and/or hypertension. The culmination of cerebrovascular disease is a *stroke*. Common symptoms of a stroke include darkening of vision, numbness, tingling or weakness on one side of the body, and a staggering gait.

Review Exercises

Multiple Choice

1. As compared to arteries, veins (*a*) contain more muscle, (*b*) appear more rounded, (*c*) stretch more, (*d*) are under a greater pressure.

2. Return of blood to the heart is *not* facilitated by (*a*) venous valves, (*b*) the skeletal-muscle pump, (*c*) skeletal-muscle groups, (*d*) venous pressure.

3. Resistive vessels of the circulatory system are (*a*) large arteries, (*b*) large veins, (*c*) small arteries and arterioles, (*d*) small veins and venules.

4. Discontinuous, or fenestrated, capillaries are found in (*a*) muscles, (*b*) adipose tissue, (*c*) the CNS, (*d*) the small intestine.

5. Compared to veins, arteries contain a thicker (*a*) endothelium, (*b*) tunica intima, (*c*) tunica media, (*d*) tunica adventitia.

6. The blood vessels that are under the greatest pressure are (*a*) large arteries, (*b*) small arteries, (*c*) veins, (*d*) capillaries.

7. Capillaries provide a total surface area of (*a*) 50 ft^2, (*b*) 700 m^2, (*c*) 7500 ft^2, (*d*) 1 square mile.

8. Interstitial fluid enters capillaries at the venular end through the action of (*a*) negative pressure, (*b*) colloid osmotic pressure, (*c*) active transport, (*d*) capillary pores.

9. A hormone that is significantly involved in regulation of blood volume is (*a*) ACTH, (*b*) osmoretic hormone, (*c*) ADH, (*d*) LH.

10. Edema is *not* caused by (*a*) high blood pressure, (*b*) increased plasma protein concentration, (*c*) leakage of plasma proteins into interstitial fluid, (*d*) obstruction of lymphatic drainage.

11. A person with a blood pressure of 135/75 has a pulse pressure of (*a*) 60, (*b*) 80, (*c*) 105, (*d*) 210.

12. Arteries are (*a*) strong, rigid vessels that are adapted for carrying blood under high pressure; (*b*) thin, elastic vessels that are adapted for transporting blood through areas of low pressure; (*c*) elastic blood vessels that form the connection between arterioles and venules, (*d*) strong, elastic vessels that are adapted for carrying blood under high pressure.

13. The innermost layer of an artery is composed of (*a*) stratified squamous epithelium, (*b*) simple cuboidal epithelium, (*c*) simple columnar epithelium, (*d*) endothelium.

14. The tunica externa is relatively thin and consists chiefly of (*a*) collagenous fibers, (*b*) elastic fibers, (*c*) loose connective tissue, (*d*) epithelium.

15. The vasa vasorum are minute vessels within (*a*) the tunica adventitia, (*b*) the tunica intima, (*c*) the tunica media, (*d*) the metaarterioles.

16. Sympathetic impulses to the smooth muscles in the walls of arteries and arterioles produce (*a*) vasodilation only, (*b*) vasodilation and vasoconstriction, (*c*) vasomotor inhibition, (*d*) arteriosclerosis.

17. Substances exchanged at the capillary level move through the capillary walls primarily by (*a*) diffusion, (*b*) filtration, (*c*) osmosis, (*d*) active transport.

18. In the brain, the endothelial cells of the capillary walls are more tightly fused than they are in other body regions. This permits the effective operation of (*a*) the precapillary sphincters, (*b*) the astrocytes, (*c*) the blood-brain barrier, (*d*) the impermeable membrane region.

19. The substances in the blood that help to maintain the osmotic pressure are (*a*) lipids, (*b*) plasma proteins, (*c*) lipid-soluble vitamins, (*d*) histamines.

20. Which venous layer is poorly developed? (*a*) tunica adventitia, (*b*) tunica intima, (*c*) tunica media

21. The accumulation of soft masses of fatty materials, particularly cholesterol, on the inside of the arterial wall is known as (*a*) ischemia, (*b*) atherosclerosis, (*c*) arteriosclerosis, (*d*) phlebitis.

22. In the measurement of blood pressure, the cuff of the sphygmomanometer usually surrounds (*a*) the radial artery, (*b*) the dorsalis pedis artery, (*c*) the brachiocephalic trunk, (*d*) the subclavian artery, (*e*) the brachial artery.

23. If the blood pressure of an individual is measured at 125 over 81, the approximate mean arterial pressure would be (*a*) 206, (*b*) 44, (*c*) 103, (*d*) 96.

24. Arterial blood pressure is independent of (*a*) blood volume, (*b*) heart rate, (*c*) peripheral resistance, (*d*) blood viscosity, (*e*) an influx of calcium ions.

25. Identify the *true* statement(s):
(*a*) An increased cardiac output is reflected in an elevated diastolic pressure.
(*b*) An increased cardiac output is reflected in a decreased diastolic pressure.
(*c*) An increase in the force of ventricular contraction produces an elevated systolic pressure.
(*d*) An increase in the force of ventricular contraction produces a decreased systolic pressure.

True or False

____ **1.** All capillaries have the same fluid exchange rate because they have similar patterns of fenestration in the endothelium.

____ **2.** The tunics of both arteries and veins consist of three layers.

____ **3.** To facilitate a high metabolic rate, the capillaries in the brain are characterized by a fenestrated endothelium.

____ **4.** Factors influencing capillary exchange include surface area, fenestrations, capillary pressure, and blood osmotic pressure.

____ **5.** The superior and inferior phrenic arteries serve the diaphragm.

____ **6.** The internal jugular veins drain blood from the brain and meninges.

____ **7.** During exercise, the diastolic pressure is greater than the systolic pressure.

____ **8.** Pulse pressure is the difference between the systolic and diastolic pressures.

____ **9.** Baroreceptors monitor changes in blood oxygen and carbon dioxide levels.

____ **10.** Hypertension is classified as being alpha (essential) and beta (nonessential).

Completion

1. The hepatic, splenic, and left gastric arteries arise from the ________________ trunk.

2. Branching from the common iliac arteries, the ________________ ________________ arteries serve the external reproductive organs and the gluteal muscles.

3. Three major vessels arise from the aortic arch: the ________________ trunk, the ________________ ________________ ________________ artery, and the ________________ ________________ artery.

4. Venous blood returning from the arm passes through the brachial vein to the ________________ vein, and then to the subclavian vein.

5. ________________ are varicose veins in the rectum.

6. The ________________ ________________ vein is the preferred site for venipuncture.
7. Systolic pressure minus the diastolic pressure is the ________________ pressure.
8. ________________ is the sustained elevation of the systemic arterial pressure.
9. Renal ________________ is caused by a narrowing of the renal arteries.
10. The tunica ________________ is the outer connective tissue layer of blood vessels.

Labeling

Label the arteries indicated on the figure to the right.

1. ______________________
2. ______________________
3. ______________________
4. ______________________
5. ______________________
6. ______________________

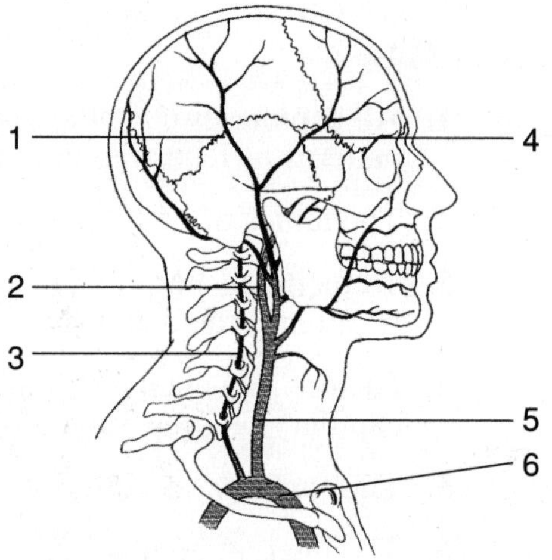

Answers and Explanations for Review Exercises

Multiple Choice

1. (*c*) The thin tunics of veins enable them to distend.
2. (*d*) The blood pressure in veins is near zero.
3. (*c*) Smooth muscles in small arteries and arterioles regulate blood flow to specific parts of the body during physiological adaptation to changing conditions.
4. (*d*) Discontinuous, or fenestrated, capillaries along the GI tract permit absorption of nutrients.
5. (*c*) The tunica media is much thicker in arteries than in veins. It is the autonomic contraction of the smooth muscles in this layer that is responsible for diastolic pressure.
6. (*a*) Blood pressure is greatest as blood leaves the heart and enters the large arteries. The pressure drops as blood passes through the remaining vessels. Blood pressure is near zero as it returns to the heart.
7. (*b*) It is estimated that the total length of capillaries if put end to end would be about 60,000 miles.
8. (*b*) The osmotic pressure is greater than the fluid hydrostatic pressure at the venular end of capillaries; thus, interstitial fluid enters the capillaries.
9. (*c*) ADH regulates total body fluid and thus blood volume by regulating the amount of urine formed.
10. (*b*) Decreased plasma protein (blood osmotic pressure) will lead to edema.
11. (*a*) Pulse pressure is the difference between the systolic and diastolic pressures.
12. (*d*) The tunica media of arteries is characterized by smooth muscle tissue and abundant elastic and collagenous fibers.

13. (*d*) The innermost layer of blood vessels is composed of simple squamous epithelium, which is referred to as *endothelium.*
14. (*c*) The tunica externa consists mainly of loose connective tissue that protects the blood vessel and anchors it to surrounding structures.
15. (*a*) The vasa vasorum are specialized microscopic vessels that provide blood to the tunics of large vessels.
16. (*b*) Sympathetic stimulation may initiate both vasodilation and vasoconstriction. For example, in a "fight-or-flight" response, blood flow to the skeletal muscles increases (vasodilation), whereas blood flow the GI tract decreases (vasoconstriction).
17. (*a*) Most exchange in capillaries takes place by diffusion, although some substances move across the capillary wall by other transport mechanisms.
18. (*c*) The blood-brain barrier inhibits (as a protective mechanism) the movement of some substances into brain tissue.
19. (*b*) Plasma proteins (mainly albumin) play a major role in the regulation of blood osmotic pressure.
20. (*c*) The tunica media is very thin in veins (hence, the lumina of veins are larger than those of arteries); the other two tunics are similar in arteries and veins.
21. (*b*) Atherosclerosis is the accumulation of soft masses of fatty material on the inside of the arteries, whereas arteriosclerosis is a generalized degenerative disorder that results in a thickening and hardening of vessels.
22. (*e*) The brachial artery provides an accessible pressure for the cuff of a sphygmomanometer. By using the same site, valid comparisons can be made to a standard norm.
23. (*c*) $125+81 = 206/2 = 103$
24. (*e*) Physiological changes in Ca^{2+} do not affect blood pressure.
25. (*a* and *c*) As the heart contracts more forcefully (increased cardiac output), there is an increase in both the systolic and diastolic pressures.

True or False

1. False; the amount of capillary fenestration varies with the function of the tissue or organ being served
2. True
3. False; as part of the blood-brain barrier, the capillaries in the brain lack fenestrations
4. True
5. True
6. True
7. False; blood pressure is an expression of the higher systolic pressure over the lower diastolic pressure
8. True
9. False; baroreceptors respond to changes in blood pressure
10. False; hypertension is classified as primary and secondary

Completion

1. celiac
2. internal iliac
3. brachiocephalic, left common carotid, left subclavian
4. axillary
5. Hemorrhoids
6. median cubital
7. pulse
8. Hypertension
9. ischemia
10. adventitia

Labeling

1. Posterior temporal
2. Internal carotid
3. Vertebral
4. Anterior temporal
5. Right common carotid
6. Brachiocephalic trunk

Lymphatic System and Body Immunity 17

Objective A To describe the functional relationship between the *lymphatic system* and the cardiovascular system.

Functioning together with the cardiovascular system, the lymphatic system (1) transports interstitial (tissue) fluid, called *lymph* in lymph vessels, from the tissues back to the blood, where it contributes to blood plasma; (2) assists in fat absorption in the small intestine; and (3) plays a key role in protecting the body from bacterial invasion via the blood.

17.1 What is edema?

Much of the fluid of the body (approximately 11%) surrounds the cells in body tissues as interstitial fluid. Excessive accumulation of interstitial fluid is known as *edema*.

17.2 Which of the following could *not* be a contributing cause of edema?

(*a*) Obstruction of lymphatic drainage
(*b*) Increased intravascular volume
(*c*) Leakage of plasma proteins into the interstitial fluid (or a decrease in plasma protein concentration by some other means)
(*d*) Allergy

All of the above are potential causes of edema. (*a*) The condition *elephantiasis* is caused by a tropical nematode parasite that blocks lymphatic drainage. (*b*) Increased intravascular volume (caused by excessive salt and water intake) produces elevated hydrostatic pressure in the venous system. (*c*) Lowered concentrations of plasma proteins—perhaps because of liver or kidney disease—provokes osmosis of blood plasma into the interstitial fluid. (*d*) Chemical mediators associated with allergic reactions cause leaking of capillary fluid and proteins.

Congestive heart failure is a major cause of edema in elderly people. In the healthy individual, the heart is able to pump the entire intravascular volume through the circulatory system without any pooling in the veins or lymph vessels. A diseased or damaged heart (e.g., from heart attack, chronic hypertension, or valvular disease) cannot generate sufficient cardiac output to push the total volume of blood through the circuit of arteries, capillaries, and veins. Venous and lymphatic circulation becomes congested as hydrostatic pressure rises, leading to serum extravasation into the extravascular space. This occurs most commonly in the gravity-dependent lower limbs, and results in swelling, or edema, in the feet and ankles. The goals of therapy are to decrease functional intravascular volume by reducing salt intake and to remove excess leaked fluids by using medications (diuretics) that increase urine output.

Objective B To specify the *routes of fluid transport* in the lymphatic system.

Interstitial fluid enters the lymphatic system through the walls of *lymph capillaries,* composed of simple squamous epithelium. From merging lymph capillaries, the lymph is carried into large *lymph ducts.* Interconnecting lymph ducts eventually empty into one of two principal vessels: the **thoracic duct** and the **right lymphatic duct**. These drain into the left subclavian vein and right subclavian vein, respectively (fig. 17.1).

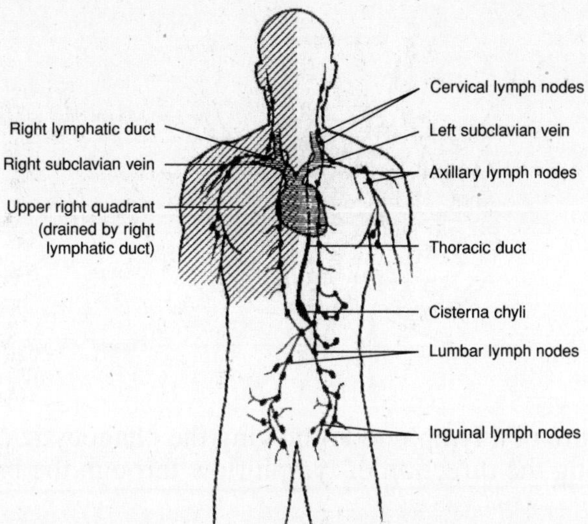

Figure 17.1 Lymph from the right upper extremity (shaded) drains through the right lymphatic duct into the right subclavian vein. Lymph from the remainder of the body drains into the thoracic duct and into the left subclavian vein.

17.3 Which two body regions are drained by the two principal lymphatic vessels?

The *right lymphatic duct* drains lymph from the upper right quadrant of the body (shaded area in fig. 17.1). The larger *thoracic duct* drains lymph from the remainder of the body.

17.4 Compare lymph ducts and veins with regard to structure.

Although thinner, the walls of lymph ducts are similar to those of veins in that they have the same three tunics (layers) and contain valves to prevent backflow.

17.5 What is the cisterna chyli, and how does it relate to lacteals?

The **cisterna chyli** is a saclike enlargement of the thoracic duct in the abdominal region. **Lacteals** are specialized lymph capillaries within the villi of the small intestine (see fig. 19.11); they transport certain products of fat absorption out of the GI tract into the cisterna chyli.

17.6 What causes lymph to flow through lymph vessels?

Involuntary contraction of skeletal muscles (tonus), intestinal peristalsis, and skeletal muscle contraction during body movement massage the lymph vessels. Gravity also aids the flow of lymph.

Metastasis of *cancer* frequently uses the route of the lymphatic system. Because of this, the path of lymph flow is clinically important. When cancer is detected, the surrounding lymph nodes are generally biopsied to determine the extent of metastasis. Once in the lymphatic system, cancer spreads rapidly to other body organs causing secondary cancerous sites.

Objective C To describe the structure and function of *lymph nodes*.

Lymph nodes are small oval bodies enclosed in fibrous capsules (fig. 17.2). They contain phagocytic *cortical tissue* (reticular tissue) adapted to filter lymph. Specialized bands of connective tissue, called *trabeculae*, divide the lymph node. *Afferent lymphatic vessels* carry lymph into the node, where it is circulated through the cortical sinuses. The filtered lymph leaves the node through the *efferent lymphatic vessels*, which merge through the concave *hilum*.

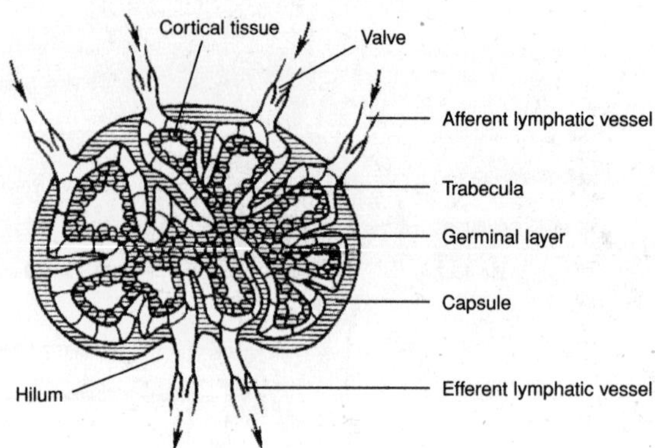

Figure 17.2 The structure of a lymph node showing the phagocytic cortical tissue. (Note the arrows indicating the direction of lymph flow through the lymph node.)

17.7 What is the function of the germinal layer of a lymph node?

The *germinal layer* harbors **lymphocytes**. Lymphocytes are leukocytes (white blood cells) that are responsible for body immunity. They have large nuclei, long life spans, and account for about one-fourth of all leukocytes.

17.8 What are macrophages?

Macrophages are large phagocytic cells found in lymphatic cortical tissue. They engulf and destroy foreign substances, damaged cells, and cellular debris before these materials can enter the blood. Thus, two major functions of lymph nodes are the harboring of lymphocytes and the macrophagic cleansing of lymph.

Lymphoid leukemia is a form of cancer characterized by an uncontrolled production of lymphocytes that remain immature. These leukemic cells eventually appear in such great numbers that they crowd out the normal, functioning cells. Chemotherapeutic drugs are fairly effective in treating lymphoid leukemia.

Objective D To chart the distribution of lymph nodes.

Lymph nodes usually occur in clusters or chains (fig. 17.3). Some of the principal groupings are the *popliteal* and *inguinal nodes* of the lower extremity, the *lumbar nodes* of the pelvic region, the *cubital* and *axillary nodes* of the upper extremity, and the *cervical nodes* of the neck. Clusters of *mesenteric nodes* (Peyer's patches) are associated with the small intestine.

During a physical examination, the physician will palpate (feel with firm pressure) the cervical and axillary lymph nodes. In conducting a breast self-examination, a woman palpates for abnormal lumps and tender areas. A detected lump may be an enlarged lymph node.

17.9 Are the tonsils lymph nodes?

The three pairs of **tonsils**—pharyngeal (adenoids), palatine, and lingual—are not specifically lymph nodes, but are lymphatic organs of the pharyngeal region. The function of the tonsils is to combat infections of the ear, nose, and throat regions. Swollen tonsils may interfere with breathing and make swallowing difficult. Children who are constant mouth–breathers frequently have enlarged pharyngeal tonsils.

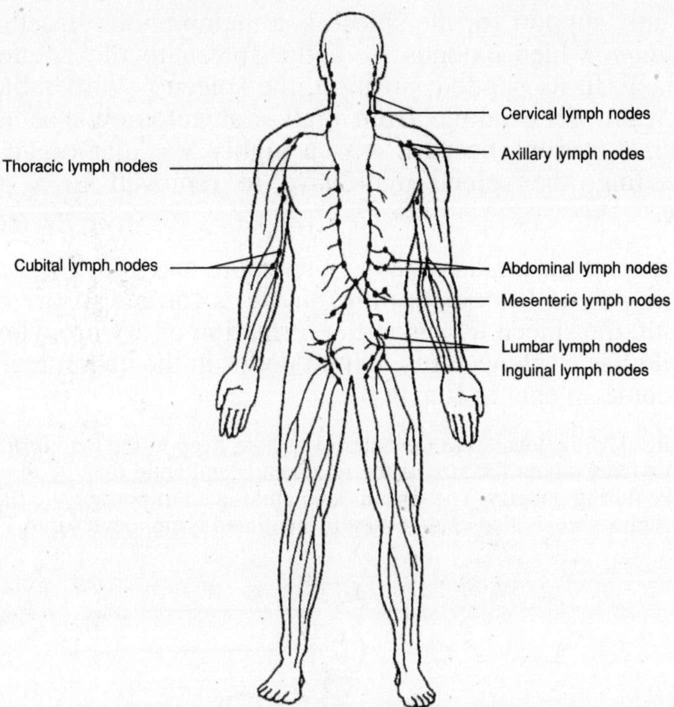

Figure 17.3 The location of the principal groups (clusters) of lymph nodes.

The tonsils tend to become swollen and inflamed after persistent infections. They may have to be surgically removed when they become so overrun with pathogens after repeated infections that they themselves become the prime sources of infection. When this happens, removal of certain tonsils may be necessary. The removal of the palatine tonsils is called a *tonsillectomy*, whereas the removal of the pharyngeal tonsils is called an *adenoidectomy*.

17.10 Characterize the spleen and the thymus as lymphoid organs.

The **spleen** (fig. 17.4) is located in the upper left portion of the abdominal cavity, beneath the diaphragm, and is suspended from the stomach. The spleen is not a vital organ in an adult, but it does assist other body organs in producing lymphocytes, filtering blood, and destroying old and worn erythrocytes (red blood cells). In addition, the spleen is a reservoir for erythrocytes.

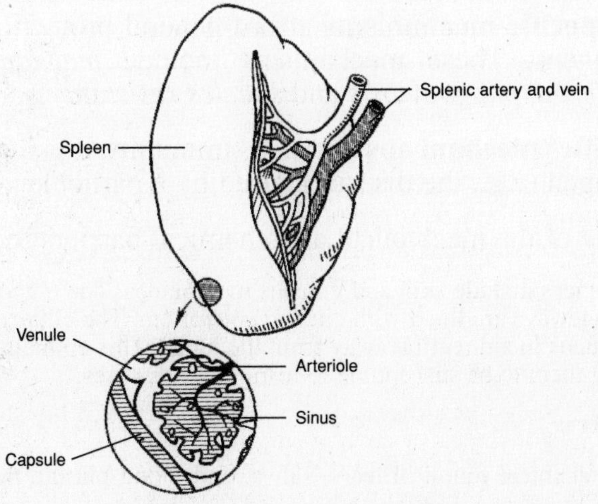

Figure 17.4 The spleen.

The only support for the spleen is a membranous structure called the *lesser omentum*, which extends from the spleen to the greater curvature of the stomach. In its pendent position, the spleen is vulnerable to trauma, which may result, for example, from a fall or an automobile accident. Trauma to the spleen is serious because it is a highly vascular organ. To avoid profuse internal bleeding, the spleen may have to removed in a procedure called a *splenectomy*.

The spleen is a vital organ in a newborn and a prepubescent child. Until the hemopoietic (blood-forming) tissue is formed in the bone marrow of an adult, the spleen assists in the formation of erythrocytes. Interestingly, if a child has a splenectomy, lymph nodes in the abdominal cavity enlarge and become splenic in function.

The **thymus** (fig. 17.5) is located in the anterior thorax, deep to the manubrium of the sternum. It is much larger in a fetus (about the size of the fetal heart) and child than in an adult because it regresses in size during puberty. The thymus of a child is an important site of immunity and is a reservoir of lymphocytes. It also changes undifferentiated lymphocytes into T lymphocytes.

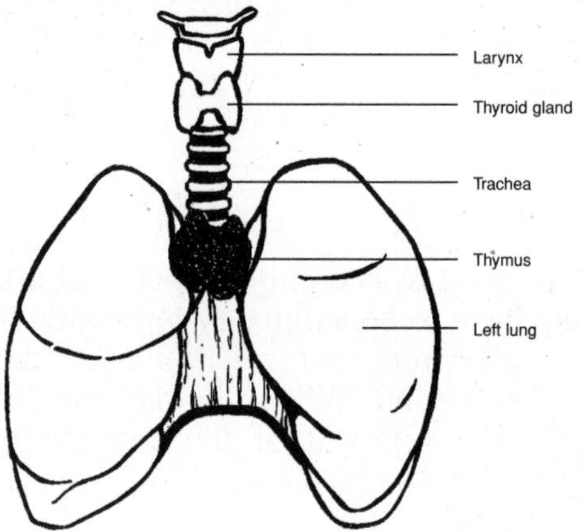

Figure 17.5 The location of the thymus.

Objective E To distinguish between *specific* and *nonspecific defenses against infection* and to describe some *barriers to infection*.

Nonspecific mechanisms afford general protection against many types of pathogens. These mechanisms include *mechanical barriers, enzymes, interferon, phagocytosis* and *species resistance*.

Specific mechanisms furnish immunity to the effect of a particular pathogen (e.g., the disease caused by a particular virus).

17.11 What are some of the mechanical and chemical barriers to infection?

Mechanical barriers include skin and mucous membranes. The mucous membranes in the respiratory passageways are lined with ciliated epithelium. The cilia continuously move particles trapped in the mucus in a direction away from the lungs. This epithelium is eventually destroyed in smokers, causing them to be susceptible to respiratory diseases.

Chemical barriers:

Lysozyme — A chemical found in tears, saliva, and blood plasma that breaks down bacterial cell walls.

Pepsin — An enzyme in the stomach that lyses (disintegrates) many microorganisms.

Hydrochloric acid — Secreted by the parietal cells in the stomach, it creates a low pH that is lethal to many pathogens.

Complement — A series of enzymatic proteins that are activated by both specific and nonspecific mechanisms.

Interferon – Any of a group of proteins that are produced by virus-infected cells and some immune system cells, inhibiting viral growth.

17.12 What kind of cells provide a second line of nonspecific defense if the mechanical and chemical barriers have been breached?

Phagocytes, which include *neutrophils, monocytes,* and *macrophages* are all cells that provide a second line of defense. **Natural killer cells** also help by releasing enzymes that punch holes in cell membranes.

Objective F To define *specific immunity* and to explain how it may be acquired.

Specific immunity refers to the resistance of the body to specific foreign agents (antigens). These include microorganisms, viruses, and their toxins, as well as foreign tissue and other substances.

17.13 What are the two ways in which immunity may be acquired by the body?

Antibody-mediated immunity. An antigen stimulates the body to produce special proteins, called *antibodies*, which can lead to the destruction of a particular antigen through an *antigen-antibody reaction*. The antibodies serve as the main weapon against invasion.

Cell-mediated immunity. Lymphocytes may become sensitized to an antigen, attach themselves to that antigen, and destroy it. In this case, cells provide the main defensive strategy.

17.14 Why is it important that the immune system be able to discriminate "self' from "nonself" antigens?

In order for the immune system to effectively rid the body of foreign invaders and yet not harm normal body cells, it must distinguish between what is and is not self. Failure to make this distinction, or the inability to launch a "nonself only" immune response, may result in an autoimmune disease (see problem 17.29).

17.15 What are the chemical characteristics of antigens?

Antigens are usually large (MW > 10^4), complex molecules, for example, proteins, polysaccharides, and mucopolysaccharides. These antigens can be found on bacterial cell walls, on cell membranes, and on viruses, or they can be free-floating. Introduction of a foreign antigen often, but not always, stimulates an immune response.

17.16 What are the chemical characteristics of antibodies?

Antibodies are *gamma globulins* composed of four interlinked polypeptide chains, two short (*light*) chains and two long (*heavy*) chains (fig. 17.6). All antibodies have portions that are structurally similar, called *constant regions*, and portions that are highly variable, called *variable regions*. The antibody's antigen-binding sites are located on the variable portion of the antibody. Small variations in the variable regions make each antibody highly specific for one particular antigen. Binding of the antigen to its specific antibody induces production of more antibodies specific to that antigen.

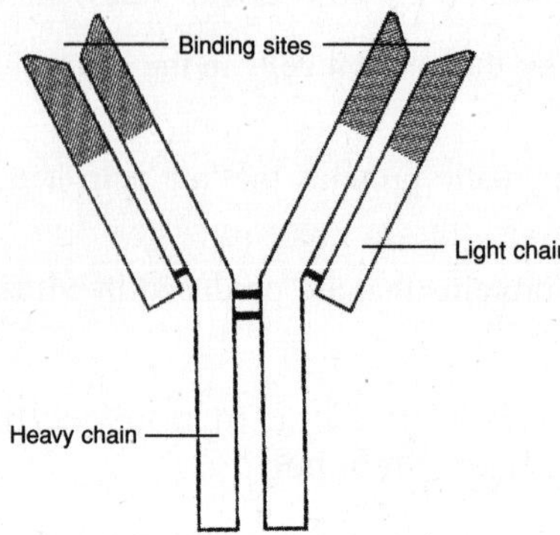

Figure 17.6 A simple model of an antibody showing binding sites and light and heavy chains. Shaded areas indicate variable regions.

17.17 What are the five main classes or isotypes of antibodies (immunoglobulins) produced by the immune system?

IgG The most abundant class and very specific for its complementary antigens; can cross the placenta.

IgM Found in higher numbers when the body first encounters an antigen; the largest immunoglobulin class (structurally a pentamer), but not as specific as IgG.

IgA Inhibits the entrance of antigens into the body; found in nasal, salivary, lacrimal, bronchial, intestinal, and vaginal secretions.

IgE Aids in immunity against parasitic worms and other parasites; also mediates allergic responses and causes degranulation of mast cells, with release of heparin, histamine, and vasoactive substances.

IgD Still of uncertain function.

17.18 Does vaccination against a disease confer active or passive immunity?

Active immunity is conferred when the body manufactures antibodies in response to direct contact with an antigen. When an individual is again exposed to the antigen, the body "remembers" it and mounts a quicker and more specific antibody response to that antigen. Active immunity can be conferred by exposure to the whole antigen (such as the chicken pox virus), or by vaccination with dead or weakened pathogens or altered toxins.

Passive immunity is conferred by the transfer of antibodies from one person to another; the recipient does not produce his or her own antibodies. For example, a gamma globulin shot (another individual's antibodies) can confer passive immunity against hepatitis A. As another example, a fetus receives IgG across the placenta from the mother. This passive immunity helps the newborn to fight disease before its own immune system has developed.

Objective G To identify the *components of the immune system* and to describe cell-mediated immunity.

The immune system is composed of lymphocytes (*T lymphocytes* and *B lymphocytes*), substances released from lymphocytes (antibodies and *cytokines*), complement, macrophages, and various other cell types and substances. Figure 17.7 shows the development of the two kinds of lymphocytes, and fig. 17.8 is a scheme of the immune system as a whole.

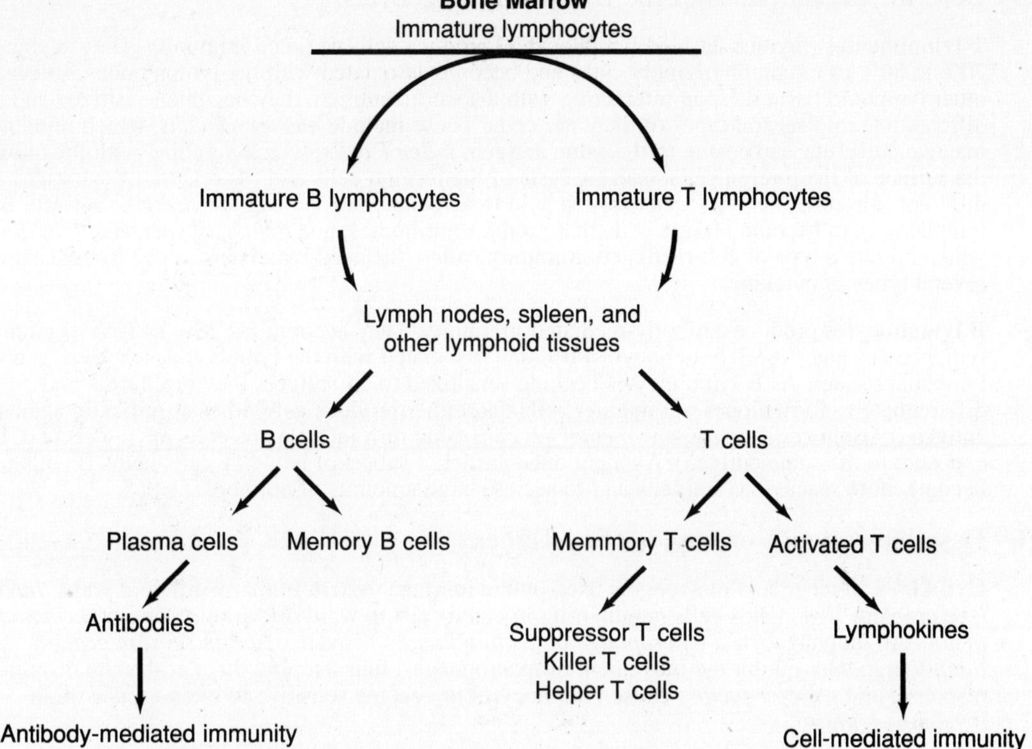

Figure 17.7 Development of T and B lymphocytes.

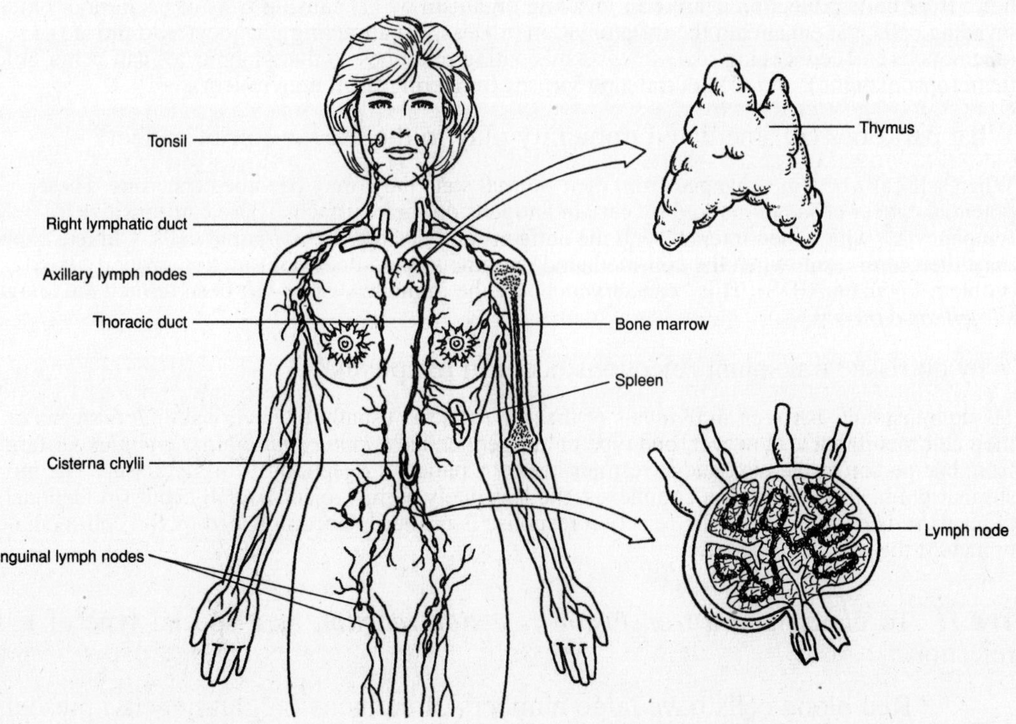

Figure 17.8 Representation of the immune system as a whole.

17.19 Describe the functions of the B and T lymphocytes.

T lymphocytes (thymus-derived lymphocytes) produce cell-mediated immunity. They account for 70% to 80% of circulating lymphocytes and become associated with the lymph nodes, spleen, and other lymphoid tissues. Upon interacting with a specific antigen, they become sensitized and differentiate into several types of daughter cells. These include *memory T cells*, which remain inactive until future exposure to the same antigen; *killer T cells*, which combine with the antigen on the surface of the foreign cells, causing lysis of the foreign cells and the release of cytokines; different subsets of *helper T cells*, which help to activate other T lymphocytes or to activate B lymphocytes to become plasma cells that produce antibodies; and *delayed-hypersensitivity T cells*, which initiate a type of cell-mediated immunity called "delayed hypersensitivity" by releasing several types of cytokines.

B lymphocytes produce antibody-mediated immunity. They account for 20% to 30% of circulating lymphocytes and (like T lymphocytes) become associated with the lymph nodes, spleen, and other lymphoid tissues. As B lymphocytes become sensitized to an antigen, they proliferate and differentiate to form clones of daughter cells that either produce antibodies specifically against that antigen (*plasma cells*) or become *memory B cells* (that turn into plasma cells upon a second, later exposure to the same antigen). As mentioned earlier, a subset of helper T cells helps B cells to become more reactive to antigens and to secrete large amounts of antibodies.

17.20 Describe and give examples of cytokines.

Cytokines are chemical messengers used by the immune system in many different ways. *Interferon* (see problem 17.11) aids cells neighboring infected cells to ward off viral infection; *chemotactic factors* attract phagocytes; *macrophage-activating factors* activate macrophages; *migration-inhibiting factors* inhibit the movement of macrophages, thus keeping them at the site of immune response; and *transfer factors* cause lymphocytes to become sensitive to the presence of an invading organism.

17.21 What role does complement play in the immune response?

The activated complement system, composed of several enzyme precursors (see problem 17.11), helps to provide protection against an invading organism by (1) causing lysis of bacteria or other invading cells, (2) enhancing the inflammation process, (3) attracting phagocytes to the area (chemotaxis), (4) enhancing *phagocytosis* by coating microbes so that phagocytes can better hold them (opsonization), and (5) neutralizing viruses (rendering them nonvirulent).

17.22 What part does cell-mediated immunity play in the prevention of cancer?

When self cells become changed from their normal state they may become cancerous. These potential cancer cells are marked by certain antigens on their surface. These can sensitize T lymphocytes, which then interact with the antigens and destroy the abnormal cells. Clinical cancer may, therefore, result when the cell-mediated immune system does not function properly (see problem 17.32 on AIDS). This "cancer watch" by the immune system has been termed the *immune surveillance theory*.

17.23 Why do tissue transplant rejections occur in recipients?

Tissue transplants between individuals of the same species usually have very similar antigens on their cell membranes. However, one type of antigen, called a *histocompatibility complex,* differs from one person to the next and is responsible for a patient's rejection of foreign tissue. The more similar two histocompatibility complexes, the less likely a graft rejection will occur (in identical twins there is no rejection). The rejection response is primarily accomplished by the cell-mediated branch of the immune system.

Objective H To understand a *transfusion rejection reaction* as a special type of tissue rejection.

Red blood cells have large numbers of antigens (agglutinogens) present on their cell membrane; these can initiate antibody (agglutinin) production, and therefore antigen-antibody reactions. One of the groups of antigens most likely to cause blood transfusion reactions is the **ABO system**.

Table 17.1 The ABO Antigen System of Blood

Genotype	*Blood group*	*Antigens (agglutinogens)*	*Antibodies (agglutinins)*
OA or AA	A	A	Anti-B
OB or BB	B	B	Anti-A
AB	AB	A and B	*none*
OO	O	*none*	Anti-A and Anti-B

17.24 Antigens of the ABO system are inherited factors and are present on the red blood cell membranes at the time of birth. Can the same be said of the corresponding antibodies?

No. The antibodies begin to appear about 3 to 8 months after birth and reach maximum concentrations at about 10 years of age. This phenomenon is not yet completely understood.

17.25 What happens if the recipient of a blood transfusion and the donor(s) are improperly matched?

An antigen-antibody reaction (*transfusion reaction*) occurs in the recipient, causing the red cells to clump together (*agglutinate*). The clumps may block or occlude small vessels throughout the body, hindering the flow of blood. The red cells may also rupture (*hemolyze*) and release hemoglobin into the plasma. A severe transfusion reaction will raise plasma bilirubin levels and lead to jaundice. In extreme cases, renal tubular damage, anuria, and death are possible.

17.26 How are blood groups matched to reduce the likelihood of a transfusion reaction?

See table 17.2.

Table 17.2 Preferred and Permissible Blood Types for Transfusions (AB = Universal Recipient; O = Universal Donor)*

Recipient's blood type	*Preferred donor's blood type*	*Permissible donor's blood type*
A	A	O
B	B	O
AB	AB	A, B, O
O	O	(*only* O)

* It should be noted that the mixing of type AB blood or type O blood should be made with caution and in small amounts. One reason is that blood groups other than the ABO group must be matured for compatability.

Objective I To describe the events leading to *erythroblastosis fetalis*.

Erythroblastosis fetalis is a hemolytic disease of the newborn resulting from an antigen-antibody reaction associated with the *Rh system* of the blood. Rh antigens (first found in the rhesus monkey, and thus designated "Rh") are present on the red blood cell membranes of about 85% of the population. These people are classed as Rh positive (Rh^+). The remaining 15% are classed as Rh negative (Rh^-). Rh^- individuals do not develop antibodies against Rh antigens until they are exposed to Rh^+ blood.

17.27 List the steps in the development of erythroblastosis fetalis.

1. An Rh^- mother and an Rh^+ father have an Rh^+ baby.
2. At birth, some of the Rh^+ red cells from the baby enter the mother's circulation, usually as a result of a tear in the placenta.
3. As the Rh antigens are foreign to the mother, she begins to produce anti-Rh antibodies (a primary response).
4. The mother becomes pregnant again, and the fetus is Rh^+.
5. The anti-Rh antibodies cross the placenta and enter the fetal blood.
6. The anti-Rh antibodies from the mother react with the Rh antigens on the fetal red cells, causing agglutination and hemolysis.

17.28 How is erythroblastosis fetalis prevented?

The Rh^- mother receives an injection of anti-Rh antibodies (*Rhogam injection)* within 72 hours after delivery or abortion. These anti-Rh antibodies will tie up and destroy any absorbed Rh^+ cells and thus prevent the mother from being sensitized and producing her own anti-Rh antibodies. The antibodies the mother receives by passive immunity only last for several months; therefore, the fetus of the next pregnancy is protected.

Objective J To give examples of *autoimmune diseases* and to explain, in general terms, why these diseases occur.

A variety of autoimmune disorders affects many different tissues in the body. These disorders may be caused by an immune cell's inability to correctly distinguish between self and nonself or by a nonspecific hyperactive response from the immune system.

17.29 What are some examples of autoimmune diseases?

Rheumatoid arthritis. An immune response that frequently results in inflammation of the joints.

Systemic lupus erythematosis **(SLE).** A systemic immune response that affects organs throughout the body.

Insulin-dependent diabetes mellitus **(IDDM).** A disease caused by an autoimmune attack on the beta cells of the pancreas.

Grave's disease. Antibody-induced stimulation of the thyroid gland.

Multiple sclerosis **(MS).** An immune response against myelin in the CNS.

17.30 Why do autoimmune diseases occur?

Since it has been demonstrated that there are circulating lymphocytes in a healthy individual that react to the body's own antigens (for example, B cells that can bind thyroglobulin or DNA, and T cells that can respond to myelin protein or collagen), the question arises, Why don't autoimmune diseases occur as the rule rather than as the exception? Several theories have been postulated to explain this.

Sequestration. Many antigens in the body are effectively "hidden" from the immune system during a person's life. For example, the lens protein of the eye and the antigens of the spermatozoa are considered sequestered antigens that evoke an immune response only when they are introduced into the bloodstream by trauma or some other means.

Immunoregulation. T cells are suppressor agents against autoimmune processes. In experiments where radiation is used to eliminate the suppressor T cells in animals, autoimmune diseases occur with greater frequency.

Cross-reactive antigens. Some viruses and bacteria express antigens on their surfaces that are structurally similar to those normally occurring in body tissue. As the body mounts a response to the invading organism, it begins to recognize native antigens as nonself and destroy its own tissue.

Genetic predisposition. There is clearly a linkage between autoimmune disease and certain HLA haplotypes that results in familial tendency toward diseases such as rheumatoid arthritis and systemic lupus erythematosus (SLE). This mechanism is poorly understood.

17.31 What is AIDS and how does HIV cause it?

AIDS (acquired immuno-deficiency syndrome) is a disease that severely impairs the immune response. It is caused by HIV (acquired immunodeficiency virus), which has a particular affinity for helper T cells (one of the central cell types in the immune system). As the helper T cells are gradually destroyed or deactivated, both cell-mediated and humoral immune responses weaken.

17.32 What are the symptoms of an HIV infection?

An HIV-infected individual passes through various stages correlated with decreasing helper T cell counts. At first, mild flulike symptoms develop, which are often overlooked. In the next stage, numerous symptoms include persistent weight loss, fever, fatigue, night sweats, and enlarged lymph nodes. The diseases that follow in time are signs of full-blown AIDS. Among them are multiple and uncommon cancers (e.g., Kaposi's sarcoma) due to dysfunctional immune surveillance. Some patients develop severe dementia, and most finally succumb to cancer or overwhelming infection.

Review Exercises

Multiple Choice

1. The immune system is involved in (*a*) destruction of abnormal or mutant cell types that arise within the body, (*b*) allergic reactions, (*c*) rejection of organ transplants, (*d*) all of the preceding.

2. Active immunity is (*a*) borrowed from an active disease case, (*b*) developed in direct response to a disease agent, (*c*) the product of borrowed antibodies, (*d*) passive immunity that is activated.

3. In the cell-mediated immune response, T lymphocytes divide and secrete (*a*) antigens, (*b*) plasmogens, (*c*) collagens, (*d*) cytokines.

4. B lymphocytes are primarily involved in (*a*) humoral immunity, (*b*) autoimmune disorders, (*c*) graft rejection, (*d*) cell-mediated immunity.

5. Plasma cells are (*a*) responsible for specific immunity, (*b*) derived from B cells, (*c*) involved in the production of antibodies, (*d*) described by all of the preceding.

6. Transfusing a person with blood plasma proteins from a person or animal that has been actively immunized against a specific antigen provides (*a*) active immunity, (*b*) passive immunity, (*c*) autoimmunity, (*d*) anti-immunity.

7. A person with type AB blood has (*a*) both anti-A and anti-B antibodies, (*b*) only anti-O antibodies, (*c*) neither anti-A nor anti-B antibodies, (*d*) no antigens.

8. When an Rh^- mother and an Rh^+ father produce an Rh^- baby, (*a*) the mother may develop Rh antibodies unless she is treated with Rhogam within 72 hours after birth of the baby, (*b*) the baby will be born with a yellowish color, (*c*) the mother will not develop any Rh antibodies, (*d*) the baby will most likely have congenital defects.

9. Substances against which the body launches an immune response are called (*a*) antibodies, (*b*) antigens, (*c*) anticlines, (*d*) agglutinins.

10. The antibodies produced and secreted by B lymphocytes are soluble proteins called (*a*) immunoglobulins, (*b*) immunosuppressants, (*c*) lymphokines, (*d*) histones.

11. Which of the following is *not* a major organ of the lymphatic system? (*a*) lymph nodes, (*b*) thymus, (*c*) kidney, (*d*) spleen

12. Which is the proper order of events in cell-mediated immunity?
(*a*) Antigen enters tissue, macrophages engulf antigen, antigen presented to members of a clone of lymphocytes, sensitized T lymphocytes attack antigen-bearing agents
(*b*) Antigen enters tissues, antigen passed to members of a clone of lymphocytes, lymphocytes sensitized, macrophages engulf antigen, T lymphocytes attack antigen-bearing agents
(*c*) Antigen enters tissues, macrophages engulf antigen, antigen passed to members of a clone of lymphocytes, lymphocytes sensitized, B lymphocytes secrete antibodies that react with antigen-bearing agents
(*d*) Antigen enters tissues, lymphocytes sensitized, antigen passed to members of a clone of lymphocytes, macrophages engulf antigen, T lymphocytes attack antigen-bearing agents

13. A dilation of the lymphatic duct in the lumbar region that marks the beginning of the thoracic duct is (*a*) the cisterna chyli, (*b*) the right lymphatic duct, (*c*) the hilum, (*d*) the mesenteric lymph node.

14. The spleen does *not* (*a*) house lymphocytes; (*b*) filter foreign particles, damaged red blood cells, and cellular debris from the blood; (*c*) contain phagocytes; (*d*) change undifferentiated lymphocytes into T lymphocytes.

15. An Rh^- mother and an Rh^+ father are preparing for the birth of their first child.
(*a*) They should arrange for the mother to receive a Rhogam injection.
(*b*) They should expect no problem with this pregnancy.
(*c*) They should expect no problems with future pregnancies.
(*d*) All of the above apply.

True or False

____ **1.** Valves are present in lymphatic vessels.

____ **2.** A person with type B blood has B antibodies.

____ **3.** The polypeptide chains of antibodies have portions that are constant and portions that are variable. It is the constant region that is responsible for binding with the antigen.

____ **4.** When antigenically stimulated, B lymphocytes proliferate and form plasma cells.

____ **5.** A metastasis of cancer frequently uses the route of the lymphatic system.

____ **6.** A person who encounters a pathogen and who has a primary immune response develops passive immunity.

____ **7.** There are five major types of immunoglobulins: IgG, IgA, IgD, IgL, and IgE.

____ **8.** The interaction of antigen with antibody is highly specific.

____ **9.** Antigens are small lipid molecules that stimulate the immune response.

____ **10.** If a child has a splenectomy, lymph nodes in the abdominal cavity enlarge and become splenic in function.

____ **11.** Passive immunity is the transfer of antibodies developed in one individual into the body of another.

____ **12.** T and B lymphocytes may cooperate in response to a particular antigen.

Completion

1. Specialized bands of connective tissue, called ________________, divide the lymph nodes.
2. Clusters of ________________ ________________ (Peyer's patches) are associated with the small intestine.
3. The ________________ is located in the anterior thorax, near the manubrium of the sternum.
4. ________________ is an enzyme in tears, saliva, and blood plasma that breaks down bacterial cell walls.
5. The immunoglobulin that aids in immunity against parasitic worms and other parasites is ________________.
6. ________________ ________________ is conferred when the body manufactures antibodies in response to direct contact with an antigen.
7. ________________ ________________ is a hemolytic disease of the newborn resulting from an antigen-antibody reaction associated with the Rh system of blood.
8. ________________ is a disease whose victims develop a severely impaired immune response through transmission of HIV.

Labeling

Label the structures indicated on the figure to the right.

1. ________________
2. ________________
3. ________________
4. ________________
5. ________________

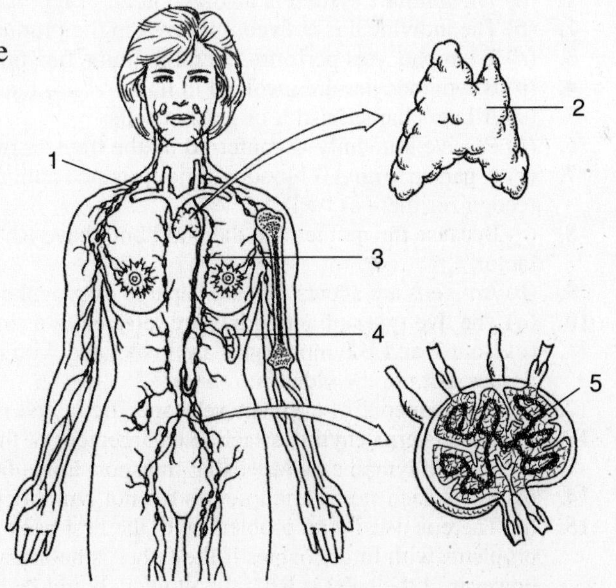

1
2
3
4
5

Matching

Match the term with its appropriate description or action.

____	**1.** Helper T cells	(*a*)	cancer characterized by an uncontrolled production of lymphocytes
____	**2.** AB blood	(*b*)	any of a group of proteins produced by virus-infected cells
____	**3.** Rhogam	(*c*)	activate other T lymphocytes or activate B lymphocytes to become plasma cells
____	**4.** A blood	(*d*)	combine with the antigen on the surface of the foreign cells, causing lysis and the release of cytokines
____	**5.** Plasma cells	(*e*)	blood type referred to as the "universal recipient"
____	**6.** O blood	(*f*)	blood type referred to as the "universal donor"
____	**7.** Lymphoid leukemia erythrematosis	(*g*)	injected into a mother with Rh^- blood giving birth to a Rh^+ child
____	**8.** Systemic lupus	(*h*)	autoimmune disease that affects many body systems
____	**9.** Killer T cells	(*i*)	blood type with anti-B antibodies in the plasma
____	**10.** Interferon	(*j*)	cells that are active in the production of antibodies

Answers and Explanations for Review Questions

Multiple Choice

1. (*d*) The immune system is involved in all of the functions listed.
2. (*b*) The individual is actively involved in the production of antibodies.
3. (*d*) T lymphocytes perform several immune functions including the release of cytokines.
4. (*a*) B lymphocytes are involved in humoral immunity by producing antibodies.
5. (*d*) All are characteristics of plasma cells.
6. (*b*) Passive immunity is conferred by the transfer of antibodies from one individual to another.
7. (*c*) A person with AB blood will not produce antibodies against A or B antigens since the body recognizes them as "self."
8. (*c*) Because the mother and the child both have Rh^- blood, there will be no problem related to the Rh factor.
9. (*b*) Antigens are agents that are capable of provoking an immune response.
10. (*a*) The five types of antibodies are referred to as immunoglobulins.
11. (*c*) Both T and B lymphocytes are associated with the lymph nodes, spleen, and other lymphoid tissues, but not the kidneys.
12. (*a*) The sequence of events is very specific in cell-mediated immunity.
13. (*a*) The cisterna chyli is a saclike enlargement of the thoracic duct in the abdominal area. Lacteals are specialized lymph capillaries that transport lipids from the intestines to the cisterna chyli.
14. (*a*) The spleen stores erythrocytes but not lymphocytes.
15. (*d*) There is usually no problem with the first baby in relation to the Rh factor, and there may not be problems with future babies if the father is heterozygous with respect to the Rh factor alleles; however, if the baby is Rh^+, the mother should be given a Rhogam injection.

True and False

1. True
2. False; a person with B blood has A antibodies in the plasma
3. False; the antibody's antigen-binding sites are located on the variable portion of the antibody
4. True
5. True
6. False; the person develops active immunity
7. False; IgM not IgL
8. True
9. False; antigens are usually large molecules, such as proteins or polysaccharides
10. True
11. True
12. True

Completion

1. trabeculae
2. mesenteric nodes
3. thymus
4. Lysozyme
5. IgE
6. Active immunity
7. Erythroblastosis fetalis
8. AIDS

Labeling

1. Right lymphatic duct
2. Thymus
3. Thoracic duct
4. Spleen
5. Lymph node

Matching

1. (*c*)
2. (*e*)
3. (*g*)
4. (*i*)
5. (*j*)
6. (*f*)
7. (*a*)
8. (*h*)
9. (*d*)
10. (*b*)

Respiratory System 18

Objective A To define *respiration.*

All cells require a continuous supply of oxygen (O_2) and must continuously eliminate a metabolic waste product, carbon dioxide (CO_2). On the macroscopic level, the term *respiration* simply means ventilation, or "breathing." On the cellular level, it refers to the processes by which cells utilize O_2, produce CO_2, and convert energy into useful forms.

18.1 Distinguish between external *respiration, internal respiration,* and *cellular respiration.*

External respiration is the process by which gases are exchanged between the blood and the air. **Internal respiration** is the process by which gases are exchanged between the blood and the cells. **Cellular respiration** is the process by which cells use O_2 for metabolism and give off CO_2 as a waste.

Objective B To identify the *basic components of the respiratory system.*

The major passages of the respiratory system are the **nasal cavity, pharynx, larynx,** and **trachea** (fig. 18.1). Within the **lungs**, the trachea branches into **bronchi, bronchioles,** and finally, **pulmonary alveoli** (see fig. 18.6).

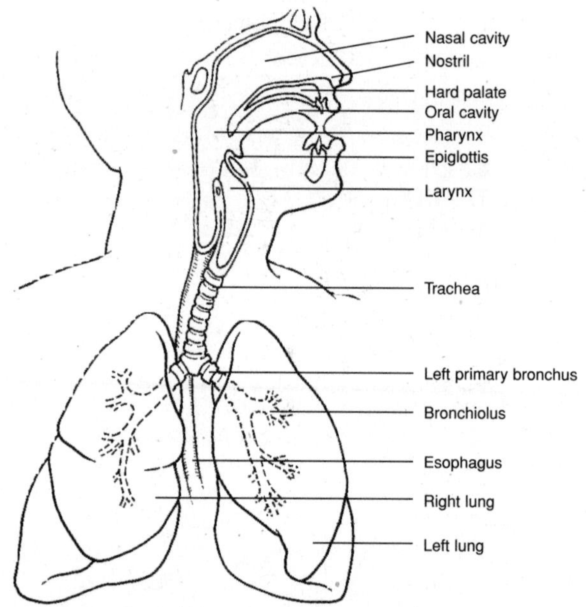

Figure 18.1 The respiratory system.

18.2 Distinguish between the conducting division and the respiratory division of the respiratory system.

The **conducting division** includes all cavities and structures that transport gases to and from the microscopic air pockets (pulmonary alveoli) in the lungs. The pulmonary alveoli constitute the **respiratory division**.

18.3 What physical requirements must be satisfied for the respiratory system to function effectively?

1. The membranes through which gases are exchanged with the circulatory system must be thin and differentially permeable so that diffusion can occur easily.
2. These membranes must be kept moist so that O_2 and CO_2 can be dissolved.
3. A rich blood supply must be present.
4. The surfaces for gas exchange must be located deep in the body so that the incoming air can be sufficiently warmed, moistened, and filtered.
5. There must be an effective pumping mechanism to constantly replenish the air.

Objective C To list the *functions of the respiratory system.*

1. *Gaseous exchange* for the cellular respiratory process
2. *Sound production* (vocalization) as expired air passes over the vocal cords
3. Assistance in *abdominal compression* during micturition (urination), defecation (passing of the feces), and parturition (childbirth)
4. *Coughing* and *sneezing* (self-cleaning reflexes)

18.4 What are the two phases of breathing?

Breathing, or pulmonary ventilation, consists of an **inspiration** (*inhalation*) **phase** and an **expiration** (*exhalation*) **phase**.

Objective D To describe the *nose, nasal cavity,* and *paranasal sinuses* as respiratory structures.

The **nose** includes an external portion that juts out from the face, and an internal **nasal cavity** for the passage of air. The **paranasal sinuses** (see problem 6.16) help, in a small way, to warm and moisten inspired air.

18.5 Describe the anatomy of the nasal cavity.

The **nasal cavity** is divided into two lateral halves, each referred to as a **nasal fossa**, by the **nasal septum**. The **vestibule** is the anterior expanded portion of a nasal fossa. In the lateral walls of either fossa are three shell-like concavities—the **superior**, **middle**, and **inferior conchae** (fig. 18.2). Air passageways, or **meatuses**, connect the conchae. The **nostrils** (*external nares*) open anteriorly into the nasal cavity, and the **choanae** (*posterior nares*) communicate posteriorly with the **nasopharynx**.

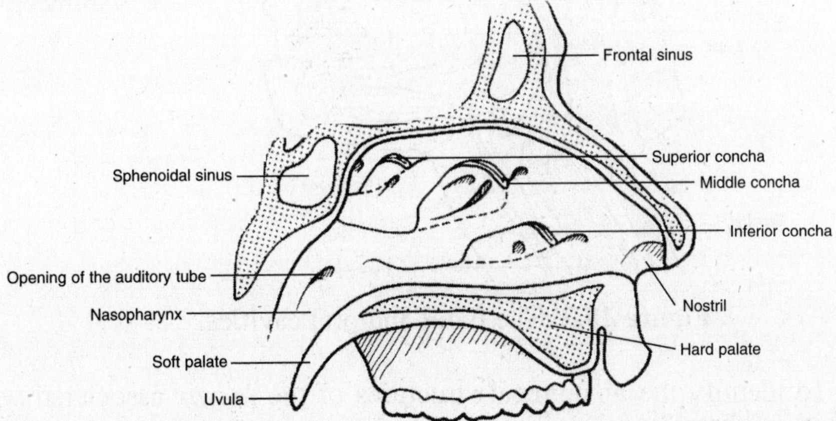

Figure 18.2 The nasal cavity and surrounding structures.

18.6 What types of tissues line the nasal cavity?

The vestibules are lined with *nonkeratinized stratified squamous epithelium* (see table 4.2); this epithelium divides rapidly and supports protective nasal hairs, or *vibrissae.* The conchae of the nasal fossae are lined with *pseudostratified ciliated columnar epithelium* (table 4.1), which secretes mucus to trap dust, pollen, smoke, and other inspired airborne particles. Specialized columnar epithelium, called *olfactory epithelium,* lines the upper medial portion of the nasal cavity, where it responds to odors.

18.7 Why are nosebleeds common?

The nasal epithelia are extensive and highly vascular, with the capillaries located close to the surface. This makes us susceptible to *epistaxes* (nosebleeds).

Objective E To describe the regions of the *pharynx*.

Survey

The *pharynx* is divided on the basis of location and function into a **nasopharynx**, an **oropharynx**, and a **laryngopharynx**. The *auditory* (eustachian) *canals, uvula,* and *pharyngeal tonsils* are in the nasopharynx; the *palatine* and *lingual tonsils* are in the oropharynx (fig. 18.3). The oropharynx and laryngopharynx have respiratory and digestive functions, while the nasopharynx serves only the respiratory system.

18.8 How does the uvula serve both the respiratory and digestive systems?

The pendulous **uvula** hangs from the middle of the lower border of the soft palate. During swallowing, the soft palate and uvula are elevated, closing off the nasal cavity so that food or fluid cannot enter.

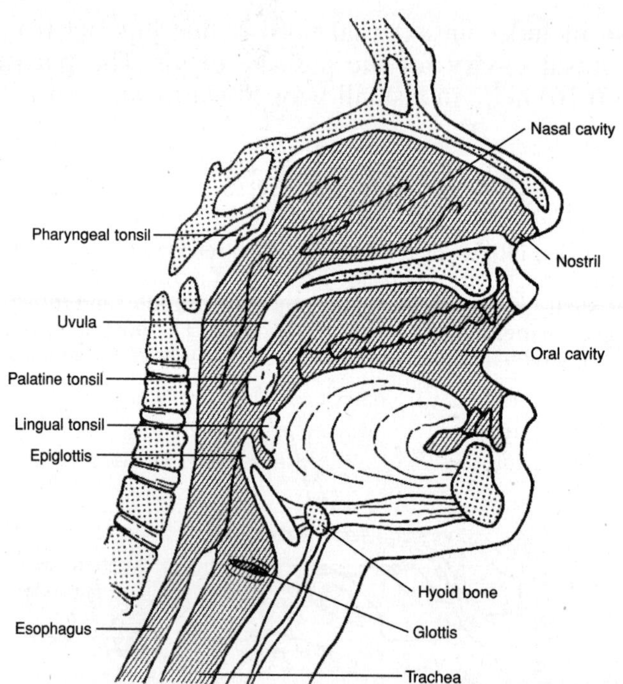

Figure 18.3 The nasal and oral cavities.

Objective F To identify the anatomical structures of the *larynx* associated with sound production and breathing.

The **larynx** (*voice box*) forms the entrance into the trachea. A primary function of the larynx is to prevent food or fluid from entering the trachea and lungs during swallowing, while permitting the passage of air into the trachea at other times. A secondary function is to produce sound vibrations.

18.9 Which cartilages of the larynx are paired and which are unpaired? Which cartilage is the largest and most prominent?

The *larynx* is a roughly triangular box composed of nine hyaline cartilages; three are large single structures and six are smaller paired structures (fig. 18.4). The **anterior thyroid cartilage** ("Adam's apple") is the largest. The spoon-shaped **epiglottis** has a cartilaginous framework. The lower portion of the larynx is formed by the ring-shaped **cricoid cartilage**. The three paired cartilages are the **arytenoid cartilages**, which support the vocal cords, and the **cuneiform** and **corniculate cartilages**, which aid the arytenoid cartilages. The **glottis** is the opening into the larynx (see fig. 18.5).

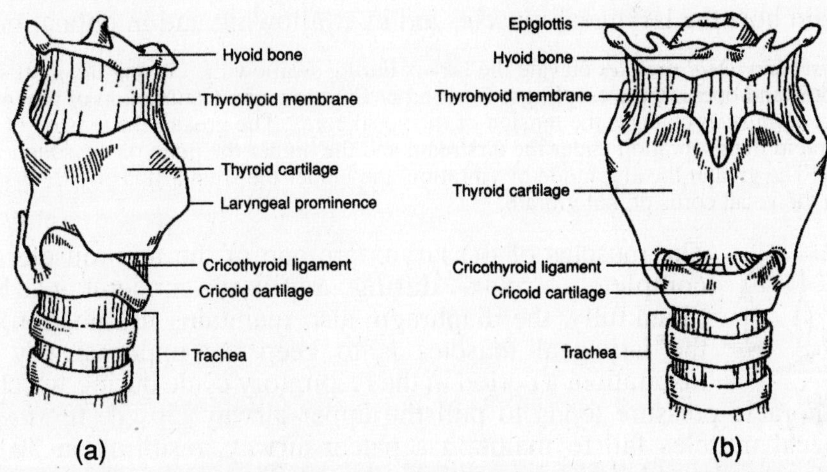

Figure 18.4 The hyoid bone and larynx. (*a*) A lateral view and (*b*) an anterior view.

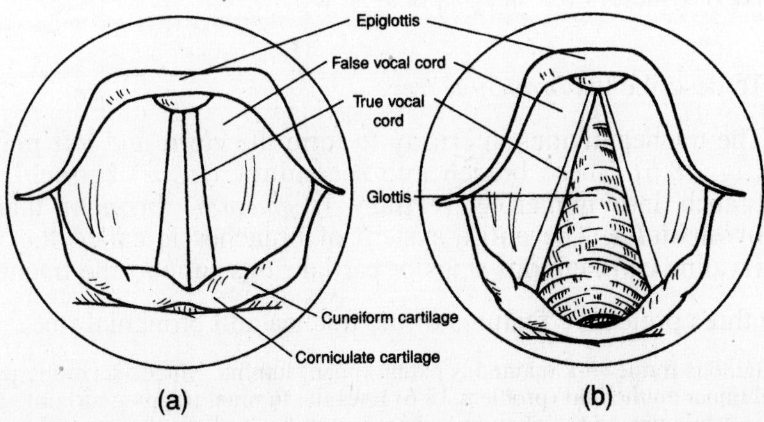

Figure 18.5 A superior view of (*a*) the opened and (*b*) the closed glottis.

18.10 Why not "Eve's apple"?

During puberty, the male sex hormone testosterone causes accelerated growth of the larynx, especially the thyroid cartilage. The larger larynx accounts for the deeper voice of males.

18.11 Explain the functional relationship between the glottis and epiglottis during swallowing.

During the final sequence of swallowing, the larynx is pulled superiorly, closing the glottis against the epiglottis. You can feel this movement by cupping your fingers lightly over the larynx and then swallowing.

With the glottis sealed, fluid or food enters the esophagus rather than the larynx and trachea. However, fluid or food may enter the glottis if it is not closed during swallowing as it should be. Fluid entering the trachea reflexively causes violent coughing in an attempt to force it out. Food entering the glottis may become lodged between the vocal cords. In this case, the *abdominal thrust* (Heimlich) *maneuver* can be used to prevent suffocation.

18.12 Explain how the laryngeal muscles aid in swallowing and in phonation (speech).

Extrinsic laryngeal muscles elevate the larynx during swallowing, closing the glottis over the epiglottis and opening the esophagus for food or fluid to enter. Contraction of the *intrinsic laryngeal muscles* changes the tension of the vocal cords. The greater the tension of the cords, the more rapid their vibration under the airstream and the higher the pitch of the sound (see problem 12.17). The greater the amplitude of vibration, the louder the sound. *Whispering* is phonation in which the vocal cords do not vibrate.

The muscles of the larynx are one of the few muscle groups to not completely relax during rapid-eye-movement (REM) sleep (thankfully, the diaphragm also maintains its function). The role of the laryngeal muscles is to keep the upper airway open during inspiration, a period in the respiratory cycle during which the negative intrathoracic pressure tends to pull the upper airway closed. In *sleep apnea*, the laryngeal muscles fail to maintain a patent airway, resulting in 30-second to 1-minute episodes of breathing cessation. Sleep apnea may be associated with snoring, daytime somnolence (due to lack of REM sleep at night), nighttime headaches (due to hypoxia), and lethargy. Pulmonary vascular hypertension due to hypoxia may ultimately lead to right ventricular heart failure and death. Obesity is one of the few recognized risk factors for sleep apnea.

Objective G To describe the *bronchial tree.*

The trachea divides inferiorly to form the **right** and **left primary bronchi** (fig. 18.6). These branch into **secondary** (*lobar*) **bronchi**, which in turn branch into numerous **tertiary** (*segmental*) **bronchi** that terminate in **bronchioles**. The entire system of branches is called the *bronchial tree.* Hyaline cartilaginous rings or partial rings support the trachea and the tree.

18.13 Describe three protective features of the trachea and bronchial tree.

The cartilaginous framework maintains patent (open) lumina. Mucus-secreting pseudostratified ciliated columnar epithelium (problem 18.6) lines the lumina, trapping airborne particles and moving this debris toward the pharynx, where it may be swallowed. Irritation to the epithelial lining of the trachea or bronchial tree elicits a violent coughing reflex that cleanses the respiratory tract.

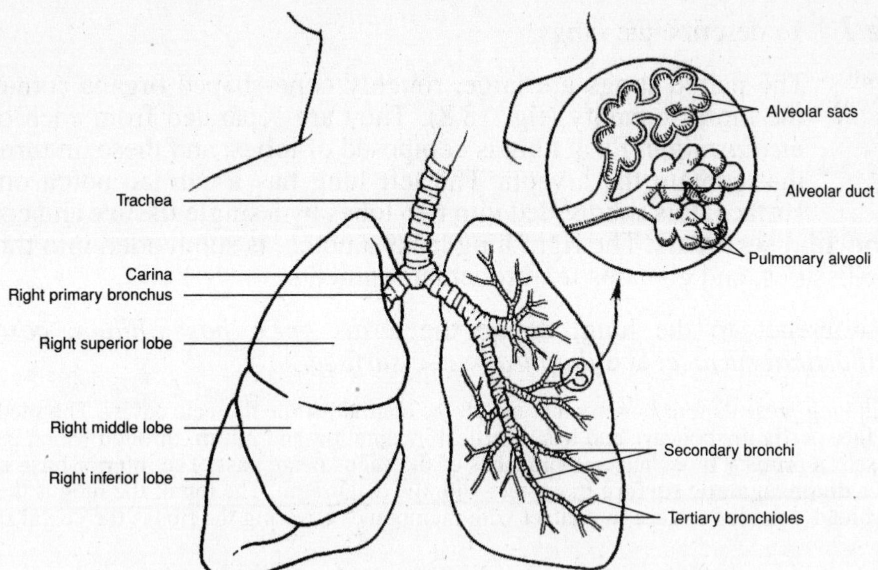

Figure 18.6 The trachea, lungs, and bronchial tree.

Objective H To describe the *respiratory division structures.*

Bronchioles end in **terminal bronchioles**, which branch into many **alveolar ducts** that lead directly into **alveolar sacs**, themselves clusters of many microscopic **pulmonary alveoli**. Alveolar ducts are lined with simple cuboidal epithelium, whereas pulmonary alveoli are lined with simple squamous epithelium. Gas exchange with the blood of the circulatory system occurs through the thin-walled, moistened pulmonary alveoli (see problem 18.2).

18.14 What are the functions of septal cells and alveolar macrophages?

Small **septal cells**, dispersed among the cells of the simple squamous epithelium lining a pulmonary alveolus, secrete a phospholipid *surfactant* that lowers the surface tension. Also found in the alveolar wall are phagocytic **alveolar macrophages** (*dust cells*) that remove dust particles or other debris from the pulmonary alveolus.

18.15 Diagram the process of external respiration (see problem 18.1) on the alveolar level.

See figure 18.7.

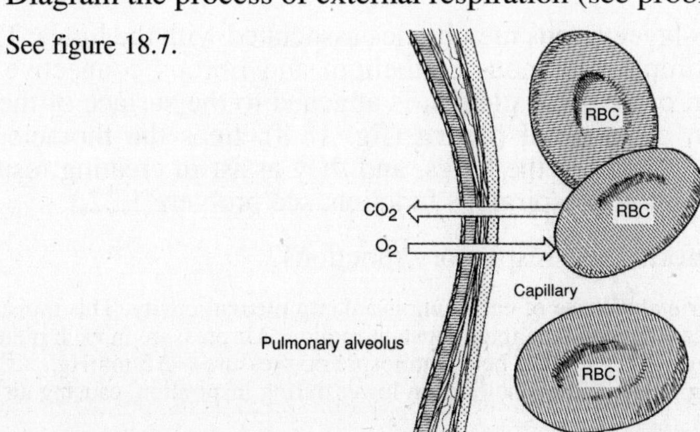

Figure 18.7 External respiration between a pulmonary alveolus and the blood within a capillary.

Objective I To describe the *lungs*.

The paired **lungs** are large, roughly cone-shaped organs contained within the thoracic cavity (fig. 18.8). They are separated from each other by the *mediastinum*. Each lung is composed of **lobes**, and these, in turn, of **lobules** that contain the alveoli. The left lung has a cardiac notch on its medial surface. It is subdivided into two lobes by a single **fissure** and contains eight **bronchial segments**. The right lung lacks a notch, is subdivided into three lobes by two fissures, and contains ten bronchial segments.

18.16 In reference to the lung, define the terms *apex, base, hilum, costal surface, mediastinal surface,* and *diaphragmatic surface.*

Each lung presents four borders that match the contour of the thoracic cavity. The **mediastinal surface** is slightly concave and has a vertical indentation, the **hilum**, through which pulmonary vessels, a primary bronchus, and branches of the vagus nerve pass. The inferior **base of the lung** has a **diaphragmatic surface** in contact with the diaphragm. The top of the lung is the **apex**, and the broad, rounded surface in contact with membranes covering the ribs is the **costal surface.**

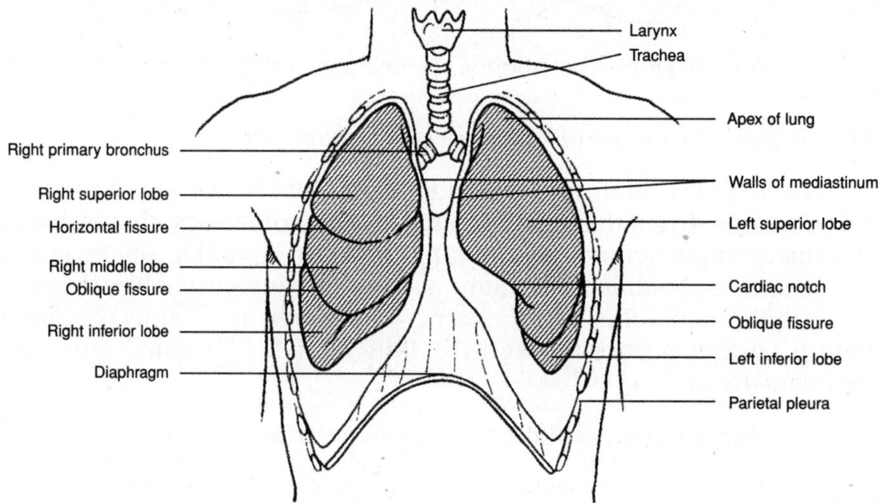

Figure 18.8 The lungs within the thoracic cavity.

Objective J To describe the *pleurae* and to explain their respiratory significance.

Pleurae are two-layer serous membranes associated with the lungs. They are composed of simple squamous epithelium and fibrous connective tissue. The inner layer, or **visceral pleura**, is attached to the surface of the lungs; the outer layer, or **parietal pleura** (fig. 18.8), lines the thoracic cavity. Pleurae serve to lubricate the lungs, and they assist in creating respiratory pressure. (For an important nonrespiratory function, see problem 1.22.)

18.17 How do the pleurae perform their respiratory functions?

Between the visceral and parietal pleurae of each lung is a damp **pleural cavity**. This moist environment serves to lubricate the lungs in their constant motion. Air pressure in each pleural cavity (the *intrathoracic pressure)* is slightly below atmospheric pressure (–2.5 mmHg, approximately) in the resting lungs, and becomes even lower during inspiration, causing air to inflate the lungs.

Pleurisy, or inflammation of the pleurae, may be secondary to some other respiratory disease or due to an autoimmune reaction associated with viral infections or autoimmune diseases. Chest pain is felt most keenly when breathing deeply or lying down. Anti-inflammatories, such as aspirin, ibuprofen, and corticosteroids, are used in treating pleurisy.

The *pleural cavity* is a potential space only. Under normal conditions the visceral and parietal pleurae are pressed tightly against one another due to the relative negative pressure in the space. This negative pressure is critical for the thoracic cavity to "pull out" on the lungs causing them to inflate. Air in the pleural space (from a hole in the chest wall or a hole in the visceral pleura) disturbs this vital homeostasis so that the lung collapses despite active expansion of the chest wall. This condition is called *pneumothorax*.

Objective K To examine the *mechanics of breathing*.

Inspiration (see problem 18.4) occurs when contraction of the respiratory muscles (diaphragm, internal intercostals; see table 8.4) causes an increase in thoracic volume, with expansion of the lungs and a decrease in intrathoracic and *intrapulmonic* (alveolar) pressures. Air enters the lungs when intrapulmonic pressure falls below atmospheric pressure (760 mmHg at sea level). **Expiration** follows passively as thoracic volume decreases and intrapulmonic pressure rises above atmospheric, with recoil of the rib cage and contraction of the lungs.

18.18 Describe the changes in shape of the thorax during inspiration and expiration.

Contraction of the dome-shaped diaphragm downward increases the thoracic vertical dimension. A simultaneous contraction of the external intercostal muscles (see fig. 8.4) increases the side-to-side and front-to-back dimensions. During deep inspiration or forced breathing, the scalenes and sternocleidomastoid muscles (see fig. 8.3), as well as the pectoralis minor muscles (see table 8.8), become involved. During forced expiration, the internal intercostal muscles are contracted, depressing the rib cage. Contracting the abdominal muscles (see table 8.6) also forces air from the lungs by elevating the diaphragm.

Objective L To identify the various *respiratory air volumes*.

The *total lung capacity* may be expressed as the sum of four volumes (fig. 18.9). These are the **tidal volume**, the volume of air moved into and out of the lungs during normal breathing; **inspiratory reserve**, the maximum volume beyond the tidal volume that can be inspired in one deep breath; **expiratory reserve**, the maximum volume beyond the tidal volume that can be forcefully exhaled following a normal expiration; and **residual volume**, the air that remains in the lungs following a forceful expiration. Respiratory air volumes are measured with the *spirometer*.

18.19 Account for the variability of respiratory air volumes.

Clinically speaking, it is important to know the amount of air that is breathed in at a given time and to be aware of difficulty in breathing. The amount of air exchanged during pulmonary ventilation varies from person to person according to age, sex, activity, and general health.

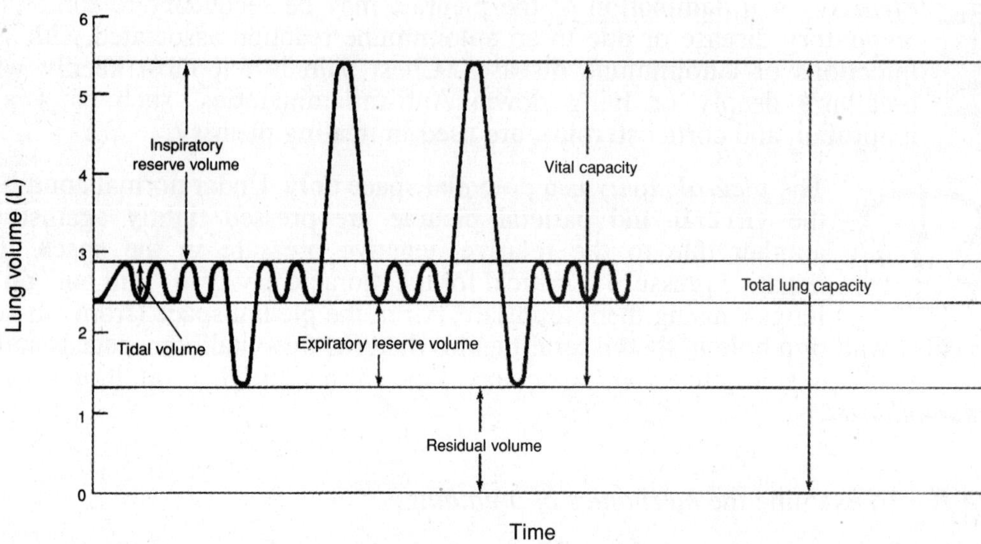

Figure 18.9 Respiratory air volumes.

18.20 Calculate the minute respiratory volume of an individual who has a tidal volume of 500 ml and a respiratory rate of 12 breaths per minute.

Minute respiratory volume is the volume of air moved in normal ventilation in one minute. Therefore,

$$\text{Minute respiratory volume} = (\text{tidal volume}) \times (\text{respiratory rate})$$
$$= (0.500 \text{ L}) \times (12 \text{ min}^{-1}) = 6 \text{ L/min}$$

The preceding is a typical example.

18.21 Define *alveolar ventilation.*

Alveolar ventilation is the volume of air exchanged in 1 minute in the pulmonary alveoli (for transport to the cells). Thus,

$$\text{Alveolar ventilation} = [(\text{tidal volume}) - (\text{dead space})] \times (\text{respiratory rate})$$

Dead space is defined as the volume of air between the mouth/nose and the pulmonary alveoli, or that space that is not involved in gas exchange. This is typically 150 ml in an adult.

Objective M To describe the transport of gases between the lungs and the cells.

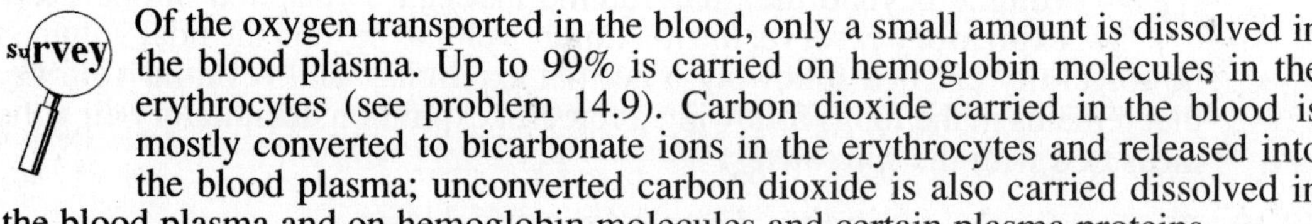

Of the oxygen transported in the blood, only a small amount is dissolved in the blood plasma. Up to 99% is carried on hemoglobin molecules in the erythrocytes (see problem 14.9). Carbon dioxide carried in the blood is mostly converted to bicarbonate ions in the erythrocytes and released into the blood plasma; unconverted carbon dioxide is also carried dissolved in the blood plasma and on hemoglobin molecules and certain plasma proteins.

18.22 Use an equation to define the oxygenation of hemoglobin.

Hemoglobin is converted from bluish-red *deoxyhemoglobin* (Hb) to scarlet *oxyhemoglobin* (HbO_2) according to the reaction

$$Hb + O_2 \longleftrightarrow HbO_2$$

18.23 What is meant by partial pressure?

In a mixture of gases, each component gas exerts a *partial pressure* that is proportional to its concentration in the mixture. For example, since air is 21% O_2, this gas is responsible for 21% of the atmospheric pressure. Since 21% of 760 mmHg is equal to 160 mmHg, the partial pressure of O_2, symbolized by P_{O_2}, in atmospheric air is 160 mmHg. Similarly,

$$P_{CO_2} = 0.3 \text{ mmHg}$$

The partial pressures of oxygen and carbon dioxide are not the same in the alveoli as they are in the atmosphere due to the high CO_2 contribution from the venous blood. Alveolar P_{O_2} is 101 mmHg and P_{CO_2} is 40 mmHg at sea level.

18.24 Explain respiratory diffusion in terms of partial pressure differences.

The difference between the P_{O_2} in the alveolus and in the pulmonary capillary ($P_{O_2} = 40$) is about 60 mmHg and therefore favors diffusion of oxygen from the alveolus into the blood. A similar calculation of the difference between the P_{CO_2} in the pulmonary capillary ($P_{CO_2} = 45$) and in the alveolus demonstrates a gradient of 5 mmHg, favoring diffusion of carbon dioxide from the blood to the alveolar air.

18.25 What factors precipitate release of O_2 from hemoglobin to body tissues?

1. A decreased concentration of O_2 in the blood plasma
2. A decreased blood pH (i.e., an increased H^+ concentration)
3. An increased body temperature

Objective N To describe the role of the respiratory system in the *acid-base balance* of the body.

The presence of the enzyme **carbonic anhydrase** in the erythrocytes causes about 67% of the CO_2 in blood to combine quickly with water to form carbonic acid, most of which dissociates into bicarbonate and hydrogen ions:

$$CO_2 + H_2O \leftrightarrow H_2CO_3 \leftrightarrow HCO_3^- + H^+$$

18.26 Define *alkali reserve* and *chloride shift*.

Bicarbonate ions (HCO_3^-), which make up a large part of the blood buffer system, constitute the **alkali reserve.** As these ions leave erythrocytes, they cause an excess of negative charge, which is relieved by the diffusion of chloride ions (Cl^-) from the blood into the cells. This movement of chloride ions is termed the **chloride shift.**

18.27 Define *respiratory acidosis* and *respiratory alkalosis*.

Respiratory acidosis, or a process that drives blood pH below 7.35, occurs when CO_2 is not eliminated from the body at a normal rate, thus increasing vascular P_{CO_2}. Lung disease or decreased mental status resulting in hypoventilation may cause respiratory acidosis. **Respiratory alkalosis**, a process that drives blood pH above 7.45, occurs when CO_2 is eliminated too rapidly, thus decreasing vascular P_{CO_2}. Either hyperventilation or the action of certain drugs (such as excessive aspirin) on the respiratory control center of the brain may produce respiratory alkalosis. The effects of hyperventilation are rapidly reversed when the person breathes into a paper bag, inhaling expired air and so causing the vascular P_{CO_2} to rise.

Objective O To describe *neural* and *chemical regulation of respiration.*

Survey

The locations of the respiratory centers in the CNS are shown in fig. 18.10 (see also fig. 10.7). The rhythmicity area of the medulla oblongata is actually composed of separate **expiratory** and **inspiratory centers**. The medulla oblongata also contains chemoreceptors concerned with respiration (fig. 18.10), as do the **carotid bodies** in the neck and **aortic bodies** in the thorax (fig. 18.11).

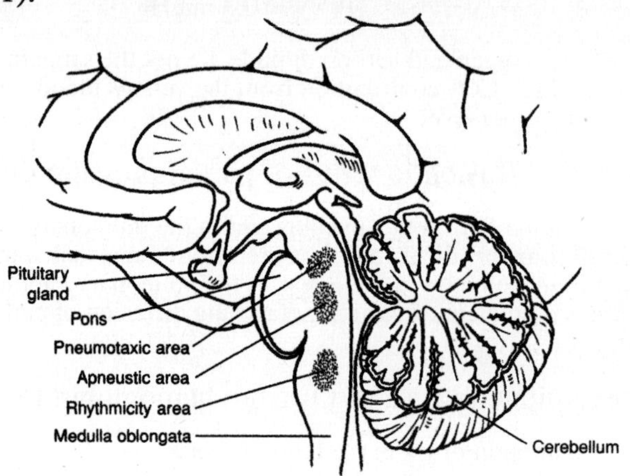

Figure 18.10 Respiratory control centers.

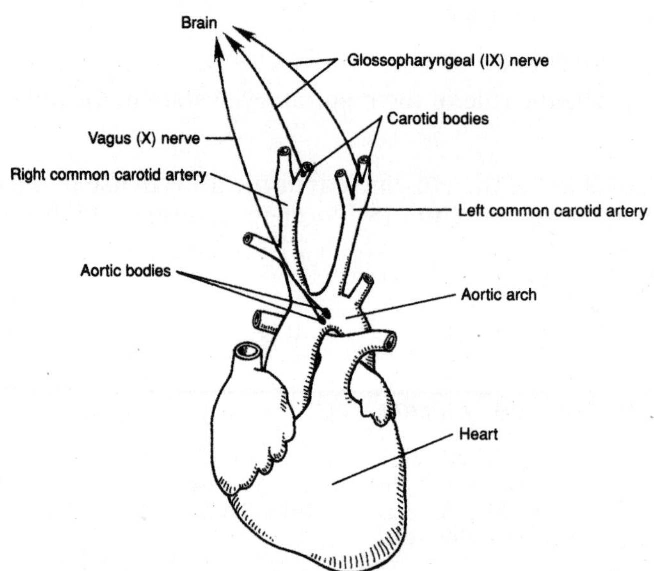

Figure 18.11 The carotid and aortic bodies.

18.28 How do the neural respiratory centers operate?

The rhythmicity area of the medulla oblongata consists of two intermingled groups of neurons. When the *inspiratory group* is excited (via the apneustic center), the respiratory muscles are signaled to accomplish inbreathing; at the same time, the *expiratory group* is inhibited. After about 2 seconds, the reciprocal process occurs: the pneumotaxic center stimulates the expiratory group to signal for exhaling, with simultaneous inhibition of the inspiratory group.

18.29 How do the respiratory chemoreceptors operate?

Central chemoreceptors, on the surface of the medulla oblongata, respond to increased P_{CO_2} or decreased pH of cerebrospinal fluid by initiating increased inspiration. Peripheral chemoreceptors, in the carotid and aortic bodies, respond to decreased P_{CO_2} by initiating increased inspiration.

The ability of the body to regulate acid/base status is one of the most elegant homeostatic mechanisms known. Almost immediately, an increase in pH caused by a loss of acid (through vomiting stomach secretions, for example) will result in a slowing of breathing, while a drop in pH (possibly due to diabetic ketoacidosis) will promote hyperventilation. Through this mechanism, the body is able to quickly reestablish normal pH.

Key Clinical Terms

Anoxia A severe shortage of oxygen in tissues and organs. Anoxia of the brain results in cell destruction within 30 seconds and in death generally within 5 to 10 minutes.

Apnea A temporary cessation of breathing that may follow hyperventilation.

Asphyxia Suffocation.

Asthma A disease characterized by recurrent attacks of dyspnea and wheezing. It may be aggravated by inhaled allergens, viral or bacterial upper respiratory infections, cold air, or exercise. The attacks are provoked by constriction of the airways and inflammation of the bronchial mucosa.

Bronchitis, acute Inflammation of the mucous membrane lining the bronchial tubes. Viral and bacterial infections, air pollution, and allergies may be causative or contributing factors.

Bronchitis, chronic Excessive mucus production in the bronchial tubes that leads to productive cough, shortness of breath, and lung damage. It is caused almost exclusively by cigarette smoking.

Cleft lip A genetic developmental disorder in which the two sides of the upper lip fail to fuse; also referred to as *harelip*.

Cleft palate A developmental deformity of the hard palate, resulting in a persistent opening between the oral and nasal cavities. The condition may be hereditary or a complication of some disease (e.g., German measles) contracted by the mother during pregnancy.

Dyspnea Difficult breathing.

Emphysema A breakdown of the alveolar walls that decreases the alveolar surface area and increases the size of air spaces distal to the terminal bronchioles. It is a frequent cause of death among heavy cigarette smokers and may also result from severe air pollution.

Epistaxis Nosebleed.

Hiccup A spasmodic contraction of the diaphragm causing a rapid, involuntary inhalation that is stopped by the sudden closure of the glottis and accompanied by a distinctive sound; also spelled *hiccough*.

Hyperventilation Excessive inhalation and exhalation.

Laryngitis Inflammation of the larynx.

Pleurisy Inflammation of the pleurae.

Pneumonia Acute infection and inflammation of the lungs, with exudation of fluids into, and consolidation (collapse) of, lung tissue.

Rhinitis Inflammation of the nasal mucosa.

Sinusitis Inflammation of the mucous membrane of one or another of the paranasal sinuses—usually secondary to a nasal infection.

Tuberculosis An inflammatory disease of the lungs, caused by the tubercle bacillus, in which the tissue caseates (becomes cheesy) and ulcerates. The disease is usually contracted by inhaling air sneezed or coughed by someone with an active case.

Review Exercises

Multiple Choice

1. Which is *not* a structure of the respiratory system? (*a*) the pharynx, (*b*) the bronchus, (*c*) the larynx, (*d*) the hyoid (*e*) the trachea

2. The roof of the nasal cavity is formed primarily by (*a*) the hard palate, (*b*) the cribriform plate of the ethmoid bone, (*c*) the superior concha, (*d*) the vomer, (*e*) the sphenoid bone.

3. The cartilages upon which the vocal cords are attached are (*a*) the thyroid and arytenoid cartilages, (*b*) the thyroid and cricoid cartilages, (*c*) the cuneiform and cricoid cartilages, (*d*) the thyroid and corniculate cartilages.

4. Pulmonary vessels, nerves, and a bronchus enter or leave the lung at (*a*) the cardiac notch, (*b*) the apex, (*c*) the capsule, (*d*) the hilum, (*e*) the base.

5. Neither the trachea nor the bronchi contain (*a*) hyaline cartilage, (*b*) ciliated columnar epithelium, (*c*) goblet cells, (*d*) simple squamous epithelium.

6. Pharyngeal tonsils are located in (*a*) the nasopharynx, (*b*) the oral cavity, (*c*) the nasal cavity, (*d*) the oropharynx, (*e*) the lingualopharynx.

7. Which of the following is a *false* statement?
(*a*) Slacker vocal cords produce higher sounds.
(*b*) During swallowing, the epiglottis is depressed to cover the glottis.
(*c*) In whispering, the vocal cords do not vibrate.
(*d*) Testosterone secretion influences laryngeal development during puberty.

8. The serous membrane in contact with the lung is (*a*) the parietal pleura, (*b*) the pulmonary mesentery, (*c*) the pulmonary peritoneum, (*d*) the visceral pleura.

9. Most of the CO_2 is transported in the blood as (*a*) carboxyhemoglobin, (*b*) HCO_3^-, (*c*) carbaminohemoglobin, (*d*) dissolved CO_2.

10. Peripheral chemoreceptors are located in (*a*) lung tissue, (*b*) the pons and medulla oblongata, (*c*) aortic and carotid bodies, (*d*) the myocardium.

11. As CO_2 produced in the tissues combines with H_2O in the blood, (*a*) carbonic acid is formed, (*b*) Cl^- enters the blood, (*c*) most of the HCO_3^- from the carbonic acid leaves the RBCs for the blood plasma, (*d*) all of the preceding occur.

12. When blood CO_2 levels rise, (*a*) only the rate of breathing decreases, (*b*) respiratory acidosis may occur, (*c*) peripheral pressure receptors respond, (*d*) both the rate and depth of breathing decrease.

13. The amount of air that is moved in and out of the lungs during normal quiet breathing is called (*a*) the vital capacity, (*b*) the tidal volume, (*c*) the residual volume, (*d*) the vital volume, (*e*) the inspiratory volume.

14. Which of the following is *not* a structural feature of the left lung? (*a*) superior lobe, (*b*) cardiac notch, (*c*) inferior lobe, (*d*) middle lobe

15. Which combination of muscles contraction causes inspiration? (*a*) internal intercostals–diaphragm, (*b*) diaphragm–abdominal complex, (*c*) external intercostals–diaphragm, (*d*) external–internal intercostals

16. The maximum amount of air that can be expired after a maximum inspiration is called (*a*) the forced expiratory volume, (*b*) the maximum expiratory flow, (*c*) the tidal volume, (*d*) the vital capacity.

17. Surfactant (*a*) reduces the surface tension in pulmonary alveoli (*b*) increases the P_{CO_2} levels in blood, (*c*) is a mucus secreted by goblet cells, (*d*) reduces friction in the pleural cavity.

18. The basic inspiratory and expiratory centers are located in (*a*) the lungs, (*b*) the medulla oblongata, (*c*) the carotid and aortic bodies, (*d*) the pons.

19. Factors determining the extent to which O_2 will combine with hemoglobin are (*a*) P_{CO_2} in blood, (*b*) body temperature, (*c*) blood H^+ concentration, (*d*) all of the preceding.

20. The cells formed in the alveolar wall that remove foreign particles from the pulmonary alveoli are called (*a*) Kupffer cells, (*b*) pulmonary reticulocytes, (*c*) surfactant cells, (*d*) alveolar macrophage cells.

True or False

____ **1.** The nasal septum divides the nose into right and left nasal cavities.

____ **2.** The pleural cavities are closed separate cavities within the thorax.

____ **3.** In normally functioning lungs, the intrathoracic pressure is always greater than the intrapulmonic pressure.

____ **4.** Expiration is usually passive and occurs with the cessation of inspiratory contractions.

____ **5.** Active transport mechanisms effect exchange of gases between the respiratory and circulatory systems.

____ **6.** An increase or decrease in blood P_{CO_2} is always accompanied by a change in plasma H^+ concentration (plasma pH).

____ **7.** In the respiratory system, simple squamous epithelium is restricted to the pulmonary alveoli.

____ **8.** The vomer and sphenoid bone form the bony framework of the nasal septum.

____ **9.** Hyperventilation may cause body fluids to become more acidic.

____ **10.** The auditory canals open into the nasopharynx.

____ **11.** An elevation in pH causes peripheral chemoreceptors to increase the rate and depth of breathing.

____ **12.** The partial pressure of a gas is directly proportional to its molecular weight.

____ **13.** An increased body temperature increases the ability of hemoglobin to supply O_2 to tissues.

____ **14.** The release of oxygen from hemoglobin is termed *oxygenation*.

____ **15.** Minute respiratory volume is the total amount of air ventilated during relaxed breathing in a single minute (tidal volume x respiratory rate).

____ **16.** An increased concentration of O_2 in the blood plasma contributes to the release of O_2 from hemoglobin to body tissues.

____ **17.** As bicarbonate ions attach to erythrocytes, the balance between positively and negatively charged ions is disturbed, causing the chloride shift.

____ **18.** The respiratory system may indirectly assist micturition, defecation, and parturition.

____ **19.** The larynx contains nine separate cartilages, the largest of which is the thyroid cartilage.

____ **20.** Apnea is difficult breathing.

Completion

1. The ________________ ________________ partitions the nasal cavity into two nasal fossa.
2. The ________________ hangs into the oropharynx from the center of the posterior border of the soft palate.
3. The ________________ is the opening into the larynx.
4. Respiratory passageways are maintained patent (open) by the cartilage within the tunicas (walls) of the tubular organs. Histological, these supportive structures consist of ________________ cartilage.
5. Alveolar epithelium secretes a phospholipid ________________ that lowers the surface tension within the pulmonary alveoli.
6. Each lung is located in a moistened ________________ cavity between the visceral and parietal pleurae.
7. ________________ ________________ is the volume of air moved in and out of the lungs during normal breathing.
8. Hemoglobin with an attached oxygen molecule is called ________________.

9. Chemoreceptors are contained in the ________________ bodies in the neck and the ________________ bodies in the thorax.
10. A severe shortage of oxygen in tissues is referred to as ________________.

Labeling

Label the structures indicated on the figure to the right.

1. ____________________
2. ____________________
3. ____________________
4. ____________________
5. ____________________
6. ____________________
7. ____________________
8. ____________________

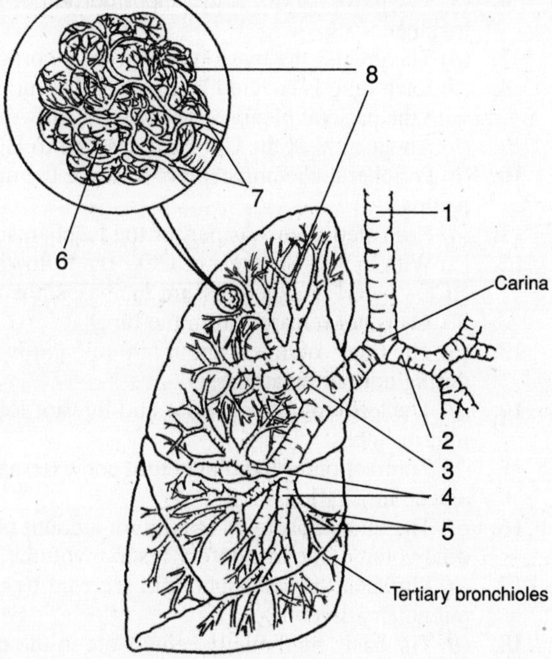

Matching

Match the respiratory air volumes with the values indicated with capital letters in the figure.

____ 1. Total lung capacity
____ 2. Expiratory reserve volume
____ 3. Vital capacity
____ 4. Residual volume
____ 5. Tidal volume
____ 6. Inspiratory reserve volume

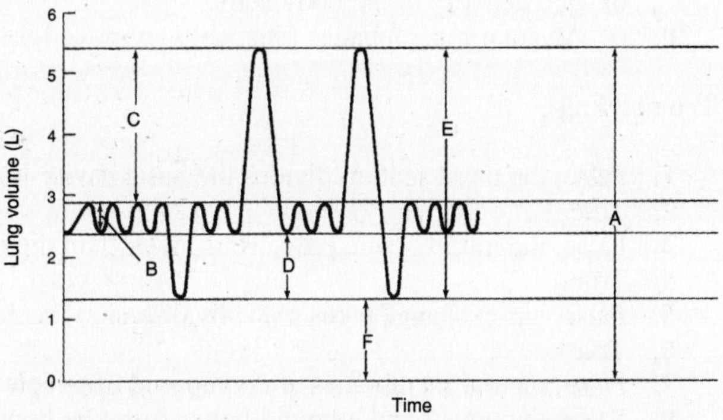

Answers and Explanations for Review Exercises

Multiple Choice

1. (*d*) The hyoid is a bone of the skeletal system that supports the larynx and tongue.
2. (*b*) The cribriform plate of the ethmoid bone contains numerous foramina for the passage of olfactory nerves into the olfactory bulb.

3. (*a*) The paired vocal cords span either side of the glottis from the thyroid cartilage to the paired arytenoid cartilages.
4. (*d*) The hilum is a vertical depression on the mediastinal surface of the lung, where structures enter and exit the lung.
5. (*d*) Simple squamous epithelium in the respiratory system is restricted to the pleural membranes and the lining of the pulmonary alveoli.
6. (*a*) The nasopharynx is the upper portion of the pharynx, where the nasal and oral cavities come together.
7. (*a*) The greater the tension of the vocal cords, the more rapid the vibration and the higher the pitch.
8. (*d*) Each lung is covered by a visceral pleura. The thoracic body wall surrounding the lung is lined with the parietal pleura. The pleural cavity is the space between the two pleural membranes.
9. (*b*) About 67% of the CO_2 is converted to HCO_3^- for transport in the blood.
10. (*c*) Peripheral chemoreceptors (outside the medulla oblongata) are located in the aortic and carotid bodies.
11. (*d*) Each step occurs as part of the mechanism by which CO_2 is transported in the blood.
12. (*b*) With increased levels of CO_2, the following reaction occurs: $CO_2 + H_2O \longleftrightarrow H_2CO_3 \longleftrightarrow HCO_3^- + H^+$. As the pressure increases, the rise in H^+ may lead to acidosis if CO_2 is not released from the lungs.
13. (*b*) The tidal volume is about 500 ml, and is the amount of air that moves into and out of the lungs during normal breathing.
14. (*d*) The left lung has superior and inferior lobes only. The right lung has superior, middle, and inferior lobes.
15. (*c*) Contraction of the diaphragm and external intercostal muscles expand the thoracic cavity, causing a deep inspiration.
16. (*d*) The vital capacity is maximum amount of air that can be forcefully exhaled. It is the sum of the tidal volume, the inspiratory reserve volume, and the expiratory reserve volume.
17. (*a*) Surfactant is a phospholipid secreted by the septal cells that lowers the surface tension in the pulmonary alveoli.
18. (*b*) The basic rhythmicity centers are in the medulla oblongata. However, there are other respiratory centers in the pons that indirectly regulate the centers in the medulla oblongata.
19. (*d*) The several factors that influence the extent to which O_2 binds to hemoglobin ensure adequate oxygen delivery to the body cells.
20. (*d*) Alveolar macrophages (dust cells) remove foreign debris from the pulmonary alveoli.

True or False

1. False; the nasal septum divides the nasal cavity into right and left nasal fossae.
2. True
3. False; the intrathoracic pressure is lower than the intrapulmonic pressure
4. True
5. False; gas exchange takes place by diffusion mechanisms
6. True
7. False; pleural membranes are composed of simple squamous epithelium
8. False; the vomer and ethmoid bones form the bony framework of the nasal septum
9. False; hyperventilation may cause body fluids to become more alkaline because of the removal of large quantities of CO_2
10. True
11. True
12. False; the partial pressure of a gas is directly proportional to its concentration
13. True
14. False; deoxyhemoglobin
15. True
16. False; a decrease in blood plasma O_2 will increase the release of O_2 from hemoglobin to body tissues
17. False; as the bicarbonate ions leave the erythrocytes, there is a decrease in the negative charge, which is relieved by the diffusion of Cl^- into the erythrocytes.
18. True

19. True
20. False; apnea is the temporary cessation of breathing

Completion

1. nasal septum
2. uvula
3. glottis
4. hyaline
5. surfactant
6. pleural
7. Tidal volume
8. oxyhemoglobin
9. carotid, aortic
10. anoxia

Labeling

1. Trachea
2. Left primary bronchus
3. Secondary bronchi
4. Segmental bronchi
5. Bronchioles
6. Pulmonary vessels
7. Pulmonary capillaries
8. Alveolar sac

Matching

1. (*A*)
2. (*D*)
3. (*E*)
4. (*F*)
5. (*B*)
6. (*C*)

Digestive System 19

Objective A To define *digestion* as a mechanical and chemical process.

The food we eat is utilized at the cellular level in chemical reactions involving synthesis of proteins, carbohydrates, hormones, and enzymes; cellular division, growth, and repair; and production of heat. To become usable by the cells, most food must first be mechanically and chemically reduced to forms that can be absorbed through the intestinal wall and transported to the cells by the blood.

19.1 Define as a mechanical and/or a chemical process: *ingestion*, *mastication*, *deglutition*, *peristalsis*, *absorption*, and *defecation.*

Ingestion. Taking of food into the gastrointestinal (GI) tract by way of the mouth (a mechanical process).

Mastication. Chewing movements to pulverize food (a mechanical process) and mix it with saliva (salivary action, a chemical process).

Deglutition. Swallowing of food (a mechanical process).

Peristalsis. Rhythmic, wavelike contractions that move food through the GI tract (a mechanical process).

Absorption. Passage of food molecules through the mucous membrane of the small intestine and into the circulatory or lymphatic systems for distribution to cells (mechanical and chemical processes).

Defecation. Discharge of indigestible wastes, called feces, from the GI tract (a mechanical process).

Objective B To describe the *structural* and *functional organization of the digestive system.*

The digestive system can be divided into a tubular **gastrointestinal tract** (*digestive tract*) and **accessory digestive organs** (see fig. 19.1). The GI tract of an adult is about 9 m (30 ft) long and extends from the oral cavity (mouth) to the anus. The regions or organs of the GI tract include the *oral cavity* (also called the *buccal cavity*), *pharynx, esophagus, stomach, small intestine,* and *large intestine*. The *rectum* and *anal canal* are located at the terminal end of the large intestine. The accessory digestive organs include the *teeth, tongue, salivary glands, liver, gallbladder,* and *pancreas*.

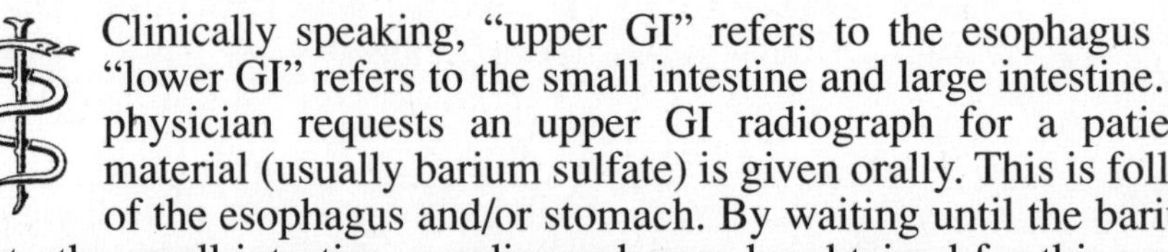

Clinically speaking, "upper GI" refers to the esophagus and stomach and "lower GI" refers to the small intestine and large intestine. For example, if a physician requests an upper GI radiograph for a patient, a radiopaque material (usually barium sulfate) is given orally. This is followed by an X ray of the esophagus and/or stomach. By waiting until the barium works its way into the small intestine, a radiograph may be obtained for this part of the lower GI tract. If a radiograph of the large intestine is needed, the patient is administered a barium enema.

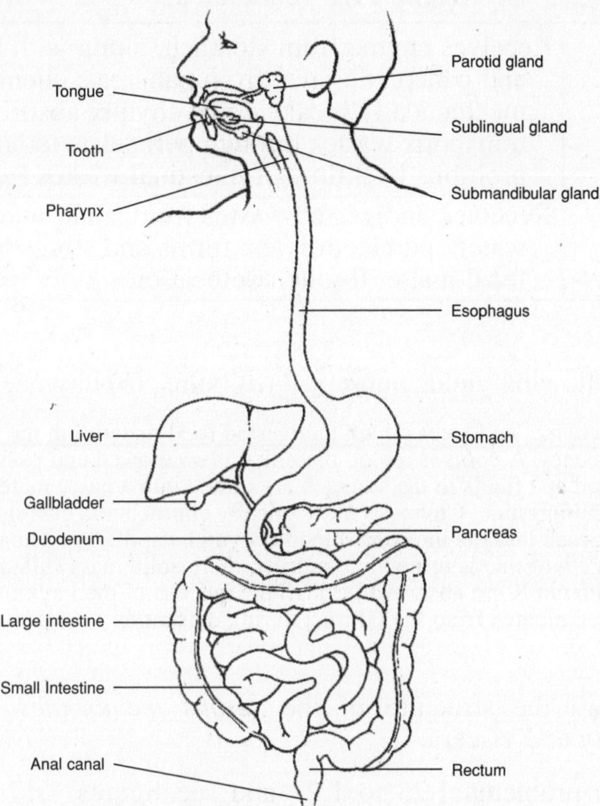

Figure 19.1 The digestive system.

19.2 Distinguish between the terms *viscera* and *gut*.

Although **viscera** is frequently used in reference to the abdominal organs of digestion, the term actually applies to all thoracic and abdominal organs. **Gut** has reference to the developing GI tract.

19.3 What are the basic functions of the organs or regions of the digestive system?

See Table 19.1.

Table 19.1

Organs or regions	*Functions*
Oral cavity	Ingests food; grinds food and mixes it with saliva (mastication); initiates digestion of carbohydrates; forms bolus (food mass); swallows bolus (deglutition)
Pharynx	Receives bolus from oral cavity; autonomically continues deglutition of bolus to esophagus
Esophagus	Transports bolus to stomach by peristalsis; esophageal sphincter restricts backflow of food
Stomach	Receives bolus from esophagus; churns bolus with gastric juice to from chyme; initiates digestion of proteins; carries out limited absorption; moves chyme into duodenum and prohibits backflow

Table 19.1 (continued)

Small intestine	Receives chyme from stomach, along with bile from liver and pancreatic juice from pancreas; chemically and mechanically breaks down chyme; absorbs nutrients; transports wastes through peristalsis to large intestine; prohibits backflow of intestinal wastes from large intestine
Large intestine	Receives undigested wastes from small intestine; absorbs water and electrolytes; forms and stores feces and expels fecal matter through defecation

19.4 Which of the following yields nutrients? (*a*) bolus, (*b*) chyme, (*c*) feces

(*b*) chyme. Bolus, chyme, and feces all are undigested food material in the GI tract but differ in location and consistency. A *bolus* is a food mass that is swallowed and passed from the oral cavity to the stomach. Food and fluids in the stomach are turned into a pasty material called *chyme* that is moved into the small intestine. Chyme is mechanically churned and chemically altered, enabling nutrients to be absorbed into the intestinal blood or lymph capillaries. The undigested chyme that is moved into the large intestine is turned into a dryer, more solid mass called *feces*. Water, electrolytes, and vitamin K are absorbed through the mucosa of the large intestine as feces are formed. Feces are eliminated from the GI tract during defecation.

Objective C To detail the structure of the *serous membranes* associated with the abdominal cavity and viscera.

Review problems 1.25 to 1.27 and see figures 1.17 and 19.2. A **serous membrane** is an epithelial and connective tissue membrane that lines body cavities and covers visceral organs within these cavities. Serous membranes also secrete a lubricative serous fluid. A serous membrane covering a visceral organ is sometimes called *serosa*. The serous membranes within the abdominal cavity are specifically referred to as the *peritoneum*. The **parietal peritoneum** lines the wall of the abdominal cavity. It comes together posteriorly to form the double-layered *mesentery* that supports the lower GI tract. The peritoneal covering continues around the abdominal viscera as the **visceral peritoneum**.

Extensions of the parietal peritoneum serve specific functions. The *falciform ligament* attaches the liver to the diaphragm and anterior abdominal wall. The *lesser omentum* extends between the liver and the lesser curvature of the stomach. The *greater omentum* extends from the greater curvature of the stomach to the transverse colon, forming an apronlike structure over the small intestine. The omentum stores fat, cushions visceral organs, supports lymph nodes, and protects against the spread of infection.

19.5 Define *mesentery* and *mesocolon.*

As stated in the survey of Objective C, the **mesentery** is a double fold of the peritoneum that supports the intestines. The **mesocolon** (see fig. 19.2) is the specific portion of the mesentery that supports the large intestine. The mesentery provides support and yet permits peristaltic digestive movements. Being a double-folded structure, the mesentery also encloses the vessels and nerves that go to and from the intestines.

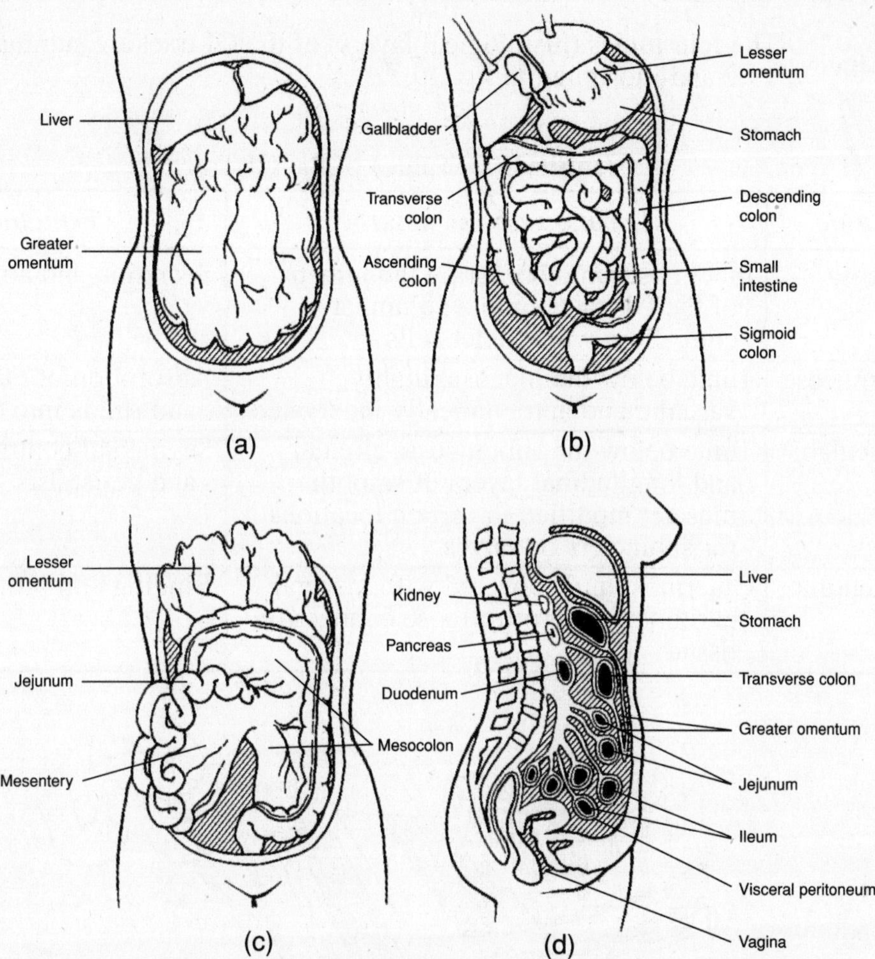

Figure 19.2 Serous membranes of the abdominal cavity. (*a*) The intact appearance of the greater omentum covering the abdominal viscera; (*b*) the abdominal viscera with the greater omentum removed; (*c*) the mesentery as seen with the greater omentum elevated; and (*d*) the viscera and serous membranes seen in sagittal view.

19.6 Distinguish between the abdominal and the peritoneal cavities.

The **abdominal cavity** is the space within the confines of the abdominal wall. The **peritoneal cavity** is the space between the *parietal* and *visceral portions* of the *peritoneum.* Most of the abdominal viscera (see problem 1.20) are located within both cavities; a few, such as the retroperitoneal organs are located within the abdominal cavity only.

The moist peritoneal cavity is an aseptic environment. *Peritonitis* is an inflammation of the peritoneum caused by bacterial contamination of the peritoneal cavity. Entry of infectious pathogens may occur through trauma (such as a stab wound), rupture of a visceral organ (such as the appendix), ectopic pregnancy, or contamination during surgery. Untreated peritonitis is generally fatal.

Objective D To identify the *basic tunics of the GI tract.*

The four tunics (histological layers) of the GI tract are summarized in table 19.2 and illustrated in fig. 19.3.

Table 19.2 Tunics of the GI Tract

Tunic	*Location and structure*	*Function*
Mucosa	Innermost tunic bordering the lumen of the GI tract; simple columnar epithelium with goblet cells	Secretion and absorption
Submucosa	Tunic below the mucosa; highly vascular and autonomically innervated	Absorption of nutrients and fluids into capillaries
Muscularis	Tunic below the submucosa; circular and longitudinal layers of smooth muscle; modified in certain locations for sphincters or valves	Segmental contractions and peristalsis
Adventitia	Outermost tunic covered with visceral peritoneum (serosa); loose connective tissue	Binding and protection

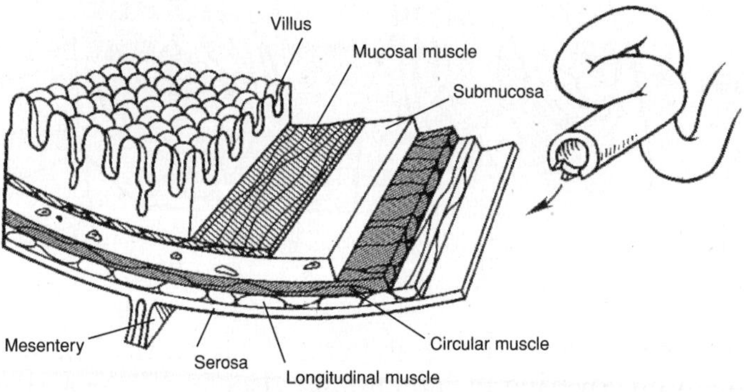

Figure 19.3 Tunics of the small intestine.

19.7 Describe the autonomic innervation of the lower GI tract.

The **submucosal plexus**, or *plexus of Meissner*, provides autonomic innervation to the *muscularis mucosae* (a thin layer of smooth muscle of the tunica mucosa). The **myenteric plexus**, or *plexus of Auerbach*, located between the longitudinal and circular muscle layers of the tunica muscularis, provides the principal innervation of the GI tract and includes fibers and ganglia from both sympathetic and parasympathetic autonomic divisions.

Objective E To describe the anatomy of the *oral cavity*, *pharynx*, and *esophagus*, and to state the digestive events that take place in these regions of the GI tract.

The **oral cavity** (mouth), or *buccal cavity*, formed by the *lips, cheeks, hard palate* and *soft palate,* and *tongue* (fig. 19.4), serves as a receptacle for food; initiates digestion through mastication by the teeth; participates in swallowing; and forms words in speech. The **pharynx**, which is posterior to the oral cavity, is a common passageway for the respiratory and digestive systems. The **esophagus** transports food and fluid from the pharynx to the stomach.

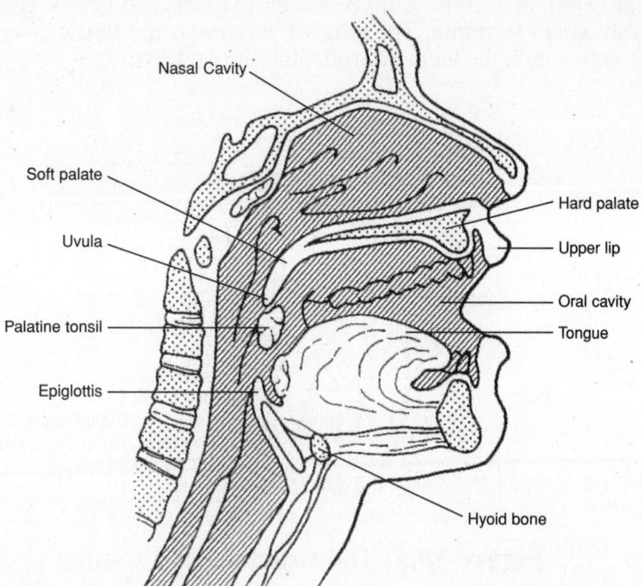

Figure 19.4 The oral cavity, pharynx, and esophagus seen in a sagittal section of the head.

19.8 Which of the following is *not* one of the four basic kinds of teeth? (*a*) incisors, (*b*) canines, (*c*) premolars, (*d*) wisdom teeth, (*e*) molars.

(*d*). The wisdom teeth are third molars. If they erupt, it is generally between the ages of 17 and 25. The teeth present in both the upper and lower jaw are listed below and their locations can be noted in fig. 6.8.

Incisors. There are four upper and four lower chisel-shaped anteromost incisors, adapted for cutting and shearing food.

Canines. There are two upper and two lower cone-shaped canines (eye teeth), adapted for holding and tearing.

Premolars. There are four upper and four lower premolars (bicuspids) with roughened cusps, adapted for crushing and grinding food.

Molars. There are six upper and six lower molars (the posteromost four of which are the wisdom teeth); adapted for crushing and grinding food.

If the wisdom teeth erupt through the gums, it is generally in a person's late teens or early twenties. There is, however, great variation in the time of eruption of the wisdom teeth. An *impacted tooth* is one that cannot emerge through the gum, and so becomes rotated, displaced, or tilted. An impacted wisdom tooth is common because the jaws are formed and the other teeth are in place long before a wisdom tooth tries to emerge. An impacted tooth can cause pain and occasionally becomes infected. Extraction of the tooth is the usual treatment.

19.9 Define *heterodontia* and *diphyodontia*.

Heterodontia refers to the differentiation of teeth for different tasks (see problem 19.8). **Diphyodontia** refers to the development of two sets of teeth in a lifetime; in humans, there are 20 **deciduous** (*baby*) **teeth**, or *milk teeth*, and 32 **permanent teeth**.

19.10 Diagram a typical tooth and state the functions of the principal structures.

The exposed **crown** of a tooth is covered with protective **enamel** (fig. 19.5). The **cusp** of the tooth is the chewing surface of the crown. The **dentin** provides structural support to the tooth; it surrounds the **pulp**, which contains nerves and blood and lymph vessels. A thin layer of **cementum** fastens the tooth to the **periodontal ligament**, which borders a *tooth socket* called a **dental**

alveolus. A **root canal**, in each root, for passage of vessels and nerves communicates with the pulp cavity through an **apical foramen**. The **gingiva**, or *gum*, is the fleshy covering over the mandible and maxilla through which the teeth protrude into the oral cavity.

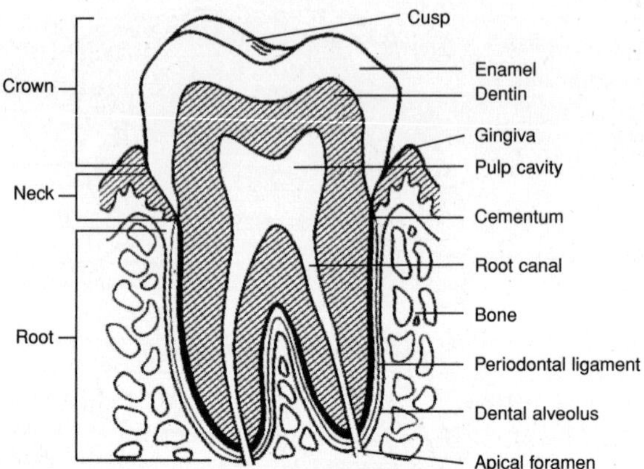

Figure 19.5 The structure of a tooth.

19.11 Where are the salivary glands located, and what are the functions of saliva?

Collectively, the three pairs of salivary glands produce 1000 to 1400 ml of saliva daily, in response to parasympathetic stimulation. The saliva cleanses the teeth, initiates carbohydrate digestion through the action of amylase (see table 19.5), helps form the bolus, lubricates the oral cavity and pharynx, and dissolves food chemicals so that they can be tasted. The location of the salivary glands is shown in fig. 19.6. The features of the glands are summarized in table 19.3.

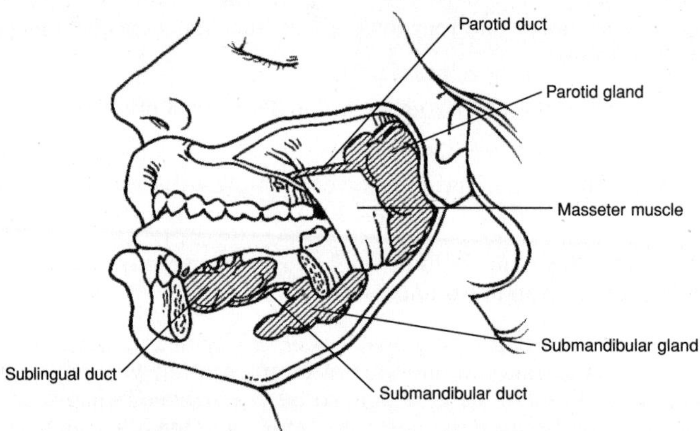

Figure 19.6 The salivary glands and their ducts.

19.12 What are the functions of the tongue?

The **tongue** moves food around in the mouth during mastication and assists in swallowing. Movement of the tongue also cleanses the teeth. The tongue is extremely important in forming sounds during speaking. *Taste buds* (see fig. 12.1) along the surface of the tongue respond to food chemicals (flavors), provoking the secretion of saliva, and to a limited extent, the flow of gastric juice (see problem 19.17).

Table 19.3 The Salivary Glands

Salivary gland	*Location*	*Salivary duct*	*Entry into oral cavity*	*Type of secretion*
Parotid gland	Anteroinferior to auricle of ear; subcutaneous over masseter muscle	Parotid duct	Lateral to upper second molar	Watery serous fluid containing salts and enzymes
Submandibular gland	Inferior to base of tongue	Submandibular duct	At papilla, lateral to lingual frenulum	Watery serous fluid containing some mucus
Sublingual gland	Anterior to submandibular gland under the tongue	Several small sublingual ducts	With submandibular duct	Mostly thick, stringy mucus containing salts and enzymes

19.13 Describe the location and structure of the palate.

The **palate** is the roof of the oral cavity, consisting anteriorly of the bony **hard palate** and posteriorly of the **soft palate**. Transverse ridges, called **palatal rugae**, are located along the mucous membranes of the hard palate, where they serve as friction bands against which the tongue is placed during swallowing. The **uvula** is suspended from the soft palate. During swallowing, the soft palate and uvula are drawn upward, closing the nasopharynx and preventing food and fluid from entering the nasal cavity.

19.14 Distinguish between the pharynx and the esophagus.

The funnel-shaped pharynx is a passageway approximately 13 cm (5 in.) long that connects the oral and nasal cavities to the esophagus and trachea (see fig. 19.1). It serves the digestive system (with the passage of food and fluid) and the respiratory system (with the passage of air).

The esophagus (see fig. 19.1) is that portion of the upper GI tract that connects the pharynx to the stomach. It is approximately 25 cm (10 in.) in length and is located behind the trachea in the thorax. The upper third of the esophagus contains skeletal muscle; the middle third contains skeletal and smooth muscle; and the lower third contains smooth muscle only.

19.15 Is deglutition voluntary or involuntary?

Only the *first stage of deglutition* (swallowing) is voluntary. This includes chewing of the food and the formation of a bolus that is forced against the soft palate with the elevated tongue. The *second stage*, the involuntary deglutition reflex, begins when pharyngeal sensory receptors are stimulated. During this stage, the uvula is elevated, sealing off the nasal cavity; the hyoid bone and larynx are elevated, so that food or fluid is less likely to enter the trachea; and the esophagus is opened. During the *third stage*, the bolus or fluid enters the esophagus and is transported to the stomach by peristalsis.

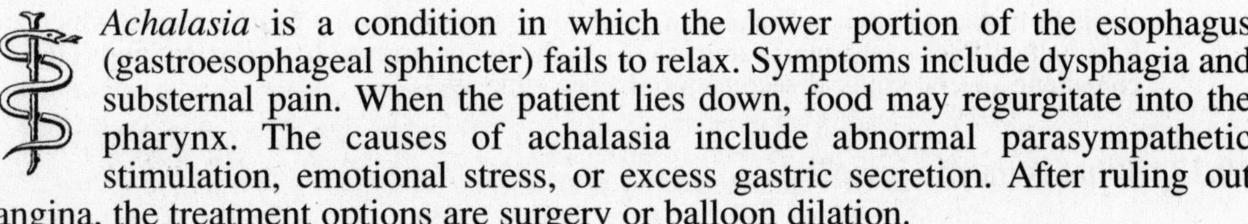

Achalasia is a condition in which the lower portion of the esophagus (gastroesophageal sphincter) fails to relax. Symptoms include dysphagia and substernal pain. When the patient lies down, food may regurgitate into the pharynx. The causes of achalasia include abnormal parasympathetic stimulation, emotional stress, or excess gastric secretion. After ruling out angina, the treatment options are surgery or balloon dilation.

Objective F To describe the structure and function of the *stomach.*

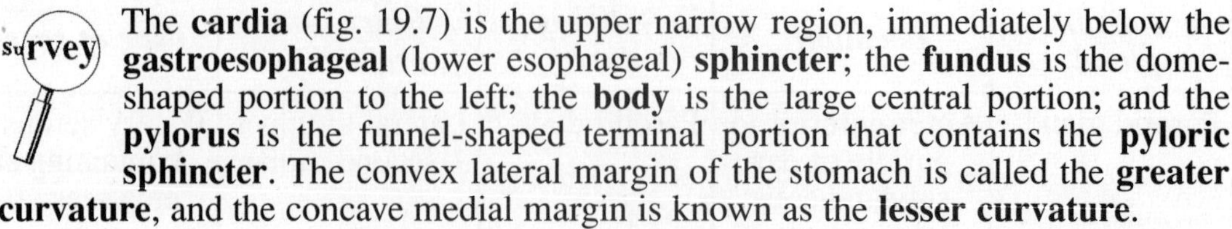

The **cardia** (fig. 19.7) is the upper narrow region, immediately below the **gastroesophageal** (lower esophageal) **sphincter**; the **fundus** is the dome-shaped portion to the left; the **body** is the large central portion; and the **pylorus** is the funnel-shaped terminal portion that contains the **pyloric sphincter**. The convex lateral margin of the stomach is called the **greater curvature**, and the concave medial margin is known as the **lesser curvature**.

The functions of the stomach are to store food as it is mechanically churned with gastric secretions (see problem 19.17) in the formation of *chyme*; to initiate the digestion of proteins; to carry out limited absorption; and to move chyme into the small intestine. In addition, the low pH of gastric juice (pH of about 2.0) helps to kill bacteria that may have been ingested with food or fluid. The mucosal wall is permeable enough to allow some absorption (for example, alcohol readily absorbs through the stomach mucosa).

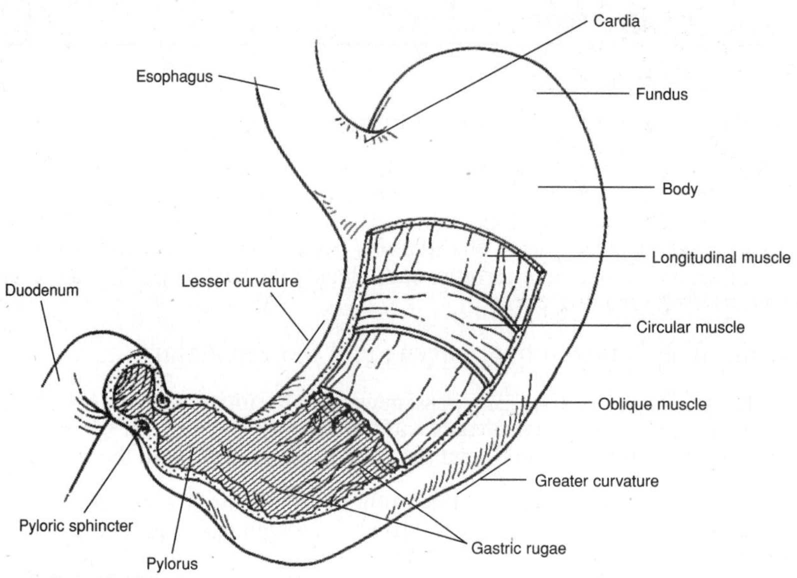

Figure 19.7 The stomach. The tunics are stripped away to expose the gastric rugae bordering the lumen of the stomach.

19.16 Explain the specializations of the tunics of the stomach.

In addition to the **longitudinal** and **circular muscle layers** (see fig. 19.7) of the tunica muscularis, the stomach has an **oblique layer**. These three smooth muscle layers enable churning movements in the formation of chyme. Distension (stretching) is permitted by **gastric rugae**, which are longitudinal folds of the tunica mucosa. Also, the tunica mucosa has **gastric glands** (fig. 19.8) containing several kinds of secretory cells (see table 19.4).

19.17 What is gastric juice?

Gastric juice is the combined secretory product of the mucous, chief, parietal, and argentaffin cells (table 19.4). Hormones secreted from G cells into the bloodstream are not part of the gastric juice.

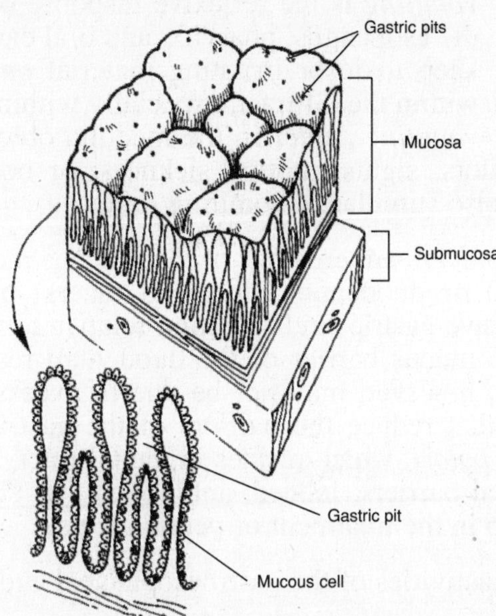

Figure 19.8 Gastric pits and gastric glands of the mucosa.

Table 19.4 Secretory Products of the Stomach

Component	*Source*	*Function*
Hydrochloric acid (HCl)	Parietal cells	Strong acid for killing pathogens; conversion of pepsinogen to pepsin
Pepsinogen	Chief cells	Inactive form of pepsin
Pepsin	Formed from pepsinogen in presence of HCl	Protein-splitting enzyme
Mucus	Goblet cells	Viscous, alkaline, protective coating of mucosa
Intrinsic factor	Parietal cells	Aids absorption of vitamin B_{12}
Serotonin and histamine	Argentaffin cells	Autocrine regulators
Gastrin	G cells	Stimulates secretion of HCl and pepsin

19.18 Describe the phases of gastric secretion.

Cephalic phase. In response to sight, taste, smell, or even certain thoughts, parasympathetic impulses in the vagus nerves initiate the secretion of 50 to 150 ml of gastric juice.

Gastric phase. Food-induced distension of the tunica mucosa, along with the chemical breakdown of protein, stimulates the release of gastrin (see table 19.4).

Intestinal phase. Chyme entering the duodenum stimulates the release of gastrin, which leads to the production of additional small quantities of gastric juice.

Vomiting is the reflexive response of emptying the stomach through the esophagus, pharynx, and oral cavity. It is a protective response to keep toxic or irritating material out of the lower GI tract. Stimuli within the GI tract, especially within the duodenum, may activate the *vomiting center* in the medulla oblongata of the brain stem, as may nauseating odors, sights, motion sickness, or body stress. Certain drugs, called *emetics*, can also stimulate a vomiting reflex.

Peptic ulcers are erosions of the mucous membranes of the stomach (gastric ulcers) or duodenum (duodenal ulcers) produced by the action of HCl. Excessive gastric acid secretion relative to the degree of protection afforded by the mucus barrier of the duodenum results in duodenal ulcers. Gastric ulcers, however, may not be due to excessive acid secretion, but rather to mechanisms that reduce the barriers of the gastric mucosa to self-digestion. The bacterium *H. pylori*, which resides in the GI tract, may contribute to the weakening of the mucosal barriers; indeed, antibiotics that eliminate this infection have been shown to help in the treatment of peptic ulcers.

19.19 Compare the activities of the gastroesophageal and pyloric sphincters.

The **gastroesophageal sphincter** (*lower esophageal sphincter*) is located at the junction of the esophagus and the stomach. It is a specialized portion of the circular layer of the tunica muscularis that constricts after food or fluid passes into the stomach. It is not a true sphincter, however, because during reflexive vomiting it opens to permit flow of the regurgitated matter into the esophagus toward the oral cavity.

Located at the terminal part of the stomach, the **pyloric sphincter** is also a specialization of the circular layer of the tunica muscularis. The pyloric sphincter regulates the movement of chyme into the duodenum. Generally, it prohibits backflow of chyme into the stomach, but with prolonged and extremely forceful vomiting it may be forced open, permitting some bile to be regurgitated.

Objective G To identify the *regions of the small intestine* and to discuss the process of *food absorption.*

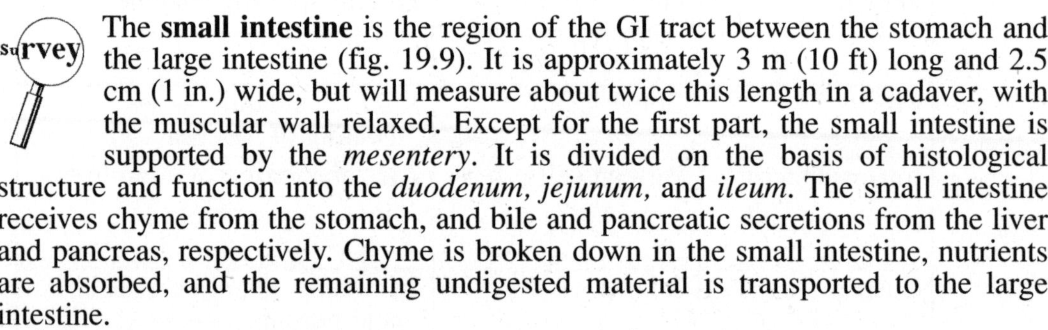

The **small intestine** is the region of the GI tract between the stomach and the large intestine (fig. 19.9). It is approximately 3 m (10 ft) long and 2.5 cm (1 in.) wide, but will measure about twice this length in a cadaver, with the muscular wall relaxed. Except for the first part, the small intestine is supported by the *mesentery*. It is divided on the basis of histological structure and function into the *duodenum, jejunum,* and *ileum*. The small intestine receives chyme from the stomach, and bile and pancreatic secretions from the liver and pancreas, respectively. Chyme is broken down in the small intestine, nutrients are absorbed, and the remaining undigested material is transported to the large intestine.

19.20 How may the three regions of the small intestine be distinguished?

The C-shaped **duodenum** measures 25 cm (10 in.) from the pyloric sphincter of the stomach to the **duodenojejunal flexure** (fig. 19.10). It receives bile secretions through the *common bile duct* from the liver and gallbladder, and pancreatic secretions through the *pancreatic duct*. Mucus-secreting **duodenal** (*Brunner's*) **glands** are numerous in the submucosa of the duodenum.

The 1-m (3-ft) **jejunum** extends from the duodenum to the ileum. It is characterized by deep folds called **plicae circulares** in the mucosa and submucosa (see problem 19.21 and figure 19.9).

The 2-m (6-ft) **ileum** joins the cecum of the large intestine at the **ileocecal valve**. Aggregates of **mesenteric lymph nodes** (*Peyer's patches*) are abundant in the ileum (see fig. 17.3).

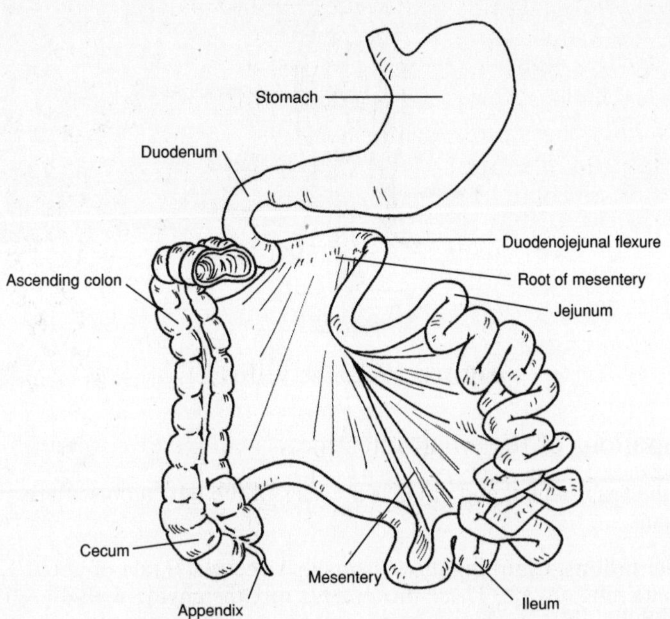

Figure 19.9 The small intestine.

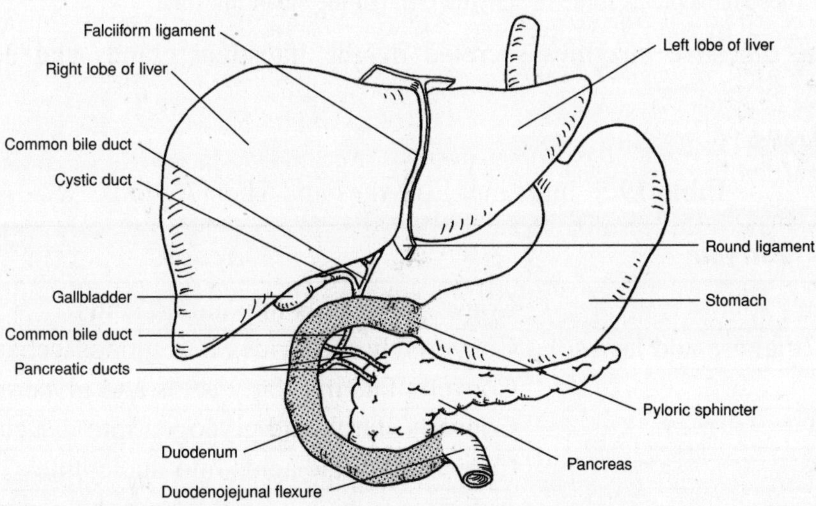

Figure 19.10 The duodenum and associated structures.

19.21 What structural modifications of the small intestine facilitate absorption?

The **plicae circulares** (see fig. 19.9) increase the absorptive area. These folds, in turn, are covered by numerous fingerlike projections of the mucosa called **villi**. Each villus (fig. 19.11) contains a capillary network, smooth muscle, and a specialized lymph vessel called a **lacteal**. Absorption is accomplished as food molecules enter the minute vessels of the villi through **microvilli**, which are microprojections on the free surface of the cells lining the villi. At the bases of the villi are **intestinal glands** (or *crypts of Lieberkühn*) that secrete digestive enzymes.

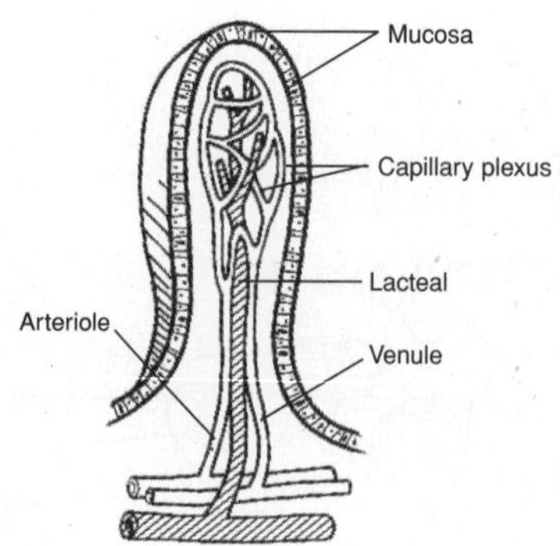

Figure 19.11 A villus.

19.22 Describe the motions of the small intestine.

Contractions of the longitudinal and circular muscles of the tunica muscularis produce three distinct types of movements.

Rhythmic segmentations. Churning movements that occur at a rate of about 12 to 16 per minute in the regions containing chyme. These movements mix the chyme with digestive juices and bring it into contact with the villi.

Pendular movements. Irregular constrictions that cause wavelike movements first in one direction, then back again. These movements further mix the chyme.

Peristalses. Rhythmic local contractions of smooth muscles that occur at a rate of 15 to 18 per minute. These movements force the chyme through the small intestine.

19.23 List the digestive enzymes secreted by the intestinal glands and describe their actions.

See table 19.5.

Table 19.5 Intestinal Enzymes and Their Actions

Enzyme	*Action*
Peptidase	Converts proteins into amino acids
Sucrase (maltase and lactase)	Converts disaccharides into monosaccharides
Lipase	Converts fats into fatty acids and glycerol
Amylase	Converts starch and glycogen into disaccharides
Nuclease	Converts nucleic acids into nucleotides
Enterokinase	Activates trypsin secreted from the pancreas

Objective H To describe the *large intestine* and to discuss its functions.

survey

The **large intestine** is the region of the GI tract that extends from its union with the small intestine at the ileocecal valve to the anus (fig. 19.12). It is approximately 1.5 m (5 ft) long and 6.5 cm (2.5 in.) in diameter. The large intestine is structurally divided into the *cecum*, *colon*, *rectum*, and *anal canal*. Other than absorbing water, electrolytes, and small amounts of vitamin K, the large intestine plays a minor digestive role. It's principal function is to form, store, and expel feces from the GI tract.

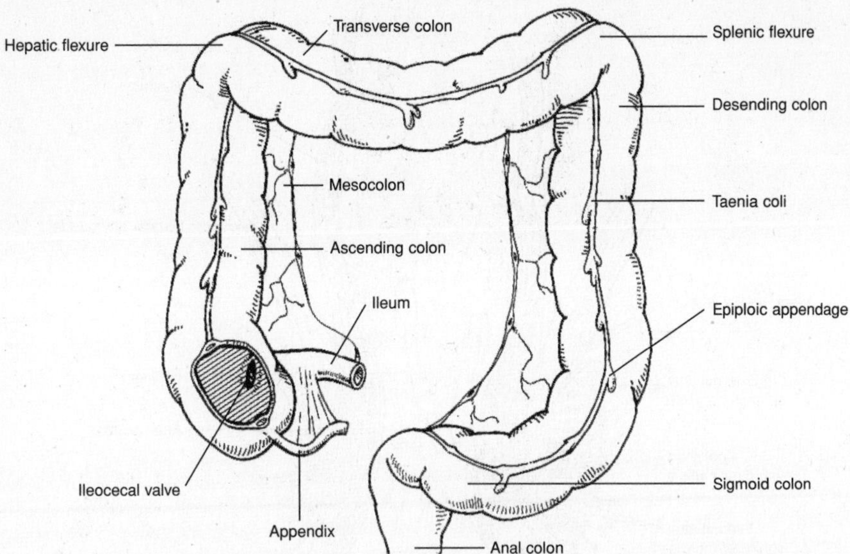

Figure 19.12 The large intestine.

19.24 What is the cecum?

The **cecum** (fig. 19.12), which resembles a dilated pouch, is the first portion of the large intestine. It receives chyme through the **ileocecal valve**, a fold of mucous membrane at the junction of the small intestine and large intestine that prohibits the backflow of chyme.

19.25 Does the appendix have a digestive function?

No. The **appendix** is an 8-cm (3-in.) fingerlike attachment to the inferomedial margin of the cecum (fig. 19.12). It has an abundance of lymphatic tissue, which may serve to resist infection.

19.26 What are the four portions of the colon?

The **colon** consists of *ascending, transverse, descending,* and *sigmoid portions* (see fig. 19.12). The **ascending colon** extends superiorly from the cecum to the level of the liver, where it bends sharply to the left at the **hepatic flexure** (*right colic flexure*) and transversely crosses the upper peritoneal cavity as the **transverse colon**. Here, another right-angle bend called the **splenic flexure** (*left colic flexure*) marks the beginning of the **descending colon**. In the pelvic region, the colon angles medially into an S-shaped bend known as the **sigmoid colon**.

19.27 Describe the terminal portion of the large intestine.

The terminal 20 cm (7.5 in.) of the large intestine is the **rectum**, and the last 2 to 3 cm of the rectum is called the **anal canal** (fig. 19.13). Folds in the mucosa of the anal canal called *anal columns* permit distension during defecation. The **anus** is the external opening of the anal canal. The *internal anal sphincter* of smooth muscle tissue and the *external anal sphincter* of skeletal muscle tissue guard the anus.

Colitis is ulceration of the mucosal lining of the colon, especially in the descending and sigmoid portions. The symptoms include diarrhea (stools contain blood and mucus), loss of appetite, nausea, and abdominal tenderness. The causes may include mycobacteria, viruses, or allergy to dairy products. Treatment of colitis includes avoidance of milk products, nuts, certain fruits, and the use of anti-inflammatory drugs. Surgery may be required for severe colitis.

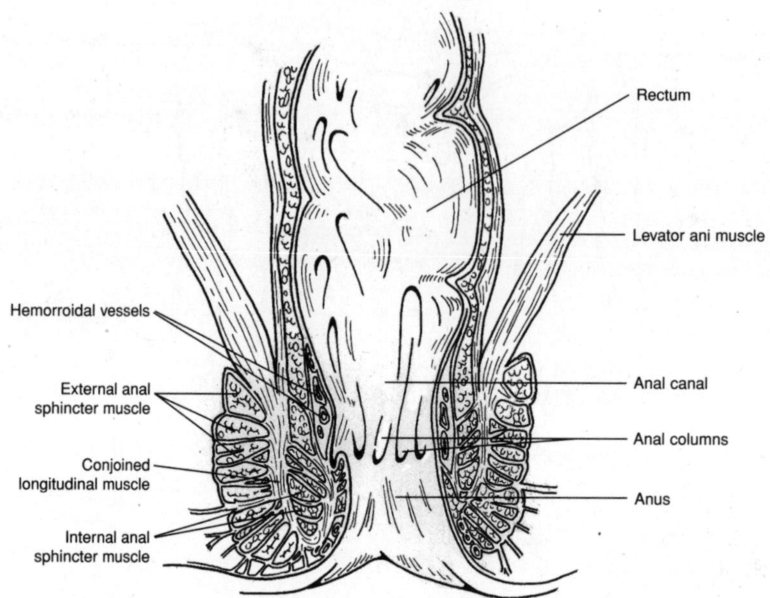

Figure 19.13 The anal canal.

19.28 Describe the tunics of the large intestine.

The mucosa and submucosa of the large intestine lack plicae, but do have sacculations, or **haustra**, along their length (see fig. 19.12). The tunica muscularis consists of longitudinal bands of smooth muscle called **taeniae coli**. In the tunica adventitia, attached superficially to the taeniae coli, are fat-filled pouches called **epiploic appendages**.

19.29 Are the movements of the large intestine the same as those of the small intestine?

Only the *peristaltic movements* of the colon are similar to those of the small intestine, although they are usually more sluggish (3 to 12 per minute). *Haustral churning* is the contraction of a haustrum stimulated by the distended tunics. *Mass movement* (two or three times a day following meals) is gross motion of fecal material, brought on by contraction of the taeniae coli.

19.30 How does the GI tract respond to parasympathetic stimulation and to sympathetic stimulation?

Parasympathetic stimulation of the GI tract and digestive organs generally increases digestive activity; specifically, glandular secretion and autonomic muscular movement. Sympathetic stimulation inhibits digestive activity. It is for this reason that excessive and prolonged stress (sympathetic stimulation) may result in GI dysfunction.

Prolonged *diarrhea* is of serious concern, especially in children and elderly people. Dehydration and loss of electrolytes present an immediate problem. Outbreaks of *cholera* are accompanied with life-threatening diarrhea. The cholera toxin stimulates the secretion of large amounts of electrolytes (sodium, chloride, and bicarbonate) and fluids in the GI tract. The loss of these body essentials causes homeostatic changes that result in death within hours or a few days. Treatment of patients with cholera includes administration of saline and glucose.

Objective I To describe the location, structure, and functions of the *liver* and *gallbladder*.

The reddish-brown **liver** is positioned beneath the diaphragm, in the epigastric and right hypochondriac regions of the abdominal cavity. Weighing 1.7 kg (3.5 to 4.0 lb) in an adult, the liver is the largest visceral organ. The **falciform ligament** separates the **right lobe** from the **left lobe** (fig. 19.14). The **caudate lobe** is located near the inferior vena cava and the **quadrate lobe** is sandwiched between the left lobe and the **gallbladder**. The **porta of the liver** is the concavity on the inferior surface where the *hepatic artery, hepatic portal vein, lymphatics,* and *nerves* enter the liver and where the **hepatic ducts** exit.

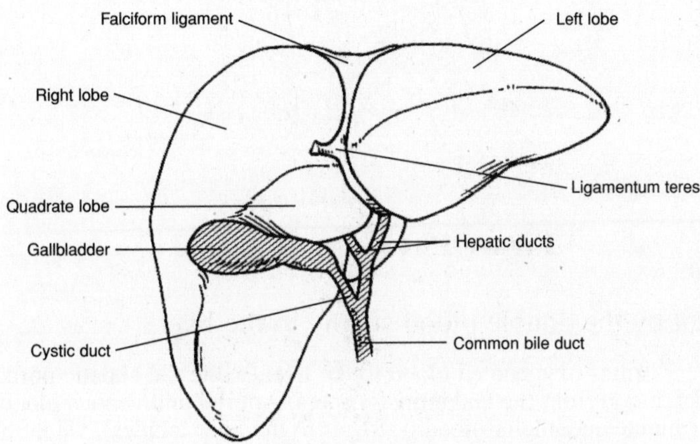

Figure 19.14 The liver and gallbladder.

The *ligamentum teres* (round ligament of the liver) extends from the falciform ligament (fig. 19.14) to the umbilicus. The ligamentum teres is the remnant of the umbilical vein of the fetus that transported oxygenated blood from the placenta. At birth, the umbilical vein collapses and becomes a supportive ligament of the liver.

19.31 Define and describe a liver lobule.

A **liver lobule** (fig. 19.15) consists of **hepatic plates** that are one to two cells thick. **Hepatocytes** are the principal cells of the hepatic plates. The hepatic plates are separated from each other by large capillary spaces called **sinusoids**. Phagocytic **Kupffer cells** line the sinusoids.

In the center of each liver lobule is a **central vein**, and at the periphery are branches of the *hepatic portal vein* and of the *hepatic artery*. These vessels open into the sinusoids. Because of the plate structure of the liver lobules, each hepatocyte is in direct contact with the blood.

Bile is produced by the hepatocytes and secreted into thin channels called **bile canaliculi** located within each hepatic plate. The bile drains toward the periphery of the hepatic plates to the **bile ducts**, which in turn drain into **hepatic ducts** that carry bile away from the liver. Since bile travels in the hepatic plates toward the periphery and blood travels in the opposite direction through the sinusoids toward the central vein, bile and blood do not mix in the liver lobules.

19.32 What are the functions of the liver?

The liver lobules are the functional structures of the liver. They carry out numerous functions, including synthesis, storage, and release of vitamins; synthesis, storage, and release of glycogen; synthesis of blood proteins; phagocytosis of worn red and white blood cells and certain bacteria; removal of toxic compounds; and production of bile.

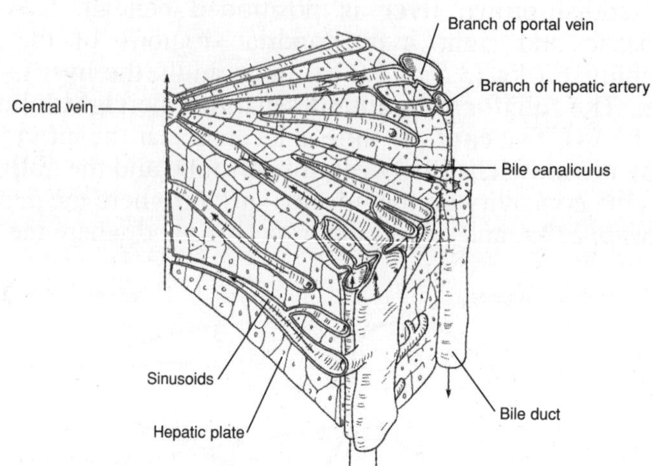

Figure 19.15 A liver lobule.

19.33 What is meant by the double blood supply to the liver?

The hepatic artery brings oxygenated blood to the liver, while the hepatic portal vein (see fig. 16.5) brings food-laden blood from the abdominal viscera. Arterial and venous blood are mixed in the liver sinusoids (minute endothelial-lined passages in the liver lobules), where oxygen, most of the nutrients, and certain toxic substances are extracted by the hepatic cells. When needed by other cells of the body, the nutrients are returned to the venous blood drainage via the hepatic veins that drain into the inferior vena cava.

19.34 Describe the gallbladder and explain the function of bile.

The **gallbladder** is a pouchlike organ attached to the inferior surface of the liver (see fig. 19.14). The mucosa lining the lumen of the gallbladder is folded into rugae similar to that of the stomach, allowing a storage capacity of about 35 to 50 ml. The function of the gallbladder is to store and concentrate bile.

Bile is continuously produced by the liver and drains through the hepatic ducts and bile duct to the duodenum. When the small intestine is empty of food, the **sphincter of ampulla** (*Oddi*) constricts, and bile is forced up the **cystic duct** to the gallbladder for storage. When food passes from the stomach into the duodenum, and under the influence of *cholecystokinin* (see table 13.4), the sphincter of ampulla relaxes and bile mixes with the chyme.

The contribution of bile to digestion is the emulsification of neutral fats and the absorption of fatty acids, cholesterol, and certain vitamins. Fatty acids are absorbed into the lacteals of the villi (see fig. 19.11) and into the lymphatic system.

Jaundice is a yellowish discoloration of the skin, mucous membranes, and sclera of the eyes produced by high blood concentrations of free or conjugated bilirubin. It is a symptom of many disorders, including liver diseases (such as hepatitis), bile duct obstructions (as produced by gallstones), and some anemias. Since free bilirubin is derived from heme, abnormally high concentrations of this pigment may be the result of an unusually high rate of red blood cell destruction. In newborns, jaundice is common and is generally of no concern; in some cases, however, it may indicate liver or bone marrow problems.

Objective J To examine the digestive role of the *pancreas*.

The **pancreas** (fig. 19.16) lies horizontally along the posterior abdominal wall, adjacent to the greater curvature of the stomach. The endocrine function of the pancreas, which involves the pancreatic islets, is discussed in problem 13.27; the exocrine function (of interest in this chapter) is discussed in problems 19.35 and 19.36.

The pancreas is about 12.5 cm (6 in.) long. It has an expanded **head**, located within the curvature of the duodenum, a centrally located **body**, and a tapering **tail**, which extends to the spleen. Pancreatic juice, produced in *acinar cells*, drains into the **pancreatic duct**, which in turn empties into the duodenum at the **duodenal papilla**.

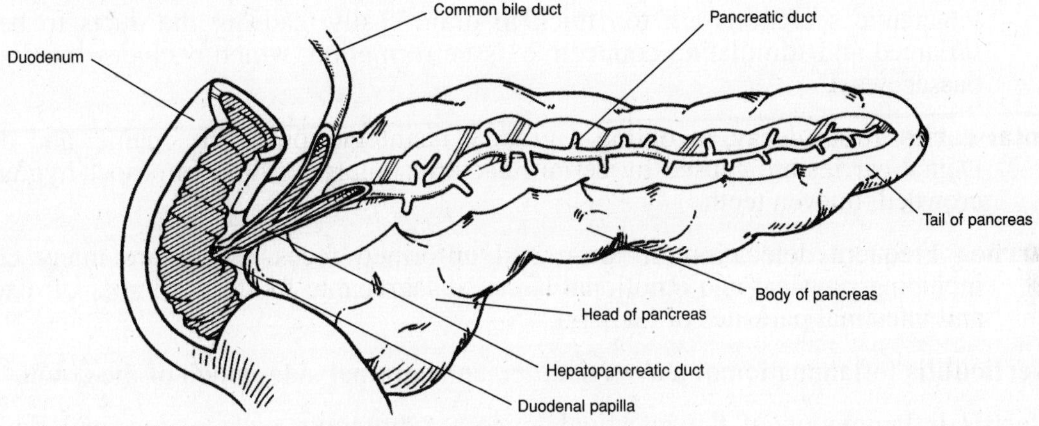

Figure 19.16 The pancreas.

19.35 Which digestive enzymes are contained in pancreatic juice?

Besides amylase, lipase, and nuclease (see table 19.5), pancreatic juice contains three peptidases—trypsin, chymotrypsin, and carboxypeptidase—that act to convert proteins into amino acids.

19.36 Describe the regulation of pancreatic exocrine secretion.

Secretin (see table 13.5) from the duodenum stimulates the release of pancreatic juice that contains a few digestive enzymes but with a high bicarbonate ion concentration. It is cholecystokinin, from the intestinal mucosa, that stimulates the release of pancreatic juice that has a high concentration of digestive enzymes.

Key Clinical Terms

Anorexia nervosa A psychological disorder characterized an inability or refusal to eat, resulting in extreme weight loss.

Appendicitis Inflammation of the appendix. The appendix is prone to infection, and so appendicitis is one of the most common surgical emergencies.

Bulimia A "binge-purge" syndrome of behavior in which uncontrollable overeating is followed by forced vomiting or use of excessive laxatives.

Cholelithiasis The presence or formation of gallstones. *Cholelithotomy* is the surgical removal of the gallstones from the gallbladder.

Cirrhosis A condition in which normal hepatocytes (liver cells) are replaced by connective tissue, causing blockage of the sinusoids. Hepatitis, alcohol, or malnutrition may cause cirrhosis.

Colostomy The surgical creation of an artificial excretory opening in the anterior abdominal wall. The colon is cut and brought to the surface of the skin. If the rectum has been removed because of cancer, the *stoma* (opening created by the colostomy) provides a permanent fecal outlet.

Constipation A condition in which the feces are retained for a longer-than-normal period of time; infrequent or difficult defecation.

Cystic fibrosis An inherited disease of the exocrine glands, particularly the pancreas. Pancreatic secretions are too thick to drain easily, causing the ducts to become inflamed and stimulating connective tissue formation, which occludes the drainage passageway.

Dental caries Tooth decay, involving a gradual disintegration of the enamel and dentin. Dental carries are caused by certain bacteria; improper diet; improper hygiene; or crowded, uneven teeth.

Diarrhea Frequent defecation of loose and unformed feces. There are many causes, including physical and emotional stress, disagreeable foods, a number of diseases, and intestinal parasites or bacteria.

Diverticulitis Inflammation of a *diverticulum*, an abnormal side-pouch of the colon.

Enteritis Inflammation of the intestinal mucosa, commonly called "intestinal flu." The symptoms of enteritis are cramps, nausea, and diarrhea. It may be caused by a virus or certain foods.

Gastric ulcer An open sore or lesion of the gastric (stomach) mucosa. Certain foods and medications, alcohol, coffee, aspirin, and overstimulation of the vagus nerves due to stress may be associated with the development of ulcers.

Gingivitis Inflammation of the gums. It may result from improper hygiene, poorly fitted dentures, improper diet, or certain body infections.

Halitosis Offensive odor of the breath. It may be due to dental caries, certain diseases, or the eating of certain foods or use of tobacco.

Hemorrhoids Varicose veins of the rectum and anus.

Hepatitis Inflammation of the liver. It is usually caused by viruses, but may also be caused by protozoa, bacteria, or the absorption of toxic materials.

Mumps (*parotitis*) A contagious viral disease resulting in inflammation of the parotid and other salivary glands. Mumps is particularly serious in adult males because it can cause testicular inflammation and sterility.

Periodontal disease A collective term for conditions characterized by degeneration of the gums, alveoli of teeth, and the associated structures.

Peritonitis Inflammation of the peritoneum.

Peptic ulcers Craterlike lesions that develop in the mucosa of the GI tract, in areas exposed to gastric juice.

Pyorrhea Discharge of pus at the base of the teeth at the gum line.

Trench mouth A contagious bacterial infection that causes inflammation, ulceration, and swelling of the floor of the mouth. Generally, it is contracted through direct contact by kissing an infected person. Trench mouth can be treated with penicillin and other medications.

Review Exercises

Multiple Choice

1. If an incision had to be made in the small intestine to remove an obstruction, which tunic would be cut first? (*a*) muscularis, (*b*) mucosa, (*c*) submucosa, (*d*) adventitia

2. The hepatic flexure of the large intestine occurs between (*a*) the transverse colon and descending colon, (*b*) the cecum and ascending colon, (*c*) the ascending colon and transverse colon, (*d*) the descending colon and rectum, (*e*) the descending colon and sigmoid colon.

3. Obstruction of the common bile duct by gallstones would most likely affect the digestion of (*a*) carbohydrates, (*b*) fats, (*c*) proteins, (*d*) nucleic acids, (*e*) none of the preceding.

4. Formation of gallstones is referred to as (*a*) jaundice, (*b*) cirrhosis, (*c*) hepatitis, (*d*) cholelithiasis.

5. Which of the following is *not* a function of saliva? (*a*) to initiate protein digestion, (*b*) to aid in cleansing the teeth, (*c*) to lubricate the pharynx, (*d*) to assist in the formation of a bolus

6. Peristalsis moves food material (*a*) in the stomach and small intestine only, (*b*) in the intestines only, (*c*) in the stomach and intestines only, (*d*) from the pharynx to the anal canal.

7. A gastrointestinal tumor involving the plicae circulares and villi might interfere with (*a*) deglutition, (*b*) absorption, (*c*) peristalsis, (*d*) defecation, (*e*) emulsification.

8. The greater omentum does *not* participate in (*a*) secretion of enzymes, (*b*) support and cushioning of the viscera, (*c*) storage of lipids, (*d*) protection against the spread of infection.

9. Most enzymes involved in protein digestion are (*a*) secreted by the pancreas, (*b*) activated by HCl, (*c*) present in the stomach, (*d*) secreted in an inactive form, (*e*) stimulated by enterokinase.

10. The terminal portion of the small intestine is (*a*) the ileum, (*b*) the cecum, (*c*) the duodenum, (*d*) the jejunum, (*e*) the colon.

11. The large intestine lacks (*a*) goblet cells, (*b*) epiploic appendages, (*c*) plicae circulares, (*d*) haustra.

12. Which of the following is *not* a major gastrointestinal hormone? (*a*) epinephrine, (*b*) secretin, (*c*) gastrin, (*d*) cholecystokinin

13. A set of permanent teeth contains (*a*) 20 teeth, (*b*) 30 teeth, (*c*) 32 teeth, (*d*) 24 teeth.

14. Teeth adapted to shear food are (*a*) premolars, (*b*) canines, (*c*) incisors, (*d*) molars.

15. A patient who has undergone a gastrectomy (removal of the stomach) may suffer from (*a*) cirrhosis, (*b*) pernicious anemia, (*c*) gastric ulcer, (*d*) inability to digest fats, (*e*) inability to digest proteins.

16. Amylase in saliva initiates digestion of (*a*) lipids, (*b*) proteins, (*c*) carbohydrates, (*d*) fats.

17. Pancreatic juice contains a protein-splitting enzyme called (*a*) trypsin, (*b*) zymogen, (*c*) pepsin (*d*) amylase, (*e*) nuclease.

18. Secretin is a hormone that (*a*) stimulates the release of pancreatic juice, (*b*) converts trypsinogen into trypsin, (*c*) activates chymotrypsin, (*d*) inhibits the action of pancreatic lipase.

19. Which of the following ducts is *not* associated with the digestive system? (*a*) cystic duct, (*b*) parotid duct, (*c*) pancreatic duct, (*d*) hepatic duct, (*e*) lacrimal duct

20. A dysfunction of the parietal cells of gastric glands would result in a decreased production of which two of the following? (*a*) mucus, (*b*) pepsinogen, (*c*) pepsin, (*d*) hydrochloric acid, (*e*) intrinsic factor

21. The part of the stomach that meets the esophagus at the gastroesophageal sphincter is (*a*) the fundus, (*b*) the cardia, (*c*) the pylorus, (*d*) the body, (*e*) the lesser curvature.

22. Which of the following statements is *false* concerning the hepatic plates?
(*a*) Each is only one or two cell layers thick.
(*b*) They are considered the functional units of the liver.
(*c*) A sinusoid separates adjacent hepatic plates.
(*d*) Arterial blood and portal venous blood mix within the sinusoids.
(*e*) Bile flows through the sinusoids of the hepatic plates.

23. The small intestine is held to the posterior abdominal wall by (*a*) the mesentery, (*b*) the falciform ligament, (*c*) the greater omentum, (*d*) the lesser omentum, (*e*) the visceral peritoneum.

24. The salivary gland located in front and slightly below the auricle of the ear is (*a*) the buccal gland, (*b*) the parotid gland, (*c*) the submandibular gland, (*d*) the sublingual gland.

25. The uvula is (*a*) a structure that guards the larynx, (*b*) a structure that extends into the lumen of the small intestine, (*c*) a fleshy extension of the soft palate, (*d*) a tonsil within the oral cavity, (*e*) a flap of the ileocecal valve.

True or False

____ 1. The principal function of the digestive system is to prepare food for cellular utilization.

____ 2. The GI tract has both sympathetic and parasympathetic innervation.

____ 3. Parasympathetic impulses to the GI tract decrease peristaltic activity.

____ 4. The tongue is a mass of smooth muscles covered by a mucous epithelial membrane.

____ **5.** Jaundice is a liver disease.

____ **6.** Pancreatic juice is secreted from acinar cells of the pancreas.

____ **7.** The falciform ligament attaches the liver to the diaphragm.

____ **8.** Intrinsic factor is necessary for the normal absorption of amino acids from the small intestine.

____ **9.** The spleen is an accessory digestive organ.

____ **10.** Deglutition is the process by which bile causes the breakdown of fat globules into smaller droplets.

____ **11.** The primary tissue of peritoneal membranes is simple squamous epithelium.

____ **12.** Intestinal rugae are folds of the mucosa within the small intestine that greatly increase the surface area for absorption.

____ **13.** Elevation of the uvula prevents food or fluid from entering the trachea.

____ **14.** The mesentery is a double-folded peritoneal membrane that supports the liver.

____ **15.** Cirrhosis is a chronic disease of the liver in which fibrous tissue replaces functional hepatic cells.

Completion

1. The ____________________ is the serous membrane that lines the wall of the abdominal cavity and covers visceral organs.
2. The ____________________ is the specific part of the mesentery that supports the large intestine.
3. Food and fluid in the stomach are consolidated into a pasty material called ____________________.
4. ____________________ refers to differentiation of the teeth for specific tasks.
5. Capillaries within the ____________________ of the small intestine are sites where nutrients and fluids are absorbed into the circulatory system.
6. A ____________________ contains a lacteal for the absorption of fats and lymph into the lymphatic system.
7. Autonomic innervation of the lower GI tract is through the ____________________ plexus to the muscularis mucosae and the ____________________ plexus to the tunica muscularis.
8. The ____________________ ligament binds the cementum of a tooth to a dental alveolus.
9. Hepatic plates within a liver lobule are separated from each other by spaces called hepatic ____________________, which permit passage of blood.
10. ____________________ from the duodenum stimulates the release of pancreatic juice.

Labeling

Label the structures indicated on the figure to the right.

1. ______________________
2. ______________________
3. ______________________
4. ______________________
5. ______________________
6. ______________________

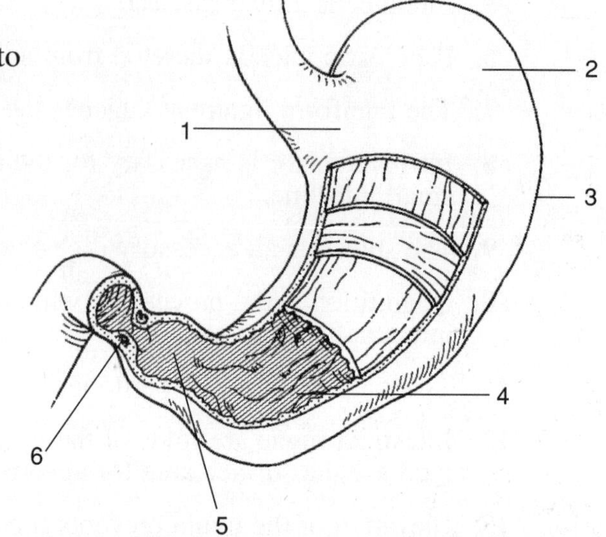

Matching

Match the digestive compound with its function.

____	**1.** Gastrin	(*a*)	emulsifies fats
____	**2.** Bile	(*b*)	converts pepsinogen to pepsin
____	**3.** Peptidase	(*c*)	converts proteins into amino acids
____	**4.** Sucrase	(*d*)	stimulates secretion of HCl and pepsin
____	**5.** Nuclease	(*e*)	converts fats into fatty acids and glycerol
____	**6.** HCl	(*f*)	converts disaccharides into monosaccharides
____	**7.** Amylase	(*g*)	activates trypsin secreted from pancreas
____	**8.** Enterokinase	(*h*)	converts nucleic acids into nucleotides
____	**9.** Lipase	(*i*)	converts starch and glycogen into disaccharides

Answers and Explanations for Review Exercises

Multiple Choice

1. (*d*) The four basic tunics (layers) of the GI tract from outermost to innermost are the adventitia, muscularis, submucosa, and mucosa. The mucosa borders the lumen of the GI tract.
2. (*c*) The ascending colon extends superiorly to the level of the liver, where it bends sharply to the left at the hepatic (right colic) flexure.
3. (*b*) Arriving at the duodenum through the common bile duct, bile is essential for the emulsification (breakdown of large droplets to small particles) of fats and the absorption of fatty acids, cholesterol, and certain vitamins.
4. (*d*) Gallstones (biliary calculi) afflict 20% of Americans over the age of 65 each year. Removal of gallstones is referred to as cholelithotomy.

5. (*a*) The digestive enzyme amylase contained in saliva initiates the digestion of carbohydrates.
6. (*d*) Autonomic peristaltic contractions move food through all regions of the GI tract.
7. (*b*) The plicae circulares, villi, and microvilli are structural features of the small intestine that increase the absorptive area.
8. (*a*) The greater omentum forms an apronlike structure over the intestines. Although it does not secrete enzymes, it stores fats, cushions visceral organs, supports lymph nodes, and protects against the spread of infection.
9. (*d*) Most enzymes involved in the digestion of proteins are secreted in an inactive form and are activated by HCl.
10. (*a*) The ileum is the last region of the small intestine, positioned between the jejunum and the ileocecal valve.
11. (*c*) The plicae circulares increase the absorptive area of the small intestine.
12. (*a*) Epinephrine (adrenaline) is a hormone secreted from the adrenal medulla during sympathetic stimulation.
13. (*c*) Human dentition includes 20 deciduous (baby) teeth and 32 permanent (adult) teeth.
14. (*c*) Shaped like chisels, the four upper and four lower incisors are adapted for cutting and shearing food, as in biting an apple.
15. (*b*) Following a gastrectomy, there is a loss of parietal cells that secrete intrinsic factor. Intrinsic factor is required for intestinal absorption of vitamin B_{12}, needed for red blood cell production. In the absence of intrinsic factor, pernicious anemia develops.
16. (*c*) Secreted in saliva, amylase initiates the digestion of carbohydrates.
17. (*a*) Pancreatic juice contains trypsin along with two other peptidases—chymotrypsin and carboxypeptidase.
18. (*a*) Secretin stimulates the release of pancreatic juice rich in bicarbonate.
19. (*e*) The lacrimal duct drains lacrimal fluid (tears) onto the surface of the eye.
20. (*d*) and (*e*) The parietal cells of the stomach secrete both hydrochloric acid and intrinsic factor.
21. (*b*) The cardia is the upper narrow region of the stomach, immediately below the gastroesophageal (lower esophageal or cardiac) sphincter.
22. (*e*) Bile flows through the bile canaliculi of the hepatic plates to the bile ducts.
23. (*a*) The mesentery is a double-layered membrane that supports the intestines.
24. (*b*) The parotid gland is the largest of the three salivary glands; it is located over the masseter muscle, anterior to the auricle of the ear.
25. (*c*) The uvula is a pendulous structure at the back of the soft palate; it is elevated while swallowing, blocking off the nasopharynx.

True or False

1. True
2. True
3. False; the parasympathetic division of the ANS increases peristaltic activity
4. False; the tongue is a mass of skeletal muscle.
5. False; jaundice is a symptom of one of several abnormal conditions in which there is an excess of bilirubin in the body fluids or tissues (e.g., liver disease, excessive breakdown of red blood cells, blocked bile duct)
6. True
7. True
8. False; intrinsic factor is necessary for the absorption of vitamin B_{12}
9. False; composed of lymphoid tissue, the spleen is considered an organ of the lymphatic system
10. False; deglutition is the mechanism of swallowing; emulsification is the breakdown of fat globules into small droplets
11. True
12. False; folds in the small intestine are called plicae circulares
13. False; elevation of the uvula prevents food or fluid from entering the nasopharynx
14. False; the mesentery supports the lower GI tract
15. True

Completion

1. peritoneum
2. mesocolon
3. chyme
4. Heterodontia
5. mucosa
6. villus
7. submucosal, myenteric
8. periodontal
9. sinusoids
10. Secretin

Labeling

1. Cardia
2. Fundus
3. Greater curvature
4. Gastric rugae
5. Pylorus
6. Pyloric sphincter

Matching

1. (*d*)
2. (*a*)
3. (*c*)
4. (*f*)
5. (*h*)
6. (*b*)
7. (*i*)
8. (*g*)
9. (*e*)

Metabolism, Nutrition, and Temperature Regulation 20

Objective A To define *metabolism* and *nutrients.*

Foods are first digested, then absorbed, and finally, metabolized. **Metabolism** refers to all the chemical reactions of the body. There are two aspects of metabolism: *catabolism*, the breaking-down process (e.g., glycolysis, in which glucose is broken down to yield products and energy), and *anabolism*, the building-up process (e.g., the biosynthesis of proteins, in which amino acids are joined together to form a protein). In terms of energy, metabolism may be thought of as a balancing of catabolism, which provides energy (stored in ATP), and anabolism, which requires energy.

20.1 What are nutrients?

Chemical substances in food that enter into metabolism are referred to as *nutrients*. They are classified as **carbohydrates, lipids** (*fats*), **proteins, vitamins, minerals,** and **water**.

All body processes, whether providing or requiring energy, are classified as metabolic processes. All metabolic reactions within the body, whether anabolic or catabolic, are catalyzed by enzymes. The number of enzymes the body employs in these processes is so great that it currently cannot even be estimated. Nevertheless, the lack of or dysfunction of even one of these enzymes can sometimes make it impossible for the body to maintain homeostasis.

Objective B To describe the major events in *carbohydrate metabolism.*

The average human diet consists largely of polysaccharide and disaccharide **carbohydrates**. When digested, these molecules are broken down into the monosaccharides glucose, fructose, and galactose. Through several biochemical reactions, the liver further converts fructose and galactose into glucose. Therefore, in the end, all carbohydrates entering the body are catabolized as glucose. The equation for glucose metabolism in the aerobic pathway is as follows:

$$\underset{\text{glucose}}{C_6H_{12}O_6} + \underset{\text{oxygen}}{6O_2} \rightarrow \underset{\text{carbon dioxide}}{6CO_2} + \underset{\text{water}}{6H_2O} + \underset{\text{adenosine triphosphate}}{\textit{energy}\ (36 \text{ or } 38 \text{ ATP})}$$

Glucose is the molecule from which the body's energy molecule (ATP) is formed. The way in which one glucose molecule is used to produce 36 or 38 ATP molecules is diagrammed in fig. 20.1.

20.2 What if oxygen is *not* present during glycolysis?

If oxygen is not present, anaerobic glucose metabolism (*glycolysis*) takes place to try to fulfill the body's energy needs. The equation for glucose metabolism in the anaerobic pathway is:

$$\underset{\text{glucose}}{C_6H_{12}O_6} \rightarrow \underset{\text{pyruvic acid}}{2C_3H_3O_3} \ \underline{\text{or}} \ \underset{\text{lactic acid}}{C_3H_6O_3} + \underset{\text{adenosine triphosphate}}{\textit{energy} \ (2\ \text{ATP})}$$

Glycolysis is much more rapid than the aerobic pathway, but it can only supply 2 ATP molecules per glucose molecule and it produces lactic acid as a by-product. Lactic acid causes early fatigue and even tissue damage, which is why anaerobic exercises (such as wind sprints and weight lifting) can be performed only for a short period of time without rest.

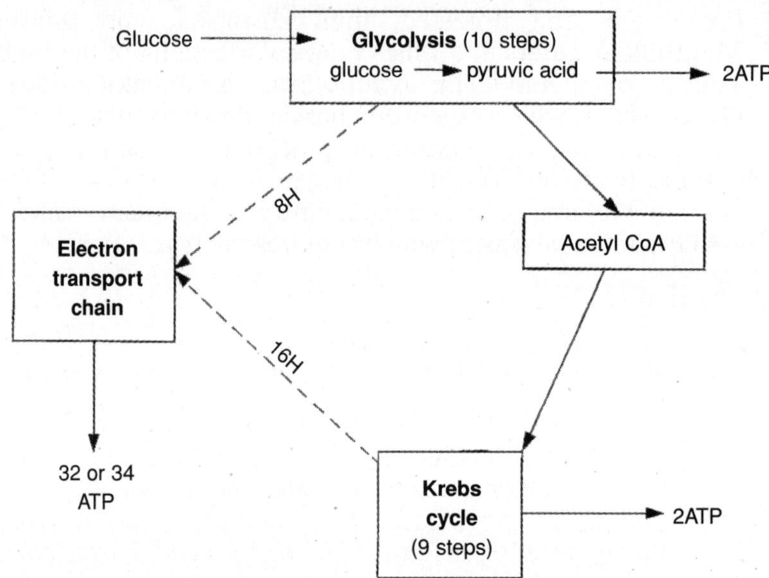

Figure 20.1 Aerobic production of ATP.

20.3 Does glucose readily move into and out of the body cells?

No. Insulin is required for the rapid movement of glucose into most cells of the body. Upon entry into a cell, glucose combines with a phosphate group, and this *phosphorylation* serves to capture the glucose molecule in the cell. Several cell types (liver, kidney tubule, intestinal epithelial) contain *phosphatase*, which removes the phosphate group and thereby allows the glucose molecule to pass out of the cell.

20.4 What is oxygen's role in metabolic processes?

Oxygen's primary role is to accept electrons in the last step of the electron transport chain during aerobic production of ATP. One oxygen atom or half an oxygen molecule (one-half O_2) accepts two electrons and then combines with hydrogen to form water (H_2O). If O_2 is not available to accept electrons, oxidative phosphorylation is no longer possible and the cell must rely instead on anaerobic metabolism for ATP production. Oxygen must be present if the catabolism of glucose is to be completed via the *Krebs cycle* and the electron transport chain, but glycolysis can occur under anaerobic conditions (such as during strenuous exercise).

20.5 In what parts of the cell does catabolism of glucose take place?

The 10 steps of glycolysis, each step mediated by a different enzyme, take place in the cytoplasm of the cell (see problem 3.5). The Krebs cycle (9 steps, 8 enzymes; also called the *citric acid* or *tricarboxylic acid cycle*) and the electron transport chain of oxidation-reduction reactions both take place in mitochondria.

20.6 Is all the energy released during glucose metabolism utilized to form ATP?

No. Less than half (about 40%) of the energy released is captured by ATP; the balance is given off as heat.

20.7 Is all of the glucose that enters body cells immediately catabolized to form energy and heat?

No. Depending on energy demands, a portion of the superfluous glucose molecules entering cell types are linked together into molecular chains called *glycogen*. The conversion of glucose to glycogen (*glycogenesis*) is instigated by the pancreatic hormone insulin (see chapter 13, Objective J). When the body needs energy, glycogen stored in liver and muscle cells is broken down and glucose is released into the blood. This inverse process, *glycogenolysis*, is spurred by the pancreatic hormone glucagon (see chapter 13, Objective J) and the adrenal hormones epinephrine and norepinephrine (see table 13.3).

Muscle glycogen supercompensation or "carbo-loading" refers to eating a carbohydrate-rich diet in order to maximize muscle glycogen levels. With maximized glycogen stores, athletes can increase their endurance. The classic method of carbo-loading is to (1) deplete glycogen stores with hard exercise, (2) eat foods high in protein and fats but low in carbohydrates for 3 days, (3) exercise again, (4) eat a 90% carbohydrate meal, (5) perform the exercise. Through this diet and exercise regimen, the body depletes its glycogen stores to such an extent that when it replaces them, it will do so at an increased level.

20.8 Can glucose be formed from noncarbohydrate sources?

Yes. Both protein and lipid molecules can be converted to glucose; the process is called *gluconeogenesis*. There are five hormones that stimulate gluconeogenesis (review chapter 13): cortisol, thyroxine, glucagon, growth hormone, and epinephrine (or norepinephrine).

Objective C To describe the major events in *lipid metabolism*.

Lipids (mainly fats) are second to carbohydrates as a source of energy for ATP synthesis. There is an increase in fat utilization when the carbohydrate level is low. Fats participate in the building of many cellular structures (e.g. membranes), and cholesterol is a precursor in the synthesis of sex hormones, adrenocorticoids, and bile salts. Lipid metabolism is diagrammed in fig. 20.2.

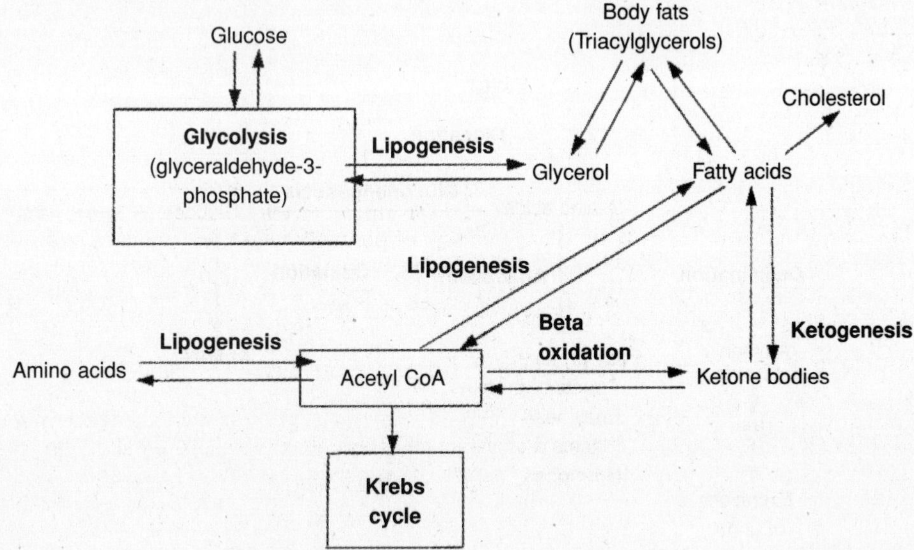

Figure 20.2 Lipid metabolism.

20.9 What happens when total food intake—whether carbohydrate, protein, or fat—exceeds the body's needs?

The excess (of that portion not already fat) is converted into fat and stored in *adipocytes* (adipose cells). These are located in subcutaneous connective tissue and in deeper supporting tissue of the viscera, mesenteries, and greater omentum. Before entering adipose cells, fats are hydrolyzed to glycerol and fatty acids by *lipoprotein lipase*, which is present in capillary endothelium. After entering the adipose cells, the glycerol and fatty acids are recombined into fats (technically, into *triacylglycerols*; see fig. 2.8).

20.10 Discuss beta oxidation and lipogenesis as inverse processes.

When stored triacylglycerols are catabolized—for energy or for synthesis of new nutrients—the glycerol components may be converted to *phosphoglyceraldehyde* (PGAL), which enters the glycolytic pathway and is utilized for energy production or synthesis of glucose. The remaining fatty acid components undergo a step-by-step breakdown of the fatty acid chain. Two carbon fragments of each chain are serially split off as acetyl coenzyme A (*acetyl CoA*) molecules. This breaking down of fatty acids to form acetyl CoA molecules is called *beta oxidation*. The oppositely directed anabolic process, leading from glucose or amino acids to lipids, is known as *lipogenesis*.

20.11 What are ketone bodies, how are they formed, and what are the clinical consequences of an excess?

As part of fatty acid catabolism, the liver condenses two molecules of acetyl CoA to form *acetoacetic acid*, which is converted to *beta-hydroxybutyric acid* and *acetone*. These three substances, collectively known as the *ketone bodies*, are normally decomposed into acetyl CoA and utilized in the Krebs cycle. If they are not—as a consequence of a meal rich in fats, a period of starvation or fasting, or a case of diabetes mellitus during which the body is relying upon fats as the primary source of energy instead of glucose—excessive amounts of ketones build up, producing the condition *ketosis*. Extreme or prolonged ketosis can lead to ketoacidosis (see problem 13.29).

Objective D To describe the major events in *protein metabolism*.

As long as the stores of carbohydrates and fats are sufficient, the body uses very little protein for energy. **Proteins** play an essential role in cellular structure and function (see table 2.7). Protein metabolism is diagrammed in fig. 20.3.

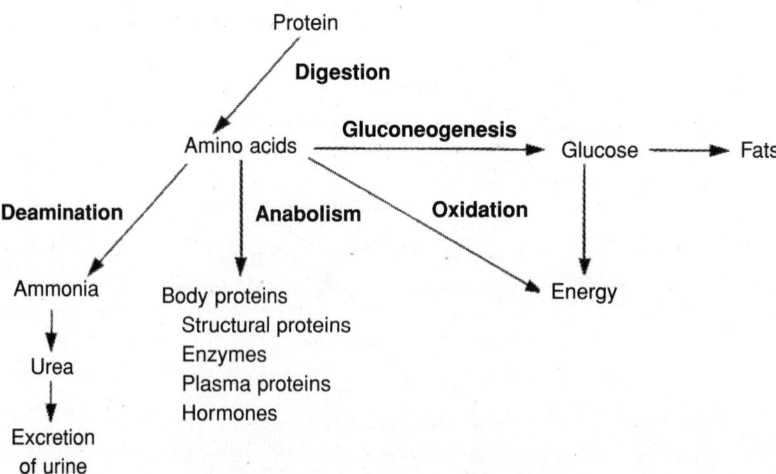

Figure 20.3 Protein metabolism.

20.12 Amino acids in proteins may be used as an energy source when other sources prove inadequate. How is this done?

Through the Krebs cycle. Depending on the amino acid, it may enter the cycle via acetyl CoA (see fig. 20.2) or it may directly enter at another stage of the cycle. The molecule is then used to create ATP molecules. Recent evidence suggests that protein can contribute from 5% to 15% of the fuel supply used during exercise, somewhat higher than the previously believed 2%.

20.13 What is meant by a negative or a positive nitrogen balance?

Negative nitrogen balance: Protein catabolism exceeds protein anabolism; occurs with a protein-poor diet or during starvation. *Positive nitrogen balance*: Protein anabolism exceeds protein catabolism; occurs during growth, recovery from illness, or following administration of anabolic steroids.

Objective E To understand how all of the different energy systems work together to provide *energy* for the entire body.

When metabolized, all three substrates (carbohydrates, lipids, and protein) can provide the energy required to perform bodily functions. The body continually uses all three of these fuels to produce ATP molecules, or energy. The intensity of exercise the body performs dictates which fuel contributes the highest percentage of energy.

20.14 Discuss the interaction of fat and carbohydrate metabolism.

As the intensity of the exercise increases, the higher the contribution of the carbohydrates in providing energy for the body. The lower the intensity, or the longer the duration of exercise at the same intensity, the higher the contribution of fat as an energy source.

20.15 Do carbohydrate, protein, and fat molecules contain the same amount of energy?

No. Carbohydrate and protein molecules contain about 4 kilocalories (kcal) of energy each. One fat molecule, on the other hand, contains about 9 kcal of energy, over twice the amount of carbohydrates and proteins.

Many a weight-conscious person has set out to take off those extra few pounds of unwanted fat. In order for someone to reduce body fat content, he or she must reduce total caloric intake as well as engage in light-intensity aerobic exercises that will allow fat stores to be used instead of glycogen or protein stores as energy sources. Also, the rate of fat metabolism increases as duration of exercise time increases. Therefore, aerobic exercises engaged in for longer periods of time help to maximize fat metabolism.

Objective F To review *hormonal regulation of metabolism.*

See table 20.1.

20.16 Briefly discuss the effects of diabetes mellitus (see problems 13.30 and 13.31) on the metabolism of carbohydrates, proteins, and lipids.

On carbohydrate metabolism. In diabetes, there is a decreased rate of entry of glucose into many tissues and an increased rate of release of glucose from the liver into the circulation, causing an extracellular glucose excess and an intracellular deficiency. With a lack of intracellular glucose, energy requirements are met by greatly increasing the rate of catabolism of fat and protein.

On protein metabolism. There is a decrease in the rate of uptake of amino acids into tissues, decreased protein synthesis, and acceleration of protein catabolism to CO_2 and H_2O and to glucose.

On lipid metabolism. There is decreased conversion of glucose to fatty acids (because of the low intracellular glucose level), and an acceleration of lipid catabolism because of the increased rate of lipid catabolism. The plasma level of free fatty acids may more than double, with a corresponding jump in the formation of acetyl CoA (see fig. 20.2) and consequent production of ketone bodies (see problem 20.11).

Table 20.1 Hormones That Affect Metabolism

Hormone	*Metabolic effects*
Insulin	Promotes glucose uptake into cells; promotes glycogenesis; promotes lipogenesis and inhibits lipolysis; promotes amino-acid uptake into cells; promotes protein synthesis
Glucagon and Epinephrine	Promotes glycogenolysis; promotes gluconeogenesis; promotes protein synthesis
Thyroxine	Promotes glycogenolysis; promotes gluconeogenesis; promotes lipolysis
Growth hormone	Promotes amino-acid uptake into cells; promotes protein synthesis; promotes glycogenolysis; promotes lipolysis
Cortisol	Promotes gluconeogenesis; promotes lipolysis; promotes the breakdown of proteins
Testosterone	Promotes protein synthesis

20.17 Explain how diabetic ketosis can lead to acidosis, and give the clinical consequences.

Like all acids, ketone bodies liberate hydrogen ions. Thus, in untreated diabetes, the blood plasma pH drops (the plasma becomes more acidic), and hydrogen ions are secreted into the kidney tubules and excreted in the urine. When the kidney's ability to remove H^+ from the plasma is exceeded, it removes Na^+ and K^+ instead. The resultant electrolyte and water losses may lead to dehydration, *hypovolemia* (decreased blood volume), depressed consciousness, and, finally, coma.

Ketone bodies are found in the body under normal conditions and are used for energy by many of the body organs. Under conditions of fasting or in the case of *diabetes mellitus*, however, there is an increase in the rate of formation of free fatty acids from adipose tissue, which results in the production of large amounts of ketone bodies. Ketone bodies in high concentrations may lead to a fruity breath odor, a state of acidosis, coma, and even death. Therefore, people with this condition must be given food in the case of fasting, or given insulin if they are diabetics, in order to reverse these symptoms.

Objective G To learn how to measure the *energy content of foods* and to know the terms used to express *body energy expenditure*.

Calorie. The basic unit of energy is the *joule* (J); but for energy as heat, one frequently uses the *calorie* (cal), where

$$1 \text{ cal} = 4.184 \text{ J}$$

The "calorie" of dieticians' charts is equal to 1000 cal, or 1 kcal.

Bomb calorimeter. The energy obtainable from a sample of food can be determined by placing it in a sealed chamber surrounded by a jacket filled with water of known volume and temperature, the whole being thermally insulated from the environment. As the food is completely oxidized (burned), heat is liberated into the water. The temperature rise of the water gives the caloric value of the food sample.

Direct calorimetry. The energy liberated by oxidation of food in the body is measured by placing a person in a chamber that is sensitive to the heat loss from the body. (Same principle as bomb calorimeter.)

Indirect calorimetry. The oxygen taken in by the way of the respiratory system is consumed in cellular oxidations. Therefore, standard tables may be used to translate the minute respiratory volume (see problem 18.20) into the rate of heat production.

Respiratory quotient (R.Q.). The ratio of the volume of carbon dioxide produced (in a given period) to the volume of oxygen consumed (in the same period) is obtained.

Metabolic rate. The total rate of body metabolism, or metabolic rate, can be measured by either the amount of heat generated by the body or by the amount of oxygen consumed by the body per minute. The *basal metabolic rate (BMR)* is the metabolic rate as measured when the person is resting but awake. The test is usually taken in the morning before rising, the person having fasted for at least 12 hours and having slept for 8 hours.

20.18 If a person has an R.Q. of 0.72, what type of food is being utilized for energy?

Fat. The R.Q. for carbohydrates (strictly for glucose) is 6/6 = 1,since in the metabolic equation,

$$C_6H_{12}O_6 + 6O_2 \rightarrow 6CO_2 + 6H_2O$$

the coefficients may be interpreted as numbers of moles or numbers of liters. The R.Q. for protein is 0.80, since protein contains less oxygen than carbohydrates and therefore requires more oxygen in burning. The R.Q. for fat is 0.70, since fat contains even less oxygen.

20.19 What factors affect the metabolic rate?

Increases in *body size, body temperature, activity, levels of thyroid hormone,* and *sympathetic stimulation* all increase metabolic rate. The metabolic rate is 10% higher in men than in women and decreases with increasing age.

Objective H To identify the *essential vitamins and minerals*, their sources in the diet, uses in the body, and the syndromes of their deficiency.

The essential vitamins and minerals necessary for body function are summarized in tables 20.2 and 20.3.

20.20 Classify vitamins according to solubility.

Fat-soluble vitamins include vitamins A, D, E, and K. They are absorbed from the GI tract along with lipids and can be stored in the body. **Water-soluble vitamins** include the B vitamins and vitamin C. They are absorbed from the small intestine along with water, and are generally not stored in the body.

20.21 Can vitamins be synthesized in the body?

Generally, no. Vitamins are obtained only from external sources, such as ingested foods. (The B vitamins and vitamin K are synthesized by bacteria in the GI tract, and vitamin D is produced in small quantities in the skin.)

Objective I To describe the body's *thermal equilibrium*.

Heat is continually being produced as a by-product of metabolism and is continually being lost to the surroundings. The body normally balances heat gain and loss.

Table 20.2 The Essential Vitamins

Vitamin	*Dietary sources*	*Major body functions*	*Deficiency syndrome*
Vitamin A (Retinol)	Green vegetables, carrots, sweet potatoes	Synthesis of rhodopsin	Night blindness; skin disorders
Vitamin B_1 (Thiamine)	Meats, grains, legumes	Coenzyme in cellular respiration	Peripheral nerve changes; beriberi
Vitamin B_2 (Riboflavin)	Meats, leafy vegetables, milk	Component of FAD	Lesions of lips, mouth, and tongue
Vitamin B_6 (Pyridoxine)	Meats, grains, vegetables	Coenzyme in amino acid metabolism	Irritability; muscle twitching; seizures
Vitamin B_{12} (Cobalamine)	Meats, eggs, dairy products	Coenzyme in nucleic acid metabolism	Pernicious anemia; nervous disorders
Niacin	Meats, grains, legumes	Component of NAD	Pellagra (lesions in skin and GI tract); nervous disorders
Pantothenic acid	Widely distributed in foods	Component of coenzyme A	None, except in controlled lab situations
Biotin	Meats, legumes, vegetables,	Component of coenzyme	No serious problems
Folic acid	Green vegetables, grains	Coenzyme in nucleic acid metabolism	Anemia; diarrhea
Vitamin C (Ascorbic acid)	Citrus fruits, tomatoes, leafy vegetables, broccoli	Formation of intercellular material in connective tissues	Scurvy
Vitamin D (Cholecal-ciferol)	Egg yolks, fortified milk	Growth of bones; absorption of calcium	Rickets in children; osteomalacia in adults
Vitamin E (Tocopherol)	Seed oils, widely distributred in foods	Antioxidant to prevent damage to cell membranes	Anemia (in premature infants)
Vitamin K (Phylloquin-one)	Meats, fruits, leafy vegetables	Synthesis of clotting factors	Hemorrhage

Table 20.3 The Essential Minerals

Mineral	*Dietary sources*	*Major body functions*	*Deficiency syndrome*
Calcium	Eggs, dairy products, vegetables	Formation of bones and teeth; clotting; nerve and muscle activity; many cellular functions	Rickets; tetany; osteoporosis
Chlorine	Table salt, most foods	Water-electrolyte balance; acid-base balance; formation of HCl in stomach	Fluid imbalance
Cobalt	Most foods	Component of vitamin B_{12}	Anemia
Copper	Most foods	Synthesis of hemoglobin; component of enzyme involved in melanin formation	Anemia
Fluorine	Seafoods, drinking water	Component of bones, teeth, and other tissues	Dental caries
Iodine	Seafoods, table salt	Component of thyroid hormones	Hypothyroidism
Iron	Meat, egg yolks, legumes, nuts, cereals	Component of hemoglobin, myoglobin, and cytochromes	Anemia
Magnesium	Many foods	Bone formation; nerve and muscle function	Tetany
Manganese	Meats	Activation of several enzymes; reproduction; lactation	Infertility
Phosphorus	Meats, dairy products, fish, poultry	Formation of bones and teeth; component of buffer system and nucleic acids	Weakness
Potassium	Meats, bananas, seafoods, milk	Nerve conduction; electrolyte balance	Skeletal and cardiac muscle weakness
Sodium	Most foods, table salt	Nerve conduction; electrolyte balance	Cramps; weakness; dehydration
Zinc	Most foods	Component of several enzymes	Reduced growth; hair loss; vomiting

20.22 Briefly describe the mechanisms of body heat loss.

Radiation. Transfer of heat, in the form of electromagnetic waves, from the surface of the body to the surrounding environment.

Evaporation. Loss of heat as water evaporates from the body surfaces (580 cal per ml of water). Water loss from the skin and lungs is normally about 600 ml per day.

Conduction. Molecule-to-molecule transfer from the surface of the body to objects in direct contact with it (e.g., clothing, water, other people).

Convection. Transfer of heat between the body and the air that overlies the body surface. A cool breeze results in movement of air over the body and a consequent loss of body heat.

Heat gain	*Heat loss*
Metabolism (see problem 20.19)	Radiation
Muscular activity (shivering)	Evaporation
	Conduction
	Convection

A *fever* is a body temperature above 99°F (37.2°C). Although fevers can result from abnormalities of the thermoregulatory mechanism or from clinical problems, they are usually caused by a bacterial or viral infections. These pathogens produce toxins that may act like, or stimulate, circulating proteins called *pyrogens*, the agents that stimulate the body's thermostat in the hypothalamus. When these pyrogens are active, the body thermostat "setting" is raised and the person feels a desire to be warmer. A slight fever may actually assist the immune system in fighting the infection, and thus help to maintain homeostasis.

Heat exhaustion and *heat stroke* are two malfunctions of the thermoregulatory system. Heat exhaustion occurs when a person is exposed to high temperatures without drinking any liquids. The elevated temperature stimulates the body's heat-loss center to secrete large amounts of sweat to provide evaporating cooling. As fluid losses increase without being replaced, the person's blood volume decreases. The resulting drop in blood pressure due to the drop in blood volume is not countered by peripheral vasoconstriction (due to the external heat), and therefore the central nervous system receives a less-than-adequate blood supply. This causes headache, nausea, and even loss of consciousness. Treatment simply includes providing water, salts, and a cooler environment. Without proper treatment, heat exhaustion may turn into heat stroke. Heat stroke occurs when the thermoregulatory center ceases to stimulate the sweat glands to produce and secrete sweat, causing body temperatures that may range from 106° to 113°F (41° to 45°C). If body temperatures are not reduced rapidly, brain, liver, and kidney cells can be destroyed.

20.23 What is the normal range of resting body temperatures?

For oral temperatures see figure. 20.4; rectal temperatures run about 1°F (.55°C) higher.

°F	°C	
102	38.8	Hard exercise
101	38.3	
100	37.7	
99	37.2	Resting
98	36.1	
97	36.1	
96	35.5	Cold weather, early morning

Figure 20.4 Range of normal body temperatures.

Objective J To outline briefly the U.S. Dietary Goals concerning nutrition.

In order to provide our bodies with the proper amounts of energy sources, we must ingest the proper amounts of nutrients. As a dietary guideline, a U.S. Senate Select Committee on Nutrition and Human Needs recruited dietary experts to formulate U.S. Dietary Goals. These goals included the following:

1. Increase carbohydrate intake to represent 55% to 60% of caloric intake.
2. Decrease fat consumption from 40% to 30% of caloric intake.
3. Decrease saturated fat intake to represent only 10% of caloric intake; increase polyunsaturated and monounsaturated fats to approximately 10% of caloric intake.
4. Reduce cholesterol intake to 300 mg per day.
5. Reduce sugar consumption to account for only 15% of caloric intake.
6. Reduce salt consumption by about 70% to approximately 3 g per day.

Review Exercises

Multiple Choice

1. Which of the following chemical reactions can occur anaerobically? (*a*) glycolysis, (*b*) Krebs cycle, (*c*) conversion of lactic acid to pyruvic acid, (*d*) conversion of pyruvic acid to acetyl CoA

2. The synthesis of glycogen molecules for cellular storage is referred to as (*a*) glycogenolysis, (*b*) beta oxidation, (*c*) glyconeogenesis, (*d*) glycogenesis.

3. Between meals, the blood glucose level is maintained by (*a*) insulin, (*b*) glycogenolysis, (*c*) lipogenesis, (*d*) glycogenesis.

4. Within the cell, Krebs cycle reactions occur in (*a*) the neurotransmitter chemicals, (*b*) the ribosomes, (*c*) the nucleolus, (*d*) the mitochondria.

5. In the absence of oxygen, how many molecules of ATP are produced by catabolism of one molecule of glucose? (*a*) 1, (*b*) 2, (*c*) 8, (*d*) 36 or 38

6. If oxygen is present, how many molecules of ATP are produced by catabolism of one molecule of glucose? (*a*) 1, (*b*) 2, (*c*) 8, (*d*) 36 or 38

7. Which of the following is *not* an energy source for cells? (*a*) glucose, (*b*) protein, (*c*) fats, (*d*) vitamins

8. During which phase of metabolism is the greatest amount of energy generated? (*a*) glycolysis, (*b*) conversion of pyruvic acid to acetyl CoA, (*c*) glycogenesis, (*d*) electron transport

9. Anabolic metabolism includes (*a*) processes by which substances are synthesized, (*b*) changes of larger molecules into smaller ones, (*c*) glycolysis, (*d*) all processes needed to maintain life.

10. Aerobic respiration increases the body's supply of (*a*) CO_2, (*b*) water, (*c*) ATP, (*d*) all of the preceding.

11. The primary role of oxygen in the body is to (*a*) help build amino acids, (*b*) allow glycolysis within the mitochondria of the cell, (*c*) accept electrons in the electron transport chain, (*d*) facilitate lipolysis within adipose cells.

12. Vitamins are essential in metabolism because they (*a*) serve as structural components, (*b*) serve as sources of energy, (*c*) act as coenzymes, (*d*) cannot be stored in the body.

13. Beriberi is caused by a deficiency of vitamin (*a*) A, (*b*) B_1, (*c*) B_{12}, (*d*) B_6.

14. What fraction of the energy released during catabolism of glucose is captured by ATP? (*a*) 25%, (*b*) 40%, (*c*) 75%, (*d*) 100%

15. The synthesis of glucose from proteins or lipids is referred to as (*a*) glycogenesis, (*b*) glucose oxidation, (*c*) glucosynthesis, (*d*) gluconeogenesis.

16. Which hormone increases the rate of amino acid uptake by cells, protein synthesis, and glycogenolysis? (*a*) cortisol, (*b*) epinephrine, (*c*) glucagon, (*d*) growth hormone

17. Which of the following hormones does *not* stimulate gluconeogenesis? (*a*) cortisol, (*b*) epinephrine, (*c*) estrogen, (*d*) thyroxine

18. The ratio of the carbon dioxide volume produced to the oxygen volume consumed is called (*a*) the bomb calorimeter, (*b*) the metabolic rate, (*c*) the direct quotient, (*d*) the respiratory quotient.

19. A respiratory quotient of 0.70 would indicate that the main source of food was (*a*) carbohydrates, (*b*) fats, (*c*) proteins, (*d*) a mixture of carbohydrates and proteins.

20. Reduced growth, hair loss, and vomiting may result from a deficiency of (*a*) iron, (*b*) copper, (*c*) potassium, (*d*) zinc.

21. Important mechanisms for heat transfer include (*a*) evaporation, (*b*) conduction, (*c*) radiation, (*d*) all of the preceding.

22. Per unit weight, which contains the most energy? (*a*) carbohydrates, (*b*) proteins, (*c*) fats, (*d*) vitamins

23. When the ambient temperature is very high, say 105° F, the body will lose heat by: (*a*) radiation, (*b*) conduction, (*c*) evaporation, (*d*) increased metabolism.

24. When the body is engaged in long-duration, low-intensity exercise, the principal energy source is (*a*) carbohydrates, (*b*) fats, (*c*) proteins, (*d*) glycogen.

True or False

____ **1.** The hormone glucagon facilitates glucose uptake by the cells.

____ **2.** Aerobic metabolism produces more ATP than does anaerobic metabolism.

____ **3.** As the intensity of exercise increases, the body begins to use glucose as its sole source of energy.

____ **4.** Vitamins are neither produced nor stored in the body.

____ **5.** Carbohydrates, as a whole, should make up about 60% of our diet.

____ **6.** Fats should constitute less than 10% of our diet.

____ **7.** Carbohydrates have more than twice the energy content of fats.

____ **8.** *All* carbohydrates ingested by the body are converted into and catabolized as glucose.

____ **9.** All processes that deal with the catabolism of glucose (glycolysis, Krebs cycle) take place in the mitochondria of the cell.

____ **10.** Both lipids and proteins can be converted into glucose.

Completion

1. Metabolism refers to all the chemical reaction in the body; ____________ is the breaking–down process and ____________ is the building–up process.

2. ____________ is the anaerobic metabolism of glucose that takes place in the cytoplasm.

3. ____________ is the conversion of glucose to glycogen.

4. The conversion of protein or lipid molecules to glucose is referred to as ____________.

5. The excessive buildup of ketones produces a condition called ____________.

6. Vitamins A, D, E, and K are referred to as the ____________ - ____________ vitamins.

7. ____________ is the transfer of heat in the form of electromagnetic waves from the surface of the body.

8. Proteins that stimulate the body's thermostat in the hypothalamus are called ____________.

9. The metabolism of one molecule of glucose forms ____________ ATPs.

10. The electron transport chain of oxidation-reduction reactions take place in ____________.

Matching

(*Set 1*) Match the vitamin with its appropriate description.

____	**1.** Vitamin A	(*a*)	deficiency most commonly produces pernicious anemia
____	**2.** Vitamin B_1	(*b*)	deficiency may result in scurvy
____	**3.** Vitamin B_6	(*c*)	involved in synthesis of rhodopsin
____	**4.** Vitamin B_{12}	(*d*)	necessary for synthesis of clotting factors
____	**5.** Vitamin C	(*e*)	also referred to as pyridoxine
____	**6.** Vitamin D	(*f*)	promotes absorption of calcium
____	**7.** Vitamin K	(*g*)	deficiency may result in beriberi

(*Set 2*) Match the mineral with its description or function.

____	**1.** Calcium	(*a*)	component of vitamin B_{12}
____	**2.** Cobalt	(*b*)	component of thyroid hormone
____	**3.** Copper	(*c*)	required for clotting, bone formation, and muscle contraction
____	**4.** Iodine	(*d*)	component of hemoglobin and myoglobin
____	**5.** Iron	(*e*)	involved in melanin formation

(*Set 3*) Match the vitamin or compound with its dietary source.

____	**1.** Ascorbic acid	(*a*)	leafy vegetables, carrots, sweet potatoes
____	**2.** Vitamin A	(*b*)	meats, fruits, leafy vegetables
____	**3.** Vitamin D	(*c*)	egg yolks, fortified milk
____	**4.** Folic acid	(*d*)	citrus fruits, tomatoes, salad greens
____	**5.** Vitamin K	(*e*)	meats, grains, legumes
____	**6.** Niacin	(*f*)	leafy vegetables, whole grains

Answers and Explanations to Review Exercises

Multiple Choice

1. (*a*) If oxygen is not present, two ATPs can be formed through anaerobic metabolism (glycolysis).
2. (*d*) Glycogenesis is the formation of glycogen from glucose. Some cells store glucose as glycogen, such as liver and muscle cells.
3. (*b*) Between meals, during fasting, and under conditions of starvation, glucose is formed from glycogen (storage) through the metabolic step called glycogenolysis.
4. (*d*) ATP is formed in mitochondria from Krebs cycle and oxidative phosphorylation reactions.
5. (*b*) Four ATP molecules are formed in glycolysis, but two ATP molecules are used in the process.

6. (*d*) Under aerobic (oxygen present) conditions, there is a net gain of 36 or 38 ATP molecules, depending on which electron transport carriers are used.
7. (*d*) Vitamins regulate chemical reactions, but do not serve as a source of energy.
8. (*d*) Most of the ATP molecules formed during the metabolism of food molecules are formed during electron transport (also called oxidative phosphorylation).
9. (*a*) Anabolic metabolism refers only to the building up process.
10. (*d*) CO_2, water, and ATP molecules are all formed as a result of aerobic metabolism.
11. (*c*) The primary role of oxygen is to accept electrons in the last step of the electron transport chain during aerobic metabolism.
12. (*c*) Vitamins are small organic molecules that act as coenzymes in metabolic reactions.
13. (*b*) Beriberi and neuritis are clinical conditions that result from a deficiency of vitamin B_1.
14. (*b*) About 40% of the energy released during glucose metabolism is captured by ATP; the balance is given off as heat.
15. (*d*) Both protein and lipid molecules can be converted to glucose via gluconeogenesis.
16. (*d*) Growth hormone stimulates amino acid uptake and protein synthesis as mechanisms to stimulate bone and tissue growth.
17. (*c*) The other hormones stimulate gluconeogenesis, whereas, estrogen has little effect on metabolic formation.
18. (*d*) The respiratory quotient (R.Q.) is an index of the body's energy expenditure.
19. (*b*) Fat has a R.Q. of 0.70, containing less oxygen than carbohydrates (R.Q. = 1.0) or protein (R.Q. = 0.80).
20. (*d*) Reduced growth, hair loss, and vomiting are classic symptoms associated with zinc deficiency.
21. (*d*) The three mechanisms listed, along with convection, are the major ways body heat is lost.
22. (*c*) Fats contain more energy (9.5 kcal/g) than do carbohydrates (4.1 kcal/g) or proteins (5.3 kcal/g). Vitamins are not a source of energy.
23. (*a*) Evaporation is the best mechanism for heat loss when the ambient temperature is high.
24. (*b*) During low-intensity, long-duration exercise, the body uses mainly fat stores instead of glycogen (carbohydrate) or protein sources.

True or False

1. False; the hormone insulin facilitates glucose uptake by the cells
2. True
3. False; although the body may use glucose to a greater extent during higher intensity exercise, all sources of ATP production are being utilized
4. False; although water-soluble vitamins are generally not stored in the body, both fat-soluble and water-soluble vitamins can be
5. True
6. False; fats should constitute less than 30% of our diet
7. False; fats contain twice the energy content of carbohydrates
8. True
9. False; all 10 steps of glycolysis take place in the cytoplasm of the cell
10. True

Completion

1. catabolism, anabolism
2. Glycolysis
3. Glycogenesis
4. gluconeogenesis
5. ketosis
6. fat-soluble
7. Radiation
8. pyrogens
9. 36 or 38
10. mitochondria

Matching

(*Set 1*)

1. (*c*)
2. (*g*)
3. (*e*)
4. (*a*)
5. (*b*)
6. (*f*)
7. (*d*)

(*Set 2*)

1. (*c*)
2. (*a*)
3. (*e*)
4. (*b*)
5. (*d*)

(*Set 3*)

1. (*d*)
2. (*a*)
3. (*c*)
4. (*f*)
5. (*b*)
6. (*e*)

Urinary System 21

Objective A To describe the components of the *urinary system* and to state their functions.

The **urinary system** (fig. 21.1) consists of the **kidneys**, which form urine; the **ureters**, which transport urine to the **urinary bladder**, where it is stored; and the **urethra**, which carries urine to the outside body.

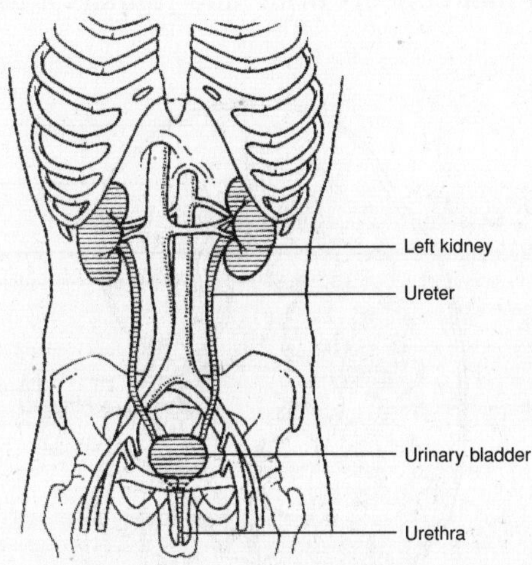

Figure 21.1 The urinary system.

21.1 What are the specific functions of the urinary system?

Like most body systems, the urinary system is involved in maintaining homeostasis. Specifically, it plays a critical role in regulating the composition of the body fluids (water balance, electrolyte balance, and acid-base balance). It also rids the body of the wastes of metabolism and phagocytosis, in addition to removing foreign chemicals, drugs, and food additives. Finally, the kidneys serve as minor endocrine organs (see fig. 13.9).

21.2 Is the name "excretory system" appropriate for the urinary system?

Although the urinary system has a major excretory function, other body systems are also excretory in nature. Carbon dioxide is eliminated through the respiratory system; excessive water, salts, and nitrogenous wastes are eliminated through the integumentary system; and digestive wastes (bile) are eliminated through the digestive system. Therefore, since other systems also eliminate wastes and since the kidneys are also involved in functions other than excretory, "urinary system" is a more precise designation.

21.3 Do *micturition* and *urination* have the same meaning?

No. **Urination** is the voiding of urine from the urinary bladder, whereas **micturition** is the physiological process of urination, which includes nerve impulses and muscular responses.

Objective B To describe the anatomy of the *kidneys* and to trace their embryonic development.

The **kidneys** are located on either side of the vertebral column in the abdominal cavity, between the levels of the twelfth thoracic and the third lumbar vertebrae. Each kidney is approximately 11.25 cm (4 in.) long, 5.5 to 7.5 cm (2 to 3 in.) wide, and 2.5 cm (1 in.) thick. The concave medial surface is called the **hilum**, and through it the **renal artery** enters and the **renal vein** and **ureter** exit (fig. 21.2). The kidney is embedded retroperitoneally in an *adipose* and *fibrous capsule*. The outer portion of the kidney, or the **renal cortex**, contains capillary tufts and convoluted tubules; the inner portion, or the **renal medulla**, is composed of a series of triangular masses, called **renal pyramids**, separated by **renal columns**. The apex of each pyramid is called a **renal papilla**, and each papilla projects into a small depression, called a **minor calyx** (plural, *calyces*). Several minor calyces unite to form a **major calyx**. In turn, the major calyces join to form the **renal pelvis** that collects urine and funnels it to the *ureter*.

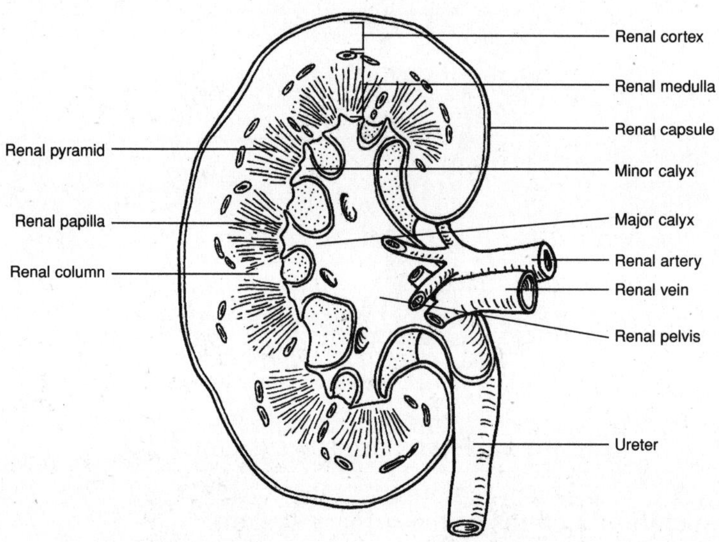

Figure 21.2 The kidney viewed in coronal section.

21.4 Describe the innervation of the kidneys.

The kidneys have an autonomic nerve supply derived from the tenth, eleventh, and twelfth thoracic nerves. Sympathetic stimulation of the renal plexus produces a vasomotor response in the kidneys that affects the circulation of the blood by regulating the diameters of arterioles.

Objective C To trace the *embryonic development* of the kidneys.

The permanent set of kidneys derives from a *metanephros*, which begins to form during the fifth week of embryonic development (fig. 21.3). The metanephros has two mesodermal sources: (1) a *ureteric bud* and (2) a *metanephrogenic mass*, which develops from the *urogenital ridge* and forms the meaty portion of the kidney.

21.5 Is the metanephros the only precursor of the kidney?

No. The most primitive kidney, the *pronephros*, develops during the fourth week and persists only through the sixth week. It never functions as a kidney, but its ductule system gives rise to portions of the *metanephros* and the *mesonephros*. The latter develops from the urogenital ridge toward the end of the fourth week and lasts through the eighth week.

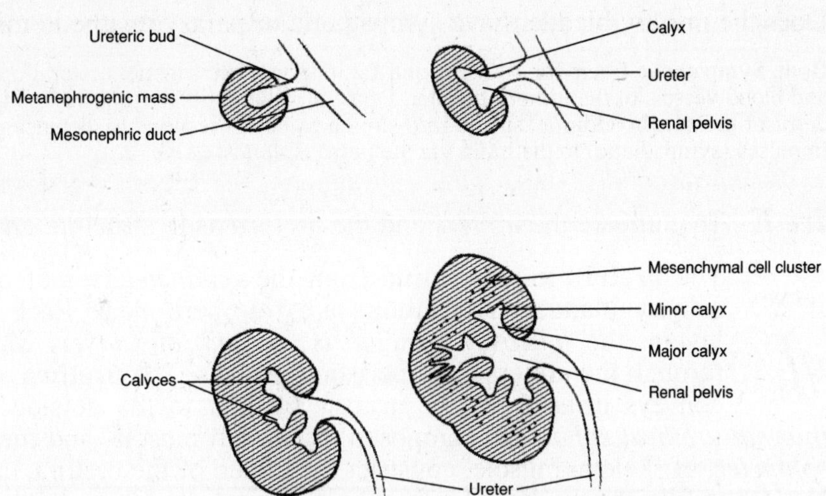

Figure 21.3 Development of the embryonic kidney.

Objective D To describe the structure and functions of the *urinary bladder*.

survey

The **urinary bladder** lies posterior to the symphysis pubis and anterior to the rectum. Its shape is determined by the volume of urine it contains. Like that of the GI tract, the wall of the urinary bladder consists of four tunics (layers): the (innermost) *mucosa*, the *submucosa*, the *muscularis*, and the *serosa* (fig. 21.4). When the urinary bladder is empty, the mucosa is folded into numerous *rugae*; the rugae disappear as the urinary bladder fills with urine. The floor of the urinary bladder is a triangular area called the **trigone**. It has an opening at each of its three angles and, lacking rugae, is smooth in appearance.

21.6 What are the functions of the tunics of the urinary bladder?

The **mucosa**, consisting of **transitional epithelium**, permits distension of the organ; the **submucosa**, consisting of vascular tissue, provides a rich blood supply; the **muscularis**, whose three interlaced smooth muscle laminas are collectively called the *detrusor muscle*, aids in micturition; and the **serosa**, consisting of simple squamous epithelium, is a physical and functional continuation of the perineum.

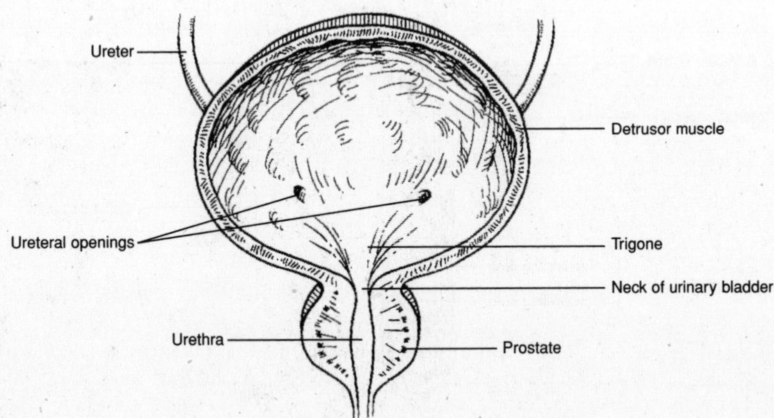

Figure 21.4 The urinary bladder.

21.7 Does the urinary bladder have sympathetic or parasympathetic innervation?

Both. Sympathetic fibers for T12, L1, and L2 innervate the trigone, ureteral openings (fig. 21.4), and blood vessels of the urinary bladder. Parasympathetic fibers from S2 to S4 innervate the detrusor muscle. In addition, specialized stretch receptors respond to distension and relay sensory impulses (sympathetic) to the brain via the *pelvic splanchnic nerve*.

Objective E To compare the *ureters* and the *urethra* as to structure and function.

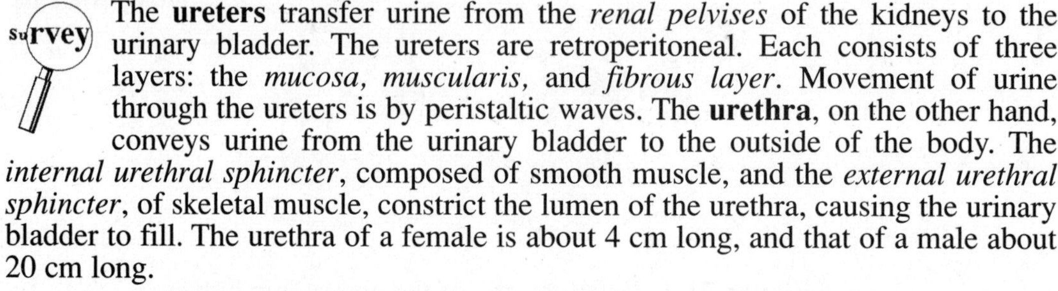

The **ureters** transfer urine from the *renal pelvises* of the kidneys to the urinary bladder. The ureters are retroperitoneal. Each consists of three layers: the *mucosa, muscularis,* and *fibrous layer*. Movement of urine through the ureters is by peristaltic waves. The **urethra**, on the other hand, conveys urine from the urinary bladder to the outside of the body. The *internal urethral sphincter*, composed of smooth muscle, and the *external urethral sphincter*, of skeletal muscle, constrict the lumen of the urethra, causing the urinary bladder to fill. The urethra of a female is about 4 cm long, and that of a male about 20 cm long.

21.8 Does the urethra convey urine only?

In the male, the urethra also conveys semen from the reproductive organs during ejaculation.

Objective F To describe the structure of a *nephron*.

Fig. 21.5 depicts the functional (urine-forming) unit of the kidney, the **nephron**. There are over 1 million nephrons per kidney. The components of a nephron are the *glomerulus, glomerular capsule* (Bowman's capsule), *proximal convoluted tubule, nephron loop* (loop of Henle), and *distal convoluted tubule*. The distal convoluted tubule empties into a collecting duct (papillary duct) which serves several nephrons.

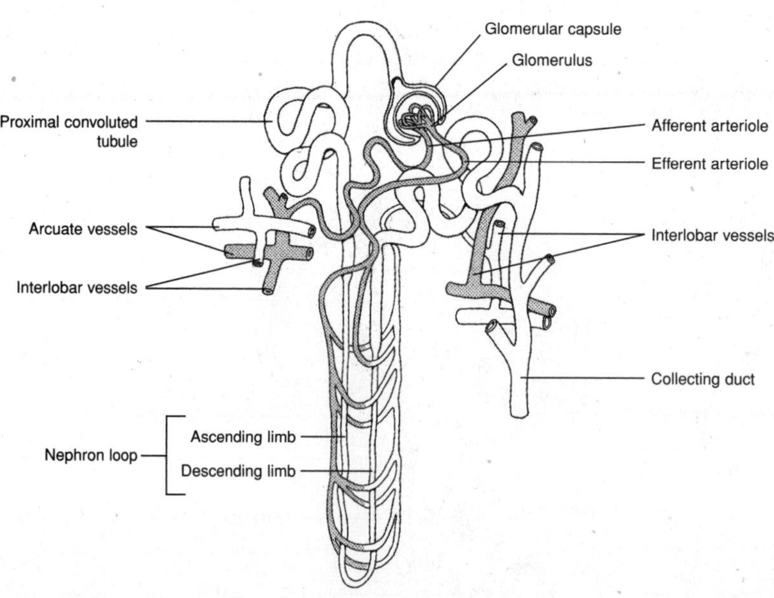

Figure 21.5 The glomerular capsule, nephron tubule, and collecting duct.

21.9 Describe the glomerulus and surrounding glomerular capsule.

The **glomerulus** (fig. 21.6) consists of a network of about 50 capillaries. The capillary endothelium lining the glomerulus has many circular *fenestrations*, or pores, whose diameters range between 50 and 100 nm. This makes the glomerulus 100 to 1000 times more permeable than typical capillaries.

The **glomerular capsule** is a double-walled cuplike structure composed of squamous epithelium. The outer parietal layer is continuous with the epithelium of the proximal tubule, whereas the inner visceral layer is composed of modified cells called **podocytes** that are closely associated with the glomerular capillaries.

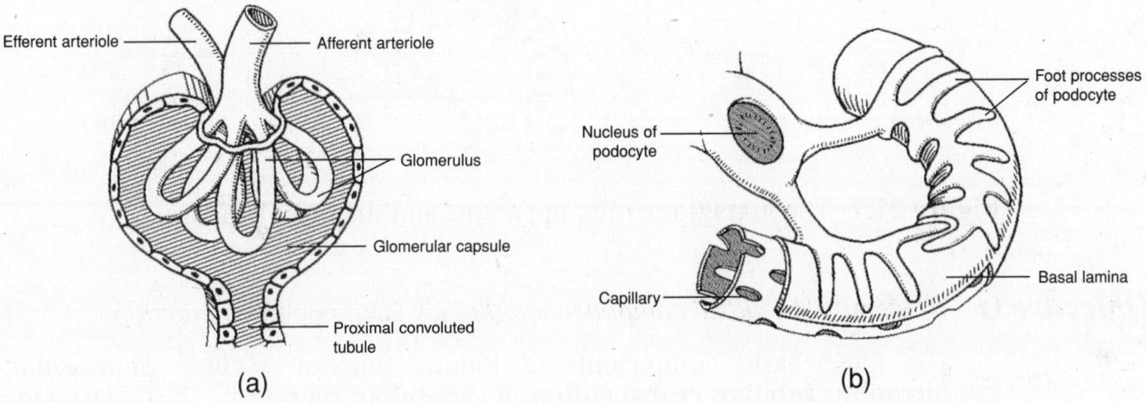

Figure 21.6 (*a*) A glomerulus and a glomerular capsule and (*b*) the associated podocyte.

21.10 Describe the structural components of the tubular segments of the nephron.

The **proximal convoluted tubule** is continuous with the parietal epithelium of the glomerular capsule; it consists of a single layer of cuboidal cells containing microvilli (as a *brush border*) that greatly increase the surface area. It terminates in the first portion of the nephron loop, called the *descending limb of the nephron loop.*

The **nephron loop** has descending and ascending thin limbs and an ascending thick portion (see fig. 21.5). The thin segments are lined with flat squamous cells that lack microvilli, as do the cuboidal cells that compose the thick segment, which runs between the afferent and efferent arterioles.

The **distal convoluted tubule** begins at the *macula densa*, a mass of specialized epithelial cells of the tubule wall, located next to the afferent arteriole (fig. 21.7). The distal convoluted tubule is shorter than the proximal convoluted tubule and has fewer microvilli. It is the last segment of the nephron and terminates as it empties into a **collecting duct** (papillary duct).

A collecting duct is formed by the confluence of several distal convoluted tubules; collecting ducts, in a renal pyramid, drain urine into the renal pelvis.

21.11 What is meant by the juxtaglomerular apparatus?

The cells of the macula densa, together with special juxtaglomerular cells of the afferent arteriole, (see fig. 21.7), compose a sensory apparatus. If the juxtaglomerular cells sense a drop in blood pressure in the afferent arteriole, or if the cells of the macula densa sense an increased sodium chloride concentration in the distal tubule, renin is released from the juxtaglomerular cells and activates the *renin-angiotensin system* (see problem 13.24).

21.12 Do the nephron loops of all nephrons extend the same distance into the renal medulla of the kidney?

No. *Cortical nephrons*, which are close to the outer surface of the kidney, have very short, thin loops, whereas *juxtaglomerular nephrons*, located deep in the renal cortex adjacent to the renal medulla, have long nephron loops that extend deep into the renal medulla.

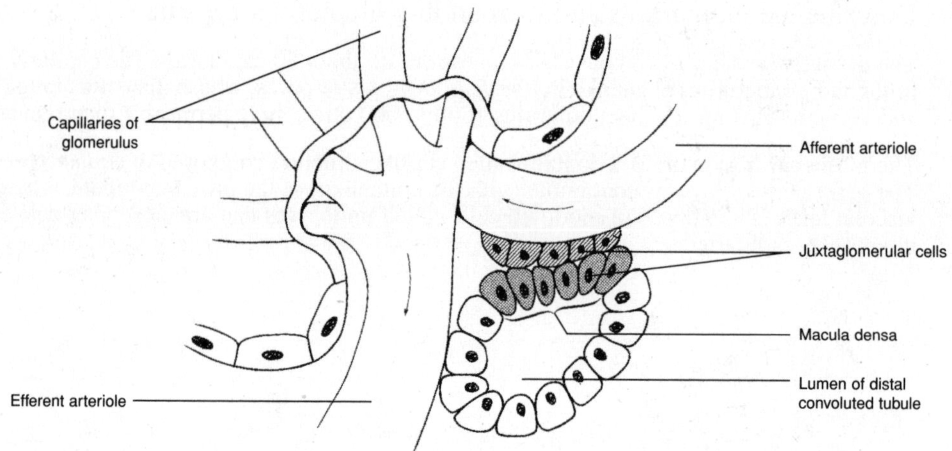

Figure 21.7 The juxtaglomerular apparatus and the macula densa.

Objective G To identify the *basic components of kidney* (i.e., nephron) *function*.

Survey

The three basic components of kidney function include **glomerular filtration**, **tubular reabsorption**, and **tubular secretion**. (1) *Glomerular filtration*: Fluid and solutes in the blood plasma of the glomerulus pass into the glomerular capsule. The portion of the blood plasma that enters the capsule is referred to as the *glomerular filtrate*; it amounts to some 180 L per day (multiple filtration). (2) *Tubular reabsorption*: Approximately 99% of the filtrate is transported actively or passively out of the lumen into the interstitial fluid, and then into the peritubular capillaries. (3) *Tubular secretion*: Noxious substances (see problem 21.19) are actively transferred from the peritubular capillaries into the interstitial fluid, and then into the tubular lumen (fig. 21.8).

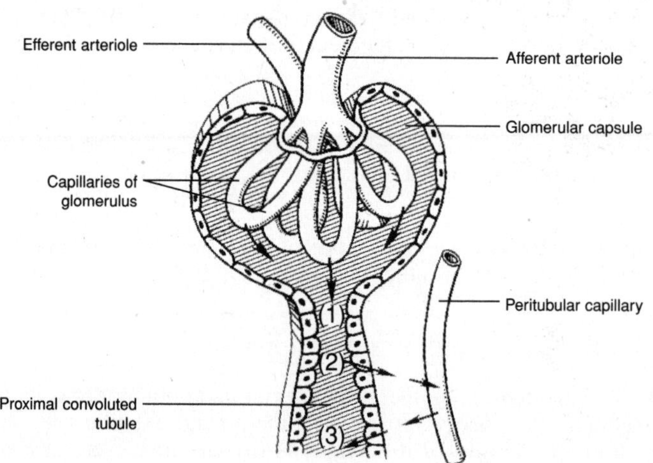

Figure 21.8 Filtration, reabsorption, and secretion in the kidney.

21.13 What is the glomerular membrane?

The membrane of the glomerular capillaries is referred to as the **glomerular membrane**. It consists of (1) the endothelial layer, (2) a basement membrane, and (3) a layer of epithelial cells that line the surface of the glomerular capsule.

21.14 Why may a larger volume of fluid move across the glomerular membrane than across the membrane of other capillaries?

Fenestrations, or pores (see problem 21.9) make the glomerular membrane more permeable. Furthermore, hydrostatic pressure within the glomerular capillaries (50 to 60 mmHg) is greater than in other capillaries (10 to 30 mmHg).

21.15 What is the composition of the glomerular filtrate?

Red and white blood cells are generally not filtered, nor are plasma proteins; therefore, the glomerular filtrate has the same composition as blood plasma, except that the filtrate has no significant amount of protein. The presence of red blood cells or protein in the urine indicates that the hydrostatic pressure in the glomerular capillaries is excessively high or that there is a defect in the glomerular membrane.

21.16 Define *glomerular filtration rate.*

The **glomerular filtration rate (GFR)** is the volume of filtrate formed by all the nephrons of both kidneys each minute. In the adult female, the GFR is about 110 ml/min; in the male, about 125 ml/min. Thus, a volume of about 7.5 L/h, or 180 L/day is formed.

21.17 What fraction of the glomerular filtrate is reabsorbed?

Approximately 99% of the filtrate is reabsorbed from the renal tubules and returned to the bloodstream, while about 1% is excreted as urine (see the average values given in table 21.1). The urine volume is regulated according to the needs of the body. Most of the solutes are reabsorbed completely or almost completely, again depending upon the body's need for that particular substance.

Table 21.1 Renal Handling of Different Substances

Substance	*Kg/day filtered*	*Kg/day excreted*	*Percentage reabsorbed*
Water	180.00	1.8	99%
Glucose	0.180	0.180	100%
Sodium	0.630	0.0032	99.5%
Urea	0.056	0.028	50%

21.18 What percentage of the glomerular filtrate water is reabsorbed in each segment of the tubule.

See fig. 21.9. About 80% of the reabsorption takes place in the proximal convoluted tubule.

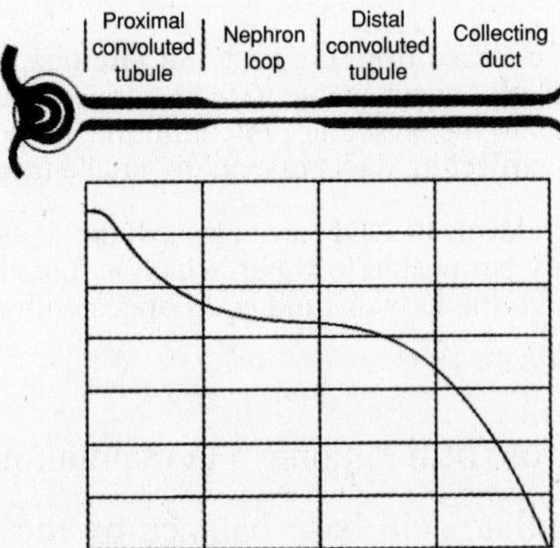

Figure 21.9 The glomerular filtrate (water) reabsorbed at progressive segments of the renal tubule.

21.19 What substances are actively transported from the peritubular capillaries to the tubule lumen?

Hydrogen, potassium, penicillin, poisons, drugs, metabolic toxins, and chemicals that are not normally present in the body.

Objective H To explain how the kidneys regulate the *concentration of urine.*

The kidneys produce either a concentrated or dilute urine depending on the operation of a **countercurrent exchange mechanism**, and the amount of circulating **antidiuretic hormone (ADH)** secreted from the posterior pituitary (see problem 13.15).

21.20 The urine-concentrating mechanism involves a combination of osmotic and diffusive transfers between the medullary interstitium (the fluid-filled space into which the tubules project), the tubules, and the vasa recta (medullary capillaries). Outline these transfers.

1. The thick portion of the ascending limb of the nephron loop actively transports negatively charged chloride ions out of the tubular fluid and into the medullary interstitium, establishing a difference in electric potential across the tubular wall (fig. 21.10). This potential causes positively charged sodium ions to pass out into the interstitium. The ascending limb is impermeable to water, and as sodium and chloride ions move out, the fluid in the ascending limb becomes more dilute as it passes toward the renal cortex.

2. Sodium and chloride ions diffuse into the descending limb, causing the fluids in the descending limb to become more concentrated. The descending limb is permeable to water, and as water diffuses out into the interstitium as a result of the osmotic gradient, the tubular fluid in the descending limb becomes more concentrated as it approaches the bend in the nephron loop.

3. Ions are actively transported into the interstitium from the collecting duct; urea passively diffuses out of the collecting duct into the interstitium.

4. Thin-walled looping vessels called the *vasa recta* parallel the course of the nephron loops. Sodium, chloride, and water diffuse into the descending vasa recta, and sodium and chloride diffuse out of the ascending vasa recta; thus, these vessels function as countercurrent exchangers. In addition, only a small quantity (1% to 2%) of the total renal blood flow passes through the vasa recta. As a result, vasa recta circulation carries only a minute amount of the medullary interstitial solutes away from the renal medulla.

21.21 How does antidiuretic hormone (ADH) participate in regulating the final urine concentration?

When low levels of ADH are released from the posterior pituitary, the distal convoluted tubules and collecting ducts become relatively impermeable to water, and, despite the high osmotic gradient, very little water is pulled out into the medullary intersititium. Therefore, fluids pass through the distal convoluted tubules and collecting ducts essentially unchanged, and a dilute urine is excreted.

When, on the other hand, high levels of ADH are released, the distal convoluted tubules and collecting ducts become highly permeable to water, which is forced by the osmotic gradient out into the interstitium. As a result, the tubular fluid equilibrates with the interstitial fluids, and a concentrated urine is excreted.

Objective I To describe the role of the kidneys in maintaining *acid-base balance.*

The kidneys regulate acid-base balance by the secretion of hydrogen ions into the tubules and the reabsorption of bicarbonate (fig. 21.11).

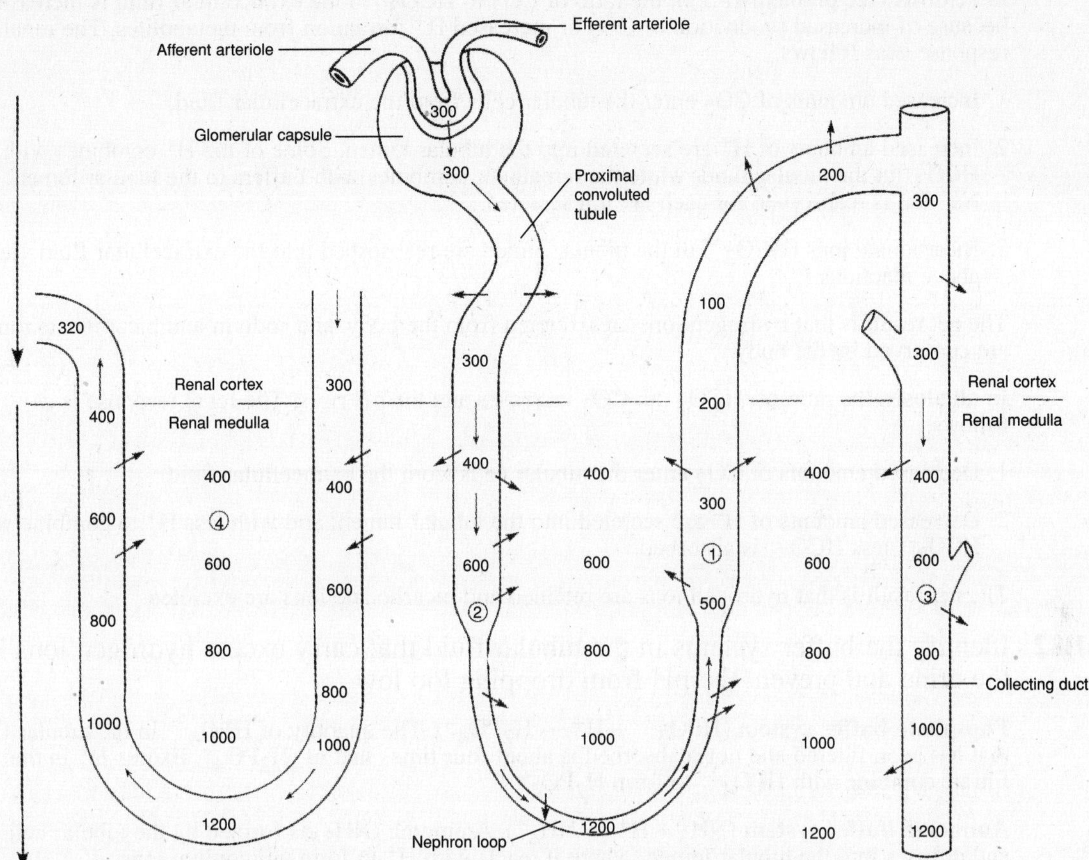

Figure 21.10 Ion concentrations and movement through the renal tubule (values represent concentrations).

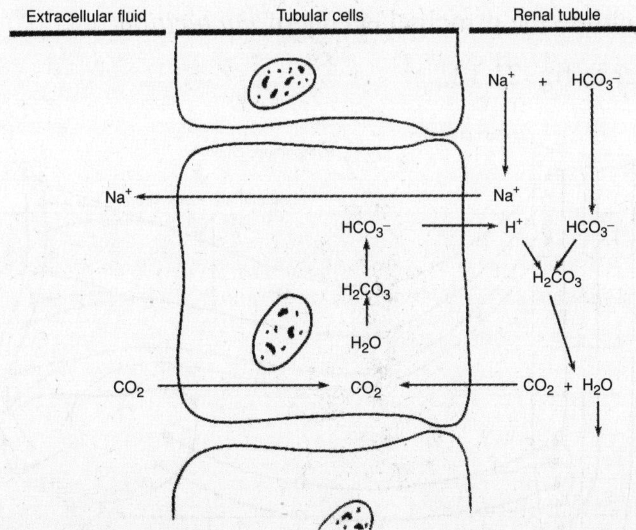

Figure 21.11 The secretion of hydrogen ions and reabsorption of bicarbonate ions in the renal tubule.

In **acidosis** (see problem 18.27), the ratio of CO_2 to HCO_3^- in the extracellular fluid is increased because of increased production of CO_2 or increased H^+ formation from metabolites. The renal response is as follows:

1. Increased amounts of CO_2 enter the tubular cells from the extracellular fluid.
2. Increased amounts of H^+ are secreted into the tubular lumen. Some of the H^+ combines with HCO_3^- in the tubular fluid, while the remainder combines with buffers in the tubular lumen. One Na^+ ion is reabsorbed for each H^+ ion secreted.
3. Bicarbonate ions (HCO_3^-) in the tubular lumen are reabsorbed into the extracellular fluid via the above reactions.

The net result is that hydrogen ions are excreted from the body, and sodium and bicarbonate ions are conserved by the body.

In **alkalosis**, the ratio of HCO_3^- to CO_2 increases and the pH rises. The renal response is as follows:

1. Decreased amounts of CO_2 enter the tubular cells from the extracellular fluid.
2. Decreased amounts of H^+ are secreted into the tubular lumen; and with less H^+ to combine with HCO_3^-, less HCO_3^- is absorbed.

The net result is that hydrogen ions are retained and bicarbonate ions are excreted.

21.22 Identify the buffer systems in the tubular fluid that carry excess hydrogen ions into the urine and prevent the pH from dropping too low.

Phosphate buffer system ($HPO_3^{2-} + H^+ \rightarrow H_2PO_4^-$). The quantity of HPO_4^{2-} in the tubular fluid that has been filtered and not reabsorbed is about four times that of $H_2PO_4^-$. Excess H^+ in the filtrate combine with HPO_4^{2-} to form $H_2PO_4^-$.

Ammonia buffer system ($NH_3 + H^+ \rightarrow NH_4^+$). Ammonia ($NH_3$) is formed by the tubular cells and diffuses into the tubular lumen, where it reacts with H^+ to form ammonium ions.

21.23 Over what range does the urinary pH change under severe acidosis or alkalosis.

4.5 to 8.0.

Objective J To outline the principal events of *micturition*.

Survey

Refer to fig. 21.12.

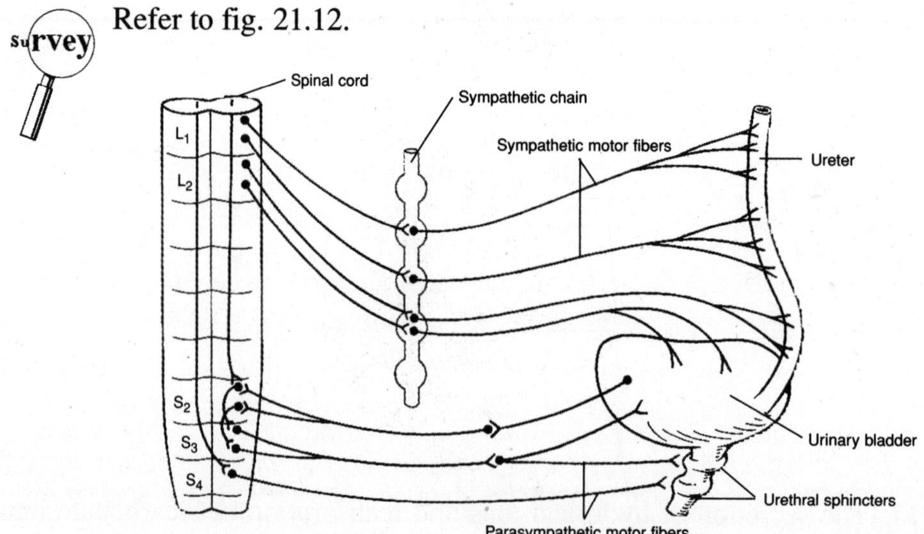

Figure 21.12 Innervation of the urinary bladder and associated structures.

1. The urinary bladder fills with urine and becomes distended.
2. Stretch receptors in the bladder wall discharge impulses via sensory neurons to the sacral spinal cord.
3. Sensory neurons send impulses up the spinal cord to the higher brain centers.
4. Parasympathetic nerve impulses stimulate the detrusor muscle and the internal urethral sphincter.
5. The detrusor muscle contracts rhythmically, and the internal urethral sphincter relaxes.
6. The need to urinate is intensified.
7. Urination can be prevented by voluntary contraction of the external urethral sphincter and by inhibition of the micturition reflex by impulses from brain centers.
8. If the decision is to urinate, the external urethral sphincter is relaxed and the micturition reflex is facilitated by impulses from brain centers.
9. The detrusor muscle contracts and urine is released through the urethra.
10. Neurons of micturition reflex centers are inactivated, the detrusor muscle relaxes, and the urinary bladder again begins to fill with urine.

21.24 Why are infants unable to contain their urine?

Voluntary inhibition of micturition requires the maturation of portions of the brain and spinal cord that is accomplished only after several years of life.

21.25 Define (adult) *incontinence*.

Incontinence is the inability to retain urine in the urinary bladder, resulting in its continuous emptying. Incontinence may be caused by central or peripheral nerve damage, various urinary diseases, or tissue damage within the urinary bladder or urethra.

Key Clinical Terms

Acute renal failure A sudden loss of kidney function, usually associated with shock or intense renal vasoconstriction, that lasts from a few days to as long as 3 weeks. In most cases the kidney damage is repairable.

Azotemia Excessive blood nitrogen compounds.

Blood urea nitrogen (BUN) An index of the accumulation of urea and other nitrogenous wastes in the urine, and thus of renal dysfunction. Normal range: 8 to 25 g/L.

Chronic renal failure A progressive destruction and shrinking of the kidneys, which become incapable of producing urine. It may be caused by chronic glomerulonephritis or pyelonephritis. Early symptoms are polyuria and nocturia; later the patient develops weakness, insomnia, loss of appetite, nausea, acidosis, and azotemia. Because of the permanent damage, the options for sustaining life are hemodialysis or kidney transplantation.

Cystitis Inflammation of the urinary bladder.

Cystoscopy Inspection of the urinary system by means of a cystoscope. Tissue and urine samples are obtained for diagnosis and for detection of obstructions.

Glomerulonephritis Inflammation of the glomeruli; generally caused by bacterial (streptococcal) infection elsewhere in the body. As toxins are given off by the streptococci, the antigen-antibody complexes accumulate in the glomeruli, producing the inflammation. If the infection is not treated, the glomeruli are replaced by fibrous tissue, and chronic renal disease may develop.

Hematuria Blood in the urine.

Hemodialysis A technique for purifying the blood outside the body.

Nephrolithiasis Renal (kidney) stones (tiny particles to large calculi) that form as a result of infections, metabolic disorders, or dehydration. They may cause obstruction and intense pain as they pass through the urinary system.

Nocturia Night urination (during sleep).

Oliguria A diminished quantity of urine.

Polyuria Excessive urine output.

Pyelography Intravenous injection of a radiopaque dye that permits X-ray examination of the kidney, ureters, and urinary bladder as the dye passes through the urinary system.

Pyelonephritis Bacterial infection and inflammation in the renal pelvis, which, if not treated, spreads progressively into the calyces and tubules of the nephrons.

Renal clearance The volume of blood plasma per minute that is cleared of a given substance. If

U = concentration of substance in urine, mg/ml
P = concentration of substance in plasma, mg/ml
F = urine flow, ml/min

then, renal clearance = UF/P.

Uremia Retention of urinary constituents in the blood, owing to kidney dysfunction.

Urethritis Inflammation of the urethra.

Urinalysis Measurement of urine volume (750 to 2000 ml/day), pH, specific gravity, protein, mucin, ketone bodies, bilirubin, glucose, blood cells, epithelial cells, and casts.

Review Exercises

Multiple Choice

1. Transitional epithelium is characteristic of (*a*) the nephron, (*b*) the glomerulus, (*c*) the urinary bladder, (*d*) the urethra.

2. The trigone is (*a*) a urine-filled cavity within the kidney, (*b*) a muscular sphincter at the neck of the urinary bladder (*c*) a smooth connective tissue region in the urinary bladder, (*d*) a tunic of the ureter.

3. Transport of urine through the ureter is by (*a*) peristalsis, (*b*) the effect of gravity, (*c*) fluid pressure, (*d*) passive transport.

4. A glomerulus is (*a*) located between a descending and ascending limb of a nephron, (*b*) composed of simple squamous epithelium, (*c*) located at the junctions of minute arteries and veins, (*d*) collapsed when not filtering urine.

5. Podocytes are specialized cells found within (*a*) the nephron loop, (*b*) the urinary bladder, (*c*) the glomerulus, (*d*) the urethra, (*e*) the glomerular capsule.

6. Regarding the kidney, which of the following is a *false* statement?
(*a*) It is drained of urine through the ureter.
(*b*) It consists of a capsule, renal cortex, and renal medulla.
(*c*) It receives blood through a renal artery and is drained of blood by a renal vein.
(*d*) It is suspended from the urinary bladder by the urinary ligament.

7. Antidiuretic hormone (ADH) is secreted by (*a*) the kidney, (*b*) the adrenal gland, (*c*) the thyroid gland, (*d*) the hypothalamus.

8. Increased permeability of the renal tubules is due to (*a*) ADH, (*b*) renin, (*c*) aldosterone, (*d*) angiotensin I.

9. The kidneys are involved in (*a*) the bicarbonate buffer system, (*b*) the phosphate buffer system, (*c*) the ammonia buffer system, (*d*) all of the preceding.

10. The epithelial cells of the proximal convoluted tubules, distal convoluted tubules, and collecting ducts all secrete (*a*) aldosterone, (*b*) ADH, (*c*) bicarbonate ions, (*d*) hydrogen ions.

11. The basic functional unit of the kidney is (*a*) the glomerulus, (*b*) the renal cortex, (*c*) the nephron, (*d*) the renal medulla.

12. Which of the following cells secretes renin? (*a*) macula densa, (*b*) glomerular epithelial cells, (*c*) juxtaglomerular cells, (*d*) basement membrane cells

13. Sodium balance is controlled by (*a*) GFR and sodium reabsorption, (*b*) GFR and renin production, (*c*) sodium reabsorption and potassium secretion, (*d*) none of the preceding.

14. To compensate for alkalosis, the kidney tubules (*a*) reabsorb more Na^+; (*b*) obtain less carbon dioxide from the blood for reaction with bicarbonate, which allows bicarbonate to pass out in the urine; (*c*) secrete more angiotensin I, which stimulates the hypothalamus to secrete more ADH to increase filtration in the glomeruli; (*d*) absorb more K^+.

15. Increased glomerular filtration results from (*a*) increased cardiac output, (*b*) a rise in environmental temperature, (*c*) decreased fluid intake, (*d*) decreased blood pressure.

16. The renal ducts of the nephron (*a*) can secrete water molecules actively into the urine, (*b*) are responsible for most of the reabsorption of water that occurs in the kidney, (*c*) determine to a large extent the final osmolality of urine, (*d*) are rendered impermeable to water by ADH.

17. Aldosterone (*a*) is produced mainly in the juxtaglomerular apparatus, (*b*) increases sodium reabsorption by the nephron, (*c*) increases potassium reabsorption by the nephron, (*d*) tends to increase the hydrogen ion concentration in the blood.

18. When a patient is treated with an aldosterone antagonist, there is likely to be a fall in (*a*) urine volume, (*b*) plasma potassium concentration, (*c*) blood viscosity, (*d*) blood volume.

19. Capillary pressure in the glomeruli (*a*) is lower than pressure in the efferent arterioles, (*b*) rises when the afferent arterioles constrict, (*c*) is higher than in most other capillaries in the body, (*d*) is reduced by about 10% when arterial pressure falls 10% below the normal level.

20. The macula densa is part of (*a*) the proximal convoluted tubule, (*b*) the afferent arteriole, (*c*) the distal convoluted tubule, (*d*) the efferent arteriole, (*e*) none of the preceding.

True or False

____ **1.** Bilirubin, a by-product of the destruction of erythrocytes, can be found in the urine.

____ **2.** Afferent arterioles bring arterial blood to the glomeruli.

____ **3.** Most water reabsorption occurs in the distal convoluted tubules.

____ **4.** The kidneys synthesize and secrete glucose during prolonged fasting.

____ **5.** The kidneys help to counter alkalosis by reabsorbing excess HCO_3^-.

____ **6.** Aldosterone increases the permeability of the distal convoluted tubules to water only.

____ **7.** The kidneys regulate glucose levels by secreting any excess in the urine.

____ **8.** One symptom of diabetes is polyuria.

____ **9.** A difference in hydrostatic pressure is one of the mechanisms for pushing blood fluid through the glomerulus to form the filtrate.

____ **10.** ADH is necessary for the reabsorption of Na^+.

Completion

1. The most primitive kidney, the ________________, begins to develop during the fourth week of the embryonic period.
2. The ________________ muscle within the wall of the urinary bladder forcefully contracts during micturition, forcing urine out of the urinary bladder.
3. The ________________ is a network of about 50 capillaries surrounded by the glomerular capsule.
4. The inner visceral layer of the glomerular capsule is composed of specialized cell called ________________.
5. Increased sodium stimulates the juxtaglomerular cells to secrete ________________.
6. Approximately ________________ % of the filtrate is reabsorbed from the renal tubules and returned to the blood.
7. The condition of blood in the urine is ________________.

8. ________________ refers to an excessive blood nitrogen compounds.

9. ________________ is the hormone that regulates water reabsorption in the distal convoluted tubule.

10. The ________________ ________________ run parallel to the nephron loops and function as countercurrent exchangers.

Labeling

Label the structures indicated on the figure to the right.

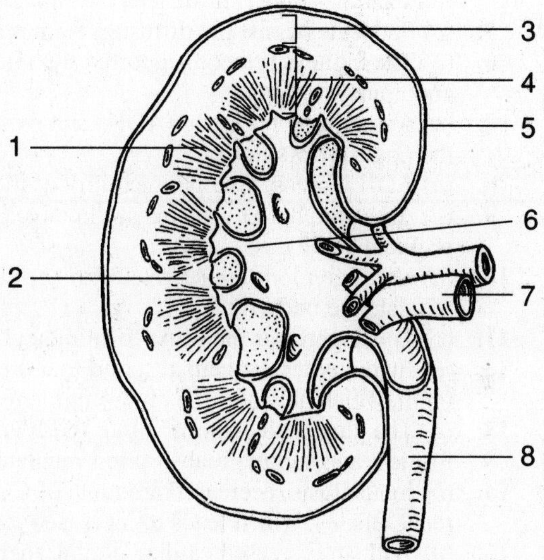

1. ______________________________
2. ______________________________
3. ______________________________
4. ______________________________
5. ______________________________
6. ______________________________
7. ______________________________
8. ______________________________

Matching

Match the structure with its description or function.

____	1. Aldosterone	(*a*)	adjacent to the macula densa of the capillaries
____	2. ADH	(*b*)	long, slender capillaries that accompany the nephron loop into the renal medulla
____	3. Juxtaglomerular cells	(*c*)	tube extending from kidney to urinary bladder
____	4. Nephron loop	(*d*)	composed of the glomerular capsule, proximal and distal convoluted tubules, nephron loop, and glomerulus
____	5. Vasa recta	(*e*)	physiological events that result in the voiding of urine
____	6. pH of the blood	(*f*)	tubule extending from the renal cortex to the renal medulla and back to the renal cortex
____	7. pH of the urine	(*g*)	7.0–7.4
____	8. Nephron	(*h*)	stimulates the zona glomerulosa of the adrenal glands
____	9. Ureter	(*i*)	4.5–6.0
____	10. Angiotensin II	(*j*)	secreted by the zona glomerulosa
____	11. Renin	(*k*)	secreted by juxtaglomerular cells
____	12. Micturition	(*l*)	synthesized in the hypothalamus

Answers and Explanations for Review Exercises

Multiple Choice

1. (*c*) The transitional epithelium lining the lumen of the urinary bladder permits distension (stretching).
2. (*c*) The trigone is a triangular sheet of connective tissue between the openings of the ureters and the urethra.
3. (*a*) Contraction of the smooth muscle within the wall of the ureter causes peristalsis, which continuously moves urine from the renal pelvis of the kidney to the urinary bladder.
4. (*b*) The simple squamous epithelium of the capillaries that form a glomerulus permits diffusion of fluid and dissolved substances into the glomerular capsule.
5. (*c*) Podocytes assist the diffusion from a glomerulus into the glomerular capsule.
6. (*d*) The kidneys are not supported by ligaments, but rather are retroperitoneal against the posterior abdominal wall.
7. (*d*) Antidiuretic hormone (ADH) and oxytocin are produced in the hypothalamus and secreted from the posterior pituitary.
8. (*a*) ADH increases the permeability of the renal tubules to water.
9. (*d*) All of the buffer systems are involved in acid-base regulation in the kidneys (see fig. 21.11 and problem 21.22).
10. (*d*) Hydrogen ions are secreted from epithelial cells throughout much of the nephron, where they regulate the body fluid pH.
11. (*c*) The nephron is the urine-forming and the blood-cleansing unit of the urinary system.
12. (*c*) Juxtaglomerular cells respond to a drop in blood pressure in the afferent arterioles and release renin, which activates the renin-angiotensin system to elevate blood pressure.
13. (*a*) The glomerular filtration rate (GFR) regulates the amount of filtrate (sodium included) that is formed, and sodium reabsorption regulates the amount of sodium that will be excreted in the urine.
14. (*b*) In alkalosis, decreased amounts of CO_2 enter the tubular cells; thus, less H^+ is secreted into the renal tubules, which leads to increased loss of bicarbonate.
15. (*a*) With an increased cardiac output, there is greater blood flow to the kidney and thus greater glomerular filtration.
16. (*c*) If large amounts of ADH are present, water will be reabsorbed from the renal tubules because of the high osmotic gradient and the urine will be concentrated; with a lack of ADH, little water will be reabsorbed in the renal tubules and the urine will be dilute.
17. (*b*) Aldosterone regulates sodium reabsorption and potassium secretion in the distal convoluted tubules and collecting ducts.
18. (*d*) If aldosterone action is blocked, there will be less sodium reabsorption; thus, more sodium will be excreted in the urine. Because of the increased sodium in the urine, it acts as an osmotic diuretic; the results are decreased body fluid and decreased blood plasma volume.
19. (*c*) Glomerular capillary pressure is higher (50 to 60 mmHg) than in most other capillaries (10 to 30 mmHg).
20. (*c*) The region of the distal convoluted tubule in contact with granular cells of the afferent arteriole is called the macula densa.

True or False

1. True
2. True
3. False; most water reabsorption occurs in the proximal convoluted tubules
4. False; the liver converts glycogen to glucose and releases it into the blood during fasting
5. False; the kidneys reabsorb HCO_3^- to help counter acidosis
6. False; aldosterone increases reabsorption of sodium, and water follows due to osmosis
7. False; excess glucose is not reabsorbed in the proximal convoluted tubules
8. True
9. True
10. False; ADH is involved in water reabsorption only

Completion

1. pronephros
2. detrusor
3. glomerulus
4. podocytes
5. renin
6. 99
7. hematuria
8. Azotemia
9. Antidiuretic hormone (ADH)
10. vasa recta

Labeling

1. Renal pyramid
2. Minor calyx
3. Renal cortex
4. Renal medulla
5. Renal capsule
6. Renal pelvis
7. Renal vein
8. Ureter

Matching

1. (*j*)
2. (*l*)
3. (*a*)
4. (*f*)
5. (*b*)
6. (*g*)
7. (*i*)
8. (*d*)
9. (*c*)
10. (*h*)
11. (*k*)
12. (*e*)

Water and Electrolyte Balance 22

Objective A To explain the *distribution of water* within the body and to list the *major functions of body water.*

survey **Water** is the most abundant substance in the human body (averaging 60% of the total body weight ([BW]), and ranging from 40% to 80%). Essentially all metabolic reactions that occur in the body require water. Body water is distributed between two major compartments: the *intracellular fluid compartment* (within the cells) and the *extracellular fluid compartment* (outside the cells) (fig. 22.1). Extracellular fluid is further divided into *blood plasma* (the fluid portion of the blood) and *interstitial fluid* (the fluid surrounding cells). Two other minor subgroups included in the extracellular fluid are *lymph* and *transcellular fluids.* Transcellular fluids include *cerebrospinal fluid* (within the central nervous system), *intraocular fluid* (within the eyes), *synovial fluid* (within joints), *pericardial fluid* (surrounding the heart), *pleural fluid* (surrounding the lungs), and *peritoneal fluid* (within the abdominal cavity).

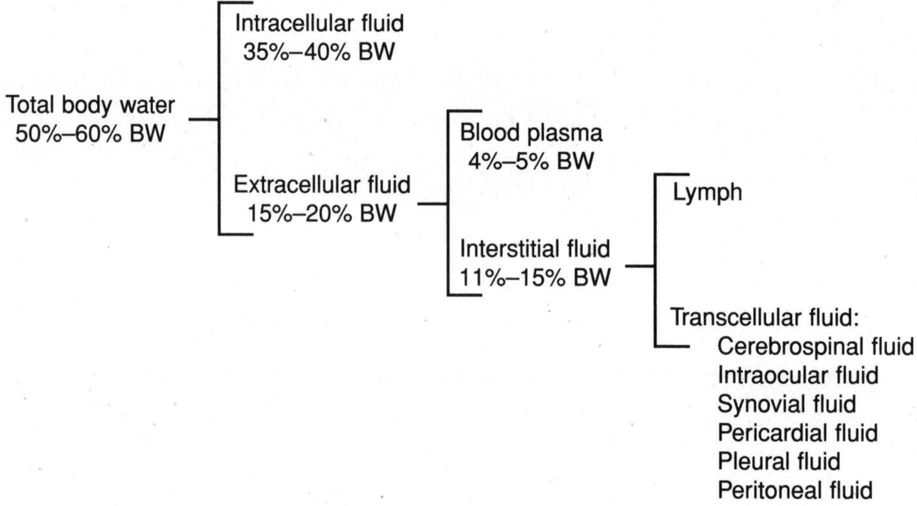

Figure 22.1 The distribution of water within the body.

22.1 List the principal functions of water in the body.

Water is a universal solvent and suspending medium. It helps to regulate body temperature, participates in hydrolysis reactions, lubricates organs, provides cellular turgidity, and helps to maintain body homeostasis.

22.2 How does total body water vary with age, sex, and body weight?

See Table 22.1.

Table 22.1 Factors That Determine the Percentage of Total Body Water

Age	Body water constitutes 75% to 80% of the body weight in infants and children. This percentage decreases with age; in elderly people, body water may constitute only 40% to 50% of the total body weight.
Sex	Women usually have less body water than men because the greater proportion of adipose tissue in women contains lesser amounts of water than other tissue types.
Weight	Obese people have less body water because of an abundance of adipose tissue.

Adults, whose percent body water is high in comparison to their skin surface, are not especially prone to severe dehydration. By contrast, infants are at high risk for dehydration when they are febrile (feverous) or lose fluids as a result of vomiting or diarrhea. A 10-kg infant has approximately 1 liter of blood plasma volume. Losing even 100 ml through the skin during a high fever constitutes a 10% decrease in total blood volume. For this reason, it is critical to administer fluids to a febrile infant to maintain body homeostasis.

Objective B To define the terms used to describe *solute concentrations* (in body fluids).

Percent solution

$$\% \text{ solution} = (\text{grams of solute})/(100 \text{ ml of solution}) = (\text{grams of solute})/(\text{dl of solution})$$

Example: If 200 ml of solution contains 5 g of dissolved NaCl, the solution is termed 2.5% NaCl.

$$\text{Mg}\% = (\text{miligrams of solute})/(100 \text{ ml of solution}) = 1000 \times (\% \text{ solution})$$

Molarity (molar concentration)

Let MW denote the molecular weight of the solute (e.g., for NaCl, MW = 23 + 35.5 = 58.5). Then *one mole* (1 mol) of solute weighs MW grams, from which

$$\text{Moles solute} = (\text{grams of solute})/\text{MW}$$

and

$$\text{Molarity} = (\text{moles solute})/(\text{liters of solution})$$

Example: The previously considered solution is 5/58.5 moles of NaCl in 0.200 L, giving a molarity of 0.427. The solution would be termed 0.427 *M* NaCl.

Fluid balance in the extracellular compartment is maintained by regulation of fluid osmolarity. For example, extracellular fluid osmolarity must be regulated to prevent shrinking or swelling of the cells. Osmolarity of a body fluid is a measure of the concentration of individual solute particles dissolved in it. The osmolarities of the extracellular and intracellular fluids are normally the same.

22.3 What is the molar concentration of a 0.9% NaCl solution?

The NaCl content is

$$0.9 \text{ g/1 dL} = 9 \text{ g/1 L}$$
$$(9 \text{ g/1 L})(1 \text{ mol/58.5 g}) = 0.154 \text{ mol/L}$$
$$\text{or } 154 \text{ mmol/L}$$

22.4 List the mean concentrations of the more important solutes in the extracellular and intracellular fluids.

See Table 22.2.

Table 22.2 Mean Concentrations of Important Body Fluid Solutes

Fluid	Na^+	K^+	Ca^{2+}	Mg^{2+}	Cl^-	Amino Acids	Glucose, mg%
Extracellular	142	4	5	3	103	5	90
Intracellular	10	140	1	58	4	40	0–20

Note: All concentrations, except those of glucose, are in miliequivalents per liter.

22.5 How are the volumes of the fluid compartments measured?

Using an *indirect dilution technique*, a specific quantity of an exogenous substance (dye, radioisotope, etc.) is introduced, which because of its chemical properties becomes evenly distributed within a certain fluid compartment. A representative sample of the fluid from that compartment is then removed and the concentration of the substance is determined.

$$\text{Compartment volume (ml)} = \frac{\text{quantity of substance introduced (mg)}}{\text{Substance conc. in compartment (mg/ml)}}$$

Compartment	*Substances used*
Total	^3H_2O (radioactive water), antipyrine
Extracellular	Thiosulfate, inulin
Blood plasma	Evans blue

The intracellular and interstitial volumes are determined indirectly, as differences.

Objective C To explain what is meant by the *fluid balance of the body.*

Under normal conditions (stable balance), fluid intake equals fluid output, so that the body maintains a constant volume. A typical water budget is given below. When body water intake exceeds water output, a positive balance exists (hydration). Conversely, when output exceeds intake, a negative balance exists (dehydration). The amount of water consumed and the amount of urine formed are the two major mechanisms by which body water content is regulated.

Water intake		***Water output***	
Ingested liquids	1400 ml	Urine	1500 ml
Solid and semisolid food	800	Skin	500
Oxidation of food	300	Feces	150
	2500		2500

Water is unconsciously regulated through the action of osmoreceptors located in the hypothalamus. These receptors sense the osmolality of the blood and determine whether more or less water is needed to maintain the correct osmolality. If more water is needed (blood is too concentrated), thirst is stimulated and we drink. Also, ADH is released from the posterior pituitary, which leads to a conservation of body fluids via the mechanism indicated in fig. 22.2. If less water is needed (blood is too dilute), thirst is suppressed and ADH release is inhibited, causing large volumes of dilute urine to be excreted.

A person may eliminate up to a liter of water over a 24-hour period without being aware of the loss. This is termed **insensible loss**. The loss occurs from the lungs and nonsweating skin. As air is breathed in, it is moistened and water is subsequently lost through evaporation. Water molecules also diffuse through the skin and evaporate without our awareness.

22.6 What is the result of a loss or gain of water that is *not* simultaneously accompanied by a loss or gain of solute?

When there is a loss of free water in the extracellular compartment, the fluid becomes too concentrated (increased osmolarity) and is referred to as being **hypertonic**. When there is a gain in free water, the fluid becomes too dilute and is referred to as being **hypotonic**.

The normal concentration of sodium in the blood plasma is 150 mEq/L. If this level drops below 120 mEq/L, the result may be lethargy, coma, or death. Unlike deficiencies of other important fluid constituents, the most common cause of this drop in sodium concentration is not a nutritional deficit of sodium, but rather an excess of water. *Gastroenteritis*, or inflammation of the stomach, can lead to vomiting and diarrhea—processes that result in a loss of fluid and solutes. If people with this common condition replace volume loss with free water instead of solute-containing beverages such as juices or sports drinks, they can easily progress to *hyponatremia*, or a state of low sodium concentration. People with psychiatric disorders urging them to binge on free water are also subject to this condition.

22.7 What are the causes and symptoms of dehydration, or hypovolemia, and what is the body's response?

Causes: Decreased intake (lack of water, psychogenic refusal to drink) and/or increased output (vomiting, diarrhea, loss of blood, drainage from burns, diabetes mellitus, diuretic use, or a lack of ADH owing to diabetes insipidus).

Symptoms: Loss of weight, rise in body temperature, increase in heart rate and cardiac output, decrease in blood pressure, sunken eyeballs.

Response: A decrease in salivary secretion and consequent drying of the mouth and pharynx produces a well-recognized symptom of thirst. Furthermore, the hypertonicity of the cerebrospinal fluid stimulates the release of ADH from the posterior pituitary which, in combination with aldosterone released from the adrenal glands, decreases urine output and stimulates drinking as indicated in fig. 22.2.

Antidiuretic hormone (*ADH*), also known as *vasopressin*, is released from the posterior pituitary. ADH plays a major role in the regulation of total body water, blood volume, and blood pressure. Dehydration or an increased blood plasma osmolality increases the secretion of ADH. ADH causes increased reabsorption of water in the kidney tubules and, therefore, decreases urine output. As a result, there is an increase in fluid, blood plasma volume, and blood pressure. This is accompanied by a decrease in plasma osmolality (see fig. 22.2).

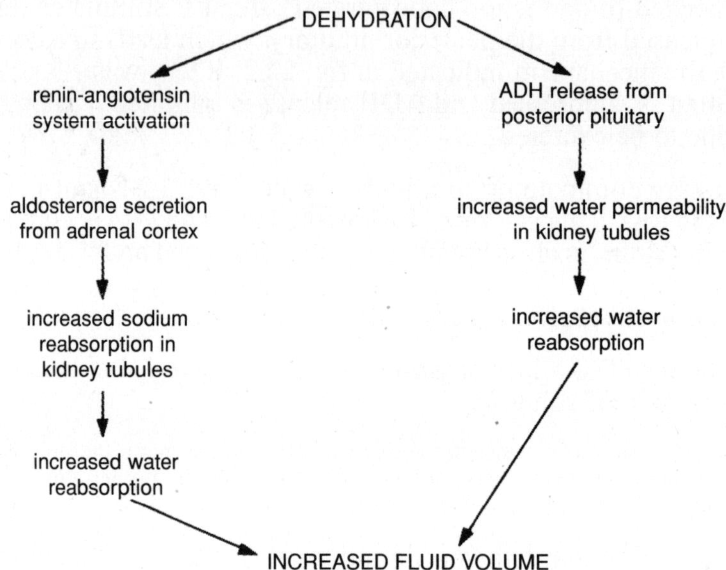

Figure 22.2 Mechanisms of aldosterone and ADH regulation of water balance.

Diabetes insipidus is a disease characterized by a deficiency of ADH secretion. Symptoms include polyuria (5 to 25 liters of urine per day), polydipsia, dehydration, fever, dry tongue, and delirium. Treatment includes the use of synthetic ADH or Diabinese (chlorpropamide, a hypoglycemic agent that increases the sensitivity of the kidney tubules to ADH).

22.8 What are the causes and symptoms of hypervolemia, and what is the body's response?

Causes: Excessive IV administration of fluids, psychogenic drinking episodes, decreased urinary output because of renal failure, or congestive heart failure.

Symptoms: Decrease in body temperature, increased blood pressure, edema, weight gain.

Response: The decrease in osmolarity of fluids in the hypothalamus causes an inhibition of thirst and a decreased release of ADH and aldosterone. With decreased ADH and aldosterone secretion, there is an increased urinary output (inverse to the process outlined in fig. 22.2).

Objective D To distinguish between *electrolytes* and *nonelectrolytes*.

Electrolytes are chemicals formed by ionic bonding that dissociate into electrically charged ions (cations and anions) when they dissolve in the body fluids. Examples of electrolytes are acids, bases, and salts. **Nonelectrolytes** are formed by covalent bonding; they do not ionize when dissolved in the body fluids. Most organic compounds are nonelectrolytes.

22.9 List the functions of electrolytes.

1. Control of the osmolarity in the fluid compartments.
2. Maintenance of the acid-base balance in body fluids.
3. Metabolization as essential minerals.
4. Participation in all cellular activities.

22.10 Identify some of the common electrolyte-imbalance disorders.

Hyponatremia (*low blood sodium level*). *Causes*: sweating, diarrhea, certain diuretics, Addison's disease, excessive water intake. *Symptoms*: muscular weakness, headache, hypotension, circulatory shock, mental confusion.

Hypernatremia (*high blood sodium level*). *Causes*: diabetes insipidus, inadequate water intake, Cushing's syndrome. *Symptoms*: nervous system disorders, mental confusion, coma.

Hypokalemia (*low blood potassium level*). *Causes*: vomiting, diarrhea, Cushing's syndrome, diuretic abuse. *Symptoms*: muscle weakness, paralysis, shallow breathing cardiac arrhythmias.

Hyperkalemia (*high blood potassium level*). *Causes*: kidney disease, Addison's disease. *Symptoms*: muscle weakness, paralysis, cardiac arrhythmias, cardiac arrest.

Hypochloremia (*low blood chloride level*). *Causes*: vomiting, diarrhea, dehydration. *Symptoms*: muscle spasms, alkalosis, depressed breathing.

Hypomagnesemia (*low blood magnesium level*). *Causes*: severe vomiting or diarrhea. *Symptoms*: hyperreflexia, sometimes leading to convulsions or tetany.

Hypermagnesemia (*high blood magnesium level*). *Causes*: impaired kidney function with reduced excretion. *Symptoms*: muscle weakness, sedation, and mental confusion.

Hypophosphatemia (*low blood phosphate level*). *Causes*: reduced intake or decreased absorption, impaired kidney function. *Symptoms*: loss of appetite, muscle and bone pain.

Review Exercises

Multiple Choice

1. Total body water would be greatest in (*a*) a newborn male, (*b*) a teenage female, (*c*) a middle-aged male, (*d*) a 100-year-old female.

2. To make 300 ml of a 10% solution, how much solute should be added? (*a*) 10 g, (*b*) 10 mg, (*c*) 3 mg, (*d*) 30 g

3. Loss of weight, elevated body temperature, decreased blood plasma, sunken eyeballs, and decreased urinary output are all symptoms of (*a*) hypervolemia, (*b*) positive water balance, (*c*) hypovolemia, (*d*) Cushing's syndrome.

4. Which of the following does not commonly contribute to hyponatremia? (*a*) inadequate sodium in the diet, (*b*) replacement of sweat losses with water, (*c*) drinking free water to replace losses due to vomiting or diarrhea, (*d*) psychogenic free-water bingeing.

5. The feeling of thirst arises from (*a*) a decreased osmolarity of the body fluids, (*b*) an increased blood plasma volume, (*c*) hypervolemia, (*d*) an increased osmolarity of body fluids and secretion of ADH.

6. The interstitial fluid is (*a*) part of the extracellular fluid, accounting for 5% of the body weight; (*b*) part of the intracellular fluid, accounting for 20% of the body weight; (*c*) part of the extracellular fluid, accounting for 15% of the body weight; (*d*) fluid confined in the cell nucleus.

7. Extracellular fluid (*a*) makes up the major proportion of total body water, (*b*) is composed mainly of transcellular fluids, (*c*) has a higher sodium/potassium ratio than intracellular fluid, (*d*) contains less glucose than intracellular fluid.

8. Hyperkalemia (*a*) causes muscle weakness, (*b*) causes cardiac arrhythmias, (*c*) causes paralysis, (*d*) may be caused by Addison's disease.

9. Kidney failure commonly causes (*a*) hyponatremia, (*b*) hyperkalemia, (*c*) hypomagnesemia, (*d*) hyperphosphatemia.

10. Which of the following has the smallest volume? (*a*) extracellular fluid, (*b*) blood plasma, (*c*) interstitial fluid, (*d*) intracellular fluid

11. The most abundant cation in the extracellular fluid is (*a*) Na^+, (*b*) Ca^{2+}, (*c*) K^+, (*d*) Mg^{2+}.

12. There is only a slight difference in the composition of: (*a*) blood plasma and intracellular fluid, (*b*) blood plasma and interstitial fluid, (*c*) intracellular and extracellular fluids, (*d*) blood plasma and water.

13. In a 200-lb male, the intracellular fluid would weigh about (*a*) 100 lb, (*b*) 60 lb, (*c*) 10 lb, (*d*) 80 lb.

14. Approximately how many milliliters of fluid are lost per day via the lungs? (*a*) 800, (*b*) 500, (*c*) 350, (*d*) 150

15. Which of the following individuals would be most seriously affected by diarrhea? (*a*) a 15-year-old male, (*b*) a 30-year-old male, (*c*) a 35-year-old male, (*d*) an 80-year-old male.

16. Potassium is the primary cation of (*a*) blood plasma, (*b*) interstitial fluid, (*c*) transcellular fluids, (*d*) intracellular fluid.

True or False

____ **1.** Infants have a low percentage of body water and adults have a high percentage.

____ **2.** Women usually have less total body water than men due to a higher percentage of body fat.

____ **3.** Amino acids are found in greater concentration in the intracellular fluid than in the extracellular fluid.

____ **4.** Through insensible fluid loss, a person may eliminate up to a liter of water in a 24-hour period and not be aware of it.

____ **5.** The amount of water consumed and the amount of urine produced are the two major mechanisms by which body water is regulated.

____ **6.** Hypernatremia is a high concentration of chloride.

Completion

1. Body fluid that is too concentrated is referred to as ________________.

2. Hypovolemia may be caused by a lack of ADH; this condition is referred to as ________________.

3. Water is autonomically regulated by ________________ located in the hypothalamus of the brain.

4. ________________ become charged ions when they dissolve in the body fluids.

5. A high level of potassium is referred to as ________________.

6. ________________ are formed by covalent bonding and do not ionize in the body fluids.

Matching

Match the substance or condition with its description

____	1. Extracellular fluid	(*a*) plasma and interstitial fluid
____	2. Evans blue	(*b*) dehydration
____	3. Diabetes insipidus	(*c*) 15%–20% body water
____	4. Intracellular fluid	(*d*) abnormal increase in fluid volume
____	5. Hypovolemia	(*e*) used to measure plasma volume
____	6. Hypervolemia	(*f*) lack of ADH

Answers and Explanations for Review Questions

Multiple Choice

1. (*a*) Total body water accounts for 75% to 90% of the total body weight in infants and young children.
2. (*d*) Percent solution = grams solute/100 ml solution.
3. (*c*) Hypovolemia is caused by a decreased intake or increase output of fluid.
4. (*a*) An excess of water, rather than inadequate sodium in the diet, contributes to hyponatremia.
5. (*d*) An increase in osmolarity or a decrease in total body water leading to ADH secretion will lead to increased thirst.
6. (*c*) The interstitial fluid is the part of the extracellular fluid that surrounds the cells.
7. (*c*) Extracellular fluid (Na^+ – 142, K^+ – 4) vs. intracellular (Na^+ – 10, K^+ – 140).
8. (*d*) Hyperkalemia may be caused by Addison's disease; hypokalemia may be caused by Cushing's disease.
9. (*b*) The kidney is the organ responsible for active excretion of excess potassium. Hyperkalemia is a common and very deadly result of kidney failure.
10. (*b*) Blood plasma makes up only about 4.5% of the total body weight.
11. (*a*) Na^+ at 142 mEq.
12. (*b*) Blood plasma and interstitial fluid constitute only about 14% of the body weight.
13. (*d*) 40% of 200 lb is 80 lb.
14. (*c*) 350 ml/day; loss of water of this type is termed *insensible loss*.
15. (*d*) The 80-year-old male is at greater risk because in an older person there is a lower percentage of total body water.
16. (*d*) The intracellular fluid contains 140 mEq of potassium.

True or False

1. False; infants have a high percentage and older people a lower percentage of body water
2. True
3. True
4. True
5. True
6. False; hypernatremia is a state of high sodium concentration

Completion

1. hypertonic
2. diabetes insipidus
3. osmoreceptors
4. Electrolytes
5. hyperkalemia
6. Nonelectrolytes

Matching

1. (*c*)
2. (*e*)
3. (*f*)
4. (*a*)
5. (*b*)
6. (*d*)

Reproductive System 23

Objective A To appreciate the biological value of *sexual reproduction*.

The male and female reproductive systems are specialized to produce offspring that have genetic diversity. This is accomplished through **sexual reproduction**, in which genes from two individuals are combined in random ways with each new generation. Genetic diversity is the basis of natural selection. As environmental conditions change over evolutionary time, the genetic traits of survival value to individuals within a population will be propagated.

Also of biological and sociological value is the fact that there are two parents to care for the young and establish a family social unit.

23.1 What are gametes?

Gametes, also called *germ cells* or *sex cells*, are the functional reproductive cells (*ova* or *spermatozoa*). They are *haploid cells*, each containing a half-complement—or 23 single chromosomes—of the genetic material. Fertilization of an **ovum** (egg cell) by a **spermatozoon** (sperm cell) produces a normal *diploid cell*, the **zygote**, in which the chromosomes of the ovum have been paired with those of the spermatozoon. In this way, genetic diversity is realized.

23.2 Define *coitus*.

Also called *copulation*, **coitus** is *sexual intercourse*. Coitus keeps ejaculated spermatozoa viable, that is, capable of fertilizing an ovum. If exposed to the air, ejaculated sperm will desiccate (dry out) and die within minutes. Discharged through the copulatory organ (penis) into the vagina during coitus, spermatozoa will remain alive up to 5 days.

23.3 How is the sex of a child determined?

One out of the 23 pairs of human chromosomes determines sex. Sex chromosomes are of two types, X and Y. A female pair of sex chromosomes consists of two X chromosomes; consequently, all female gametes, or ova, contain a single X chromosome. The male pair of sex chromosomes consists of an X chromosome and a Y chromosome; thus, an equal number of X and Y male gametes, or spermatozoa, are produced. It follows that an offspring will be female or male according to whether the fertilizing spermatozoon is X-bearing or Y-bearing. The two possibilities are equally likely.

The differentiation of the genitalia is hormonally controlled. Because they are in an environment low in androgen while in the mother's uterus, all embryos begin their development as females. The secretion of androgens (male sex hormones) early in the fetal period from the male testes masculinizes the genitalia of male embryos. In the absence of androgens, an embryo will continue to develop as female.

Objective B To understand the processes of *spermatogenesis* and *oogenesis*.

Spermatogenesis is the process by which sperm cells (which become free-swimming spermatozoa) are produced in the testes of a male. **Oogenesis** is the process by which ova are produced in the ovaries of a female. Both processes involve a special kind of cell division called *meiosis*.

23.4 Describe meiosis and explain how it differs in oogenesis and spermatogenesis.

In **meiosis** (fig. 23.1), each chromosome duplicates itself as in mitosis (see fig. 3.9). However, unlike the process of mitosis, the homologous chromosomes are attached to each other and come to lie alongside one another in pairs, producing a *tetrad* of four *chromatids*. Two *maturation divisions* are required to effect the separation of the tetrad into four daughter cells, each with one-half the original number of chromosomes. In these divisions, maternal and paternal chromosomes become freely assorted, yielding a great number of possible combinations in the haploid gametes.

The nuclear aspects of meiosis are similar in males and females. There is, however, a marked difference in the cytoplasmic aspects. The two meiotic divisions of the primary oocyte do not result in four equally mature gametes, as in the male, but rather in only one mature ovum.

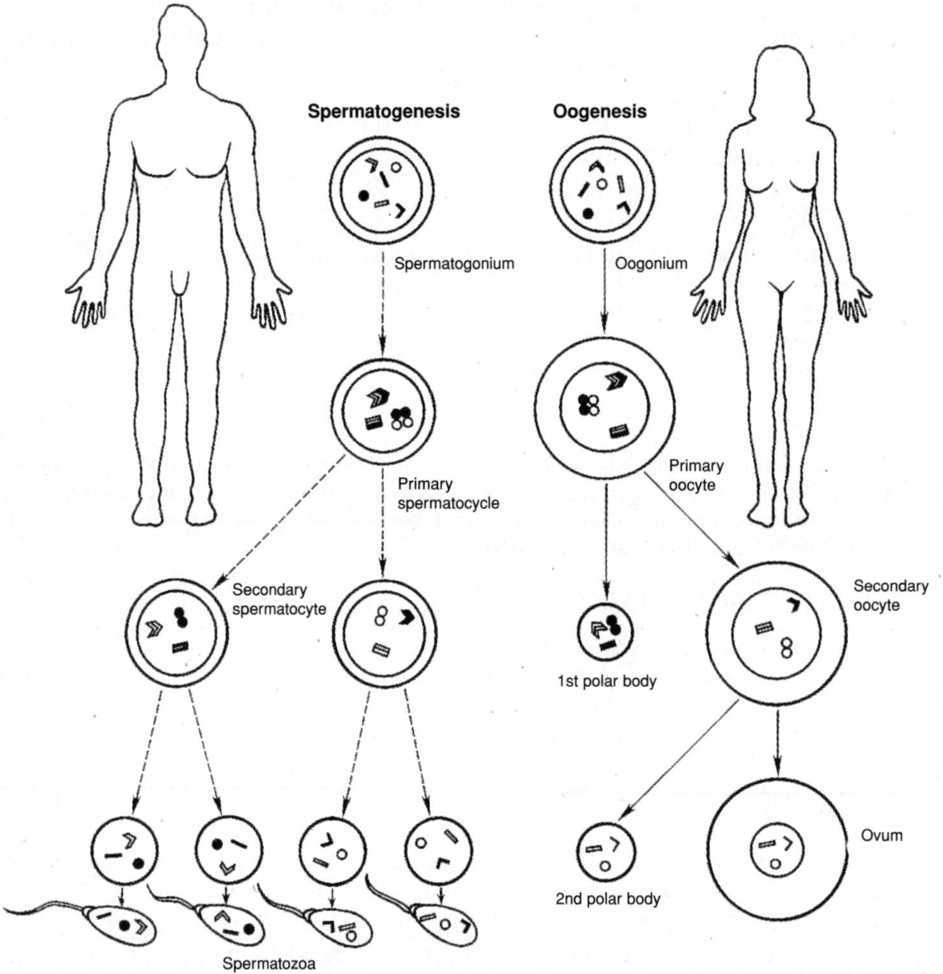

Figure 23.1 The process of meiosis in the male is referred to as spermatogenesis; in the female, as oogenesis. Spermatogenesis results in four spermatozoa, whereas oogenesis results in one ovum.

Objective C To describe the *events* and *conditions* associated with *reproduction.*

Survey

The reproductive system is characterized by *latent development*; that is, it does not become functional until a certain degree of physical growth has been attained. Body growth and the stimulation of reproductive maturation are hormonally regulated. Although formed prenatally, the reproductive organs of both sexes are underdeveloped and remain so until puberty (table 23.1) during early adolescence. *Adolescence* is the period of growth and development between childhood and adulthood. It begins around the age of 10 in girls and the age of 12 in boys. Most people have reached their adult height and are physically mature by age 20.

Table 23.1 Reproductive Events and Conditions

Event/Condition	*Definition*
Puberty	Period of development when the reproductive organs become functional
Menstruation	Discharge of blood and tissue (menses) from the uterus at the end of the monthly female reproductive cycle
Ovulation	Rupture of an ovarian (graafian) follicle, with the release of an ovum
Erection	Engorgement of erectile (vascular) tissue of reproductive organs with blood
Ejaculation	The forceful discharge of semen from the erect penis of the male
Fertility	The capacity to conceive (gravid ovum) or to induce conception (viable spermatozoan)
Pregnancy	The condition in which a female is carrying a developing offspring within her body
Gestation	Period of development of the conceptus in the uterus from the time of fertilization of the ovum until birth
Parturition	Childbirth; accompanied by a series of uterine contractions known as *labor*
Lactation	Production and secretion of milk from the mammary glands
Menopause	The end of a woman's reproductive capabilities; termination of ovulation marked by the cessation of menstrual periods

23.5 What is the length of the human gestation period and how is the "due date" determined?

Normal gestation for humans is 9 months. Knowing this and the pattern of menstruation makes it possible to determine the parturition date. In a typical reproductive cycle (see problem 23.32), a woman ovulates 14 days prior to the onset of the next menstruation and is fertile for approximately 20 to 24 hours following ovulation. Adding 9 months, or 38 weeks, to the time of ovulation gives the estimated parturition date.

Objective D To differentiate between *primary* and *secondary sex organs*, and *secondary sex characteristics.*

The **primary sex organs**, or **gonads**, are the *testes* in the male and the *ovaries* in the female. The gonads function as mixed glands in that they produce both hormones (endocrine system) and gametes (reproductive system). The **secondary**, or *accessory*, **sex organs** are those structures that mature at puberty (table 23.1) and that are essential in caring for and transporting gametes. **Secondary sex characteristics** are features that are considered sexual attractants.

23.6 What is responsible for the maturation of the secondary sex organs?

Increased release of sex hormones at puberty—*testosterone* in the male; *estradiol* and other *estrogens* in the female.

23.7 Which of the following are secondary sex organs and which are secondary sex characteristics? (*a*) uterine tubes, (*b*) pubic hair, (*c*) epididymides, (*d*) breasts

The *uterine* (*fallopian*) *tubes* of the female and the *epididymides* of the male transport ova and spermatozoa, respectively. They are **secondary sex organs** because they mature during puberty and are essential for sexual reproduction. Enlarged *breasts* in the female are considered a sexual attractant and are therefore a **secondary sex characteristic**; the same is true of *pubic hair*, in both sexes.

Objective E To identify the organs of the *male reproductive system.*

The **male sex organs** are formed prenatally under the influence of testosterone secreted by the gonads (testes). During puberty, the secondary sex organs mature and become functional. A listing of the male reproductive organs and their functions is presented in table 23.2.

Table 23.2 Organs of the Male Reproductive System

Organ(s)	*Description and location*	*Function*
Testes	Primary sex organs; posterior to the penis within the scrotum	Produce spermatozoa (gametes) and testosterone (male sex hormone)
Scrotum	Pouch of skin; posterior to the penis	Encloses and protects testes
Epididymides	Mass of tubules attached to the posterior surface of the testes	Site of sperm maturation; store spermatozoa
Ductus (vas) deferentia	Ducts extending from the epididymides to the ejaculatory ducts	Store spermatozoa; transport spermatozoa during ejaculation
Prostate	Walnut-sized gland at the base of the urinary bladder, surrounding the prostatic urethra	Secretes alkaline fluid that helps neutralize acidic environment of the vagina; enhances motility of spermatoza
Seminal vesicles	Club-shaped glands posterior to the prostate, attached to the ejaculatory ducts	Secrete alkaline fluid containing nutrients and prostaglandins
Bulbourethral glands	Pea-sized glands inferior to the prostate; empty into the membranous urethra	Secrete fluid that lubricates urethra and end of penis
Ejaculatory ducts	Short ducts between the ductus deferentia and the prostatic urethra	Receive spermatozoa and additives to produce seminal fluid
Penis	Pendant organ anterior to the scrotum and attached to the pubis	Convey urine and seminal fluid to outside of body; organ of coitus

23.8 Which male reproductive organs are singular and which are paired?

See fig. 23.2. The *scrotum*, *prostate*, *urethra*, and *penis* are unpaired male reproductive structures. The *testes* (sing., *testis*); *epididymides* (sing., *epididymis*); *ductus deferentia* (sing., *ductus deferens*); *seminal vesicles*, and *bulbourethral glands* are paired.

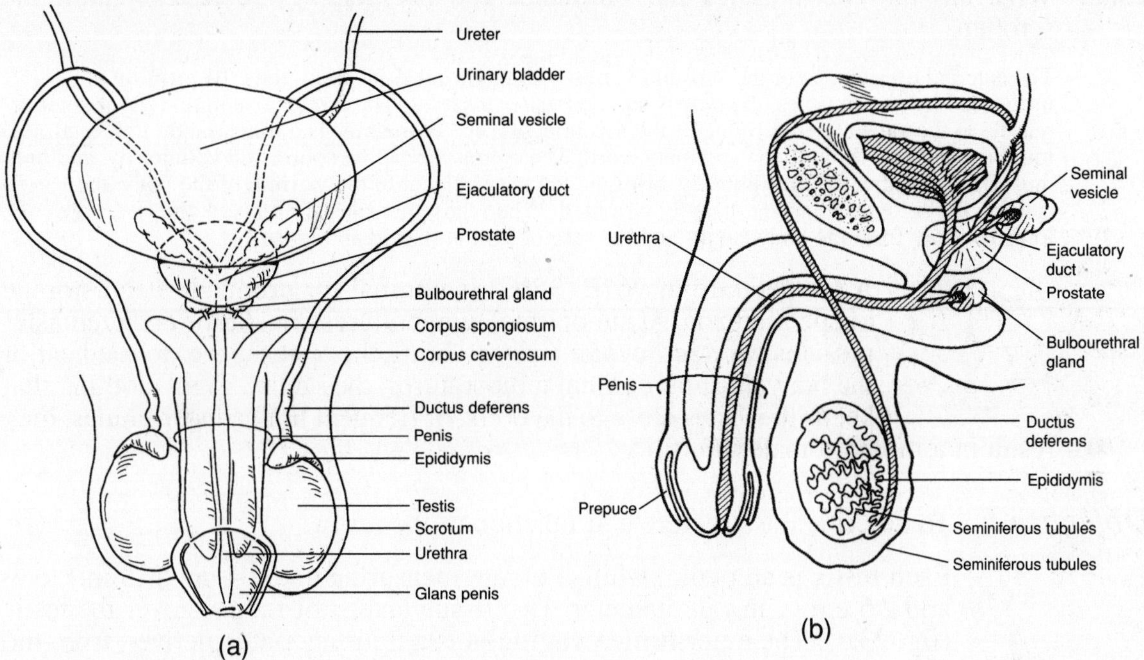

Figure 23.2 The male reproductive system. (*a*) An anterior view and (*b*) a sagittal view.

23.9 Which organs constitute the male external genitalia?

The *penis*, *scrotum*, *testes*, *epididymides*, and portions of the *ductus deferentia* constitute the external genitalia of the male. The ejaculatory ducts, seminal vesicles, prostate, and bulbourethral glands are located in the floor of the pelvic cavity. Each ductus deferens contained within a spermatic cord (see problem 23.11) enters the pelvic cavity through the inguinal canal.

23.10 Detail the interconnection of the accessory glands and the spermatic ducts that transport spermatozoa.

The **ejaculatory ducts** open into the prostatic portion of the **urethra**. An ejaculatory duct is formed by the union of the **ductus deferens** and the duct of the **seminal vesicle**. The **prostate**, which surrounds the junction of the ejaculatory duct and the urethra, drains directly into both of these ducts. The **bulbourethral glands** empty into the urethra at the base of the penis.

Because the *prostate* is subject to change that accompanies aging, a routine physical examination of the male includes rectal palpation of the prostate. A common change is *benign prostatic hyperplasia* or prostatic enlargement. In this condition, the urethra may be obstructed, causing difficult urination. A more serious condition is *prostatic carcinoma*, or cancer of the prostate. This condition is common in males over the age of 60 and is the second leading cause of death from cancer in males in the United States. Early detection and treatment is essential to achieve a successful cure.

23.11 What is the spermatic cord?

The **spermatic cord** is a structure that extends from a testis to the *inguinal ring* (see fig. 23.2). It is composed of the ductus deferens, spermatic vessels, nerves, cremaster muscle (see problem 23.12), and connective tissue.

23.12 What are the two muscles that influence the position of the testes within the scrotum?

The external appearance of the scrotum varies with environmental conditions. Its position is influenced by two muscles, the *dartos* and *cremaster muscles*. The **dartos** is a thin layer of smooth muscle in the subcutaneous tissue of the scrotum, and the **cremaster** is a thin strand of skeletal muscle extending through the spermatic cord. The cremaster can be contracted voluntarily, and both muscles are contracted involuntarily, bringing the testes closer to the warmth of the body and causing the scrotum to appear heavily wrinkled. When these muscles are relaxed, the testes are lowered away from the body cavity and the scrotal skin is flaccid and loose.

A temperature of 96°F (35°C) is optimal for production and storage of spermatozoa. At suboptimal temperatures, the dartos and cremaster muscles contract involuntarily to bring the testes closer to the heat of the body; at superoptimal temperatures, they relax. Tight clothing that keeps the testes close to the body, or frequent hot baths or saunas, may result in temporary male infertility.

Objective F To describe the structure and function of the *testes*.

Each **testis** is an ovoid, whitish organ, measuring about 4 cm (1.5 in.) long and 2.5 cm (1 in.) in diameter. Two tissue layers, or tunics, cover the testis (fig. 23.3). The outer **tunica vaginalis** is a thin sac that is derived from the peritoneum during the prenatal descent of the testis into the scrotum. The **tunica albuginea** is a tough fibrous membrane that directly encapsulates the testis. Inward extensions of the tunica albuginea partition the testis into 250 to 300 wedge-shaped **lobules**. A **scrotal septum** separates each testis into its own compartment. The testes produce spermatozoa and androgens. Androgens regulate spermatogenesis and the development and functioning of the secondary sex organs.

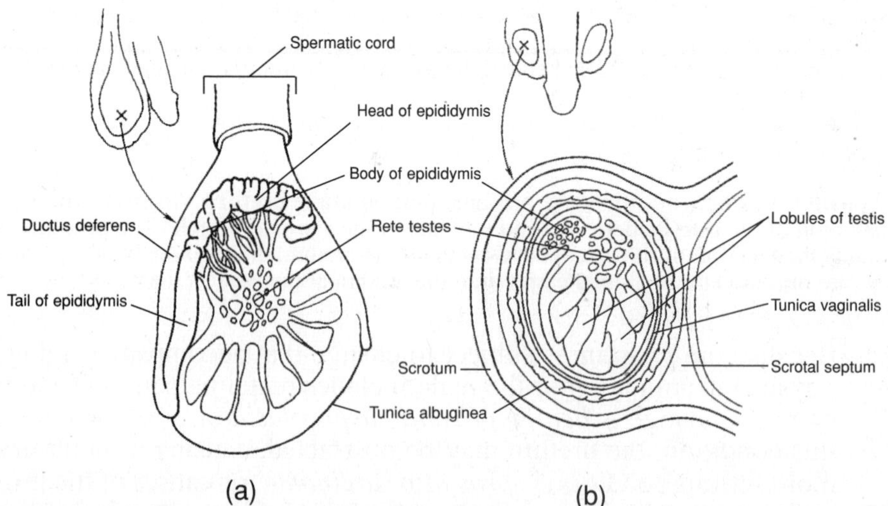

Figure 23.3 The gross structure of the testis, epididymis, and scrotum. (*a*) A sagittal view and (*b*) a coronal view.

The descent of the testes into the scrotum starts during the twenty-eighth week of prenatal development and usually is completed in the twenty-ninth week. If one or both testes are not in the scrotum at birth (a condition called *cryptorchidism*), it may be possible to induce descent by administering certain hormones. If this procedure does not work, surgery is necessary and is generally performed before age 5. Failure to correct the condition may result in sterility and/or a tumorous testis.

23.13 Where are the spermatozoa produced and stored within the testis.

Spermatogenesis occurs in the *seminiferous tubules* (fig. 23.4) Once the spermatozoa are produced, they move through the seminiferous tubules and enter the *rete testis* for further maturation. The spermatozoa are then transported out of the testis through a series of efferent ductules into the *epididymis* for the final stages of maturation. Mature spermatozoa are stored in the epididymis and also in the first portion of the ductus deferens. The entire process from production to maturation requires approximately 2 months.

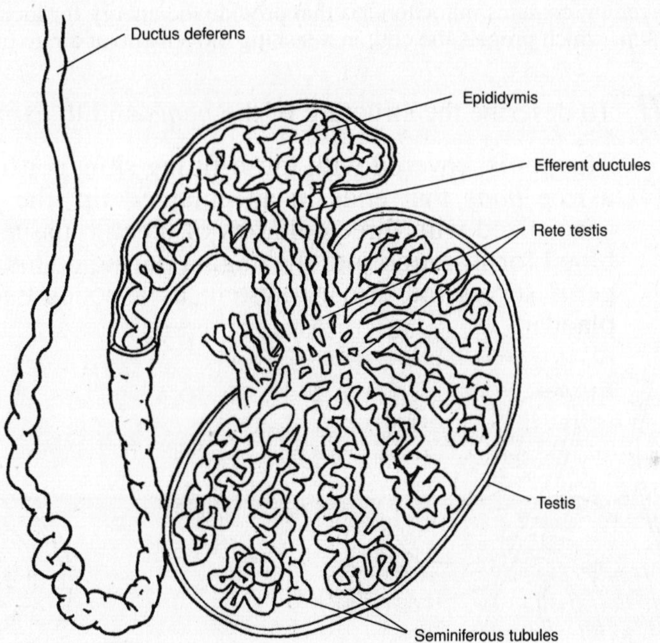

Figure 23.4 The histological structure of a testis.

23.14 Distinguish between sustentacular cells and interstitial cells.

Sustentacular cells (*Sertoli cells*) are nongerminal cells that provide many essential molecules to the developing germ cells. **Interstitial cells** (*Leydig cells*) are located between the seminiferous tubules where they produce and secrete *androgens*, the male sex hormones.

Objective G To describe the structure of a mature *spermatozoon.*

A mature **spermatozoon**, or sperm cell, is a microscopic tadpole-shaped cell about 60 μm long, consisting of an oval *head*, a cylindrical *body*, and an elongated *tail* (fig. 23.5).

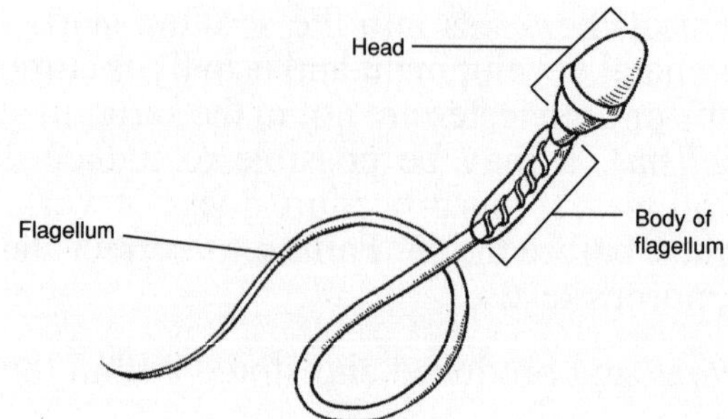

Figure 23.5 The structure of a spermatozoon.

23.15 What are the functions of the separate parts of a spermatozoon?

The *head* of a spermatozoon contains a nucleus with 23 chromosomes (haploid cell). The tip of the head (*acrosome*) contains an enzyme that helps the spermatozoon penetrate the ovum. The *body* of the flagellum contains mitochondria that provide the energy for locomotion via the *tail* of the flagellum, which propels the cell, in a lashing movement, at a rate of about 3 mm per hour.

Objective H To describe the structure of the *penis* and to explain its functions.

The **penis**, covered with loose-fitting skin, consists of an attached *root* and a free *body* that ends in an enlarged tip, the *glans penis*. The penis is specialized with three columns of erectile tissue to become engorged with blood for insertion into the vagina during coitus. The urethra of the flaccid penis serves the urinary system as a conduit for urine from the urinary bladder.

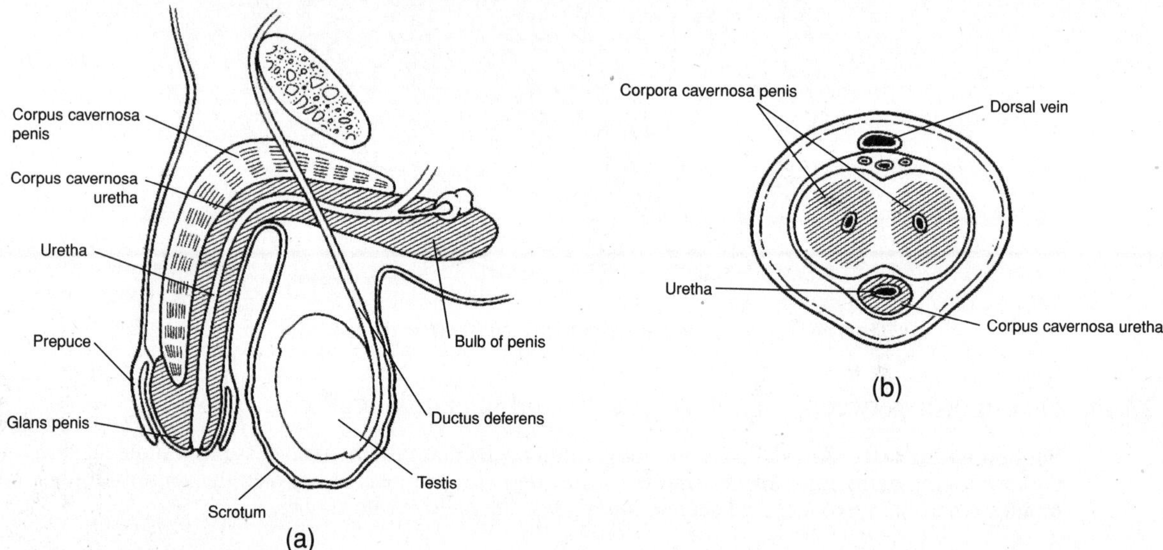

Figure 23.6 The penis. (*a*) A sagittal section and (*b*) a cross section.

23.16 How is the penis attached to the pelvic floor (perineum)?

The **root of the penis** is expanded posteriorly to form the **bulb** and the **crura** (sing., *crus*) **of the penis**. The bulb is located in the *urogenital triangle* of the perineum, where it is attached to the undersurface of the *perineal membrane* and enveloped by the *bulbospongiosus muscle*. Each crus, in turn, attaches the root of the penis to the pubic arch and to the perineal membrane, and each is enveloped by an *ischiocavernosus muscle*.

23.17 Which parts of the penis are erectile, and which are not?

Erectile parts of the penis. The **body of the penis** is composed of three cylindrical columns of erectile tissue that are bound together by fibrous tissue and covered with skin. The paired dorsally positioned masses are named the **corpora cavernosa penis.** The fibrous tissue between the two corpora forms a *median septum*. The **corpus spongiosum penis** is ventral to the other two and surrounds the urethra.

Nonerectile parts of the penis. The **glans penis** is the enlarged terminal part of the corpus spongiosum penis. In an uncircumcised male, it is covered by the retractable **prepuce**, or *foreskin*. The **corona glandis** is the prominent ridge of the glans penis. On the undersurface of the glans penis, a vertical fold of tissue called the **frenulum** attaches the prepuce to the glans penis. The prepuce protects the vascular glans penis.

Circumcision is the surgical removal of the prepuce. It is generally performed for hygienic purposes because the glans penis is easier to clean if exposed. A waxy substance, called *smegma*, is secreted along the inner surface of the prepuce. If smegma is allowed to accumulate, it becomes a nutrient source for bacteria, possibly resulting in mild inflammation and infection in this area. Cleansing the glans penis of an uncircumcised male requires retraction of the prepuce. *Phimosis* is a developmental problem in which the prepuce is so tight that it cannot be retracted. A circumcision is required to correct this condition.

23.18 What causes erection of the penis?

Erection of the penis depends on a surplus of blood entering the arteries of the penis as compared to the volume exiting through venous drainage. Normally, a constant sympathetic stimulation of the arterioles of the penis maintains a partial constriction of smooth muscles within the arteriolar walls, so that an even flow of blood is maintained throughout the penis. During sexual excitement, however, parasympathetic impulses cause marked vasodilation of the arterioles, and, with more blood entering than leaving, the penis grows turgid. A simultaneous slight vasoconstriction of the dorsal veins of the corpora cavernosa penis and the corpus spongiosum penis adds to this effect.

23.19 Define *emission* and *ejaculation*.

Continued sexual stimulation following erection of the penis causes emission. **Emission** is the movement of spermatozoa from the epididymides to the ejaculatory ducts and the secretions of the seminal vesicles and prostate in the formation of semen (see Objective I). **Ejaculation** immediately follows emission and is the expulsion of semen in a short series of spurts through the urethra of the penis. It is accompanied by orgasm, which is considered the climax of the sex act. In contrast to erection, emission and ejaculation involve sympathetic innervation of the male accessory sex organs.

Impotence is the inability of a sexually mature male to sustain an erection until ejaculation. The causes of impotency may be physical (e.g., a structural abnormality of the penis, vascular irregularities, neurological disorders, certain diseases, or conditions incidental to old age). Generally, however, the cause of impotence is psychological, and counseling by a sex therapist is usually the recommended treatment.

23.20 Are there erectile tissues in the female genitalia?

Yes. The erectile structures of the female genitalia are homologous (derived from the same embryonic tissues) to those of the male, and include the *clitoris* (see fig. 23.7) and *vestibular bulbs*. The clitoris is the homologue of the glans penis; the vestibular bulbs are homologous to the erectile tissues in the body of the penis. In addition, there is erectile tissue within the *areolae* of the breasts.

Objective I To describe the composition of *semen* and the factors involved in *male fertility*.

Semen (*seminal fluid*) is the mixture of fluids that is ejaculated from the erect penis. Generally, between 1.5 and 5.0 ml of semen are ejected during an ejaculation. Semen consists of mature sperm that were stored in the epididymides and ductus deferentia, and additives from the seminal vesicles and prostate. Over 99% of a typical discharge of semen (ejaculate) comes from the seminal vesicles and prostate. The bulk of the fluid (about 60%) is produced by the seminal vesicles, and the rest (about 40%) is contributed by the prostate. Spermatozoa constitute less than 1% of an ejaculate. Semen has a pH of about 6.5 and contains large quantities of prostaglandins (see problem 13.1).

23.21 Estimate the concentration of spermatozoa in semen (*sperm count*).

Normally, 200 to 500 million sperm are ejected during an ejaculation. This amounts to roughly 100 million per ml of ejaculate. If the sperm concentration is less than 10 million per ml, the male is likely to be infertile. A male who has had a *vasectomy* (removal of a portion of each ductus deferens) can still ejaculate semen, but it will not contain spermatozoa.

A *varicocele* exists when one or both of the testicular veins draining from the testes are swollen (varicosed), resulting in poor vascular circulation in the testes, and hence interference with spermatogenesis. Male infertility is primarily due to a varicocele. A varicocele can be surgically corrected.

Objective J To identify the organs of the *female reproductive system*.

The **primary female sex organs** are the ovaries. Female **secondary sex organs** develop prenatally as a result of the absence of testes and androgens. During puberty, the secondary sex organs mature and become functional under the influence of estrogens secreted by the ovaries. A listing of the female reproductive organs and their functions is presented in table 23.3.

23.22 Which female reproductive organs are singular and which are paired? What is the plural spelling of these terms?

See fig. 23.7. The *uterus*, *vagina*, and *clitoris* are singular female reproductive organs. The *ovaries*; *uterine tubes*; *labia majora* (sing., *labium major*); *labia minora* (sing., *labium minor*); *vestibular glands*; and *mammary glands* (or *mammae*) are paired.

23.23 Which organs constitute the female external genitalia?

The female external genitalia, collectively called the *vulva*, include the *mons pubis*, *labia majora*, *labia minora*, *clitoris*, and *vaginal orifice* (see table 23.3 and fig. 23.8).

Derived from similar embryonic tissues, many of the female reproductive organs are *homologous* to male reproductive organs. Obviously, the gonads (ovaries and testes) are homologous. Other homologous organs include the labia majora and the scrotum; the clitoris and the glans penis; the labia minora and the body of the penis; the vestibular glands and the bulbourethral glands.

Table 23.3 Organs of the Female Reproductive System

Organ(s)	***Description and location***	***Function***
Ovaries	Primary sex organ; upper pelvic cavity on both lateral sides of uterus	Produce of ova (gametes) and female sex hormones
Uterine tubes (fallopian tubes)	Open-ended tubes that extend from the ovaries to the uterus	Convey ova toward uterus; site of fertilization; convey developing blastocyst to uterus
Uterus	Hollow, musculomembranous organ shaped like an inverted pear; maintained in position within the pelvic cavity by muscles and ligaments	Site of implantation; sustains life of embryo and fetus during pregnancy; plays active role in parturition
Vagina	Hollow, musculomembranous organ positioned between the urinary bladder and urethra anteriorly and the rectum posteriorly	Conveys uterine secretion to outside of body; receives erect penis and semen during coitus; passageway for fetus during parturition
Labia majora	Two longitudinal folds of skin that extend from the mons pubis to the perineum; separated longitudinally by the pudendal cleft	Form margins of pudendal cleft; enclose and protect other external reproductive organs
Labia minora	Two longitudinal folds of skin medial to the labia majora; separated longitudinally by the vaginal vestibule	Form margins of vestibule; protect openings of vagina and urethra
Clitoris	Rounded projection at the upper part of the pudendal cleft, sheathed by a prepuce	Provides feeling of pleasure during sexual stimulation
Vestibular glands	Subcutaneous within the wall of the vaginal opening	Secrete lubricating fluid into the vestibule and vaginal opening during coitus
Mammary glands	Composed of lobes within the breasts	Produce and secrete milk for nourishment of an infant

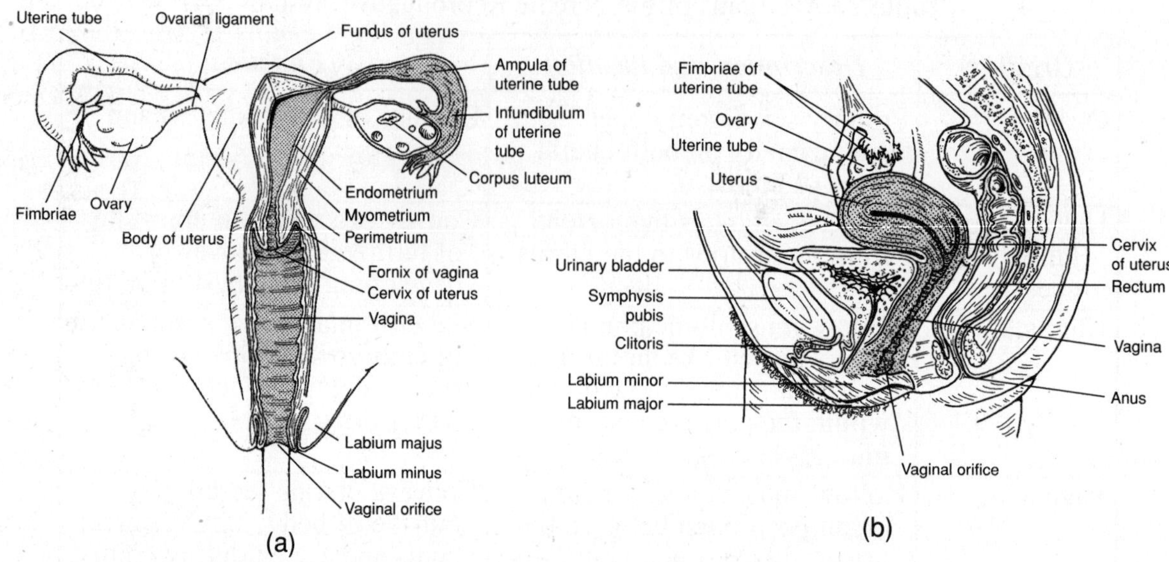

Figure 23.7 The female reproductive system. (*a*) An anterior view and (*b*) a sagittal view.

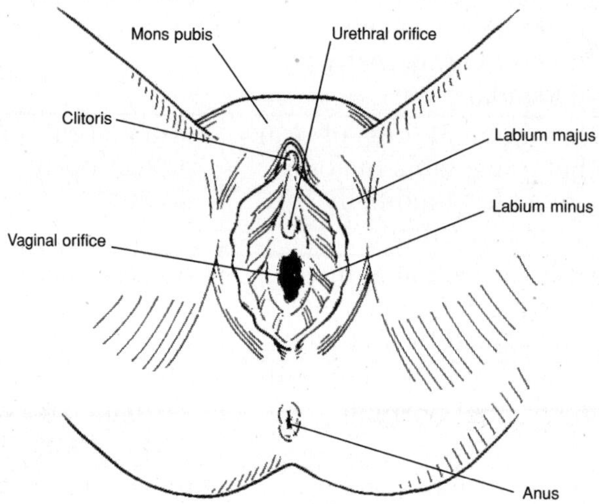

Figure 23.8 The external genitalia of the female reproductive system.

23.24 What type of epithelium lines the lumen of the vagina?

The vagina of an adult female is about 8 cm (4 in.) long but can distend to accommodate the size and shape of the erect penis during coitus. Considerable distension is permitted by the *stratified squamous epithelium*, which is folded into a series of **vaginal rugae**.

23.25 What is the pH of the vagina?

The mucosal layer of the vagina contains few glands; the acidic mucus that is present in the vagina comes primarily from glands within the uterus. The pH of the vagina is quite acidic (about 4.0). This acidity retards microbial growth but is also hostile to spermatozoa. The additives within semen, however, temporarily neutralize the acidity of the vagina to ensure the survival of the spermatozoa in the ejaculate.

The *hymen* is a thin fold of mucous membrane that may cover the vaginal orifice to some extent. It is a developmental remnant of tissue that has tremendous individual variation. The hymen may be absent in a baby girl, or it may partially, or even completely cover the vaginal orifice. If the hymen is present, it may be ruptured during childhood in the course of normal exercise, or during adolescence during the insertion of a tampon. On the other hand, a hymen may be so elastic that it persists even after coitus. Therefore, the presence of a hymen is not a reliable sign of virginity.

23.26 Describe the structure of the uterus.

The dome-shaped portion of the uterus above the entrance of the uterine tubes is the *fundus*; the tapered central region is the *body*; and the lower narrow portion that protrudes into the vagina is the *cervix* (see fig. 23.7). The interior of the body of the uterus is the *uterine cavity*, and the interior of the cervix is the *cervical canal*. The junction of the uterine cavity with the cervical canal is called the *isthmus of the uterus*, whereas the opening of the cervical canal into the cavity of the vagina is called the *uterine ostium*.

The wall of the uterus is composed of three layers: the *perimetrium*, *myometrium*, and *endometrium*. The **perimetrium** is the thin outer covering and a part of the peritoneum of the pelvic cavity. The thick **myometrium** is composed of smooth muscle layers that are thickest in the fundus and thinnest in the cervix. Premenstrual cramps are due to contractions of the myometrium; forceful contractions during labor aid parturition. Composed of two layers, the **endometrium** is the inner mucosal lining of the uterus. The superficial *stratum functionale* is composed of columnar epithelium and is the layer shed as *menses* during menstruation. The deeper *stratum basale* is highly vascular and regenerates the stratum functionale after each menstruation.

Endometriosis is a condition characterized by the presence of endometrial tissues at sites other than the lining of uterine cavity. Frequent sites of endometrial tissues are the ovaries, uterine tubes, and outer layer of the uterus. It is speculated that endometrial tissues become established at ectopic sites through a backflush of some menses during menstruation. Women with endometriosis bleed internally with each hormonally stimulated menstruation. Endometriosis can be medically treated. If untreated, it can cause infertility.

Objective K To describe the structure of an *ovary* and the cyclical development of the *ovarian follicle, ovum,* and *corpus luteum.*

survey

The **ovaries** are located in the upper pelvic cavity, one on each side of the uterus, and are held in position by several ligaments. In the outer region of each ovary are tiny masses of cells, called *primary follicles* (fig. 23.9). Each primary follicle contains an immature *egg*. As many as 20 follicles begin to develop at the beginning of a 28-day ovarian cycle; normally, however, only one follicle reaches full development and the others undergo degeneration. At about the middle of the cycle, the *mature ovarian* (*graafian*) *follicle* containing a nearly completely formed *ovum* (egg) bulges from the surface of the ovary and releases the ovum, in the process known as *ovulation*. After ovulation, the follicle cells undergo a structural change (*luteinization*) to form the *corpus luteum*.

23.27 We usually speak of the female *menstrual cycle* rather than the *ovarian cycle*. Are these distinct cycles?

No. They are dual manifestations of the basic *female hormonal cycle* (see Objective M). When "ovarian cycle" is used, the focus is on the changes in an ovary effected by the hormonal changes; when "menstrual cycle" is used, the focus is on the flow or nonflow of blood from the uterus through the vagina, resulting from the very same hormonal changes.

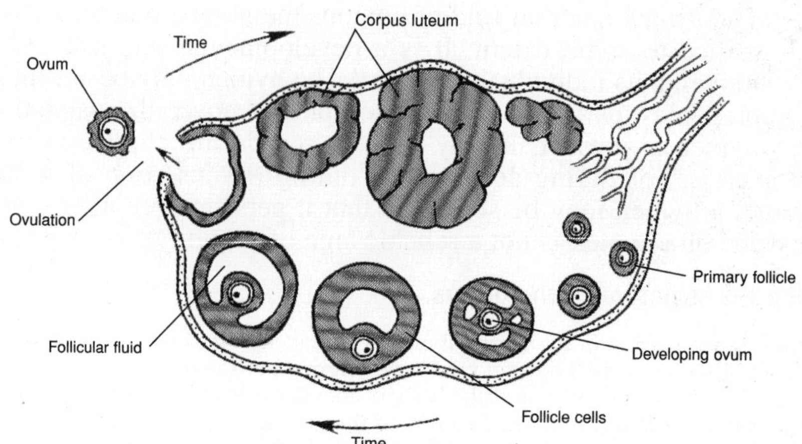

Figure 23.9 An ovary showing follicular development and ovulation.

23.28 What causes a primary follicle to mature?

Follicle-stimulating hormone (*FSH*) (see tables 13.1 and 13.3), along with subsequent small quantities of *luteinizing hormone* (*LH*). After about the fifth day of the menstrual cycle, the two ovaries, between them, contain just one maturing ovarian follicle.

23.29 What is responsible for ovulation and luteinization?

A sharp rise in LH secretion causes

1. The mature follicle to rupture, releasing its egg
2. The ruptured follicle to fill with blood
3. The clotted blood to be replaced with lipid-rich luteal cells, forming the corpus luteum

23.30 What are some of the indicators of ovulation?

Body temperature. The basal body temperature usually rises upon ovulation and remains elevated during the rest of the menstrual cycle. To detect the occurrence of ovulation, the body temperature is taken with a basal thermometer each morning and recorded.

Cervical mucus. Cervical mucus, when smeared on a glass slide, dried, and examined under a microscope, shows distinct patterns at different times during the ovulation cycle. In the middle of the cycle (when ovulation occurs), the mucus exhibits a fernlike pattern.

Abdominal pain. The rupture of the follicle at the time of ovulation may cause some hemorrhage and local inflammation, which in turn can induce low abdominal pain.

23.31 What is the length of the fertile period in the monthly ovarian cycle?

Ejaculated spermatozoa can survive up to 5 days in the female reproductive tract, whereas the egg survives only about 24 hours after ovulation. Therefore, coitus must occur no earlier than 5 days before ovulation and no later than 24 hours after ovulation if pregnancy is to result. In other words, the fertile period is approximately 6 days long.

Objective L To detail the events of the *menstrual cycle*.

Menstruation begins at puberty and menstrual cycles continue until menopause, approximately 36 to 40 years later. The day on which discharge of blood from the vagina begins is taken as the first day of the cycle; the cycle ends on the last day prior to the next menstrual flow. The cycle normally runs for about 1 lunar month, or 28 days, but it can vary from 22 to 35 days.

23.32 Describe how the 28-day menstrual cycle is divided into four phases.

See fig. 23.10.

Menstrual phase or **menses.** Days 1 to 5 (± 2 days) before the onset of the next menstruation

Proliferative or **preovulatory (follicular) phase.** Day 5 to the day of ovulation (14 ± 2 days before the onset of the next menstruation)

Secretory or **progesterone (luteal) phase.** Day of ovulation to day 28

Ischemic phase. Days 27 and 28

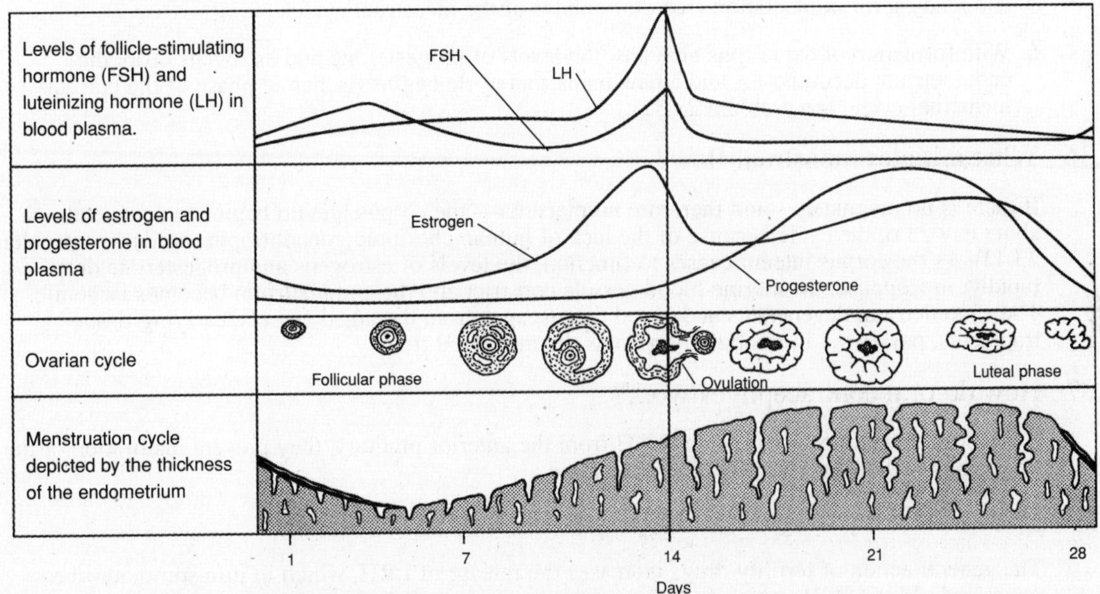

Figure 23.10 The menstrual and ovarian cycle.

23.33 Define *menarche.*

The epoch of the first menstrual flow, or *period*, is called *menarche*; it marks the beginning of a female's reproductive life and is one of the most obvious signals of the onset of puberty. Menarche usually occurs between the ages of 9 and 17 (the average age is 12.5).

23.34 Which is of more constant duration, the preovulatory phase or the secretory phase?

The preovulatory phase is quite variable, lasting longer in long (29- to 35-day) cycles and ending sooner in short (22- to 27-day) cycles. The length of the secretory phase is usually constant at about 14 days.

Objective M To describe the *female hormonal cycle.*

Under the control of the hypothalamus, the anterior pituitary and the ovaries secrete steroid hormones. Cyclic changes in these secretions regulate all female reproductive activities.

23.35 Analyze the female hormonal cycle as a progression of events. Start with the CNS and ignore feedback mechanisms.

1. The hypothalamus begins to release luteinizing-releasing hormone (LRH); the target organ is the anterior pituitary.
2. LRH stimulates the secretion of FSH and LH.
3. FSH and LH affect the ovaries as detailed in table 13.3. (This constitutes the *ovarian cycle*, which terminates in the reabsorption of the corpus luteum, leaving a white body, or *corpus albicans*).
4. The mature ovarian follicle secretes estrogens, which provoke a thickening of the endometrium (proliferative phase of the menstrual cycle).
5. The corpus luteum secretes progesterone and estrogens, which act to level off the endometrium and ready it for implantation (secretory phase of the menstrual cycle).
6. With formation of the corpus albicans, the levels of progesterone and estrogens drop, the endometrium decomposes, and a new menstrual cycle begins (ischemic phase of the current menstrual cycle; see problem 23.32).

23.36 What initiates menstrual flow?

If there is no pregnancy—and therefore no placenta— the corpus luteum begins to degenerate on about day 23 of the cycle because of the lack of human chorionic gonadotropin (hCG) (see problem 23.42). As the corpus luteum ceases to function, the levels of estrogens and progesterone drop rapidly; in response, the uterine blood vessels constrict and the endometrium becomes ischemic. The endometrium degenerates and blood that escapes from damaged vessels, carrying tissue fragments, passes out through the vagina as the menstrual flow.

23.37 How do oral contraceptives work?

By inhibiting the release of LH and FSH from the anterior pituitary, they prevent maturation of the developing follicle and thereby block ovulation.

23.38 How do fertility drugs work?

The general action of fertility drugs increases the release of LRH, which in turn stimulates the release of LH and FSH (which further ovulation).

Objective N To trace the paths of the ovum and spermatozoa through the female reproductive tract and to describe the early embryonic development of the fertilized egg.

After being deposited in the vagina, spermatozoa must move up through the cervical canal and uterine cavity into a uterine tube (the correct one), in order to encounter the ovum on its way down from the ovary. **Fertilization** (conception) normally occurs in the distal one-third of the uterine tube. Following fertilization, the **zygote** (fertilized egg) undergoes mitosis during its approximate 3-day journey down the uterine tube to the uterine cavity (see fig. 23.11). There the developing **blastocyst** remains free for about another 3 days before it begins to implant into the endometrium.

23.39 Define the terms *cleavage, morula, blastocyst,* and *implantation.*

Cleavage. Early successive divisions of the zygote.

Morula. A solid globular mass of cells that results from 16 to 72 cell divisions.

Blastocyst. A hollow ball of cells that enters into the uterine cavity between days 3 and 5 after conception.

Implantation. The embedding of the blastocyst into the endometrium.

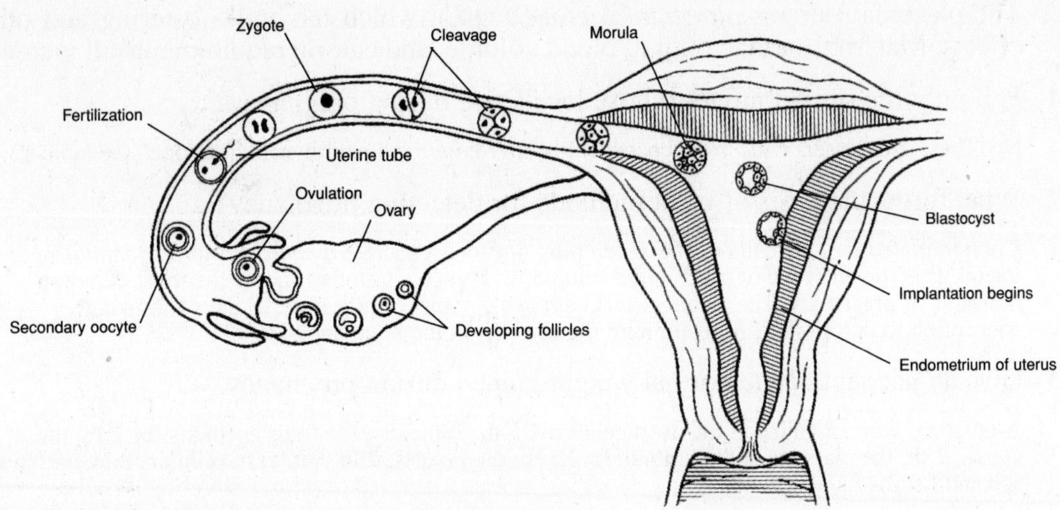

Figure 23.11 A diagrammatic representation of ovulation, fertilization, and the developmental events during the first week.

23.40 Approximately how many spermatozoa reach the uterus? The uterine tube? The ovum?

Of the 200 to 500 million spermatozoa contained within seminal fluid ejaculated into the vagina, only about 1 million reach the uterus, and just a few thousand reach the mouth of the uterine tube containing the ovulated egg. Only about 100 to 200 spermatozoa reach the upper part of the uterine tube, where fertilization takes place.

An *ectopic pregnancy* occurs when the blastocyst implants at a site other than the uterine cavity. The most frequent ectopic site is a uterine tube; one then speaks of a *tubular pregnancy*. Other ectopic sites are the cervix (*cervical pregnancy*) and the linings of abdominal viscera (*abdominal pregnancies*). An ectopic pregnancy presents a serious health risk to a woman. If it is not self-aborted (*spontaneous abortion*), a *therapeutic abortion* is usually performed.

Objective O To summarize the hormonal and other changes that occur during *pregnancy*.

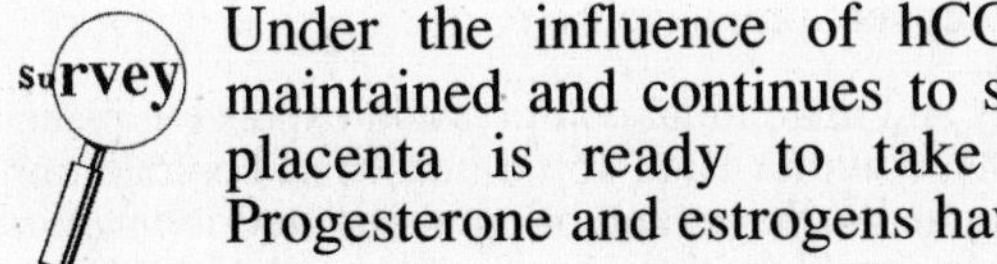

Under the influence of hCG (see table 13.4), the corpus luteum is maintained and continues to secrete progesterone and estrogens until the placenta is ready to take over the hormone-producing function. Progesterone and estrogens have the following effects during pregnancy:

1. They sustain the endometrium.
2. They inhibit the release of FSH and LH (thereby halting the menstrual cycle).
3. They stimulate the development of the mammary glands.
4. They inhibit (with progesterone) or stimulate (with estrogens) uterine contractions.
5. They increase (with estrogens) uteroplacental blood flow and also cause enlargement of the uterus, breasts, vagina, and vaginal orifice.

The placenta secretes *placental lactogen* (*PL*), which has milk-fostering and other effects. Maternal cardiac output, blood volume, and caloric requirements all increase.

23.41 Is it only the sex hormones whose levels rise during pregnancy?

No. There are (lesser) rises in glucocorticoids, thyroxine, and parathyroid hormone (see table 13.1).

23.42 What forms the basis of most methods for detecting pregnancy?

Upon implantation, the cells of the developing blastocyst secrete *human chorionic gonadotropin* (*hCG*). It is the presence of hCG in the urine of the pregnant mother that is the most common indicator of pregnancy. The levels of hCG are usually sufficiently high 10 days following conception to permit detections through the use of a home pregnancy kit.

23.43 Give an account of the normal weight gained during pregnancy.

A pregnant woman gains on the average 20 to 25 lb. Typically, the fetus accounts for 7 lb; the uterus, 2 lb; the placenta and membranes, 2.5 lb; the breasts, 2 lb. Fat, extracellular fluid, and blood account for the balance.

Objective P To describe the mechanisms of *labor* and *parturition* (childbirth).

Labor and **parturition** are the culmination of gestation. Labor consists of a sequence of physiological and physical events. The onset of labor is denoted by rhythmic and forceful contractions of the myometrium of the uterus. In *true labor* (as opposed to false labor), the pains from uterine contractions occur at regular intervals and intensify as the interval between contractions shortens. This is accompanied by cervical dilation and a discharge of blood-containing mucus from the cervical canal and out the vagina.

The uterine contractions of labor are stimulated by *oxytocin* produced in the hypothalamus and released from the posterior pituitary, and *prostaglandins* produced within the uterus itself. Induced labor can be accomplished by injections of oxytocin or by the insertion of prostaglandins into the vagina as a suppository.

23.44 What are the stages of labor?

Dilation stage. In this stage, the cervix dilates to a diameter of approximately 10 cm. The amniotic sac (bag of water) containing amniotic fluid generally ruptures. (If it does not rupture spontaneously, it is done surgically.) The dilation stage generally lasts from 8 to 24 hours.

Expulsion stage. Forceful uterine contractions and abdominal compression expels the fetus. This stage is the actual period of parturition. The expulsion stage may require 30 minutes or more in a first pregnancy but only a few minutes in subsequent pregnancies.

Placental stage. Forceful uterine contractions and abdominal compression expels the placenta from the uterus. This stage follows the expulsion stage within 10 to 15 minutes. The contractions constrict uterine vessels, thus greatly reducing hemorrhage, and bring about uterine *involution* (shrinkage to the former size). In a normal delivery, blood loss does not exceed 350 ml.

In about 5% of vaginal births, the buttocks are expelled first, rather than the head, in what is called a *breech delivery*. The principal concern of a breech birth is the increased time and difficulty of the expulsion stage of labor. If an infant cannot be rotated or delivered breech, a *cesarean section* is performed. In a C-section, the baby is delivered through an incision through the abdomen and uterus.

Objective Q To describe the hormonal control of *mammary gland development* and *lactation*.

The **mammary glands** within the *breasts* are accessory reproductive organs that are specialized to produce milk after pregnancy. Mammary glands are specialized sudoriferous glands (see problem 5.27). At the onset of puberty, the ovarian hormones stimulate the mammary glands and lactiferous ducts (fig. 23.12) to develop. During pregnancy, further glandular and ductile development takes place under the influence of progesterone and estrogens, respectively. Several other hormones are necessary to prepare the mammary glands for milk production.

Prolactin is inhibited during pregnancy by the high levels of progesterone and estrogens. After parturition, however, estrogens and progesterone dwindle, and secretion of prolactin is no longer inhibited. Prolactin, as it name implies, stimulates milk production. Nursing stimulates the *nipple* and *areola*, sending sensory input via the spinal cord to the hypothalamus, which releases oxytocin. Oxytocin stimulates contraction of the myoepithelial cells, which causes the ejection, or letdown, of milk.

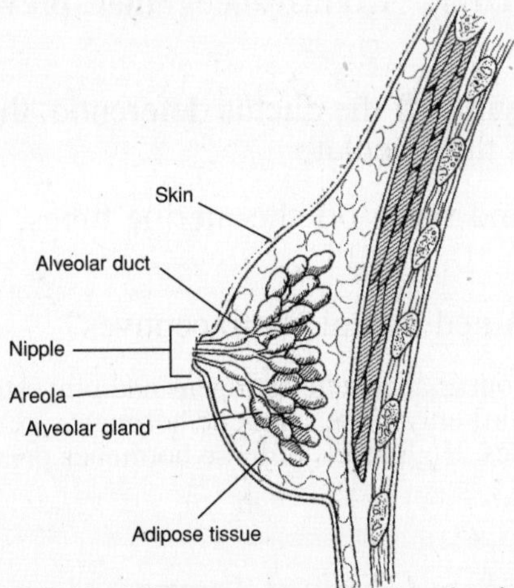

Figure 23.12 The mammary gland within the breast.

23.45 What is colostrum?

Milk is not produced during the first day or two after parturition; however, a few milliliters of a clear, fat-free, yellowish-white fluid, known as **colostrum**, is secreted at this time. It serves as the baby's first food and provides important maternal immunity to the infant.

23.46 How much milk is produced during lactation?

Up to 1.5 L of milk is secreted daily at the peak of nursing activity. This volume requires the daily metabolism of about 100 g of lactose, 50 g of fat, and 2 to 3 g of calcium phosphate.

23.47 Can a nursing mother become pregnant?

The release of prolactin during nursing usually inhibits the menstrual cycle and therefore ovulation; however, a nursing mother can become pregnant. Nursing does not constitute a reliable contraceptive method.

Objective R To list the common *methods of contraception.*

Anything that interferes with fertilization of the egg or implantation of the blastocyst can be considered a means of contraception. The following is a partial listing of some of the more common birth-control methods.

Rhythm method. Abstinence from coitus for a few days before and a few days after ovulation

Coitus interruptus. Withdrawal of the penis prior to ejaculation

Condom. A rubber or latex sheath that covers the penis during coitus, preventing semen from being deposited in the vagina

Diaphragm and **spermicidal foams**, **gels**, and **sponges.** Barriers that prevent spermatozoa from entering the cervix

Oral contraceptives ("*the pill*"). Drugs that inhibit the release of gonadotropins (LH and FSH) and, therefore, prevent ovulation

Subdermal implants. Small (2-in.) rods filled with a hormonal contraceptive drug and implanted under the skin, usually on the upper arm

Intrauterine devices (*IUDs*). Contrivances that prevent implantation of the fertilized egg

Vasectomy. Cutting and tying off the ductus deferentia, thus preventing sperm from becoming a component of the ejaculate

Tubal ligation. Cutting and tying off the uterine tubes, thus preventing ova from contacting spermatozoa

23.48 What substances are contained in oral contraceptives?

Oral contraceptives generally contain synthetic estrogens and progesterone (the Minipill contains synthetic progesterone [progestin] only). The synthetic hormones are not degraded in the GI tract and act like the natural hormones. High levels of these hormones preserve the endometrium, so that ovulation does not occur.

Key Clinical Terms

Amenorrhea A suppression or absence of menstruation. Irregular or interrupted menstruation may be a natural response to stress of various kinds and not necessarily related to pregnancy.

Cryptorchidism Arrested descent of a testis.

Dysmenorrhea Difficult or painful menstruation, generally due to obstruction, inflammation, or some disease.

Episiotomy Incision of the perineum to facilitate parturition and prevent excessive laceration.

Hysterectomy Removal of the uterus.

Laparoscopy Visual examination of the abdominal organs, especially the uterus, uterine tubes, and ovaries.

Oophorectomy Removal of an ovary.

Orchitis Inflammation of the testis; may be the result of infectious diseases such as mumps.

Papanicolaou smear A *Pap smear* is a diagnostic test for cervical or endometrial cancer. Shed epithelial cells of vaginal and uterine regions are screened for malignancy.

Prolapsed uterus Inferior displacement of the uterus, which may distend from the vagina.

Prostatitis Inflammation of the prostate.

Pruritus vulvae Severe itching of the vulva; frequently accompanies vaginitis.

Sexually transmitted diseases (STDs) Contagious diseases that affect the reproductive systems of both males and females; frequently called *venereal diseases* (*VDs*). Some of the more common STDs are as follows:

Gonorrhea. Commonly called clap, it is caused by the gonococcus bacterium, or *Neisseria gonorrhoeae*. Advanced stages involve the reproductive and urinary systems, the conjunctiva of the eyes, and the joints.

Syphilis. Less common than gonorrhea, but the more serious of the two STDs. It is caused by a spirochete, *Treponema pallidum*, and may be chronically degenerative if untreated.

Acquired immune deficiency syndrome (AIDS). An incurable STD with a high incidence among homosexual males and drug users. It is caused by the human immunodeficiency virus (HIV) and is rapidly on the increase. Blocking the spread of AIDS currently depends on education and safer sexual practices.

Chancroid. Infectious genital ulcers caused by a bacillus, *Haemophilus ducreyi.*

Genital herpes. An incurable STD caused by type II herpes simplex virus. During the recurring infectious stage, the afflicted person will have numerous clusters of painful genital blisters.

Vaginitis Inflammation of the vagina; may be due to a sexually transmitted disease or to certain bacterial or fungus infections.

Review Exercises

Multiple Choice

1. Which of the following sex chromosome combinations causes a male fetus to develop? (*a*) XX, (*b*) XY, (*c*) YY, (*d*) XO, (*e*) YO
2. Which of the following is *not* an accessory male reproductive gland? (*a*) prostate, (*b*) bulbourethral gland, (*c*) glans penis, (*d*) seminal vesicle
3. The penis is (*a*) the male primary sex organ, (*b*) composed of four longitudinal columns of erectile tissue, (*c*) homologous to the female labia majora, (*d*) a copulatory organ.
4. What is the percentage of spermatozoa in a typical ejaculate of semen? (*a*) less than 1%, (*b*) about 10%, (*c*) 50%, (*d*) over 90%
5. Secretions from the seminal vesicles account for what percentage of the additives to semen? (*a*) 10%, (*b*) 25%, (*c*) 60%, (*d*) 90%
6. The portion of the female reproductive system that is homologous to the glans penis of the male is (*a*) the labia majora, (*b*) the clitoris, (*c*) the ovary, (*e*) the vagina.
7. The vascular mucosal lining of the uterus is (*a*) the peritoneum, (*b*) the endometrium, (*c*) the mediastinum, (*d*) the synovial membrane, (*e*) the mesentery.

8. Fertilization normally occurs in (*a*) the uterine tube, (*b*) the vagina, (*c*) the uterus, (*d*) the ovary , (*e*) the peritoneal cavity.

9. The duct that transports spermatozoa during emission up over the pubic arch and to the side of the urinary bladder is (*a*) the epididymis, (*b*) the urethra, (*c*) the convoluted tubule, (*d*) the ductus deferens, (*e*) the spermatic cord.

10. Which of the following is *not* part of the female external genitalia? (*a*) clitoris, (*b*) vagina, (*c*) labia majora, (*d*) labia minora

11. Testosterone is responsible for maintenance of (*a*) a functional male reproductive system, (*b*) regular ovulation, (*c*) a mature endometrium, (*d*) cells of the interstitial spaces, (*e*) all of the preceding.

12. The function of the male secondary sex organs is to (*a*) transfer spermatozoa to the female, (*b*) regulate sperm production, (*c*) produce sperm, (*d*) produce male sex hormones, (*e*) accomplish all of the preceding.

13. The filling of venous sinuses under the influence of sexual stimulation is most closely associated with (*a*) spermatogenesis, (*b*) menstruation, (*c*) ovulation, (*d*) erection.

14. The number of gametes resulting from a single sequence of spermatogenesis is (*a*) one, (*b*) two, (*c*) four, (*d*) eight, (*e*) over one hundred.

15. The tightly convoluted tubule that lies along the posterior surface of the testis is (*a*) the seminiferous tubule, (*b*) the rete testis, (*c*) the epididymis, (*d*) the ductus deferens, (*e*) the seminal vesicle.

16. Spermatozoa are discharged through the male genital ducts in the order (*a*) epididymis, ductus deferens, ejaculatory duct, urethra; (*b*) ejaculatory duct, epididymis, ductus deferens, urethra; (*c*) epididymis, ductus deferens, urethra, ejaculatory duct; (*d*) ductus deferens, epididymis, ejaculatory duct, urethra.

17. The development of the ovarian follicle is influenced by (*a*) prolactin, (*b*) FSH, (*c*) testosterone, (*d*) insulin, (*e*) none of the preceding.

18. Which of the following secretes progesterone? (*a*) anterior pituitary, (*b*) corpus luteum, (*c*) corpus luteum and ovarian follicle, (*d*) hypothalamus, (*e*) posterior pituitary

19. Which of the following statements regarding postmenopausal blood hormone concentrations is *true*? (*a*) Estrogens remain low. (*b*) FSH remains high. (*c*) Progesterone remains low. (*d*) a, b, and c are true. (*e*) Only a and c are true.

20. Formation and maintenance of the corpus luteum is mainly effected by (*a*) FSH, (*b*) LH, (*c*) progesterone, (*d*) estrogens.

21. A woman's body temperature (*a*) rises at the onset of menstruation, (*b*) drops 2 days before ovulation, (*c*) remains higher after ovulation than before ovulation, (*d*) drops abruptly 2 days after ovulation.

22. Which of the following secretes estrogens? (*a*) anterior pituitary, (*b*) corpus luteum and ovarian follicle, (*c*) ovarian follicle only, (*d*) hypothalamus, (*e*) posterior pituitary

23. Menstruation is initiated by (*a*) a sudden release of FSH from the anterior pituitary, (*b*) a lack of estrogens and progesterone due to degeneration of the corpus luteum, (*c*) an increased release of estrogens and progesterone from the corpus luteum, (*d*) a sudden drop in LH.

24. Which hormone stimulates testosterone secretion? (*a*) LH, (*b*) progesterone, (*c*) FSH, (*d*) ACTH

25. In a normal, healthy 25-year-old woman with a menstrual cycle of 28 days, (*a*) the proliferative phase of the uterus is caused by estrogens produced by ovarian follicles, (*b*) menstruation is caused by progesterone from the corpus luteum, (*c*) injections of estrogens and/or progesterone will cause an enlargement of the ovaries and an increase in the production of mature ovarian follicles, (*d*) the concentration of estradiol in the blood plasma begins to fall prior to ovulation and continues to decrease until menstruation, (*e*) all of the preceding are true.

26. Spermatozoa are stored prior to ejaculation in (*a*) the prostatic urethra, (*b*) the prostate, (*c*) the epididymides, (*d*) the seminal vesicles, (*e*) the ejaculatory ducts.

27. Oral contraceptive treatment with mixtures of estrogens and progesterone (*a*) if given daily throughout the year would tend to prevent menstruation from occurring, (*b*) is thought to act mainly by preventing implantation of the blastocyst, (*c*) is thought to depress anterior pituitary secretion of gonadotropic hormones, (*d*) may cause a decrease in body weight.

28. An embryo with XX chromosomes develops female secondary sex organs because of (*a*) estrogens, (*b*) androgens, (*c*) absence of androgens, (*d*) absence of estrogens.

29. The milk-ejection reflex is stimulated by (*a*) oxytocin, (*b*) estrogen, (*c*) prolactin, (*d*) progesterone.

30. Once ejaculated into the vagina, spermatozoa have a life expectancy of (*a*) 10 to 12 hours, (*b*) 1 day, (*c*) 2 to 3 days, (*d*) up to 5 days.

True or False

____ **1.** Except in circumcised individuals, both the glans penis and the clitoris are covered by a prepuce.

____ **2.** The ovaries and uterus are the primary sex organs of the female.

____ **3.** Seminal vesicles, bulbourethral glands, and the prostate are all accessory glands of the male reproductive system.

____ **4.** The labia majora of the female genitalia are homologous to the scrotum in the male.

____ **5.** Meiosis is peculiar to the gonads.

____ **6.** An ectopic pregnancy results from an implantation site other than the endometrium of the uterus.

____ **7.** As a retractable sheath of skin, the prepuce of the penis has no known function.

____ **8.** Sympathetic stimulation of the arteries within the penis causes engorgement of the erectile tissue as arterial flow increases and venous drainage decreases.

____ **9.** The ejaculatory ducts store spermatozoa and additives to produce semen prior to ejaculation.

____ **10.** Mammary glands are modified sebaceous glands.

____ **11.** Prolactin causes the breasts to enlarge and the mammary glands to mature during puberty.

____ **12.** The first menstrual discharge is referred to as menarche.

____ **13.** Located at the vaginal orifice, the vestibular glands maintain the acidic pH of the vagina.

____ **14.** Interstitial cells produce spermatozoa and secrete nutrients to developing spermatozoa within the testes.

____ **15.** The secretory phase of menstruation is characterized by discharge of the menses.

Completion

1. ____________________ is a condition in which one or both testes fail to descend into the scrotum.

2. ____________________ is the discharge of semen from the erect penis.

3. The ____________________ tubes of female and the ____________________ deferens of the male transport gametes.

4. ____________________ cells are located between the seminiferous tubules of a testis where they produce and secrete androgens.

5. A ____________________ exists when one or both of the testicular veins draining blood from the testes are swollen, resulting in poor testicular circulation.

6. The ____________________ is a thin remnant of mucous membrane that may partially cover the vaginal orifice.

7. Ejaculated spermatozoa can live up to ______ days, whereas an ovulated egg can survive only about ______ hours.

8. The first menstrual period is called ____________________.

9. Mitotic divisions of a zygote are referred to as ____________________.

10. An ____________________ ____________________ occurs when a blastocyst implants at a site other than the uterine cavity.

Labeling

Label the structures indicated on the figure to the right.

1. ______________________________
2. ______________________________
3. ______________________________
4. ______________________________
5. ______________________________
6. ______________________________
7. ______________________________
8. ______________________________
9. ______________________________
10. ______________________________

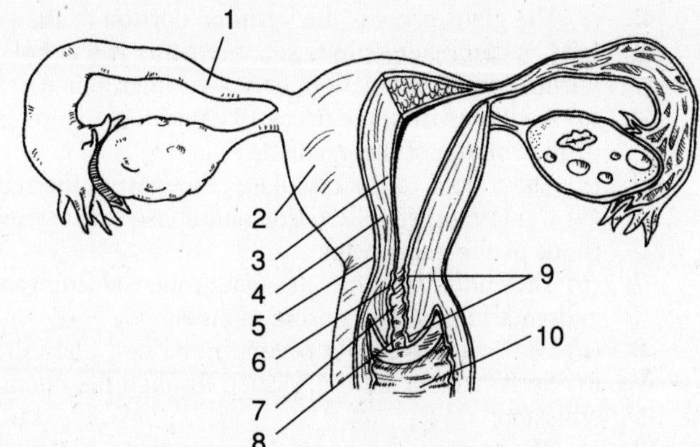

Matching

Match the structure with its function.

____	**1.** Vestibular gland	(*a*)	protective sheath
____	**2.** Prepuce	(*b*)	produces testosterone
____	**3.** Ovarian follicle	(*c*)	stores spermatozoa
____	**4.** Scrotum	(*d*)	nourishes spermatozoa
____	**5.** Sustentacular cell	(*e*)	secretes estrogens
____	**6.** Uterine tube	(*f*)	secretes a lubricant
____	**7.** Clitoris	(*g*)	encloses testes
____	**8.** Urethra	(*h*)	transports ova
____	**9.** Interstitial cell	(*i*)	contains erectile tissue
____	**10.** Epididymis	(*j*)	transports semen

Answers and Explanations for Review Exercises

Multiple Choice

1. (*b*) The genetic combination of XY = male; XX = female.
2. (*c*) The glans penis is the terminal portion of the penis; it is not a glandular structure.
3. (*d*) Containing erectile tissue, the penis is a copulatory organ that transfers spermatozoa to the vagina during coitus (copulation, or sexual intercourse).
4. (*a*) Additives from the accessory reproductive organs (prostate and seminal vesicles) account for 99% of the volume of the ejaculate.
5. (*c*) About 60% of the ejaculate comes from the seminal vesicles and 40% from the prostate.
6. (*b*) Containing erectile tissue and dense sensory innervation, the clitoris is derived from the same tissue as the glans penis.
7. (*b*) The endometrium is the inner mucosal lining of the uterus. The stratum functionale portion of the endometrium is shed during menses.
8. (*a*) Fertilization usually occurs in the first one-third of a uterine tube.
9. (*d*) Spermatozoa are transported through the ductus deferentia to the ejaculatory ducts during emission.
10. (*b*) The vagina is an internal reproductive organ and is not considered part of the vulva.
11. (*a*) Testosterone and its derivatives are responsible for initiating growth and maintaining the male reproductive system.
12. (*a*) The male secondary sex organs nurture spermatozoa and transport semen to the female during coitus.
13. (*d*) Erection of male and female reproductive organs occurs as blood fills the venous sinuses of erectile tissues.
14. (*c*) In spermatogenesis, four equally mature gametes are formed, but in oogenesis only one mature ovum is formed.
15. (*c*) The epididymis is comma-shaped mass of threadlike tubules attached to the posterior border of a testis.
16. (*a*) Emission is movement of spermatozoa through the epididymides and ductus deferentia. Ejaculation is movement of semen through the ejaculatory ducts and the urethra.
17. (*b*) Follicle-stimulating hormone (FSH), along with small quantities of luteinizing hormone (LH), stimulate maturation of a primary follicle.
18. (*b*) Progesterone is secreted by the corpus luteum of the ovary. During pregnancy, it is secreted by the placenta.
19. (*d*) Following menopause, FSH and LH continue to be released from the anterior pituitary. The ovaries no longer respond, however; thus no follicles develop and little or no estrogens or progesterone are produced.
20. (*b*) Luteinizing hormone (LH) is involved in the formation and maintenance of the corpus luteum.
21. (*c*) The basal body temperature rises at the time of ovulation and remains elevated during the last half of the ovarian cycle.
22. (*b*) Progesterone is secreted by the corpus luteum of the ovary. During pregnancy, it is secreted by the placenta.
23. (*b*) A decrease in estrogen and progesterone as the corpus luteum degenerates (in the absence of pregnancy) initiates menstruation.
24. (*a*) LH stimulates the interstitial (Leydig) cells of the testes to secrete testosterone.
25. (*a*) The proliferative phase is from about day 5 of the ovarian cycle until the time of ovulation. Growth of the endometrium during this phase is influenced by estrogen.
26. (*c*) Spermatozoa are stored in the epididymides and first portions of the ductus deferentia.
27. (*c*) Combination oral contraceptives inhibit the release of LH and FSH from the anterior pituitary, and thus prevent maturation of the developing follicle.
28. (*c*) Androgens masculinize the genitalia; in the absence of androgens, the embryos develop as females.
29. (*a*) Prolactin stimulates the milk production and oxytocin stimulates the myoepithelial cells, which cause the milk-ejection reflex.
30. (*d*) Spermatozoa in the female genital tract have a normal life span of up to 5 days.

True or False

1. True
2. False; the uterus is a secondary sex organ
3. True
4. True
5. True
6. True
7. False; the prepuce is a protective sheath over the vascular glans penis
8. False; parasympathetic impulses cause marked vasodilation of the arterioles which increase blood flow into the penis causing an erection.
9. False; spermatozoa mix with additives from the accessory reproductive glands in the ejaculatory ducts
10. False; mammary glands are modified sudoiferous (sweat) glands.
11. False; prolactin stimulates milk production
12. True
13. False; secretions from the vestibular glands moisten and lubricate the vaginal orifice during sexual arousal
14. False; interstitial cells secrete testosterone; the sustentacular cells provide nutrients to the spermatozoa
15. False; the secretory phase is from the time of ovulation until the start of the menstrual phase

Completion

1. Cryptorchidism
2. Ejaculation
3. uterine (fallopian), ductus
4. Interstitial (Leydig)
5. varicocele
6. hymen
7. 5, 24
8. menarche
9. cleavage
10. ectopic pregnancy

Labeling

1. Uterine tube
2. Body of uterus
3. Endometrium
4. Myometrium
5. Perimetrium
6. Isthmus of uterus
7. Cervical canal
8. Uterine ostium
9. Cervix
10. Vagina

Matching

1. (*f*)
2. (*a*)
3. (*e*)
4. (*g*)
5. (*d*)
6. (*h*)
7. (*i*)
8. (*j*)
9. (*b*)
10. (*c*)

Index